Classical Hypergeometric Functions and Generalizations

Contemporary Mathematics

818

Classical Hypergeometric Functions and Generalizations

Minisymposium on All Things Hypergeometric, q-series and Generalizations
16th International Symposium on Orthogonal Polynomials,
Special Functions and Applications (OPSFA-16)
June 13–17, 2022
Centre de Recherches Mathématiques, Montréal, Québec, Canada

AMS Special Session on Hypergeometric Functions and q-series
2022 AMS Fall Western Sectional Meeting
October 22–23, 2022
University of Utah, Salt Lake City, Utah

AMS Special Session on Hypergeometric Functions, q-series
and Generalizations
2023 AMS Spring Eastern Virtual Sectional Meeting
April 1–2, 2023

Howard S. Cohl
Roberto S. Costas-Santos
Robert S. Maier
Editors

American Mathematical Society

2020 *Mathematics Subject Classification.* Primary 33C05, 33C20, 33C45, 33C90, 33E05, 11G20, 11G55, 14H52.

Library of Congress Cataloging-in-Publication Data

Names: Cohl, Howard S., 1968– editor. | Costas-Santos, Roberto S., 1974– editor. | Maier, Robert S., 1968– editor. | Minisymposium on All Things Hypergeometric, q-series and Generalizations, International Symposium on Orthogonal Polynomials, Special Functions and Applications (OPSFA-16) (16th : 2022 : Montréal, Québec). | AMS Special Session on Hypergeometric Functions and q-series (2022 : Salt Lake City, Utah). | Virtual AMS Special Session on Hypergeometric Functions, q-series and Generalizations (2023 : online).

Title: Classical hypergeometric functions and generalizations / Minisymposium on All Things Hypergeometric, q-series and Generalizations, 16th International Symposium Orthogonal Polynomials, Special Functions and Applications (OPSFA-16), June 13–17, 2022, Centre de Recherches Mathématiques, Montréal, Québec, Canada; AMS Special Session on Hypergeometric Functions and q-series, October 22–23, 2022, 2022 AMS Fall Western Sectional Meeting, University of Utah, Salt Lake City, Utah, USA; AMS Special Session on Hypergeometric Functions, q-series and Generalizations, April 1–2, 2023, 2023 AMS Spring Eastern Virtual Sectional Meeting; Howard S. Cohl, Roberto S. Costas-Santos, Robert S. Maier, editors.

Description: Providence, Rhode Island : American Mathematical Society, [2025] | Series: Contemporary mathematics, 0271-4132 ; volume 818 | Includes bibliographical references.

Identifiers: LCCN 2024039112 | ISBN 9781470476533 (paperback) | ISBN 9781470479381 (ebook)

Subjects: LCSH: Hypergeometric functions–Congresses. | AMS: Special functions – Hypergeometric functions – Classical hypergeometric functions. | Special functions – Hypergeometric functions – Generalized hypergeometric series. | Special functions – Hypergeometric functions – Orthogonal polynomials and functions of hypergeometric type (Jacobi, Laguerre, Hermite, Askey scheme, etc.). | Special functions – Hypergeometric functions – Applications. | Special functions – Other special functions – Elliptic functions and integrals. | Number theory – Arithmetic algebraic geometry (Diophantine geometry) – Curves over finite and local fields. | Number theory – Arithmetic algebraic geometry (Diophantine geometry) – Polylogarithms and relations with K-theory. | Algebraic geometry – Curves – Elliptic curves.

Classification: LCC QA353.H9 C63 2025 | DDC 515/.5–dc23/eng/20250207

LC record available at https://lccn.loc.gov/2024039112

Contemporary Mathematics ISSN: 0271-4132 (print); ISSN: 1098-3627 (online)

DOI: https://doi.org/10.1090/conm/818

Visit the AMS home page at `https://www.ams.org/`

10 9 8 7 6 5 4 3 2 1 30 29 28 27 26 25

Dedicated to Richard Askey, who noticed and expounded on the utility and functionality of hypergeometric functions.

Contents

Preface

Hypergeometric functions, and the hypergeometric series by which they can be defined, are parametric mathematical objects of much interest in classical analysis. 'The' hypergeometric function is the classical function now denoted by ${}_2F_1(z)$, which was investigated by Gauss and is now named after him, though it was previously studied by Euler. For a beautiful description of the history of the hypergeometric function see Chapter 27 in [**1**] by Ranjan Roy. It satisfies a second-order differential equation with respect to z, and the coefficients of its defining power series satisfy a two-term recurrence. Beginning with Kummer, ${}_2F_1$ was generalized to ${}_rF_s$. All these functions satisfy contiguous function relations in which parameters are displaced by integers [**4**], and they have many interesting transformation properties. The nicest generalized hypergeometric functions may be the non-confluent ones, for which $r = s + 1$.

These classical functions are prominent among the many special functions that arise in mathematical physics and elsewhere. They also appear repeatedly in the theory of orthogonal polynomials. For instance, Jacobi polynomials are simply terminating hypergeometric series of the ${}_2F_1$ (Gauss) type, and the three-term recurrences that these orthogonal polynomials satisfy can be viewed as specializations of certain contiguous function relations satisfied by the parametric function ${}_2F_1$. The interested reader will find a wealth of related information in the monographs of Andrews, Askey, and Roy [**2**], and Ismail [**3**].

This is the first volume of a two-volume collection of recent research results related to hypergeometric functions, which come with full bibliographies and which should serve as entry points to the current literature. (The second volume is titled "Applications and q-Extensions of Hypergeometric Functions.") The present volume is divided into two parts. The papers in the first part deal with generalized functions ${}_rF_s$, and in many cases with the Gauss function ${}_2F_1$. Those in the second part deal with special functions that are related to ${}_2F_1$ or extend it, but differ from the now-classical ${}_rF_s$. The Lamé and Heun functions, which were first investigated in the nineteenth century, are an example. Like ${}_2F_1(z)$, they satisfy parametric second-order differential equations with respect to z, but the coefficients in their defining power series satisfy three-term rather than two-term recurrences. Other related functions that could be mentioned are elliptic integrals, elliptic functions (the inverses of same), and elliptic modular functions. An altogether different direction in which ${}_2F_1(z)$ and ${}_rF_s(z)$ can be extended involves the replacement of the classical complex field $\mathbb{C}$ by a finite field.

The first part contains five papers of varying lengths. They touch on such topics as sums involving digamma functions that reduce to hypergeometric functions;

addition theorems for Jacobi functions of the first and second kind (which are specialized Gauss functions); bounds for ratios of contiguous hypergeometric functions; trigonometric expansions of the Ferrers function of the first kind (a Gauss function that transforms well under a quadratic transformation of variables); and infinite products of incomplete Beta functions, which are particular Gauss functions.

The first paper (by A. Çetinkaya and D. Karp) is *On digamma series convertible into hypergeometric series*, which treats the differentiation of generalized hypergeometric functions with respect to parameters. The second paper (by H. S. Cohl, R. S. Costas-Santos, L. Durand, C. Montoya and G. Ólafsson) is *Double summation addition theorems for Jacobi functions of the first and second kind.* The third paper (by J. Segura) is *On bounds for ratios of contiguous hypergeometric functions.* The fourth paper (by H. Volkmer) is *Expansion of the Ferrers function of the first kind in a trigonometric series.* The fifth paper (by N. S. Witte) is *An infinite product of incomplete Beta function-type hypergeometric functions, and its probabilistic origins.*

The second part of this volume features six papers. The mathematical topics considered include eta and generalized eta functions and Dedekind sums; Apéry-like sequences and four-term recurrences; Heun, Lamé, and Lamé–Wangerin functions; hypergeometric functions over finite fields; and logarithmic integrals and hyper-Mahler measures.

The first paper (by Heng Huat Chan and Teoh Guan Chua) is *The quadratic reciprocity law and the Gauss–Schering lemma*, and is connected to hypergeometric functions through the transformation properties of the Dedekind eta function and its generalizations. The second paper (by S. Cooper) is *Apéry-like sequences defined by four-term recurrence relations*, a partially expository paper which discusses sequences arising from the theory of elliptic modular forms. The third paper (by Yifeng Huang, K. Ono and H. Saad) is *Counting matrix points in certain varieties over finite fields*, which treats Legendre elliptic curves and certain K3 surfaces. The fourth paper (by E. McSpirit and K. Ono) is *Hypergeometry and the AGM [Arithmetic–Geometric Mean] over finite fields*, which obtains new identities for Gauss's class numbers of positive definite quadratic forms. The fifth paper (by H. Volkmer) is *The Heun eigenvalue problem and some of its special cases*, which rigorously justifies certain numerical solution schemes. The sixth paper (by Yajun Zhou) is *Hyper-Mahler measures via Goncharov–Deligne cyclotomy.*

This two-volume collection follows in the tradition of an earlier volume in the AMS Contemporary Mathematics series, "Special Functions and Orthogonal Polynomials," edited by D. Dominici and R. S. Maier. The contributions to that volume by the late Prof. Diego Dominici are warmly remembered.

Howard S. Cohl
Roberto S. Costas-Santos
Robert S. Maier

Bibliography

[1] Ranjan Roy, *Sources in the development of mathematics. Infinite series and products from the fifteenth to the twenty-first century*, Cambridge University Press, Cambridge, 2011, DOI 10.1017/CBO9780511844195. MR2807493

[2] George E. Andrews, Richard Askey, and Ranjan Roy, *Special functions*, Encyclopedia of Mathematics and its Applications, vol. 71, Cambridge University Press, Cambridge, 1999, DOI 10.1017/CBO9781107325937. MR1688958

[3] Mourad E. H. Ismail, *Classical and quantum orthogonal polynomials in one variable*, Encyclopedia of Mathematics and its Applications, vol. 98, Cambridge University Press, Cambridge, 2005. With two chapters by Walter Van Assche, with a foreword by Richard A. Askey, DOI 10.1017/CBO9781107325982. MR2191786

[4] Earl D. Rainville, *Special functions*, The Macmillan Company, New York, 1960. MR107725

Hypergeometric Functions

Contemporary Mathematics
Volume **818**, 2025
https://doi.org/10.1090/conm/818/16366

On digamma series convertible into hypergeometric series

Asena Çetinkaya and Dmitrii Karp

ABSTRACT. Series containing the digamma function arise when calculating parametric derivatives of hypergeometric functions, and play a role in evaluation of Feynman diagrams. As these series are typically non-hypergeometric, a few instances when they are summable in terms of hypergeometric functions are of importance. In this paper, by employing appropriate limiting processes, we convert multi-term identities for the generalized hypergeometric functions evaluated at positive/negative unity into identities connecting them to digamma series. The resulting formulas can be viewed as hypergeometric expressions for the sum of the partial derivatives of the generalized hypergeometric function with respect to all its parameters, and seem to have no direct analogues in the literature.

1. Introduction and notation

The series containing the logarithmic derivative $\psi(z) = \Gamma'(z)/\Gamma(z)$ of Euler's gamma function $\Gamma(z)$, known as the digamma or psi function, appear in a number of contexts. First of all they may represent the parameter derivatives of hypergeometric functions, which play an important role in several areas of mathematical physics, most notably in evaluating Feynman diagrams [**15, 16**] and in problems involving fractional differential equations [**3**]. A number of other problems from mathematical physics, where hypergeometric functions and their parameter derivatives are relevant can be found in the introductory sections of [**1**] and [**18**]. It is thus not surprising that the papers mentioned above as well as [**9**] are all published in physics or mathematical physics journals. Probably the first comprehensive collection of psi series evaluations is given in Hansen's book [**14**, section 55, 360–6] (see also formulas numbered by 55.$*$ in the addendum [**5**] by Borwein).

This topic has been rather active over the last two decades with a variety of results found in [**1, 3, 6, 7, 9, 11, 12, 17–20**], and numerous references in these papers. Results up to 2008 are summarized in Brychkov's book [**6**]: see, in particular, sections 1.30.2 and 6.2. Most formulas derived in the literature convert psi series obtained by differentiating hypergeometric functions with respect to their parameters into multiple hypergeometric series, frequently of Kampé de Fériet type [**1, 7, 9, 11, 17, 18, 20**]. In more special cases, however, such series can be summed by, or rather transformed into, single hypergeometric series [**2**, Chapter 3]. Such

2020 *Mathematics Subject Classification.* Primary 33C20, 33B15; Secondary 33C99.
Key words and phrases. hypergeometric series, digamma series, summation formula, hypergeometric identity.

special cases are of importance, as the properties of hypergeometric functions and algorithms for computing them are well developed, which facilitates handling the corresponding psi series.

An elegant set of formulas of this type has been established by Cvijović in [**9**] as a follow-up of a previous work by Miller [**18**], and later used by Cvijović and Miller to establish new reduction formulas for Kampé de Fériet functions [**10**]. The series handled in these references are power series whose coefficients contain the digamma function, and the derivations are largely based on the Euler–Pfaff transformation for the Gauss hypergeometric function ${}_2F_1$ and the Kummer transformation for the confluent hypergeometric function ${}_1F_1$. Other works use series rearrangements and the hypergeometric differential equation. More recent results in this direction can be found in [**12, 13**].

Our method of proof differs from all those techniques and is based on degeneration of multi-term identities for generalized hypergeometric functions evaluated at positive or negative unity, including those recently established in [**8**]. We will use this approach to transform certain rather general series containing digamma functions into combinations of univariate hypergeometric functions evaluated at the same point.

Due to the significantly richer collection of transformations of hypergeometric functions at a fixed argument, our formulas have an arbitrary number of free parameters, unlike those due to Miller and Cvijović [**9, 10, 18**] and those due to González-Santander and Sánchez Lasheras [**12, 13**], as well as those listed in [**6**, 6.2.1.62–81].

To illustrate the flavor of the results anticipated in this paper let us introduce the following shorthand notation:

$$\begin{aligned}\Gamma(\mathbf{a})=\prod\nolimits_{i=1}^{p}\Gamma(a_i),\qquad (\mathbf{a})_k=\prod\nolimits_{i=1}^{p}(a_i)_k,\qquad \sin(\mathbf{a})=\prod\nolimits_{i=1}^{p}\sin(a_i),\\ \mathbf{a}+\beta=(a_1+\beta,\ldots,a_p+\beta),\qquad \mathbf{a}_{[k]}=(a_1,\ldots,a_{k-1},a_{k+1},\ldots,a_p),\end{aligned}\tag{1.1}$$

for any vector $\mathbf{a}=(a_1,\ldots,a_p)\in\mathbb{C}^p$ (the set of complex p-tuples) and a scalar β. Among other things, we will establish the following identities:

(1.2)
$$\begin{aligned}&\sum_{k=0}^{\infty}\frac{(\mathbf{a})_k(\pm1)^k}{\pi(\mathbf{b})_k k!}\Bigg[\sum_{b\in\mathbf{b}}\psi(b+k)+\psi(1+k)-\sum_{a\in\mathbf{a}}\psi(a+k)\Bigg]\\ &=\Bigg[\sum_{a\in\mathbf{a}}\cot(\pi a)-\sum_{b\in\mathbf{b}}\cot(\pi b)\Bigg]{}_pF_{p-1}\left(\begin{matrix}\mathbf{a}\\ \mathbf{b}\end{matrix}\middle|\pm1\right)\mp\frac{(1-\mathbf{b})_1}{\pi(1-\mathbf{a})_1}{}_{p+1}F_p\left(\begin{matrix}1,1,2-\mathbf{b}\\ 2-\mathbf{a}\end{matrix}\middle|\pm1\right)\\ &\quad+\frac{\Gamma(1-\mathbf{a})}{\Gamma(1-\mathbf{b})}\\ &\quad\times\sum_{k=1}^{p-1}\frac{[\cos(\pi b_k)]^{\{\substack{1\\0}\}}\Gamma(b_k-1)\Gamma(b_k-\mathbf{b}_{[k]})}{\sin(\pi b_k)\Gamma(b_k-\mathbf{a})}{}_pF_{p-1}\left(\begin{matrix}1-b_k+\mathbf{a}\\ 2-b_k,1-b_k+\mathbf{b}_{[k]}\end{matrix}\middle|\pm1\right)\end{aligned}$$

and

(1.3)
$$\begin{aligned}&\sum_{k=0}^{\infty}\frac{(\mathbf{a})_k}{(\mathbf{b})_k k!}\Bigg[\sum_{b\in\mathbf{b}}\psi(b+k)+\psi(1+k)-\sum_{a\in\mathbf{a}}\psi(a+k)\Bigg]\\ &=\frac{\Gamma(\mathbf{b})}{\Gamma(\mathbf{a})}\sum_{k=1}^{p}\frac{\pi\Gamma(a_k)\Gamma(\mathbf{a}_{[k]}-a_k)}{\sin(\pi a_k)\Gamma(\mathbf{b}-a_k)}{}_pF_{p-1}\left(\begin{matrix}a_k,1-\mathbf{b}+a_k\\ 1-\mathbf{a}_{[k]}+a_k\end{matrix}\right)-\frac{(1-\mathbf{b})_1}{(1-\mathbf{a})_1}{}_{p+1}F_p\left(\begin{matrix}1,1,2-\mathbf{b}\\ 2-\mathbf{a}\end{matrix}\right),\end{aligned}$$

where here and in the sequel ${}_pF_q\left(\begin{matrix}\mathbf{a}\\\mathbf{b}\end{matrix}\middle| z\right)$ stands for the generalized hypergeometric function [**4**, (2.1.2)] and an omitted argument z designates $z=1$. Note that via the elementary relations

$$\frac{d}{da}(a)_n=(a)_n(\psi(a+n)-\psi(a)),\qquad \frac{d}{db}\left[\frac{1}{(b)_n}\right]=\frac{1}{(b)_n}(\psi(b)-\psi(b+n)),$$

the expressions on the left hand sides of (1.2) and (1.3) take the form

$$\sum_{k=0}^{\infty}\frac{(\mathbf{a})_k(\pm1)^k}{(\mathbf{b})_k k!}\left[\sum_{b\in\mathbf{b}}\psi(b+k)+\psi(1+k)-\sum_{a\in\mathbf{a}}\psi(a+k)\right]$$

$$=-\mathbf{1}\cdot\nabla_{\mathbf{a},\mathbf{b},\beta}\,{}_{p+1}F_p\left(\begin{matrix}\mathbf{a},1\\\mathbf{b},\beta\end{matrix}\middle|\pm1\right)\Bigg|_{\beta=1}+\left\{\sum_{b\in\mathbf{b}\cup\{1\}}\psi(b)-\sum_{a\in\mathbf{a}}\psi(a)\right\}{}_pF_{p-1}\left(\begin{matrix}\mathbf{a}\\\mathbf{b}\end{matrix}\middle|\pm1\right),$$

where $\mathbf{1}\cdot\nabla_{\mathbf{a},\mathbf{b},\beta}f(\mathbf{a},\mathbf{b},\beta)$ denotes the sum of the first partial derivatives of f with respect to the components of $\mathbf{a}$, $\mathbf{b}$ and the variable β. Hence, formulas (1.2), (1.3) and many further identities in this paper can be viewed as expressions for the sum of the components of the gradient of the generalized hypergeometric function with respect to all its parameters.

Another way of writing the left hand sides of (1.2) and (1.3) is via Kampé de Fériet functions [**21**, (28)]. Indeed, using

$$\psi(b+k)-\psi(b)=\frac{1}{b}\sum_{j=0}^{k-1}\frac{(b)_j}{(b+1)_j}$$

and rearranging summations, we will have

$$\sum_{k=0}^{\infty}\frac{(\mathbf{a})_k(\pm1)^k}{(\mathbf{b})_k k!}\left[\sum_{b\in\mathbf{b}}\psi(b+k)+\psi(1+k)-\sum_{a\in\mathbf{a}}\psi(a+k)\right]$$

$$=\pm\frac{(\mathbf{a})_1}{(\mathbf{b})_1}\sum_{b\in\mathbf{b}\cup\{1\}}\frac{1}{b}F^{p:2:1}_{p:1:0}\left(\begin{matrix}\mathbf{a}+1 & :b,1 & :1\\\mathbf{b}+1,2 & :b+1 & :-\end{matrix}\middle|\pm1,\pm1\right)$$

$$\mp\frac{(\mathbf{a})_1}{(\mathbf{b})_1}\sum_{a\in\mathbf{a}}\frac{1}{a}F^{p:2:1}_{p:1:0}\left(\begin{matrix}\mathbf{a}+1 & :a,1 & :1\\\mathbf{b}+1,2 & :a+1 & :-\end{matrix}\middle|\pm1,\pm1\right)$$

$$+\left\{\sum_{b\in\mathbf{b}\cup\{1\}}\psi(b)-\sum_{a\in\mathbf{a}}\psi(a)\right\}{}_pF_{p-1}\left(\begin{matrix}\mathbf{a}\\\mathbf{b}\end{matrix}\middle|\pm1\right).$$

These are formal expressions and convergence needs to be discussed separately for specific values of parameters.

This paper is organized as follows. In Sections 2, 3 and 4 we will apply degeneration process to the three identities [**8**, (5.2)], [**8**, (5.1)] and [**8**, (5.4)], respectively. This results in three transformation formulas for series containing digamma functions whose combinations and particular cases yield, among other things, the above identities (1.2) and (1.3).

Another important particular case is given at the end of Section 4 in formula (4.4). Comparing the right-hand sides of (1.2) and (1.3), as well as of other similar formulas, we further get two presumably new multi-term identities for the generalized hypergeometric function evaluated at unity, which are presented in formulas (3.6) and (4.5).

2. The first transformation formula

The formulas presented in the Introduction will be derived as corollaries of several transformation formulas for series containing both digamma functions and rising factorials of the same collection of arguments. One of the denominator parameters in these formulas must be a positive integer. In this section, this integer represents the limiting value of the appropriate parameter difference in the identity [**8**, (5.2)] which will serve as the starting point for the degeneration process. It will be convenient to use the notation

$$ {}_p\phi_q\binom{\mathbf{a}}{\mathbf{b}}=\frac{\Gamma(\mathbf{a})}{\Gamma(\mathbf{b})}\,{}_pF_q\binom{\mathbf{a}}{\mathbf{b}}=\sum_{k=0}^{\infty}\frac{\Gamma(\mathbf{a}+k)}{\Gamma(\mathbf{b}+k)k!}. \tag{2.1}$$

The main result of this section is the following.

THEOREM 2.1. *Suppose* $\mathbf{c}\in(\mathbb{C}\setminus\mathbb{Z})^{p+1}$, $\mathbf{d}\in(\mathbb{C}\setminus\mathbb{Z})^{p-1}$, $m\in\mathbb{Z}_{\geqslant 0}$ *and* $\Re(\sum_{j=1}^{p-1}d_j-\sum_{j=1}^{p+1}(c_j))>-m-1$. *Then*

$$\begin{aligned}
&\sum_{k=0}^{\infty}\frac{(\mathbf{c})_k}{(\mathbf{d})_k(m+k)!k!}\Big\{\sum_{d\in\mathbf{d}}\psi(d+k)+\psi(k+1)+\psi(1+m+k)-\sum_{c\in\mathbf{c}}\psi(c+k)\Big\}\\
&=\frac{\pi}{m!}\Big\{\sum_{c\in\mathbf{c}}\cot(\pi c)-\sum_{d\in\mathbf{d}}\cot(\pi d)\Big\}\,{}_{p+1}F_p\binom{\mathbf{c}}{1+m,\mathbf{d}}\\
&\quad+\frac{(1-\mathbf{d})_1}{(1-\mathbf{c})_1(m-1)!}\,{}_{p+2}F_{p+1}\binom{1-m,2-\mathbf{d},1,1}{2-\mathbf{c}}\\
&\quad-\frac{\Gamma(1-\mathbf{c})}{\Gamma(1-\mathbf{d})}\sum_{k=1}^{p-1}\frac{(-1)^m\Gamma(d_k-m)\Gamma(d_k)\Gamma(d_k-\mathbf{d}_{[k]})}{(d_k-m-1)(d_k-1)\Gamma(d_k-\mathbf{c})}\\
&\quad\times\,{}_{p+1}F_p\binom{1+\mathbf{c}-d_k}{2-d_k,2-d_k+m,1+\mathbf{d}_{[k]}-d_k}.
\end{aligned}\tag{2.2}$$

PROOF. We start with the identity [**8**, (5.2)] written as follows:

$$\begin{aligned}
&\frac{\Gamma(\mathbf{b}_{[1]}-b_1)}{\Gamma(\mathbf{w}-b_1)}\,{}_{p+1}F_p\binom{1-\mathbf{w}+b_1}{1-\mathbf{b}_{[1]}+b_1}+\frac{\Gamma(\mathbf{b}_{[2]}-b_2)}{\Gamma(\mathbf{w}-b_2)}\,{}_{p+1}F_p\binom{1-\mathbf{w}+b_2}{1-\mathbf{b}_{[2]}+b_2}\\
&=-\sum_{k=3}^{p+1}\frac{\Gamma(\mathbf{b}_{[k]}-b_k)}{\Gamma(\mathbf{w}-b_k)}\,{}_{p+1}F_p\binom{1-\mathbf{w}+b_k}{1-\mathbf{b}_{[k]}+b_k}.
\end{aligned}\tag{2.3}$$

This is valid for $\mathbf{b},\mathbf{w}\in\mathbb{C}^{p+1}$ with the components of $\mathbf{b}$ different modulo integers. Setting $b_2=b_1+m+\varepsilon$, $m\in\mathbb{Z}_{\geqslant 0}$, using the reflection formula $\Gamma(z)\Gamma(1-z)=\pi/\sin(\pi z)$, and the elementary relation $\sin(\pi(z+m))=(-1)^m\sin(\pi z)$, we obtain for the left-hand side of (2.3):

$$\begin{aligned}
\text{LHS of (2.3)}&=\frac{(-1)^m}{\pi\sin(\pi\varepsilon)}\big(f_1(\varepsilon)-f_2(\varepsilon)\big)\\
&=\frac{(-1)^m}{\sin(\pi\varepsilon)}\big(f_1(0)+\varepsilon f_1'(0)-f_2(0)-\varepsilon f_2'(0)+O(\varepsilon^2)\big)\\
&=\frac{(-1)^m\varepsilon}{\sin(\pi\varepsilon)}\big(f_1'(0)-f_2'(0)+O(\varepsilon)\big)\to\frac{(-1)^m}{\pi}\big(f_1'(0)-f_2'(0)\big)\ \text{as}\ \varepsilon\to 0,
\end{aligned}$$

where we employed Taylor's theorem for the functions

$$f_1(\varepsilon) = \frac{\sin(\pi(\mathbf{w}-b_1))}{\pi\sin(\pi(\mathbf{b}_{[1,2]}-b_1))}{}_{p+1}\phi_p\left(\begin{matrix}1-\mathbf{w}+b_1\\ 1-m-\varepsilon, 1-\mathbf{b}_{[1,2]}+b_1\end{matrix}\right)$$

and

$$f_2(\varepsilon) = \frac{\sin(\pi(1-\mathbf{w}+b_1+\varepsilon))}{\pi\sin(\pi(1-\mathbf{b}_{[1,2]}+b_1+\varepsilon))}{}_{p+1}\phi_p\left(\begin{matrix}1-\mathbf{w}+b_1+m+\varepsilon\\ 1+m+\varepsilon, 1-\mathbf{b}_{[1,2]}+b_1+m+\varepsilon\end{matrix}\right),$$

with ϕ defined in (2.1). The penultimate equality above is due to $f_1(0) = f_2(0)$, which is rather straightforward to verify. To compute the derivatives we will need the two differentiation rules

(2.4)
$$\frac{\partial}{\partial\varepsilon}\left[\frac{\sin(\pi(\mathbf{a}+\varepsilon))}{\sin(\pi(\mathbf{d}+\varepsilon))}\right] = \pi\frac{\sin(\pi(\mathbf{a}+\varepsilon))}{\sin(\pi(\mathbf{d}+\varepsilon))}\left[\sum_{a\in\mathbf{a}}\cot(\pi(a+\varepsilon)) - \sum_{d\in\mathbf{d}}\cot(\pi(d+\varepsilon))\right]$$

and

$$\frac{\partial}{\partial\varepsilon}\left[\frac{\Gamma(\mathbf{a}+\varepsilon+k)}{\Gamma(\mathbf{d}+\varepsilon+k)}\right] = \frac{\Gamma(\mathbf{a}+\varepsilon+k)}{\Gamma(\mathbf{d}+\varepsilon+k)}\left[\sum_{a\in\mathbf{a}}\psi(a+\varepsilon+k) - \sum_{d\in\mathbf{d}}\psi(d+\varepsilon+k)\right], \tag{2.5}$$

which hold for any fixed vectors $\mathbf{a}$, $\mathbf{d}$. Hence, we get:

$$\begin{aligned} f_1'(0) &= \frac{\partial}{\partial\varepsilon}\left[\frac{\sin(\pi(\mathbf{w}-b_1))}{\pi\sin(\pi(\mathbf{b}_{[1,2]}-b_1))}{}_{p+1}\phi_p\left(\begin{matrix}1-\mathbf{w}+b_1\\ 1-m-\varepsilon, 1-\mathbf{b}_{[1,2]}+b_1\end{matrix}\right)\right]_{\varepsilon=0} \\ &= \frac{\sin(\pi(\mathbf{w}-b_1))}{\pi\sin(\pi(\mathbf{b}_{[1,2]}-b_1))}\sum_{k=0}^{m-1}\frac{\psi(1-m+k)\Gamma(1-\mathbf{w}+b_1+k)}{\Gamma(1-m+k)\Gamma(1-\mathbf{b}_{[1,2]}+b_1+k)k!} \\ &\quad + \frac{\sin(\pi(\mathbf{w}-b_1))}{\pi\sin(\pi(\mathbf{b}_{[1,2]}-b_1))}\sum_{k=m}^{\infty}\frac{\psi(1-m+k)\Gamma(1-\mathbf{w}+b_1+k)}{\Gamma(1-m+k)\Gamma(1-\mathbf{b}_{[1,2]}+b_1+k)k!}. \end{aligned}$$

In view of the reflection formula, we have

$$\begin{aligned} \frac{d}{dz}\left[\frac{1}{\Gamma(z)}\right] &= \frac{d}{dz}\left[\frac{\sin(\pi z)\Gamma(1-z)}{\pi}\right] \\ &= \cos(\pi z)\Gamma(1-z) - \frac{1}{\pi}\sin(\pi z)\psi(1-z)\Gamma(1-z) = -\frac{\psi(z)}{\Gamma(z)}, \end{aligned}$$

so that for $n = 0, 1, 2, \ldots,$

$$\frac{\psi(-n)}{\Gamma(-n)} = -\cos(\pi(-n))\Gamma(1+n) = (-1)^{n+1}n!\,, \tag{2.6}$$

and according to (2.4), the derivative $f_1'(0)$ takes the form

$$f_1'(0) = V + \frac{\sin(\pi(\mathbf{w}-b_1))}{\pi\sin(\pi(\mathbf{b}_{[1,2]}-b_1))}\sum_{k=0}^{\infty}\frac{\psi(1+k)\Gamma(1-\mathbf{w}+b_1+m+k)}{\Gamma(1+m+k)\Gamma(1-\mathbf{b}_{[1,2]}+b_1+m+k)k!},$$

where

$$V = \frac{\sin(\pi(\mathbf{w}-b_1))}{\pi\sin(\pi(\mathbf{b}_{[1,2]}-b_1))}\sum_{k=0}^{m-1}\frac{(-1)^{m-k}(m-k-1)!\Gamma(1-\mathbf{w}+b_1+k)}{\Gamma(1-\mathbf{b}_{[1,2]}+b_1+k)k!}.$$

In a similar fashion, according to (2.4) and (2.5) we have

(2.7)

$$
\begin{aligned}
&f_2'(0)\\
&=\frac{\partial}{\partial\varepsilon}\left[\frac{\sin(\pi(1-\mathbf{w}+b_1+\varepsilon))}{\pi\sin(\pi(1-\mathbf{b}_{[1,2]}+b_1+\varepsilon))}{}_{p+1}\phi_p\left(\begin{matrix}1-\mathbf{w}+b_1+m+\varepsilon\\1+m+\varepsilon,1-\mathbf{b}_{[1,2]}+b_1+m+\varepsilon\end{matrix}\right)\right]_{\varepsilon=0}\\
&=U+\frac{\sin(\pi(\mathbf{w}-b_1))}{\pi\sin(\pi(\mathbf{b}_{[1,2]}-b_1))}\sum_{k=0}^{\infty}\frac{\Gamma(1-\mathbf{w}+b_1+m+k)}{\Gamma(1+m+k)\Gamma(1-\mathbf{b}_{[1,2]}+b_1+m+k)k!}\\
&\qquad\times\left\{\sum_{w\in\mathbf{w}}\psi(1-w+b_1+m+k)-\sum_{b\in\mathbf{b}_{[1,2]}}\psi(1-b+b_1+m+k)-\psi(1+m+k)\right\},
\end{aligned}
$$

where

$$
\begin{aligned}
U=&\frac{\sin(\pi(\mathbf{w}-b_1))}{\sin(\pi(\mathbf{b}_{[1,2]}-b_1))}\left\{\sum_{b\in\mathbf{b}_{[1,2]}}\cot(\pi(b-b_1))-\sum_{w\in\mathbf{w}}\cot(\pi(w-b_1))\right\}\\
&\times{}_{p+1}\phi_p\left(\begin{matrix}1-\mathbf{w}+b_1+m\\1+m,1-\mathbf{b}_{[1,2]}+b_1+m\end{matrix}\right)
\end{aligned}
$$

and we applied the three elementary relations $\sin(\pi(x-m))=(-1)^m\sin(\pi x)$, $\cot(\pi(x\pm m))=\cot(\pi x)$, and $\cot(\pi(1-x))=-\cot(\pi x)$. Thus, we arrive at

$$
\begin{aligned}
&f_1'(0)-f_2'(0)\\
&=V-U+\frac{\sin(\pi(\mathbf{w}-b_1))}{\pi\sin(\pi(\mathbf{b}_{[1,2]}-b_1))}\sum_{k=0}^{\infty}\frac{\Gamma(1-\mathbf{w}+b_1+m+k)}{(m+k)!\,\Gamma(1-\mathbf{b}_{[1,2]}+b_1+m+k)k!}\\
&\qquad\times\Bigg\{\psi(k+1)+\psi(1+m+k)\\
&\qquad\qquad+\sum_{b\in\mathbf{b}_{[1,2]}}\psi(1-b+b_1+m+k)-\sum_{w\in\mathbf{w}}\psi(1-w+b_1+m+k)\Bigg\}.
\end{aligned}
$$

Substituting this back into (2.3) for $b_2=b_1+m$ we get

$$
\begin{aligned}
&\frac{\sin(\pi(\mathbf{w}-b_1))}{\pi\sin(\pi(\mathbf{b}_{[1,2]}-b_1))}\sum_{k=0}^{\infty}\frac{\Gamma(1-\mathbf{w}+b_1+m+k)}{\Gamma(1-\mathbf{b}_{[1,2]}+b_1+m+k)(m+k)!k!}\\
&\qquad\times\Bigg\{\psi(k+1)+\psi(1+m+k)\\
&\qquad\qquad+\sum_{b\in\mathbf{b}_{[1,2]}}\psi(1-b+b_1+m+k)-\sum_{w\in\mathbf{w}}\psi(1-w+b_1+m+k)\Bigg\}\\
&=U-V-(-1)^m\pi\sum_{k=3}^{p+1}\frac{\Gamma(\mathbf{b}_{[k]}-b_k)}{\Gamma(\mathbf{w}-b_k)}{}_{p+1}F_p\left(\begin{matrix}1-\mathbf{w}+b_k\\1-\mathbf{b}_{[k]}+b_k\end{matrix}\right),
\end{aligned}
$$

or, in view of the definitions of V and U,

$$
\begin{aligned}
&\sum_{k=0}^{\infty}\frac{(1-\mathbf{w}+b_1+m)_k}{(1-\mathbf{b}_{[1,2]}+b_1+m)_k(m+k)!k!}\\
&\qquad\times\Big\{\psi(k+1)+\psi(1+m+k)\\
&\qquad\qquad+\sum_{b\in\mathbf{b}_{[1,2]}}\psi(1-b+b_1+m+k)-\sum_{w\in\mathbf{w}}\psi(1-w+b_1+m+k)\Big\}\\
&=\frac{\pi}{m!}\Bigg[\sum_{b\in\mathbf{b}_{[1,2]}}\cot(\pi(b-b_1))-\sum_{w\in\mathbf{w}}\cot(\pi(w-b_1))\Bigg]\\
&\qquad\times{}_{p+1}F_p\left(\begin{matrix}1-\mathbf{w}+b_1+m\\1+m,1-\mathbf{b}_{[1,2]}+b_1+m\end{matrix}\right)\\
&\quad-\frac{(1-\mathbf{b}_{[1,2]}+b_1)_m}{(1-\mathbf{w}+b_1)_m}\sum_{k=0}^{m-1}\frac{(-1)^{m-k}(m-k-1)!(1-\mathbf{w}+b_1)_k}{(1-\mathbf{b}_{[1,2]}+b_1)_k k!}\\
&\quad-\frac{(-1)^m\Gamma(\mathbf{w}-b_1-m)}{\Gamma(\mathbf{b}_{[1,2]}-b_1-m)}\sum_{k=3}^{p+1}\frac{\Gamma(\mathbf{b}_{[k]}-b_k)}{\Gamma(\mathbf{w}-b_k)}{}_{p+1}F_p\left(\begin{matrix}1-\mathbf{w}+b_k\\1-\mathbf{b}_{[k]}+b_k\end{matrix}\right).
\end{aligned}
$$

Writing $1-\mathbf{w}+b_1+m=\mathbf{c}\in\mathbb{C}^{p+1}$, $1-\mathbf{b}_{[1,2]}+b_1+m=\mathbf{d}\in\mathbb{C}^{p-1}$ we arrive at (2.2) on application of the straightforward relation

$$
\begin{aligned}
\frac{(1-\mathbf{d})_m}{(1-\mathbf{c})_m}\sum_{k=0}^{m-1}\frac{(m-k-1)!(\mathbf{c}-m)_k}{(-1)^{m-k}(\mathbf{d}-m)_k k!}&=\frac{(1-\mathbf{d})_1}{(1-\mathbf{c})_1}\sum_{j=0}^{m-1}\frac{(-1)^{j+1}(2-\mathbf{d})_j j!}{(2-\mathbf{c})_j(m-1-j)!}\\
&=-\frac{(1-\mathbf{d})_1}{(1-\mathbf{c})_1(m-1)!}{}_{p+2}F_{p+1}\left(\begin{matrix}1-m,2-\mathbf{d},1,1\\2-\mathbf{c}\end{matrix}\right),
\end{aligned}\tag{2.8}
$$

where in the first equality we used

$$
\frac{(1-\mathbf{d})_m}{(\mathbf{d}-m)_k}=(-1)^{k(p-1)}(1-\mathbf{d})_1(2-\mathbf{d})_{m-k-1}. \qquad \square
$$

3. The second transformation formula

Here we will derive another hypergeometric representation for the left-hand side of (2.2). Comparing the two we will obtain a presumably new multi-term identity for the generalized hypergeometric function and the "+" case of formula (1.2).

THEOREM 3.1. *Suppose* $\mathbf{c} \in (\mathbb{C} \setminus \mathbb{Z})^{p+1}$, $\mathbf{d} \in (\mathbb{C} \setminus \mathbb{Z})^{p-1}$, $m \in \mathbb{Z}_{\geqslant 0}$ *and* $\Re(\sum_{j=1}^{p-1} d_j - \sum_{j=1}^{p+1}(c_j)) > -m-1$. *Then*

$$
\begin{aligned}
&\sum_{k=0}^{\infty} \frac{(\mathbf{c})_k}{(\mathbf{d})_k (m+k)! k!} \left\{ \sum_{d \in \mathbf{d}} \psi(d+k) + \psi(k+1) + \psi(1+m+k) - \sum_{c \in \mathbf{c}} \psi(c+k) \right\} \\
&= \frac{\pi}{m!} \Big[\sum_{c \in \mathbf{c}_{[1]}} \cot(\pi c) - \sum_{d \in \mathbf{d}_{[1]}} \cot(\pi d) \Big] {}_{p+1}F_p \left(\begin{matrix} \mathbf{c} \\ 1+m, \mathbf{d} \end{matrix} \right) \\
&+ \frac{(1-\mathbf{d})_1}{(1-\mathbf{c})_1 (m-1)!} {}_{p+2}F_{p+1} \left(\begin{matrix} 1-m, 2-\mathbf{d}, 1, 1 \\ 2-\mathbf{c} \end{matrix} \right) \\
&+ \frac{\Gamma(c_1)\Gamma(d_1)\Gamma(1-\mathbf{c}_{[1]})\Gamma(1-\mathbf{d}_{[1]}+c_1)}{(1-c_1)_m \Gamma(d_1-c_1)\Gamma(1-\mathbf{c}_{[1]}+c_1)\Gamma(1-\mathbf{d}_{[1]})} {}_{p+1}F_p \left(\begin{matrix} c_1, c_1-m, 1-\mathbf{d}+c_1 \\ 1-\mathbf{c}_{[1]}+c_1 \end{matrix} \right) \\
&- \frac{(-1)^m \Gamma(1-\mathbf{c}) \sin(\pi c_1)}{\Gamma(1-\mathbf{d}) \sin(\pi d_1)} \sum_{k=2}^{p-1} \frac{\Gamma(d_k-m)\Gamma(d_k)\Gamma(d_k-\mathbf{d}_{[k]}) \sin(\pi(d_k-d_1))}{(d_k-m-1)(d_k-1)\Gamma(d_k-\mathbf{c}) \sin(\pi(d_k-c_1))} \\
&\quad \times {}_{p+1}F_p \left(\begin{matrix} 1+\mathbf{c}-d_k \\ 2-d_k, 2-d_k+m, 1+\mathbf{d}_{[k]}-d_k \end{matrix} \right).
\end{aligned}
\tag{3.1}
$$

PROOF. We start with identity [**8**, (5.1)] written as follows

$$
\begin{aligned}
&\frac{\Gamma(a_2)\Gamma(\mathbf{a}_{[1,2]}-a_2)}{\Gamma(1-a_1+a_2)\Gamma(\mathbf{b}-a_2)} {}_{p+1}F_p \left(\begin{matrix} a_2, 1-\mathbf{b}+a_2 \\ 1-\mathbf{a}_{[2]}+a_2 \end{matrix} \right) \\
&\quad + \frac{\Gamma(a_3)\Gamma(\mathbf{a}_{[1,3]}-a_3)}{\Gamma(1-a_1+a_3)\Gamma(\mathbf{b}-a_3)} {}_{p+1}F_p \left(\begin{matrix} a_3, 1-\mathbf{b}+a_3 \\ 1-\mathbf{a}_{[3]}+a_3 \end{matrix} \right) \\
&\quad = \frac{\Gamma(\mathbf{a}_{[1]})}{\Gamma(1-a_1)\Gamma(\mathbf{b})} {}_{p+1}F_p \left(\begin{matrix} \mathbf{a} \\ \mathbf{b} \end{matrix} \right) \\
&\quad - \sum_{k=4}^{p+1} \frac{\Gamma(a_k)\Gamma(\mathbf{a}_{[1,k]}-a_k)}{\Gamma(1-a_1+a_k)\Gamma(\mathbf{b}-a_k)} {}_{p+1}F_p \left(\begin{matrix} a_k, 1-\mathbf{b}+a_k \\ 1-\mathbf{a}_{[k]}+a_k \end{matrix} \right).
\end{aligned}
\tag{3.2}
$$

If we set $a_3 = a_2 + \varepsilon + m$ here, then after a simple calculation similar to that in the proof of Theorem 2.1 we obtain

$$
\begin{aligned}
&\text{LHS of } (3.2) \\
&= \frac{(-1)^m}{\sin(\pi\varepsilon)} \{f_1(\varepsilon) + f_2(\varepsilon)\} \\
&= \frac{(-1)^m}{\sin(\pi\varepsilon)} \Big[f_1(0) + \varepsilon f_1'(0) + f_2(0) + \varepsilon f_2'(0) + O(\varepsilon^2) \Big] \\
&= \frac{(-1)^m \varepsilon}{\sin(\pi\varepsilon)} \Big(f_1'(0) + f_2'(0) + O(\varepsilon) \Big) \to \frac{(-1)^m}{\pi} \Big(f_1'(0) + f_2'(0) \Big) \text{ as } \varepsilon \to 0,
\end{aligned}
$$

where, in view of (2.1),

$$
\begin{aligned}
f_1(\varepsilon) &= \frac{\sin(\pi(a_1-a_2))\sin(\pi(\mathbf{b}-a_2))}{\pi \sin(\pi(\mathbf{a}_{[2,3]}-a_2))} {}_{p+1}\phi_p \left(\begin{matrix} a_2, 1-\mathbf{b}+a_2 \\ 1-\varepsilon-m, 1-\mathbf{a}_{[2,3]}+a_2 \end{matrix} \right), \\
f_2(\varepsilon) &= \frac{\sin(\pi(a_2-a_1+\varepsilon))\sin(\pi(1-\mathbf{b}+a_2+\varepsilon))}{\pi \sin(\pi(1-\mathbf{a}_{[2,3]}+a_2+\varepsilon))} \\
&\quad \times {}_{p+1}\phi_p \left(\begin{matrix} a_2+\varepsilon+m, 1-\mathbf{b}+a_2+\varepsilon+m \\ 1+\varepsilon+m, 1-\mathbf{a}_{[2,3]}+a_2+\varepsilon+m \end{matrix} \right),
\end{aligned}
$$

and we applied the easily verifiable identity $f_1(0) = -f_2(0)$. Then, after a simple calculation we get

$$f_1'(0) = \frac{\sin(\pi(\mathbf{b}-a_2))}{\pi\sin(\pi(\mathbf{a}_{[1,2,3]}-a_2))}\sum_{k=0}^{m-1}\frac{(-1)^{m-k}(m-k-1)!\Gamma(a_2+k)\Gamma(1-\mathbf{b}+a_2+k)}{\Gamma(1-\mathbf{a}_{[2,3]}+a_2+k)k!}$$
$$+\frac{\sin(\pi(\mathbf{b}-a_2))}{\pi\sin(\pi(\mathbf{a}_{[1,2,3]}-a_2))}\sum_{k=0}^{\infty}\frac{\Gamma(a_2+m+k)\Gamma(1-\mathbf{b}+a_2+m+k)\psi(1+k)}{\Gamma(1+m+k)\Gamma(1-\mathbf{a}_{[2,3]}+a_2+m+k)k!},$$

and in accordance with (2.4) and (2.5),

$$f_2'(0) = -\frac{\sin(\pi(\mathbf{b}-a_2))}{\sin(\pi(\mathbf{a}_{[1,2,3]}-a_2))}\Big\{\sum_{a\in\mathbf{a}_{[1,2,3]}}\cot(\pi(a-a_2)) - \sum_{b\in\mathbf{b}}\cot(\pi(b-a_2))\Big\}$$
$$\times\,{}_{p+1}\phi_p\left(\begin{matrix}a_2+m, 1-\mathbf{b}+a_2+m\\ 1+m, 1-\mathbf{a}_{[2,3]}+a_2+m\end{matrix}\right)$$
$$-\frac{\sin(\pi(\mathbf{b}-a_2))}{\pi\sin(\pi(\mathbf{a}_{[1,2,3]}-a_2))}\sum_{k=0}^{\infty}\frac{\Gamma(a_2+m+k)\Gamma(1-\mathbf{b}+a_2+m+k)}{\Gamma(1+m+k)\Gamma(1-\mathbf{a}_{[2,3]}+a_2+m+k)k!}$$
$$\times\Big[\psi(a_2+m+k)+\sum_{b\in\mathbf{b}}\psi(1-b+a_2+m+k)$$
$$-\sum_{a\in\mathbf{a}_{[2,3]}}\psi(1-a+a_2+m+k)-\psi(1+m+k)\Big].$$

Substituting $(f_1'(0)+f_2'(0))/\pi$ in place of the left-hand side of (3.2) after some tedious but elementary transformations we obtain

$$\sum_{k=0}^{\infty}\frac{(a_2+m)_k(1-\mathbf{b}+a_2+m)_k}{(1+m)_k(1-\mathbf{a}_{[2,3]}+a_2+m)_k k!}$$
$$\times\Big\{\sum_{a\in\mathbf{a}_{[2,3]}}\psi(1-a+a_2+m+k)+\psi(1+m+k)+\psi(1+k)-\psi(a_2+m+k)$$
$$-\sum_{b\in\mathbf{b}}\psi(1-b+a_2+m+k)\Big\}$$
$$=\hat{U}-\hat{V}+\frac{(-1)^m\Gamma(\mathbf{b}-a_2-m)\Gamma(1-a_1+a_2+m)m!\,\Gamma(\mathbf{a}_{[1,2]})}{\Gamma(\mathbf{a}_{[1,2,3]}-a_2-m)(a_2)_m\Gamma(1-a_1)\Gamma(\mathbf{b})}\,{}_{p+1}F_p\left(\begin{matrix}\mathbf{a}\\ \mathbf{b}\end{matrix}\right)$$
$$-\frac{(-1)^m\Gamma(\mathbf{b}-a_2-m)\Gamma(1-a_1+a_2+m)}{\Gamma(\mathbf{a}_{[1,2,3]}-a_2-m)\Gamma(a_2+m)}$$
$$\times\sum_{k=4}^{p+1}\frac{m!\,\Gamma(a_k)\Gamma(\mathbf{a}_{[1,k]}-a_k)}{\Gamma(1-a_1+a_k)\Gamma(\mathbf{b}-a_k)}\,{}_{p+1}F_p\left(\begin{matrix}a_k, 1-\mathbf{b}+a_k\\ 1-\mathbf{a}_{[k]}+a_k\end{matrix}\right),$$

where

$$\hat{U}=\pi\Big[\sum_{a\in\mathbf{a}_{[1,2,3]}}\cot(\pi(a-a_2))-\sum_{b\in\mathbf{b}}\cot(\pi(b-a_2))\Big]\,{}_{p+1}F_p\left(\begin{matrix}a_2+m, 1-\mathbf{b}+a_2+m\\ 1+m, 1-\mathbf{a}_{[2,3]}+a_2+m\end{matrix}\right)$$

and

$$\hat{V}=\frac{(1-\mathbf{a}_{[2,3]}+a_2)_m m!}{(1-\mathbf{b}+a_2)_m(a_2)_m}\sum_{k=0}^{m-1}\frac{(-1)^{m-k}(m-k-1)!(a_2)_k(1-\mathbf{b}+a_2)_k}{(1-\mathbf{a}_{[2,3]}+a_2)_k k!}.$$

Writing $\mathbf{c} = (a_2 + m, 1 - \mathbf{b} + a_2 + m) \in \mathbb{C}^{p+1}$, $\mathbf{d} = (1 - \mathbf{a}_{[2,3]} + a_2 + m) \in \mathbb{C}^{p-1}$ and applying (2.8), we arrive at (3.1). □

Next, we derive the "+" case of (1.2) from the above theorem.

COROLLARY 3.2. *Suppose* $\mathbf{c} \in (\mathbb{C} \setminus \mathbb{Z})^p$, $\mathbf{d} \in (\mathbb{C} \setminus \mathbb{Z})^{p-1}$ *and* $\Re(\sum_{j=1}^{p-1} d_j - \sum_{j=1}^{p}(c_j)) > 0$. *Then*

(3.3)

$$\sum_{k=0}^{\infty} \frac{(\mathbf{c})_k}{(\mathbf{d})_k k!} \Big\{ \sum_{d\in\mathbf{d}} \psi(\mathbf{d}+k) + \psi(1+k) - \sum_{c\in\mathbf{c}} \psi(\mathbf{c}+k) \Big\}$$

$$= \pi \left[\sum_{c\in\mathbf{c}} \cot(\pi c) - \sum_{d\in\mathbf{d}} \cot(\pi d) \right] {}_pF_{p-1}\left(\begin{matrix}\mathbf{c}\\ \mathbf{d}\end{matrix}\right) - \frac{(1-\mathbf{d})_1}{(1-\mathbf{c})_1} {}_{p+1}F_p\left(\begin{matrix}1,1,2-\mathbf{d}\\ 2-\mathbf{c}\end{matrix}\right)$$

$$- \frac{\pi\Gamma(1-\mathbf{c})}{\Gamma(1-\mathbf{d})} \sum_{k=1}^{p-1} \cot(\pi d_k) \frac{\Gamma(d_k-1)\Gamma(d_k-\mathbf{d}_{[k]})}{\Gamma(d_k-\mathbf{c})} {}_pF_{p-1}\left(\begin{matrix}1-d_k+\mathbf{c}\\ 2-d_k, 1-d_k+\mathbf{d}_{[k]}\end{matrix}\right).$$

PROOF. Setting $c_1 = 1 + m$ in (3.1) we obtain

(3.4)

$$\sum_{k=0}^{\infty} \frac{(\mathbf{c}_{[1]})_k}{(\mathbf{d})_k k!} \left[\sum_{d\in\mathbf{d}} \psi(d+k) + \psi(1+k) - \sum_{c\in\mathbf{c}_{[1]}} \psi(c+k) \right]$$

$$= \pi \left[\sum_{c\in\mathbf{c}_{[1]}} \cot(\pi c) - \sum_{d\in\mathbf{d}_{[1]}} \cot(\pi d) \right] {}_pF_{p-1}\left(\begin{matrix}\mathbf{c}_{[1]}\\ \mathbf{d}\end{matrix}\right) - m!\,\hat{V}$$

$$+ \frac{(-1)^m m!\,\Gamma(d_1)\Gamma(1-\mathbf{c}_{[1]})\Gamma(1-\mathbf{d}_{[1]}+m+1)}{\Gamma(d_1-m-1)\Gamma(1-\mathbf{c}_{[1]}+m+1)\Gamma(1-\mathbf{d}_{[1]})} {}_{p+1}F_p\left(\begin{matrix}m+1,1,1-\mathbf{d}+c_1\\ 1-\mathbf{c}_{[1]}+c_1\end{matrix}\right)$$

$$+ \frac{\Gamma(d_1)\Gamma(1-\mathbf{c}_{[1]})}{\Gamma(1-\mathbf{d}_{[1]})}$$

$$\times \sum_{k=2}^{p-1} \frac{\Gamma(d_k-1)\Gamma(d_k)\Gamma(1-d_k)\Gamma(d_k-\mathbf{d}_{[1,k]})}{\Gamma(d_k-\mathbf{c}_{[1]})\Gamma(1-d_k+d_1)} {}_pF_{p-1}\left(\begin{matrix}1-d_k+\mathbf{c}_{[1]}\\ 2-d_k, 1-d_k+\mathbf{d}_{[k]}\end{matrix}\right),$$

where

$$\hat{V} = \frac{(1-\mathbf{d})_m}{(1-\mathbf{c})_m} \sum_{k=0}^{m-1} \frac{(-1)^{m-k}(m-k-1)!(\mathbf{c}-m)_k}{(\mathbf{d}-m)_k k!} = \frac{(1-\mathbf{d})_1}{m!\,(1-\mathbf{c}[1])_1} \sum_{j=0}^{m-1} \frac{j!(2-\mathbf{d})_j}{(2-\mathbf{c}_{[1]})_j}$$

(we used that $c_1 = m+1$ in the last equality). It is easy to see that for $c_1 = m+1$, in view of $(a+m)_j = (a)_{m+j}/(a)_m$ we will have

$$\begin{aligned} &- m!\,\hat{V} + \frac{(-1)^m m!\,\Gamma(d_1)\Gamma(1-\mathbf{c}_{[1]})\Gamma(1-\mathbf{d}_{[1]}+m+1)}{\Gamma(d_1-m-1)\Gamma(1-\mathbf{c}_{[1]}+m+1)\Gamma(1-\mathbf{d}_{[1]})} \\ &\times {}_{p+1}F_p\left(\begin{matrix}m+1,1,1-\mathbf{d}+c_1\\ 1-\mathbf{c}_{[1]}+c_1\end{matrix}\right) = -\frac{(1-\mathbf{d})_1}{(1-\mathbf{c}_{[1]})_1} {}_{p+1}F_p\left(\begin{matrix}1,1,2-\mathbf{d}\\ 2-\mathbf{c}_{[1]}\end{matrix}\right). \end{aligned} \tag{3.5}$$

Substituting (3.5) into (3.1) and changing $\mathbf{c}_{[1]} \to \mathbf{c}$ we obtain

$$\sum_{k=0}^{\infty} \frac{(\mathbf{c})_k}{(\mathbf{d})_k k!} \left[\sum_{d\in\mathbf{d}} \psi(d+k) + \psi(1+k) - \sum_{c\in\mathbf{c}} \psi(\mathbf{c}+k) \right]$$

$$
\begin{aligned}
&= \pi\left[\sum_{c\in\mathbf{c}}\cot(\pi c) - \sum_{d\in\mathbf{d}_{[1]}}\cot(\pi d)\right]{}_pF_{p-1}\left(\begin{matrix}\mathbf{c}\\ \mathbf{d}\end{matrix}\right) - \frac{(1-\mathbf{d})_1}{(1-\mathbf{c})_1}\,{}_{p+1}F_p\left(\begin{matrix}1,1,2-\mathbf{d}\\ 2-\mathbf{c}\end{matrix}\right)\\
&\quad + \frac{\pi\Gamma(1-\mathbf{c})}{\Gamma(1-\mathbf{d})}\\
&\quad\times\sum_{k=2}^{p-1}\frac{\sin[\pi(d_k-d_1)]\Gamma(d_k-1)\Gamma(d_k-\mathbf{d}_{[k]})}{\sin(\pi d_1)\sin(\pi d_k)\Gamma(d_k-\mathbf{c})}\,{}_pF_{p-1}\left(\begin{matrix}1-d_k+\mathbf{c}\\ 2-d_k, 1-d_k+\mathbf{d}_{[k]}\end{matrix}\right).
\end{aligned}
$$

Using

$$
\frac{\sin(\pi(d_k-d_1))}{\sin(\pi d_1)\sin(\pi d_k)} = \cot(\pi d_1) - \cot(\pi d_k)
$$

and expressing as in [**8**, (5.3)],

$$
\begin{aligned}
&\sum_{k=2}^{p-1}\frac{\Gamma(d_k-1)\Gamma(d_k-\mathbf{d}_{[k]})}{\Gamma(d_k-\mathbf{c})}\,{}_pF_{p-1}\left(\begin{matrix}1-d_k+\mathbf{c}\\ 2-d_k, 1-d_k+\mathbf{d}_{[k]}\end{matrix}\right)\\
&= -\frac{\Gamma(1-\mathbf{d})}{\Gamma(1-\mathbf{c})}\,{}_pF_{p-1}\left(\begin{matrix}\mathbf{c}\\ \mathbf{d}\end{matrix}\right) - \frac{\Gamma(d_1-1)\Gamma(d_1-\mathbf{d}_{[1]})}{\Gamma(d_1-\mathbf{c})}\,{}_pF_{p-1}\left(\begin{matrix}1-d_1+\mathbf{c}\\ 2-d_1, 1-d_1+\mathbf{d}_{[1]}\end{matrix}\right),
\end{aligned}
$$

we finally arrive at (3.3). □

Equating (2.2) and (3.1), after some rearrangements and trigonometric manipulations we have

$$
\begin{aligned}
&\frac{\Gamma(a_1)\Gamma(d_1)\Gamma(1-\mathbf{d}_{[1]}+a_1)}{(1-a_1)_m\Gamma(d_1-a_1)\Gamma(1-\mathbf{a}_{[1]}+a_1)}\,{}_{p+1}F_p\left(\begin{matrix}a_1, a_1-m, 1-\mathbf{d}+a_1\\ 1-\mathbf{a}_{[1]}+a_1\end{matrix}\right)\\
&\quad+\frac{\Gamma(d_1-\mathbf{d}_{[1]})}{\Gamma(d_1-\mathbf{a})}\\
&\quad\times\frac{\Gamma(d_1-m-1)}{(-1)^m\Gamma(1-d_1)}\Gamma(1-a_1)\Gamma(d_1-1)\,{}_{p+1}F_p\left(\begin{matrix}1-d_1+\mathbf{a}\\ 2-d_1, 2-d_1+m, 1-d_1+\mathbf{d}_{[1]}\end{matrix}\right)\\
&\quad-\frac{\pi\Gamma(1-\mathbf{d}_{[1]})}{m!\,\Gamma(1-\mathbf{a}_{[1]})}\left(\cot(\pi a_1)-\cot(\pi d_1)\right){}_{p+1}F_p\left(\begin{matrix}\mathbf{a}\\ 1+m, \mathbf{d}\end{matrix}\right)\\
&= -(-1)^m\sum_{k=2}^{p-1}\frac{\Gamma(d_k-m-1)\Gamma(d_k-\mathbf{d}_{[k]})\Gamma(d_1)\Gamma(1-a_1)\Gamma(1-d_k+a_1)}{\Gamma(2-d_k)\Gamma(a_1-d_1)\Gamma(1-a_1+d_1)\Gamma(d_k-\mathbf{a}_{[1]})}\\
&\quad\times{}_{p+1}F_p\left(\begin{matrix}1-d_k+\mathbf{a}\\ 2-d_k, 2-d_k+m, 1-d_k+\mathbf{d}_{[k]}\end{matrix}\right).
\end{aligned}
$$

This leads to

$$
\begin{aligned}
&\frac{\Gamma(a_1)\Gamma(d_1)\Gamma(1-\mathbf{d}_{[1]}+a_1)}{(1-a_1)_m\Gamma(d_1-a_1)\Gamma(1-\mathbf{a}_{[1]}+a_1)}\,{}_{p+1}F_p\left(\begin{matrix}a_1, a_1-m, 1-\mathbf{d}+a_1\\ 1-\mathbf{a}_{[1]}+a_1\end{matrix}\right)\\
&\quad-\frac{\pi\Gamma(1-\mathbf{d}_{[1]})}{m!\,\Gamma(1-\mathbf{a}_{[1]})}\left(\cot(\pi a_1)-\cot(\pi d_1)\right){}_{p+1}F_p\left(\begin{matrix}\mathbf{a}\\ 1+m, \mathbf{d}\end{matrix}\right)\\
&\quad= -(-1)^m\sum_{k=1}^{p-1}\frac{\Gamma(d_1)\Gamma(d_k-m-1)\Gamma(d_k-\mathbf{d}_{[k]})\Gamma(1-a_1)\Gamma(1-d_k+a_1)}{\Gamma(2-d_k)\Gamma(a_1-d_1)\Gamma(1-a_1+d_1)\Gamma(d_k-\mathbf{a}_{[1]})}\\
&\quad\times{}_{p+1}F_p\left(\begin{matrix}1-d_k+\mathbf{a}\\ 2-d_k, 2-d_k+m, 1-d_k+\mathbf{d}_{[k]}\end{matrix}\right).
\end{aligned}
$$

Rearranging the above formula, we obtain the following presumably new $(p+1)$-term identity.

COROLLARY 3.3. *The following identity holds for* $m \in \mathbb{N}$ *and the values of parameters making the hypergeometric series involved converge:*

$$
\begin{aligned}
&-\frac{\Gamma(1-\mathbf{d})}{m!\,\Gamma(1-\mathbf{a})}{}_{p+1}F_p\left(\begin{matrix}\mathbf{a}\\1+m,\mathbf{d}\end{matrix}\right)\\
&+\frac{\sin(\pi a_1)\Gamma(a_1)\Gamma(1-\mathbf{d}+a_1)}{\pi\Gamma(1-a_1+m)\Gamma(1-\mathbf{a}_{[1]}+a_1)}{}_{p+1}F_p\left(\begin{matrix}a_1,a_1-m,1-\mathbf{d}+a_1\\1-\mathbf{a}_{[1]}+a_1\end{matrix}\right)\\
&=(-1)^m\sum_{k=1}^{p-1}\frac{\Gamma(d_k-m-1)\Gamma(d_k-\mathbf{d}_{[k]})}{\Gamma(2-d_k)\Gamma(d_k-\mathbf{a})}\\
&\quad\times\frac{\sin(\pi a_1)}{\sin(\pi(d_k-a_1))}{}_{p+1}F_p\left(\begin{matrix}1-d_k+\mathbf{a}\\2-d_k,2-d_k+m,1-d_k+\mathbf{d}_{[k]}\end{matrix}\right).
\end{aligned}\tag{3.6}
$$

This formula can be viewed as the degenerate case of [**8**, (5.3)] when one of the bottom parameters is a positive integer. We emphasize that computing the corresponding limit in [**8**, (5.3)] will lead to psi series, not to the above, formula which appears to be significantly less trivial.

4. The third transformation formula

Suppose $0 \leqslant n \leqslant p$, $0 \leqslant s \leqslant p$ are integers that satisfy $s+n \geqslant p$, and $\mathbf{a} \in \mathbb{C}^n$, $\mathbf{b} \in \mathbb{C}^s$, $\mathbf{c} \in \mathbb{C}^{p-n}$, $\mathbf{d} \in \mathbb{C}^{p-s}$ are complex vectors such that

$$
\Re\Big(\sum\nolimits_{j=1}^{n} a_j+\sum\nolimits_{j=1}^{p-n} c_j-\sum\nolimits_{j=1}^{s} b_j-\sum\nolimits_{j=1}^{p-s} d_j\Big)>0.
$$

Under these conditions we have recently established the identity [**8**, Theorem 5.1]

$$
\begin{aligned}
&\sum_{k=1}^{s}\frac{A_k}{b_k}{}_{p+1}F_p\left(\begin{matrix}1-\mathbf{a}+b_k,1-\mathbf{c}+b_k,b_k\\1-\mathbf{b}_{[k]}+b_k,1-\mathbf{d}+b_k,b_k+1\end{matrix}\middle|(-1)^{p-s-n}\right)\\
&+\sum_{k=1}^{n}\frac{B_k}{(1-a_k)}{}_{p+1}F_p\left(\begin{matrix}1+\mathbf{b}-a_k,1+\mathbf{d}-a_k,1-a_k\\1+\mathbf{a}_{[k]}-a_k,1+\mathbf{c}-a_k,2-a_k\end{matrix}\middle|(-1)^{p-s-n}\right)\\
&=\frac{\Gamma(1-\mathbf{a})\Gamma(\mathbf{b})}{\Gamma(\mathbf{c})\Gamma(1-\mathbf{d})},
\end{aligned}\tag{4.1a}
$$

where

$$
A_k=\frac{\Gamma(\mathbf{b}_{[k]}-b_k)\Gamma(1-\mathbf{a}+b_k)}{\Gamma(\mathbf{c}-b_k)\Gamma(1-\mathbf{d}+b_k)},\qquad B_k=\frac{\Gamma(a_k-\mathbf{a}_{[k]})\Gamma(1+\mathbf{b}-a_k)}{\Gamma(a_k-\mathbf{d})\Gamma(1+\mathbf{c}-a_k)},\tag{4.1b}
$$

and it is assumed that no numerator gamma factor becomes infinite. If the latter condition is violated it is still possible to compute the corresponding limit, leading naturally to psi series. Using this approach we will establish the following.

THEOREM 4.1. *Suppose* $0 \leqslant n \leqslant p$, $0 \leqslant s \leqslant p-1$, $\kappa := s-n \geqslant -2$, *and*

$$
\begin{aligned}
&\mathbf{a}_1 \in (\mathbb{C}\setminus\mathbb{Z})^n, && \mathbf{a}_2 \in \mathbb{C}^{p-n}, && \mathbf{a}=(\mathbf{a}_1,\mathbf{a}_2)\in\mathbb{C}^p,\\
&\mathbf{b}_1 \in (\mathbb{C}\setminus\mathbb{Z})^s, && \mathbf{b}_2 \in \mathbb{C}^{p-1-s}, && \mathbf{b}=(\mathbf{b}_1,\mathbf{b}_2)\in\mathbb{C}^{p-1}.
\end{aligned}
$$

Assume further that $\Re\left(\sum_{b\in\mathbf{b}} b - \sum_{a\in\mathbf{a}} a\right) > 0$. *Then*

(4.2)

$$\sum_{k=0}^{\infty}\frac{(\mathbf{a})_k[(-1)^{\kappa}]^k}{\pi(\mathbf{b})_k k!}\left[\sum_{b\in\mathbf{b}}\psi(b+k)+\psi(1+k)-\sum_{a\in\mathbf{a}}\psi(a+k)\right]$$
$$=\left\{\sum_{a\in\mathbf{a}_1}\cot(\pi a)-\sum_{b\in\mathbf{b}_1}\cot(\pi b)\right\}{}_pF_{p-1}\left(\begin{matrix}\mathbf{a}\\ \mathbf{b}\end{matrix}\middle|(-1)^{\kappa}\right)$$
$$-\frac{(-1)^{\kappa}(1-\mathbf{b})_1}{\pi(1-\mathbf{a})_1}{}_{p+1}F_p\left(\begin{matrix}1,1,2-\mathbf{b}\\ 2-\mathbf{a}\end{matrix}\middle|(-1)^{\kappa}\right)$$
$$+\frac{\Gamma(1-\mathbf{a}_1)\Gamma(\mathbf{b}_2)}{\Gamma(1-\mathbf{b}_1)\Gamma(\mathbf{a}_2)}$$
$$\times\sum_{a_k\in\mathbf{a}_2}\frac{\Gamma(a_k)\Gamma(\mathbf{a}_{2[k]}-a_k)\Gamma(1-\mathbf{b}_1+a_k)}{\sin(\pi a_k)\Gamma(\mathbf{b}_2-a_k)\Gamma(1-\mathbf{a}_1+a_k)}{}_pF_{p-1}\left(\begin{matrix}a_k,1-\mathbf{b}+a_k\\ 1-\mathbf{a}_{[k]}+a_k\end{matrix}\middle|(-1)^{\kappa}\right)$$
$$+\frac{\Gamma(1-\mathbf{a}_1)\Gamma(\mathbf{b}_2)}{\Gamma(1-\mathbf{b}_1)\Gamma(\mathbf{a}_2)}\sum_{b_k\in\mathbf{b}_1}\frac{\Gamma(b_k-1)\Gamma(b_k-\mathbf{b}_{1[k]})\Gamma(1-b_k+\mathbf{a}_2)}{\sin(\pi b_k)\Gamma(b_k-\mathbf{a}_1)\Gamma(1-b_k+\mathbf{b}_2)}$$
$$\times{}_pF_{p-1}\left(\begin{matrix}1-b_k+\mathbf{a}\\ 2-b_k,1-b_k+\mathbf{b}_{[k]}\end{matrix}\middle|(-1)^{\kappa}\right).$$

PROOF. Setting $b_2=b_1+m+\varepsilon$ the first two terms on the left-hand side of (4.1a) with coefficients A_1, A_2 take the form (put $\varkappa=p-s-n$ for brevity):

$$\frac{\Gamma(\varepsilon+m)\Gamma(\mathbf{b}_{[1,2]}-b_1)\Gamma(1-\mathbf{a}+b_1)}{b_1\Gamma(\mathbf{c}-b_1)\Gamma(1-\mathbf{d}+b_1)}$$
$$\times{}_{p+1}F_p\left(\begin{matrix}1-\mathbf{a}+b_1,1-\mathbf{c}+b_1,b_1\\ 1-m-\varepsilon,1-\mathbf{b}_{[1,2]}+b_1,1-\mathbf{d}+b_1,b_1+1\end{matrix}\middle|(-1)^{\varkappa}\right)$$
$$+\frac{\Gamma(-\varepsilon-m)\Gamma(\mathbf{b}_{[1,2]}-b_1-m-\varepsilon)\Gamma(1-\mathbf{a}+b_1+m+\varepsilon)}{(b_1+m+\varepsilon)\Gamma(\mathbf{c}-b_1-m-\varepsilon)\Gamma(1-\mathbf{d}+b_1+m+\varepsilon)}$$
$$\times{}_{p+1}F_p\left(\begin{matrix}1-\mathbf{a}+b_1+m+\varepsilon,1-\mathbf{c}+b_1+m+\varepsilon,b_1+m+\varepsilon\\ 1+m+\varepsilon,1-\mathbf{b}_{[1,2]}+b_1+m+\varepsilon,1-\mathbf{d}+b_1+m+\varepsilon,b_1+m+\varepsilon+1\end{matrix}\middle|(-1)^{\varkappa}\right).$$

Simple transformations of this expression lead to

$$\frac{(-1)^m}{\pi^{\varkappa+1}\sin(\pi\varepsilon)}\{f_1(\varepsilon)-f_2(\varepsilon)\}=\frac{(-1)^m}{\pi^{\varkappa+1}\sin(\pi\varepsilon)}\left[f_1(0)+\varepsilon f_1'(0)-f_2(0)-\varepsilon f_2'(0)+O(\varepsilon^2)\right]$$

as $\varepsilon\to 0$, where ϕ is defined in (2.1) and

$$f_1(\varepsilon)=\frac{\sin(\pi(\mathbf{c}-b_1))}{\sin(\pi(\mathbf{b}_{[1,2]}-b_1))}{}_{p+1}\phi_p\left(\begin{matrix}1-\mathbf{a}+b_1,1-\mathbf{c}+b_1,b_1\\ 1-m-\varepsilon,1-\mathbf{b}_{[1,2]}+b_1,1-\mathbf{d}+b_1,b_1+1\end{matrix}\middle|(-1)^{\varkappa}\right),$$
$$f_2(\varepsilon)=\frac{(-1)^{m\varkappa}\sin(\pi(1-\mathbf{c}+b_1+\varepsilon))}{\sin(\pi(1-\mathbf{b}_{[1,2]}+b_1+\varepsilon))}$$
$$\times{}_{p+1}\phi_p\left(\begin{matrix}1-\mathbf{a}+b_1+m+\varepsilon,1-\mathbf{c}+b_1+m+\varepsilon,b_1+m+\varepsilon\\ 1+m+\varepsilon,1-\mathbf{b}_{[1,2]}+b_1+m+\varepsilon,1-\mathbf{d}+b_1+m+\varepsilon,b_1+m+\varepsilon+1\end{matrix}\middle|(-1)^{\varkappa}\right).$$

Note that $f_1(0)=f_2(0)$, because $\sin(\pi(1-\mathbf{c}+b_1))=\sin(\pi(\mathbf{c}-b_1))$ and

$${}_{p+1}\phi_p\left(\begin{matrix}1-\mathbf{a}+b_1,1-\mathbf{c}+b_1,b_1\\ 1-m,1-\mathbf{b}_{[1,2]}+b_1,1-\mathbf{d}+b_1,b_1+1\end{matrix}\middle|(-1)^{\varkappa}\right)$$
$$=\sum_{k=0}^{\infty}\frac{\Gamma(1-\mathbf{a}+b_1+k)\Gamma(1-\mathbf{c}+b_1+k)\Gamma(b_1+k)[(-1)^{\varkappa}]^k}{\Gamma(1-m+k)\Gamma(1-\mathbf{b}_{[1,2]}+b_1+k)\Gamma(1-\mathbf{d}+b_1+k)\Gamma(b_1+1+k)k!}$$

$$
\begin{aligned}
&= \sum_{k=m}^{\infty} \frac{\Gamma(1-\mathbf{a}+b_1+k)\Gamma(1-\mathbf{c}+b_1+k)\Gamma(b_1+k)[(-1)^{\varkappa}]^k}{\Gamma(1-m+k)\Gamma(1-\mathbf{b}_{[1,2]}+b_1+k)\Gamma(1-\mathbf{d}+b_1+k)\Gamma(b_1+1+k)k!}\\
&= \sum_{k=0}^{\infty} \frac{\Gamma(1-\mathbf{a}+b_1+m+k)\Gamma(1-\mathbf{c}+b_1+m+k)\Gamma(b_1+m+k)[(-1)^{\varkappa}]^k}{k!\Gamma(1-\mathbf{b}_{[1,2]}+b_1+m+k)\Gamma(1-\mathbf{d}+b_1+m+k)\Gamma(b_1+1+m+k)(m+k)!}\\
&= (-1)^{m\varkappa}\\
&\quad\times {}_{p+1}\phi_p\left(\begin{matrix}1-\mathbf{a}+b_1+m, 1-\mathbf{c}+b_1+m, b_1+m\\ 1+m, 1-\mathbf{b}_{[1,2]}+b_1+m, 1-\mathbf{d}+b_1+m, b_1+m+1\end{matrix}\middle|(-1)^{\varkappa}\right).
\end{aligned}
$$

Hence,

$$
\begin{aligned}
&\frac{(-1)^m}{\pi^{\varkappa+1}\sin(\pi\varepsilon)}\{f_1(\varepsilon)-f_2(\varepsilon)\}\\
&= \frac{(-1)^m\varepsilon}{\pi^{\varkappa+1}\sin(\pi\varepsilon)}\Big(f_1'(0)-f_2'(0)+O(\varepsilon)\Big) \to \frac{(-1)^m}{\pi^{\varkappa+2}}\Big(f_1'(0)-f_2'(0)\Big) \text{ as } \varepsilon\to 0.
\end{aligned}
$$

In view of $\frac{d}{d\varepsilon}\left[\Gamma(1-m-\varepsilon)\right]^{-1} = \psi(1-m-\varepsilon)/\Gamma(1-m-\varepsilon)$ and formula (2.6), we get by separating summations over $k=0,\dots,m-1$ and $k=m,m+1,\dots$:

$$
\begin{aligned}
f_1'(0) &= \frac{\sin(\pi(\mathbf{c}-b_1))}{\sin(\pi(\mathbf{b}_{[1,2]}-b_1))}\\
&\quad\times\sum_{k=0}^{m-1}\frac{(-1)^{m-k}(m-k-1)![(-1)^{\varkappa}]^k\Gamma(1-\mathbf{a}+b_1+k)\Gamma(1-\mathbf{c}+b_1+k)\Gamma(b_1+k)}{\Gamma(1-\mathbf{b}_{[1,2]}+b_1+k)\Gamma(1-\mathbf{d}+b_1+k)\Gamma(b_1+1+k)k!}\\
&\quad+\frac{\sin(\pi(\mathbf{c}-b_1))}{\sin(\pi(\mathbf{b}_{[1,2]}-b_1))}\\
&\quad\times\sum_{k=0}^{\infty}\frac{[(-1)^{\varkappa}]^{k+m}\Gamma(1-\mathbf{a}+b_1+m+k)\Gamma(1-\mathbf{c}+b_1+m+k)\Gamma(b_1+m+k)\psi(1+k)}{\Gamma(1-\mathbf{b}_{[1,2]}+b_1+m+k)\Gamma(1-\mathbf{d}+b_1+m+k)\Gamma(b_1+1+m+k)\Gamma(1+m+k)k!},
\end{aligned}
$$

or

$$
\begin{aligned}
f_1'(0) &= \frac{\pi^{\varkappa+2}\Gamma(1-\mathbf{a}+b_1)\Gamma(\mathbf{b}_{[1,2]}-b_1)}{\Gamma(\mathbf{c}-b_1)\Gamma(1-\mathbf{d}+b_1)}\\
&\quad\times\sum_{k=0}^{m-1}\frac{(-1)^{m-k}(m-k-1)![(-1)^{\varkappa}]^k(1-\mathbf{a}+b_1)_k(1-\mathbf{c}+b_1)_k}{(1-\mathbf{b}_{[1,2]}+b_1)_k(1-\mathbf{d}+b_1)_k(b_1+k)k!}\\
&\quad+\frac{\pi^{\varkappa+2}\Gamma(\mathbf{b}_{[1,2]}-b_1-m)\Gamma(1-\mathbf{a}+b_1+m)}{\Gamma(\mathbf{c}-b_1-m)\Gamma(1-\mathbf{d}+b_1+m)}\\
&\quad\times\sum_{k=0}^{\infty}\frac{(-1)^{\varkappa k}(1-\mathbf{a}+b_1+m)_k(1-\mathbf{c}+b_1+m)_k\psi(1+k)}{(1-\mathbf{b}_{[1,2]}+b_1+m)_k(1-\mathbf{d}+b_1+m)_k(b_1+m+k)(m+k)!k!}.
\end{aligned}
$$

Similarly, by applying (2.4) and (2.5) and in view of $\sin(\pi(1-\alpha))=\sin(\pi\alpha)$ and $\cot(\pi(1-\alpha))=-\cot(\pi\alpha)$, we get [2]

$$
\begin{aligned}
f_2'(0) &= \frac{(-1)^{m\varkappa}\pi\sin(\pi(\mathbf{c}-b_1))}{\sin(\pi(\mathbf{b}_{[1,2]}-b_1))}\left[\sum_{b\in\mathbf{b}_{[1,2]}}\cot(\pi(b-b_1))-\sum_{c\in\mathbf{c}}\cot(\pi(c-b_1))\right]\\
&\quad\times {}_{p+1}\phi_p\left(\begin{matrix}1-\mathbf{a}+b_1+m, 1-\mathbf{c}+b_1+m, b_1+m\\ 1+m, 1-\mathbf{b}_{[1,2]}+b_1+m, 1-\mathbf{d}+b_1+m, b_1+m+1\end{matrix}\middle|(-1)^{\varkappa}\right)\\
&\quad+\frac{(-1)^{m\varkappa}\sin(\pi(\mathbf{c}-b_1))}{\sin(\pi(\mathbf{b}_{[1,2]}-b_1))}\\
&\quad\times\sum_{k=0}^{\infty}\frac{[(-1)^{\varkappa}]^k\Gamma(1-\mathbf{a}+b_1+m+k)\Gamma(1-\mathbf{c}+b_1+m+k)\Gamma(b_1+m+k)\Psi}{\Gamma(1-\mathbf{b}_{[1,2]}+b_1+m+k)\Gamma(1-\mathbf{d}+b_1+m+k)\Gamma(b_1+1+m+k)(m+k)!k!}
\end{aligned}
$$

$$
\begin{aligned}
&= \frac{\pi^{\varkappa+3}\Gamma(\mathbf{b}_{[1,2]}-b_1-m)\Gamma(1-\mathbf{a}+b_1+m)}{\Gamma(\mathbf{c}-b_1-m)\Gamma(1-\mathbf{d}+b_1+m)(b_1+m)m!} \\
&\quad\times\left[\sum_{b\in\mathbf{b}_{[1,2]}}\cot(\pi(b-b_1))-\sum_{c\in\mathbf{c}}\cot(\pi(c-b_1))\right] \\
&\quad\times {}_{p+1}F_p\left(\begin{matrix}1-\mathbf{a}+b_1+m,1-\mathbf{c}+b_1+m,b_1+m\\ 1+m,1-\mathbf{b}_{[1,2]}+b_1+m,1-\mathbf{d}+b_1+m,b_1+m+1\end{matrix}\middle|(-1)^{\varkappa}\right) \\
&+\frac{\pi^{\varkappa+2}\Gamma(\mathbf{b}_{[1,2]}-b_1-m)\Gamma(1-\mathbf{a}+b_1+m)}{\Gamma(\mathbf{c}-b_1-m)\Gamma(1-\mathbf{d}+b_1+m)} \\
&\quad\times\sum_{k=0}^{\infty}\frac{[(-1)^{\varkappa}]^k(1-\mathbf{a}+b_1+m)_k(1-\mathbf{c}+b_1+m)_k\Psi}{(1-\mathbf{b}_{[1,2]}+b_1+m)_k(1-\mathbf{d}+b_1+m)_k(b_1+m+k)(m+k)!k!},
\end{aligned}
$$

where

$$
\begin{aligned}
\Psi=&\sum_{a\in\mathbf{a}}\psi(1-a+b_1+m+k)+\sum_{c\in\mathbf{c}}\psi(1-c+b_1+m+k)+1/(b_1+m+k)\\
&-\sum_{b\in\mathbf{b}_{[1,2]}}\psi(1-b+b_1+m+k)-\sum_{d\in\mathbf{d}}\psi(1-d+b_1+m+k)-\psi(1+m+k).
\end{aligned}
$$

Combining the above expressions we obtain

$$
\begin{aligned}
&\frac{(-1)^m}{\pi^{\varkappa+2}}\Big(f_1'(0)-f_2'(0)\Big)=\frac{(-1)^m\Gamma(1-\mathbf{a}+b_1)\Gamma(\mathbf{b}_{[1,2]}-b_1)}{\Gamma(\mathbf{c}-b_1)\Gamma(1-\mathbf{d}+b_1)} \\
&\quad\times\sum_{k=0}^{m-1}\frac{(-1)^{m-k}(m-k-1)![(-1)^{\varkappa}]^k(1-\mathbf{a}+b_1)_k(1-\mathbf{c}+b_1)_k}{(1-\mathbf{b}_{[1,2]}+b_1)_k(1-\mathbf{d}+b_1)_k(b_1+k)k!} \\
&-\frac{\pi(-1)^m\Gamma(\mathbf{b}_{[1,2]}-b_1-m)\Gamma(1-\mathbf{a}+b_1+m)}{\Gamma(\mathbf{c}-b_1-m)\Gamma(1-\mathbf{d}+b_1+m)(b_1+m)m!} \\
&\quad\times\left[\sum_{b\in\mathbf{b}_{[1,2]}}\cot(\pi(b-b_1))-\sum_{c\in\mathbf{c}}\cot(\pi(c-b_1))\right] \\
&\quad\times {}_{p+1}F_p\left(\begin{matrix}1-\mathbf{a}+b_1+m,1-\mathbf{c}+b_1+m,b_1+m\\ 1+m,1-\mathbf{b}_{[1,2]}+b_1+m,1-\mathbf{d}+b_1+m,b_1+m+1\end{matrix}\middle|(-1)^{\varkappa}\right) \\
&-\frac{(-1)^m\Gamma(\mathbf{b}_{[1,2]}-b_1-m)\Gamma(1-\mathbf{a}+b_1+m)}{\Gamma(\mathbf{c}-b_1-m)\Gamma(1-\mathbf{d}+b_1+m)} \\
&\quad\times\sum_{k=0}^{\infty}\frac{[(-1)^{\varkappa}]^k(1-\mathbf{a}+b_1+m)_k(1-\mathbf{c}+b_1+m)_k\Psi}{(1-\mathbf{b}_{[1,2]}+b_1+m)_k(1-\mathbf{d}+b_1+m)_k(b_1+m+k)(m+k)!k!}.
\end{aligned}
$$

Substitution back into (4.1) then yields: [2]

$$
\begin{aligned}
&\frac{\Gamma(1-\mathbf{a}+b_1)\Gamma(\mathbf{b}_{[1,2]}-b_1)}{(-1)^m\Gamma(\mathbf{c}-b_1)\Gamma(1-\mathbf{d}+b_1)}\sum_{k=0}^{m-1}\frac{(m-k-1)![(-1)^{\varkappa}]^k(1-\mathbf{a}+b_1)_k(1-\mathbf{c}+b_1)_k}{(-1)^{m-k}(1-\mathbf{b}_{[1,2]}+b_1)_k(1-\mathbf{d}+b_1)_k(b_1+k)k!} \\
&-\frac{\pi(-1)^m\Gamma(\mathbf{b}_{[1,2]}-b_1-m)\Gamma(1-\mathbf{a}+b_1+m)}{\Gamma(\mathbf{c}-b_1-m)\Gamma(1-\mathbf{d}+b_1+m)(b_1+m)m!}\left[\sum_{b\in\mathbf{b}_{[1,2]}}\cot(\pi(b-b_1))-\sum_{c\in\mathbf{c}}\cot(\pi(c-b_1))\right] \\
&\quad\times {}_{p+1}F_p\left(\begin{matrix}1-\mathbf{a}+b_1+m,1-\mathbf{c}+b_1+m,b_1+m\\ 1+m,1-\mathbf{b}_{[1,2]}+b_1+m,1-\mathbf{d}+b_1+m,b_1+m+1\end{matrix}\middle|(-1)^{\varkappa}\right) \\
&\quad+\frac{(-1)^m\Gamma(\mathbf{b}_{[1,2]}-b_1-m)\Gamma(1-\mathbf{a}+b_1+m)}{\Gamma(\mathbf{c}-b_1-m)\Gamma(1-\mathbf{d}+b_1+m)} \\
&\quad\times\sum_{k=0}^{\infty}\frac{[(-1)^{\varkappa}]^k(1-\mathbf{a}+b_1+m)_k(1-\mathbf{c}+b_1+m)_k\Psi}{(1-\mathbf{b}_{[1,2]}+b_1+m)_k(1-\mathbf{d}+b_1+m)_k(b_1+m+k)(m+k)!k!}
\end{aligned}
$$

$$
=\frac{\Gamma(1-\mathbf{a})\Gamma(\mathbf{b})}{\Gamma(\mathbf{c})\Gamma(1-\mathbf{d})}-\sum_{k=3}^{m'}\frac{A_k}{b_k}{}_{p+1}F_p\left(\begin{matrix}1-\mathbf{a}+b_k,1-\mathbf{c}+b_k,b_k\\1-\mathbf{b}_{[k]}+b_k,1-\mathbf{d}+b_k,b_k+1\end{matrix}\middle|(-1)^{\varkappa}\right)
$$
$$
-\sum_{k=1}^{n}\frac{B_k}{(1-a_k)}{}_{p+1}F_p\left(\begin{matrix}1+\mathbf{b}-a_k,1+\mathbf{d}-a_k,1-a_k\\1+\mathbf{a}_{[k]}-a_k,1+\mathbf{c}-a_k,2-a_k\end{matrix}\middle|(-1)^{\varkappa}\right),
$$

where $\varkappa=p-m'-n$ and the numbers A_k, B_k are defined in (4.1b) with $b_2=b_1+m$ on the right-hand side. Simple transformations of the above equality lead to

$$
\sum_{k=0}^{\infty}\frac{[(-1)^{\varkappa}]^k(1-\mathbf{a}+b_2)_k(1-\mathbf{c}+b_2)_k\Psi}{(1-\mathbf{b}_{[1,2]}+b_2)_k(1-\mathbf{d}+b_2)_k(b_2+k)(m+k)!k!}
$$
$$
=-\frac{(-1)^{m\varkappa}(1-\mathbf{b}_{[1,2]}+b_1)_m(1-\mathbf{d}+b_1)_m}{(1-\mathbf{c}+b_1)_m(1-\mathbf{a}+b_1)_m}
$$
$$
\times\sum_{k=0}^{m-1}\frac{(-1)^{m-k}(m-k-1)![(-1)^{\varkappa}]^k(1-\mathbf{a}+b_1)_k(1-\mathbf{c}+b_1)_k}{(1-\mathbf{b}_{[1,2]}+b_1)_k(1-\mathbf{d}+b_1)_k(b_1+k)k!}
$$
$$
+\frac{\pi}{b_2m!}\left[\sum_{b\in\mathbf{b}_{[1,2]}}\cot(\pi(b-b_1))-\sum_{c\in\mathbf{c}}\cot(\pi(c-b_1))\right]
$$
$$
\times{}_{p+1}F_p\left(\begin{matrix}1-\mathbf{a}+b_2,1-\mathbf{c}+b_2,b_2\\1+m,1-\mathbf{b}_{[1,2]}+b_2,1-\mathbf{d}+b_2,b_2+1\end{matrix}\middle|(-1)^{\varkappa}\right)
$$
$$
+\frac{(-1)^m\Gamma(\mathbf{c}-b_2)\Gamma(1-\mathbf{d}+b_2)}{\Gamma(\mathbf{b}_{[1,2]}-b_2)\Gamma(1-\mathbf{a}+b_2)}\left\{\frac{\Gamma(1-\mathbf{a})\Gamma(\mathbf{b})}{\Gamma(\mathbf{c})\Gamma(1-\mathbf{d})}-\sum_{k=3}^{m'}\frac{\Gamma(\mathbf{b}_{[k]}-b_k)\Gamma(1-\mathbf{a}+b_k)}{b_k\Gamma(\mathbf{c}-b_k)\Gamma(1-\mathbf{d}+b_k)}\right.
$$
$$
\times{}_{p+1}F_p\left(\begin{matrix}1-\mathbf{a}+b_k,1-\mathbf{c}+b_k,b_k\\1-\mathbf{b}_{[k]}+b_k,1-\mathbf{d}+b_k,b_k+1\end{matrix}\middle|(-1)^{\varkappa}\right)
$$
$$
\left.-\sum_{k=1}^{n}\frac{\Gamma(a_k-\mathbf{a}_{[k]})\Gamma(1+\mathbf{b}-a_k)}{(1-a_k)\Gamma(a_k-\mathbf{d})\Gamma(1+\mathbf{c}-a_k)}{}_{p+1}F_p\left(\begin{matrix}1+\mathbf{b}-a_k,1+\mathbf{d}-a_k,1-a_k\\1+\mathbf{a}_{[k]}-a_k,1+\mathbf{c}-a_k,2-a_k\end{matrix}\middle|(-1)^{\varkappa}\right)\right\}.
$$

Denote

$$
\boldsymbol{\alpha}=1-\mathbf{a}+b_1+m,\quad \boldsymbol{\beta}=1-\mathbf{c}+b_1+m,\quad \boldsymbol{\gamma}=1-\mathbf{b}_{[1,2]}+b_1+m,\quad \boldsymbol{\delta}=1-\mathbf{d}+b_1+m
$$

and recall that $b_2=b_1+m$, so that

$$
\mathbf{a}=1-\boldsymbol{\alpha}+b_2,\quad \mathbf{b}=(b_2-m,b_2,1-\boldsymbol{\gamma}+b_2),\quad \mathbf{c}=1-\boldsymbol{\beta}+b_2,\quad \mathbf{d}=1-\boldsymbol{\delta}+b_2.
$$

Then, the latter identity takes the form:

$$
\frac{1}{b_2m!}\sum_{k=0}^{\infty}\frac{(\boldsymbol{\alpha})_k(\boldsymbol{\beta})_k(b_2)_k[(-1)^{\varkappa}]^k}{(\boldsymbol{\gamma})_k(\boldsymbol{\delta})_k(b_2+1)_k(1+m)_kk!}\Psi_1
$$
$$
=\frac{\pi}{m!\,b_2}\left[\sum_{\beta\in\boldsymbol{\beta}}\cot(\pi\beta)-\sum_{\gamma\in\boldsymbol{\gamma}}\cot(\pi\gamma)\right]{}_{p+1}F_p\left(\begin{matrix}\boldsymbol{\alpha},\boldsymbol{\beta},b_2\\1+m,\boldsymbol{\gamma},\boldsymbol{\delta},b_2+1\end{matrix}\middle|(-1)^{\varkappa}\right)
$$
$$
+\frac{(-1)^m\Gamma(1-\boldsymbol{\beta})\Gamma(\boldsymbol{\delta})}{\Gamma(1-\boldsymbol{\gamma})\Gamma(\boldsymbol{\alpha})}
$$
$$
\times\left\{\frac{\Gamma(\boldsymbol{\alpha}-b_2)\Gamma(b_2-m)\Gamma(b_2)\Gamma(1-\boldsymbol{\gamma}+b_2)}{\Gamma(1-\boldsymbol{\beta}+b_2)\Gamma(\boldsymbol{\delta}-b_2)}\right.
$$
$$
-\sum_{k=1}^{m'-2}\frac{\Gamma(\gamma_k-m-1)\Gamma(\gamma_k-1)\Gamma(\gamma_k-\boldsymbol{\gamma}_{[k]})\Gamma(1-\gamma_k+\boldsymbol{\alpha})}{(1-\gamma_k+b_2)\Gamma(\gamma_k-\boldsymbol{\beta})\Gamma(1-\gamma_k+\boldsymbol{\delta})}
$$
$$
\times{}_{p+1}F_p\left(\begin{matrix}1-\gamma_k+\boldsymbol{\alpha},1-\gamma_k+\boldsymbol{\beta},1-\gamma_k+b_2\\2+m-\gamma_k,2-\gamma_k,1-\gamma_k+\boldsymbol{\gamma}_{[k]},1-\gamma_k+\boldsymbol{\delta},2-\gamma_k+b_2\end{matrix}\middle|(-1)^{\varkappa}\right)
$$

$$-\sum_{k=1}^{n}\frac{\Gamma(\boldsymbol{\alpha}_{[k]}-\alpha_k)\Gamma(\alpha_k-m)\Gamma(\alpha_k)\Gamma(1-\boldsymbol{\gamma}+\alpha_k)}{(\alpha_k-b_2)\Gamma(\boldsymbol{\delta}-\alpha_k)\Gamma(1-\boldsymbol{\beta}+\alpha_k)}$$

$$\times {}_{p+1}F_p\left(\begin{matrix}\alpha_k-m,\alpha_k,1-\boldsymbol{\gamma}+\alpha_k,1-\boldsymbol{\delta}+\alpha_k,\alpha_k-b_2\\ 1-\boldsymbol{\alpha}_{[k]}+\alpha_k,1-\boldsymbol{\beta}+\alpha_k,1+\alpha_k-b_2\end{matrix}\middle|(-1)^{\varkappa}\right)\Bigg\}$$

(4.3)

$$-\frac{(1-\boldsymbol{\gamma})_m(1-\boldsymbol{\delta})_m}{(1-\boldsymbol{\alpha})_m(1-\boldsymbol{\beta})_m}\sum_{k=0}^{m-1}\frac{(m-k-1)![(-1)^{\varkappa-1}]^{k+m}(\boldsymbol{\alpha}-m)_k(\boldsymbol{\beta}-m)_k}{(\boldsymbol{\gamma}-m)_k(\boldsymbol{\delta}-m)_k(b_2-m+k)k!},$$

where

$$\Psi_1=\sum_{\gamma\in\boldsymbol{\gamma}}\psi(\gamma+k)+\sum_{\delta\in\boldsymbol{\delta}}\psi(\delta+k)+\psi(1+m+k)+\psi(1+k)$$
$$-\sum_{\alpha\in\boldsymbol{\alpha}}\psi(\alpha+k)-\sum_{\beta\in\boldsymbol{\beta}}\psi(\beta+k)+\frac{1}{b_2+k}.$$

In view of (2.8) we have

$$\frac{(1-\boldsymbol{\gamma})_m(1-\boldsymbol{\delta})_m}{(1-\boldsymbol{\alpha})_m(1-\boldsymbol{\beta})_m}\sum_{k=0}^{m-1}\frac{(m-k-1)![(-1)^{\varkappa-1}]^{k+m}(\boldsymbol{\alpha}-m)_k(\boldsymbol{\beta}-m)_k}{(\boldsymbol{\gamma}-m)_k(\boldsymbol{\delta}-m)_k(b_2-m+k)k!}$$
$$=\frac{(-1)^{\varkappa-1}(1-\boldsymbol{\gamma})_1(1-\boldsymbol{\delta})_1}{(m-1)!(1-\boldsymbol{\alpha})_1(1-\boldsymbol{\beta})_1}\sum_{j=0}^{m-1}\frac{(1-m)_j(2-\boldsymbol{\gamma})_j(2-\boldsymbol{\delta})_j[(1)_j]^2(-1)^{\varkappa j}}{(2-\boldsymbol{\beta})_j(2-\boldsymbol{\alpha})_j(b_2-1-j)j!}$$
$$=\frac{(-1)^{\varkappa-1}(1-\boldsymbol{\gamma})_1(1-\boldsymbol{\delta})_1}{(m-1)!(1-\boldsymbol{\alpha})_1(1-\boldsymbol{\beta})_1(b_2-1)}{}_{p+2}F_{p+1}\left(\begin{matrix}1-m,2-\boldsymbol{\gamma},2-\boldsymbol{\delta},1,1,1-b_2\\ 2-\boldsymbol{\alpha},2-\boldsymbol{\beta},2-b_2\end{matrix}\middle|(-1)^{\varkappa}\right).$$

Next, we put $\boldsymbol{\alpha}=(1+m,b_2+1,\boldsymbol{\alpha}')$ and substitute it into the large identity above:

$$\sum_{k=0}^{\infty}\frac{(\boldsymbol{\alpha}')_k(\boldsymbol{\beta})_k(b_2)_k[(-1)^{\varkappa}]^k}{(\boldsymbol{\gamma})_k(\boldsymbol{\delta})_k k!}$$
$$\times\left\{\sum_{\gamma\in\boldsymbol{\gamma}}\psi(\gamma+k)+\sum_{\delta\in\boldsymbol{\delta}}\psi(\delta+k)+\psi(1+k)\right.$$
$$\left.-\sum_{\alpha\in\boldsymbol{\alpha}'}\psi(\alpha+k)-\sum_{\beta\in\boldsymbol{\beta}}\psi(\beta+k)-\psi(b_2+k)\right\}$$

$$
\begin{aligned}
&= \pi\left[\sum_{\beta\in\boldsymbol{\beta}}\cot(\pi\boldsymbol{\beta}) - \sum_{\gamma\in\boldsymbol{\gamma}}\cot(\pi\boldsymbol{\gamma})\right]{}_{p-1}F_{p-2}\left(\left.\begin{matrix}\boldsymbol{\alpha}',\boldsymbol{\beta},b_2\\ \boldsymbol{\gamma},\boldsymbol{\delta}\end{matrix}\right|(-1)^{\varkappa}\right)\\
&\quad+\frac{(-1)^m\Gamma(1-\boldsymbol{\beta})\Gamma(\boldsymbol{\delta})}{\Gamma(1-\boldsymbol{\gamma})\Gamma(\boldsymbol{\alpha}')\Gamma(b_2)}\left\{\frac{\pi\Gamma(\boldsymbol{\alpha}'-b_2)\Gamma(b_2)\Gamma(1-\boldsymbol{\gamma}+b_2)}{(-1)^m\sin(\pi b_2)\Gamma(1-\boldsymbol{\beta}+b_2)\Gamma(\boldsymbol{\delta}-b_2)}\right.\\
&\qquad\times{}_{p-1}F_{p-2}\left(\left.\begin{matrix}b_2,1-\boldsymbol{\gamma}+b_2,1-\boldsymbol{\delta}+b_2\\ 1-\boldsymbol{\alpha}'+b_2,1-\boldsymbol{\beta}+b_2\end{matrix}\right|(-1)^{\varkappa}\right)\\
&\quad-\sum_{k=1}^{m'-2}\frac{\pi\Gamma(\gamma_k-1)\Gamma(\gamma_k-\boldsymbol{\gamma}_{[k]})\Gamma(1-\gamma_k+\boldsymbol{\alpha}')\Gamma(1-\gamma_k+b_2)}{(-1)^m\sin(\pi(\gamma_k-1))\Gamma(\gamma_k-\boldsymbol{\beta})\Gamma(1-\gamma_k+\boldsymbol{\delta})}\\
&\qquad\times{}_{p-1}F_{p-2}\left(\left.\begin{matrix}1-\gamma_k+\boldsymbol{\alpha}',1-\gamma_k+\boldsymbol{\beta},1-\gamma_k+b_2\\ 2-\gamma_k,1-\gamma_k+\boldsymbol{\gamma}_{[k]},1-\gamma_k+\boldsymbol{\delta}\end{matrix}\right|(-1)^{\varkappa}\right)\\
&\quad+\sum_{k=1}^{n-2}\frac{\pi\Gamma(b_2-\alpha_k')\Gamma(\boldsymbol{\alpha}_{[k]}'-\alpha_k')\Gamma(\alpha_k')\Gamma(1-\boldsymbol{\gamma}+\alpha_k')}{(-1)^m\sin(\pi\alpha_k')\Gamma(\boldsymbol{\delta}-\alpha_k')\Gamma(1-\boldsymbol{\beta}+\alpha_k')}\\
&\qquad\left.\times{}_{p-1}F_{p-2}\left(\left.\begin{matrix}\alpha_k',1-\boldsymbol{\gamma}+\alpha_k',1-\boldsymbol{\delta}+\alpha_k'\\ 1-b_2+\alpha_k',1-\boldsymbol{\alpha}_{[k]}'+\alpha_k,1-\boldsymbol{\beta}+\alpha_k'\end{matrix}\right|(-1)^{\varkappa}\right)\right\}\\
&\quad-\frac{(-1)^{\varkappa}(1-\boldsymbol{\gamma})_1(1-\boldsymbol{\delta})_1}{(1-b_2)(1-\boldsymbol{\alpha}')_1(1-\boldsymbol{\beta})_1}{}_pF_{p-1}\left(\left.\begin{matrix}1,1,2-\boldsymbol{\gamma},2-\boldsymbol{\delta}\\ 2-b_2,2-\boldsymbol{\alpha}',2-\boldsymbol{\beta}\end{matrix}\right|(-1)^{\varkappa}\right),
\end{aligned}
$$

where we applied the easily verifiable relations

$$
\begin{aligned}
&\frac{(-1)^m\Gamma(1-\boldsymbol{\beta})\Gamma(\boldsymbol{\delta})}{\Gamma(1-\boldsymbol{\gamma})\Gamma(\boldsymbol{\alpha}')\Gamma(b_2)}\frac{\Gamma(\boldsymbol{\alpha}'-m-1)\Gamma(b_2-m)m!\,\Gamma(2-\boldsymbol{\gamma}+m)}{(m+1-b_2)\Gamma(\boldsymbol{\delta}-m-1)\Gamma(2-\boldsymbol{\beta}+m)}\\
&\qquad\times{}_pF_{p-1}\left(\left.\begin{matrix}1,m+1,2-\boldsymbol{\gamma}+m,2-\boldsymbol{\delta}+m\\ 2-\boldsymbol{\alpha}'+m,2-\boldsymbol{\beta}+m,2-b_2+m\end{matrix}\right|(-1)^{\varkappa}\right)\\
&\quad+\frac{(-1)^m(1-\boldsymbol{\gamma})_m(1-\boldsymbol{\delta})_m}{(1-b_2)_m(1-\boldsymbol{\alpha}')_m(1-\boldsymbol{\beta})_m}\\
&\qquad\times\sum_{k=0}^{m-1}\frac{(m-k-1)!(1)_k(b_2-m)_k(\boldsymbol{\alpha}'-m)_k(\boldsymbol{\beta}-m)_k}{[(-1)^{\varkappa-1}]^{k+m}(\boldsymbol{\gamma}-m)_k(\boldsymbol{\delta}-m)_k k!}\\
&\quad=\frac{(-1)^{\varkappa}(1-\boldsymbol{\gamma})_1(1-\boldsymbol{\delta})_1}{(1-b_2)(1-\boldsymbol{\alpha}')_1(1-\boldsymbol{\beta})_1}{}_pF_{p-1}\left(\left.\begin{matrix}1,1,2-\boldsymbol{\gamma},2-\boldsymbol{\delta}\\ 2-b_2,2-\boldsymbol{\alpha}',2-\boldsymbol{\beta}\end{matrix}\right|(-1)^{\varkappa}\right)
\end{aligned}
$$

and

$$
\begin{aligned}
&\frac{\pi\Gamma(\boldsymbol{\alpha}'-b_2)\Gamma(b_2)\Gamma(1-\boldsymbol{\gamma}+b_2)}{(-1)^m\sin(\pi b_2)\Gamma(1-\boldsymbol{\beta}+b_2)\Gamma(\boldsymbol{\delta}-b_2)}\\
&\qquad+\frac{\pi\Gamma(\boldsymbol{\alpha}'-b_2-1)\Gamma(b_2+1)\Gamma(2-\boldsymbol{\gamma}+b_2)}{(-1)^m\sin(\pi b_2)\Gamma(\boldsymbol{\delta}-b_2-1)\Gamma(2-\boldsymbol{\beta}+b_2)}\\
&\qquad\times{}_pF_{p-1}\left(\left.\begin{matrix}b_2+1,2-\boldsymbol{\gamma}+b_2,2-\boldsymbol{\delta}+b_2,1\\ 2-\boldsymbol{\alpha}'+b_2,2-\boldsymbol{\beta}+b_2,2\end{matrix}\right|(-1)^{\varkappa}\right)\\
&\quad=\frac{\pi\Gamma(\boldsymbol{\alpha}'-b_2)\Gamma(b_2)\Gamma(1-\boldsymbol{\gamma}+b_2)}{(-1)^m\sin(\pi b_2)\Gamma(1-\boldsymbol{\beta}+b_2)\Gamma(\boldsymbol{\delta}-b_2)}\\
&\qquad\times{}_{p-1}F_{p-2}\left(\left.\begin{matrix}b_2,1-\boldsymbol{\gamma}+b_2,1-\boldsymbol{\delta}+b_2\\ 1-\boldsymbol{\alpha}'+b_2,1-\boldsymbol{\beta}+b_2\end{matrix}\right|(-1)^{\varkappa}\right).
\end{aligned}
$$

Hence, writing $\boldsymbol{\alpha}=(b_2,\boldsymbol{\alpha}')\in\mathbb{C}^{n-1}$ and recalling that $\boldsymbol{\beta}\in\mathbb{C}^{p-n}$, $\boldsymbol{\gamma}\in\mathbb{C}^{m'-2}$, $\boldsymbol{\delta}\in\mathbb{C}^{p-m'}$, $\varkappa=p-n-m'$ and $m'+n\geqslant p$, we get

$$\begin{aligned}
&\sum_{k=0}^{\infty}\frac{(\boldsymbol{\alpha})_k(\boldsymbol{\beta})_k[(-1)^{\varkappa}]^k}{(\boldsymbol{\gamma})_k(\boldsymbol{\delta})_k k!}\Big\{\sum_{\gamma\in\boldsymbol{\gamma}}\psi(\gamma+k)+\sum_{\delta\in\boldsymbol{\delta}}\psi(\delta+k)+\psi(1+k)\\
&\qquad\qquad-\sum_{\alpha\in\boldsymbol{\alpha}}\psi(\alpha+k)-\sum_{\beta\in\boldsymbol{\beta}}\psi(\beta+k)\Big\}\\
&=\pi\left[\sum_{\beta\in\boldsymbol{\beta}}\cot(\pi\beta)-\sum_{\gamma\in\boldsymbol{\gamma}}\cot(\pi\gamma)\right]{}_{p-1}F_{p-2}\left(\begin{matrix}\boldsymbol{\alpha},\boldsymbol{\beta}\\ \boldsymbol{\gamma},\boldsymbol{\delta}\end{matrix}\middle|(-1)^{\varkappa}\right)\\
&-\frac{(-1)^{\varkappa}(1-\boldsymbol{\gamma})_1(1-\boldsymbol{\delta})_1}{(1-\boldsymbol{\alpha})_1(1-\boldsymbol{\beta})_1}{}_pF_{p-1}\left(\begin{matrix}1,1,2-\boldsymbol{\gamma},2-\boldsymbol{\delta}\\ 2-\boldsymbol{\alpha},2-\boldsymbol{\beta}\end{matrix}\middle|(-1)^{\varkappa}\right)\\
&+\frac{\Gamma(1-\boldsymbol{\beta})\Gamma(\boldsymbol{\delta})}{\Gamma(1-\boldsymbol{\gamma})\Gamma(\boldsymbol{\alpha})}\sum_{k=1}^{n-1}\frac{\pi\Gamma(\boldsymbol{\alpha}_{[k]}-\alpha_k)\Gamma(\alpha_k)\Gamma(1-\boldsymbol{\gamma}+\alpha_k)}{\sin(\pi\alpha_k)\Gamma(\boldsymbol{\delta}-\alpha_k)\Gamma(1-\boldsymbol{\beta}+\alpha_k)}\\
&\times{}_{p-1}F_{p-2}\left(\begin{matrix}\alpha_k,1-\boldsymbol{\gamma}+\alpha_k,1-\boldsymbol{\delta}+\alpha_k\\ 1-\boldsymbol{\alpha}_{[k]}+\alpha_k,1-\boldsymbol{\beta}+\alpha_k\end{matrix}\middle|(-1)^{\varkappa}\right)\\
&+\frac{\Gamma(1-\boldsymbol{\beta})\Gamma(\boldsymbol{\delta})}{\Gamma(1-\boldsymbol{\gamma})\Gamma(\boldsymbol{\alpha})}\sum_{k=1}^{m'-2}\frac{\pi\Gamma(\gamma_k-1)\Gamma(\gamma_k-\boldsymbol{\gamma}_{[k]})\Gamma(1-\gamma_k+\boldsymbol{\alpha})}{\sin(\pi\gamma_k)\Gamma(\gamma_k-\boldsymbol{\beta})\Gamma(1-\gamma_k+\boldsymbol{\delta})}\\
&\times{}_{p-1}F_{p-2}\left(\begin{matrix}1-\gamma_k+\boldsymbol{\alpha},1-\gamma_k+\boldsymbol{\beta}\\ 2-\gamma_k,1-\gamma_k+\boldsymbol{\gamma}_{[k]},1-\gamma_k+\boldsymbol{\delta}\end{matrix}\middle|(-1)^{\varkappa}\right).
\end{aligned}$$

Introducing the notation

$$\begin{aligned}
&\boldsymbol{\beta}=\mathbf{a}_1\in\mathbb{C}^n, &&\boldsymbol{\alpha}=\mathbf{a}_2\in\mathbb{C}^{p-n}, &&\mathbf{a}=(\mathbf{a}_1,\mathbf{a}_2)\in\mathbb{C}^p,\\
&\mathbf{b}_1=\boldsymbol{\gamma}\in\mathbb{C}^m, &&\mathbf{b}_2=\boldsymbol{\delta}\in\mathbb{C}^{p-1-m}, &&\mathbf{b}=(\mathbf{b}_1,\mathbf{b}_2)\in\mathbb{C}^{p-1},
\end{aligned}$$

we finally arrive at (4.2) under the conditions $0\leqslant n\leqslant p$, $0\leqslant m\leqslant p-1$ and $\kappa:=m-n\geqslant-2$. □

A particular case of this identity for $n=p$, $m=p-1$ takes the form:

(4.4)
$$\begin{aligned}
&\sum_{k=0}^{\infty}\frac{(\mathbf{a})_k(-1)^k}{\pi(\mathbf{b})_k k!}\left[\sum_{b\in\mathbf{b}}\psi(b+k)+\psi(1+k)-\sum_{a\in\mathbf{a}}\psi(a+k)\right]\\
&=\left[\sum_{a\in\mathbf{a}}\cot(\pi a)-\sum_{b\in\mathbf{b}}\cot(\pi b)\right]{}_pF_{p-1}\left(\begin{matrix}\mathbf{a}\\ \mathbf{b}\end{matrix}\middle|-1\right)\\
&+\frac{(1-\mathbf{b})_1}{\pi(1-\mathbf{a})_1}{}_{p+1}F_p\left(\begin{matrix}1,1,2-\mathbf{b}\\ 2-\mathbf{a}\end{matrix}\middle|-1\right)\\
&+\frac{\Gamma(1-\mathbf{a})}{\Gamma(1-\mathbf{b})}\sum_{b_k\in\mathbf{b}}\frac{\Gamma(b_k-1)\Gamma(b_k-\mathbf{b}_{[k]})}{\sin(\pi b_k)\Gamma(b_k-\mathbf{a})}{}_pF_{p-1}\left(\begin{matrix}1-b_k+\mathbf{a}\\ 2-b_k,1-b_k+\mathbf{b}_{[k]}\end{matrix}\middle|-1\right).
\end{aligned}$$

Combining this formula with (3.3) we obtain:

$$\sum_{k=0}^{\infty}\frac{(\mathbf{a})_k(\pm1)^k}{\pi(\mathbf{b})_kk!}\Big[\sum_{b\in\mathbf{b}}\psi(b+k)+\psi(1+k)-\sum_{a\in\mathbf{a}}\psi(a+k)\Big]$$
$$=\Big[\sum_{a\in\mathbf{a}}\cot(\pi a)-\sum_{b\in\mathbf{b}}\cot(\pi b)\Big]{}_pF_{p-1}\left(\begin{matrix}\mathbf{a}\\ \mathbf{b}\end{matrix}\Big|\pm1\right)$$
$$\mp\frac{(1-\mathbf{b})_1}{\pi(1-\mathbf{a})_1}{}_{p+1}F_p\left(\begin{matrix}1,1,2-\mathbf{b}\\ 2-\mathbf{a}\end{matrix}\Big|\pm1\right)$$
$$+\frac{\Gamma(1-\mathbf{a})}{\Gamma(1-\mathbf{b})}$$
$$\times\sum_{b_k\in\mathbf{b}}\frac{[\cos(\pi b_k)]^{\left\{\substack{1\\0}\right\}}\Gamma(b_k-1)\Gamma(b_k-\mathbf{b}_{[k]})}{\sin(\pi b_k)\Gamma(b_k-\mathbf{a})}{}_pF_{p-1}\left(\begin{matrix}1-b_k+\mathbf{a}\\ 2-b_k,1-b_k+\mathbf{b}_{[k]}\end{matrix}\Big|\pm1\right)$$

which is precisely formula (1.2) given in the Introduction. Formula (1.3), on the other hand, is obtained by choosing $m=n=0$ in (4.2):

$$\sum_{k=0}^{\infty}\frac{(\mathbf{a})_k}{\pi(\mathbf{b})_kk!}\Big[\sum_{b\in\mathbf{b}}\psi(b+k)+\psi(1+k)-\sum_{a\in\mathbf{a}}\psi(a+k)\Big]$$
$$=\frac{\Gamma(\mathbf{b})}{\Gamma(\mathbf{a})}\sum_{a_k\in\mathbf{a}}\frac{\Gamma(a_k)\Gamma(\mathbf{a}_{[k]}-a_k)}{\sin(\pi a_k)\Gamma(\mathbf{b}-a_k)}{}_pF_{p-1}\left(\begin{matrix}a_k,1-\mathbf{b}+a_k\\ 1-\mathbf{a}_{[k]}+a_k\end{matrix}\right)$$
$$-\frac{(1-\mathbf{b})_1}{\pi(1-\mathbf{a})_1}{}_{p+1}F_p\left(\begin{matrix}1,1,2-\mathbf{b}\\ 2-\mathbf{a}\end{matrix}\right).$$

Combining it with the "+" case of (1.2) we also get:

(4.5)
$$\Big[\sum_{a\in\mathbf{a}}\cot(\pi a)-\sum_{b\in\mathbf{b}}\cot(\pi b)\Big]{}_pF_{p-1}\left(\begin{matrix}\mathbf{a}\\ \mathbf{b}\end{matrix}\right)$$
$$=\frac{\Gamma(\mathbf{b})}{\Gamma(\mathbf{a})}\sum_{k=1}^{p}\frac{\Gamma(a_k)\Gamma(\mathbf{a}_{[k]}-a_k)}{\sin(\pi a_k)\Gamma(\mathbf{b}-a_k)}{}_pF_{p-1}\left(\begin{matrix}a_k,1-\mathbf{b}+a_k\\ 1-\mathbf{a}_{[k]}+a_k\end{matrix}\right)$$
$$-\frac{\Gamma(1-\mathbf{a})}{\Gamma(1-\mathbf{b})}\sum_{k=1}^{p-1}\cot(\pi b_k)\frac{\Gamma(b_k-1)\Gamma(b_k-\mathbf{b}_{[k]})}{\Gamma(b_k-\mathbf{a})}{}_pF_{p-1}\left(\begin{matrix}1-b_k+\mathbf{a}\\ 2-b_k,1-b_k+\mathbf{b}_{[k]}\end{matrix}\right)$$

—another multi-term hypergeometric identity which appears to be new and combines the features of [**8**, (5.1)] and [**8**, (5.3)].

References

[1] L. U. Ancarani and G. Gasaneo, *Derivatives of any order of the hypergeometric function* ${}_pF_q(a_1,\dots,a_p;b_1,\dots,b_q;z)$ *with respect to the parameters* a_i *and* b_i, J. Phys. A **43** (2010), no. 8, 085210, 11, DOI 10.1088/1751-8113/43/8/085210. MR2592353

[2] A. Apelblat, *Bessel and related functions. Vol. 1. Theoretical aspects*, De Gruyter STEM, De Gruyter, Berlin, 2020. Mathematical operations with respect to the order. MR4201828

[3] A. Apelblat and F. Mainardi, Differentiation of the Wright functions with respect to parameters and other results. Appl. Sci. 12 (2022), 12825.

[4] G. E. Andrews, R. Askey, and R. Roy, *Special functions*, Encyclopedia of Mathematics and its Applications, vol. 71, Cambridge University Press, Cambridge, 1999, DOI 10.1017/CBO9781107325937. MR1688958

[5] J. Borwein, Addendum to Hansen's Table of Series and Products, available via https://carma.newcastle.edu.au/resources/jon/Preprints/Books/Other/hansen07.pdf

[6] Yu. A. Brychkov, Special Functions. Derivatives, Integrals, Series and Other Formulas. Chapman & Hall/CRC, 2008.

[7] Yu. A. Brychkov and K. O. Geddes, *Differentiation of hypergeometric functions with respect to parameters*, Abstract and applied analysis, World Sci. Publ., River Edge, NJ, 2004, pp. 15–28. MR2094371

[8] A. Çetinkaya, D. Karp, and E. Prilepkina, *Hypergeometric functions at unit argument: simple derivation of old and new identities*, SIGMA Symmetry Integrability Geom. Methods Appl. **17** (2021), Paper No. 098, 25, DOI 10.3842/SIGMA.2021.098. MR4335019

[9] D. Cvijović, *Closed-form summations of certain hypergeometric-type series containing the digamma function*, J. Phys. A **41** (2008), no. 45, 455205, 7, DOI 10.1088/1751-8113/41/45/455205. MR2515836

[10] D. Cvijović and A. R. Miller, *A reduction formula for the Kampé de Fériet function*, Appl. Math. Lett. **23** (2010), no. 7, 769–771, DOI 10.1016/j.aml.2010.03.006. MR2639876

[11] B. Xh. Fejzullahu, *Parameter derivatives of the generalized hypergeometric function*, Integral Transforms Spec. Funct. **28** (2017), no. 11, 781–788, DOI 10.1080/10652469.2017.1362635. MR3724503

[12] J. González-Santander, F. Sánchez Lasheras, Finite and infinite hypergeometric sums involving the digamma function. Mathematics 10 (2022), 2990.

[13] J. González-Santander, F. Sánchez Lasheras, Sums involving the digamma function connected to the incomplete beta function and the Bessel functions, Mathematics 2023, 11(8), 1937; `https://doi.org/10.3390/math11081937`.

[14] E. R. Hansen, A Table of Series and Products. Englewood Cliffs, NJ: Prentice-Hall, 1975.

[15] M. Yu. Kalmykov and B. A. Kniehl, All order-ϵ expansions of hypergeometric functions of one variable. Physics of Particles and Nuclei, 41 (2010), no. 6, 942–945.

[16] M. Kalmykov, V. Bytev, B. A. Kniehl, S.-O. Moch, B. F. L. Ward, and S. A. Yost, *Hypergeometric functions and Feynman diagrams*, Anti-differentiation and the calculation of Feynman amplitudes, Texts Monogr. Symbol. Comput., Springer, Cham, [2021] ©2021, pp. 189–234, DOI 10.1007/978-3-030-80219-6_9. MR4368097

[17] H. Kang and C. An, *Differentiation formulas of some hypergeometric functions with respect to all parameters*, Appl. Math. Comput. **258** (2015), 454–464, DOI 10.1016/j.amc.2015.02.017. MR3323082

[18] A. R. Miller, *Summations for certain series containing the digamma function*, J. Phys. A **39** (2006), no. 12, 3011–3020, DOI 10.1088/0305-4470/39/12/010. MR2212070

[19] T. M. Rassias and H. M. Srivastava, *Some classes of infinite series associated with the Riemann zeta and polygamma functions and generalized harmonic numbers*, Appl. Math. Comput. **131** (2002), no. 2-3, 593–605, DOI 10.1016/S0096-3003(01)00172-2. MR1920248

[20] P. C. Sofotasios and Yu. A. Brychkov, *On derivatives of hypergeometric functions and classical polynomials with respect to parameters*, Integral Transforms Spec. Funct. **29** (2018), no. 11, 852–865, DOI 10.1080/10652469.2018.1504042. MR3854517

[21] H. M. Srivastava and P. W. Karlsson, *Multiple Gaussian hypergeometric series*, Ellis Horwood Series: Mathematics and its Applications, Ellis Horwood Ltd., Chichester; Halsted Press [John Wiley & Sons, Inc.], New York, 1985. MR834385

Istanbul Kültür University, Bakırköy Campus, Ataköy Building, Bakırköy 34158, Istanbul, Turkey

Email address: `asnfigen@hotmail.com`

Department of Mathematics, Holon Institute of Technology, 52 Golomb Street, POB 305 Holon 5810201, Israel and School of Economics and Management, Far Eastern Federal University, 10 Ajax Bay, Vladivostok, Russia

Email address: `dimkrp@gmail.com`

Contemporary Mathematics
Volume **818**, 2025
https://doi.org/10.1090/conm/818/16367

Double summation addition theorems for Jacobi functions of the first and second kind

Howard S. Cohl, Roberto S. Costas-Santos, Loyal Durand, Camilo Montoya, and Gestur Ólafsson

Dedicated to Dick Askey whose favorite function was the Jacobi polynomial.

ABSTRACT. In this paper, we review and derive hyperbolic and trigonometric double summation addition theorems for Jacobi functions of the first and second kind. In connection with these addition theorems, we perform a full analysis of the relation between (i) Jacobi functions with symmetric, antisymmetric, and half odd integer parameter values, and (ii) certain Gauss hypergeometric functions that satisfy a quadratic transformation, including associated Legendre, Gegenbauer and Ferrers functions of the first and second kind. We also introduce Olver normalizations of the Jacobi functions, which are particularly useful in the derivation of expansion formulas when the parameters are integers. We apply the addition theorems for Jacobi functions of the second kind to separated eigenfunction expansions of fundamental solutions of Laplace–Beltrami operators on compact and noncompact rank-one symmetric spaces.

1. Introduction

Jacobi polynomials (hypergeometric polynomials) were introduced by the German mathematician Carl Gustav Jacob Jacobi (1804–1851). These polynomials first appear in an article by Jacobi, which was published posthumously in 1859 by Heinrich Eduard Heine [**29**]. Jacobi polynomials, $P_n^{(\alpha,\beta)}$, which for $\Re\alpha, \Re\beta > -1$ are orthogonal on the real segment $[-1,1]$ [**30**, (9.8.2)], and can be defined in terms of a terminating sum by

$$\begin{aligned} P_n^{(\alpha,\beta)}(\cos\theta) := &\frac{\Gamma(\alpha+1+n)}{\Gamma(n+1)\,\Gamma(\alpha+\beta+1+n)} \\ &\times \sum_{k=0}^{n}(-1)^k\binom{n}{k}\frac{\Gamma(\alpha+\beta+1+n+k)}{\Gamma(\alpha+1+k)}\sin^{2k}(\tfrac12\theta), \end{aligned} \tag{1.1}$$

2020 *Mathematics Subject Classification.* Primary 54C40, 14E20; Secondary 46E25, 20C20.
Key words and phrases. Addition theorems, Jacobi function of the first kind, Jacobi function of the second kind, Jacobi polynomials, ultraspherical polynomials.
LD would like to thank the Aspen Center for Physics, supported by The National Science Foundation grant PHY-2210452, for its hospitality while parts of this work were done.

where $\Gamma(\cdot)$ is the gamma function [**11**, (5.2.1)] and $\binom{n}{k}$ is the binomial coefficient [**11**, (1.2.1)]. This definition of the Jacobi polynomial is equivalent to the Gauss hypergeometric representation [**11**, (18.5.7)]

$$P_n^{(\alpha,\beta)}(x) = \frac{(\alpha+1)_n}{n!}\, {}_2F_1\left(\begin{matrix} -n, n+\alpha+\beta+1 \\ \alpha+1 \end{matrix}; \frac{1-x}{2}\right), \tag{1.2}$$

where $x = \cos\theta$. We will return to the notations used in (1.2) in the following section.

Ultraspherical polynomials, traditionally defined by [**11**, (18.7.1)]

$$C_n^\mu(\cos\theta) := \frac{\Gamma(\mu+\frac12)\Gamma(2\mu+n)}{\Gamma(2\mu)\Gamma(\mu+\frac12+n)} P_n^{(\mu-\frac12,\mu-\frac12)}(\cos\theta), \tag{1.3}$$

are symmetric ($\alpha = \beta$) Jacobi polynomials. These polynomials are commonly referred to as Gegenbauer polynomials after the Austrian mathematician Leopold Gegenbauer (1849–1903). However the Czech (Austrian) astronomer and mathematician Moriz Allé discovered and used many of their fundamental properties including their generating function *and* addition theorems [**1**] almost a decade prior to Gegenbauer [**19, 20**] and Heine [**24**, p. 455]. See the nice discussion of the history of the addition theorem for ultraspherical polynomials by Koornwinder in [**39**, p. 383]. The addition theorem for ultraspherical polynomials is [**11**, (18.18.8)]

$$\begin{aligned} &C_n^\mu(\cos\theta_1\cos\theta_2 \pm \sin\theta_1\sin\theta_2\cos\phi) \\ &\quad = \frac{n!}{(2\mu)_n}\sum_{k=0}^{n} \frac{(\mp1)^k(\mu)_k(2\mu)_{2k}}{(-n)_k(\mu-\frac12)_k(2\mu+n)_k} \\ &\qquad \times (\sin\theta_1\sin\theta_2)^k C_{n-k}^{\mu+k}(\cos\theta_1)C_{n-k}^{\mu+k}(\cos\theta_2)C_k^{\mu-\frac12}(\cos\phi). \end{aligned} \tag{1.4}$$

This result is quite important by itself. In the special case $\mu = \frac12$, it becomes one way of writing the addition theorem for spherical harmonics on the two-dimensional sphere (Legendre polynomials), namely

$$\begin{aligned} &P_n(\cos\theta_1\cos\theta_2 \pm \sin\theta_1\sin\theta_2\cos\phi) \\ &\quad = \sum_{k=-n}^{n} (\pm1)^k \frac{(n-k)!}{(n+k)!} \mathsf{P}_n^k(\cos\theta_1)\mathsf{P}_n^k(\cos\theta_2)\cos(k\phi), \end{aligned} \tag{1.5}$$

where the P_n^k are Ferrers functions of the first kind [**11**, (14.3.1)]. See the foreword of Willard Miller (1977) [**45**] written by Richard Askey for a beautiful discussion (on pp. xix–xx) of addition theorems (see also [**28**, §2.7]).

Given the addition theorem for ultraspherical polynomials (1.4), it was a natural problem to extend it to Jacobi polynomials for $\alpha \neq \beta$. It was a good match when Richard Askey, on sabbatical at the Mathematical Centre in Amsterdam during 1969–1970, met Tom Koornwinder there, who had some experience with group-theoretical methods and was looking for a good subject for a Ph.D. thesis. Askey suggested to Koornwinder the problem of finding an addition theorem for Jacobi polynomials, and he also arranged that Koornwinder could attend a special year at the Mittag-Leffler Institute in Sweden. There Koornwinder obtained the desired result [**35–38**]. He later found that his group-theoretic method and the resulting addition theorem in a special case had been anticipated by two papers in Russian. Vilenkin and Šapiro [**51**] had realized that disk polynomials [**11**, (18.37.1)] and in particular the Jacobi polynomials $P_n^{(\alpha,0)}$, for α an integer, can be interpreted as spherical functions on the complex projective space $\mathrm{SU}(\alpha+2)/\mathrm{U}(\alpha+1)$ [**5, 6**], or as spherical functions on the complex unit sphere $\mathrm{U}(\alpha+2)/\mathrm{U}(\alpha+1)$ in $\mathbb{C}^{\alpha+2}$, a homogeneous space of the unitary group $\mathrm{U}(\alpha+2)$. (See references by Ikeda, Kayama and Seto in [**35, 36**].) Šapiro obtained from that observation the addition theorem for Jacobi polynomials in the $\beta=0$ case [**52**].

Koornwinder initially presented his addition theorem for Jacobi polynomials in a series of three papers in 1972 [**35, 37, 38**]. Koornwinder gave four different proofs of the addition formula for Jacobi polynomials. His first proof focused on the spherical functions of the Lie group $\mathrm{U}(d)/\mathrm{U}(d-1)$, $d\geqslant 2$ an integer, and appeared in [**37, 38**]; his second proof, which used ordinary spherical harmonics, appeared in [**31**]; his third proof was an analytic proof and appeared in [**3, 32, 33**]; and a short fourth proof using orthogonal polynomials in three variables appeared in [**34**].

Let us consider the trigonometric context of Koornwinder's addition theorem for the Jacobi polynomials. Let $n\in\mathbb{N}_0$, $\alpha>\beta>-\frac{1}{2}$, $\cos\theta_1=\frac{1}{2}(\mathrm{e}^{i\theta_1}+\mathrm{e}^{-i\theta_1})$, $\cos\theta_2=\frac{1}{2}(\mathrm{e}^{i\theta_2}+\mathrm{e}^{-i\theta_2})$, and $w\in(-1,1)$, $\phi\in[0,\pi]$. Then Koornwinder's addition theorem for Jacobi polynomials is

$$\begin{aligned}
&P_n^{(\alpha,\beta)}\left(2|\cos\theta_1\cos\theta_2\pm\mathrm{e}^{i\phi}w\sin\theta_1\sin\theta_1|^2-1\right)\\
&\quad=\frac{n!\,\Gamma(\alpha+1)}{\Gamma(\alpha+n+1)}\sum_{k=0}^{n}\frac{(\alpha+1)_k(\alpha+\beta+n+1)_k}{(\alpha+k)(\beta+1)_k(-n)_k}\\
&\qquad\times\sum_{l=0}^{k}(\mp1)^{k-l}\frac{(\alpha+k+l)(-\beta-n)_l}{(\alpha+n+1)_l}(\cos\theta_1\cos\theta_2)^{k-l}\left(\sin\theta_1\sin\theta_2\right)^{k+l}\\
&\qquad\times P_{n-k}^{(\alpha+k+l,\beta+k-l)}\left(\cos(2\theta_1)\right)P_{n-k}^{(\alpha+k+l,\beta+k-l)}\left(\cos(2\theta_2)\right)\\
&\qquad\times w^{k-l}P_l^{(\alpha-\beta-1,\beta+k-l)}(2w^2-1)\frac{\beta+k-l}{\beta}C_{k-l}^{\beta}(\cos\phi).
\end{aligned}\tag{1.6}$$

As we will see in Section 3 below, this addition theorem and its various counterparts for Jacobi functions of the first and second kind are deeply connected to a system of 2-variable orthogonal polynomials, sometimes referred to as parabolic biangle polynomials and denoted by $\mathcal{P}_{k,l}^{(\alpha,\beta)}(w,\phi)$.

In the case of ultraspherical and Jacobi polynomials, the sum is terminating, as one would expect since the object of study is a polynomial. However, as Flensted-Jensen and Koornwinder realized [**17**], the addition theorem for Jacobi polynomials can be extended to Jacobi functions of the first kind by formally taking the outer sum limit to infinity. While Jacobi polynomials $P_n^{(\alpha,\beta)}$ have n discrete, the Jacobi functions $\varphi_\lambda^{(\alpha,\beta)}$ have λ continuous (see (3.6) below). When one starts to consider Jacobi functions, many new questions arise which must be understood for a full theory of separated eigenfunction expansions. First of all, one must consider two separate contexts: the trigonometric context, where the arguments of the functions are analytically continued from the segment $(-1,1)$; and the hyperbolic context, where the arguments of the functions are analytically continued from the segment $(1,\infty)$. On top of that, one must consider the expansions of Jacobi functions of the second kind. Gegenbauer and Jacobi functions are solutions of second-order ordinary differential equations. Therefore there are two linearly independent solutions, namely the functions of the first kind and those of the second kind. The separated eigenfunction expansions of the Gegenbauer functions of the first and second kind were treated quite extensively in a paper by Durand, Fishbane and Simmons (1976) [**14**]. Since one should refer to ultraspherical polynomials as opposed to Gegenbauer polynomials, it would be prudent in future publications to refer instead to ultraspherical functions. Durand extended Koornwinder's addition theorem to Jacobi (and other) functions of the second kind in [**13**].

The study of multi-summation addition theorems for Jacobi functions of the first and second kind seems not to have moved forward since the advances by Durand and by Flensted-Jensen and Koornwinder. In the remainder of this paper, we give the full multi-summation expansions of Jacobi functions of the second kind and bring the full theory of the expansions of Jacobi functions to a circle.

Addition theorems, such as the ones for Jacobi polynomials, are intimately related to separated eigenfunction expansions of spherical functions (reproducing kernels) on Riemannian symmetric spaces. For instance, the argument on the left-hand side of Gegenbauer's addition theorem (1.4) is easily expressible in terms of the geodesic distance between two arbitrary points on the d-dimensional real hypersphere. One is often interested in eigenfunction expansions of a fundamental solution of Laplace's or Poisson's equation, because this allows one to perform a multipole expansion of arbitrarily shaped mass distributions. When one performs a global analysis of the Laplacian (Laplace–Beltrami operator) on rank-one symmetric spaces, one finds that a fundamental solution is given in terms of radial solutions of a second-order differential equation, and since the solutions are singular at the origin, the solutions are given in terms of functions of the second kind. For rank-one symmetric spaces beyond the real case (complex, quaternionic, and octonionic), the fundamental solutions are given in terms of Jacobi functions of the second kind. This is the motivation for the work in the present paper. We were originally motivated by the symmetric spaces. They show up in the solution to the problem we were trying to solve, and the project evolved from a study of the singular part which arises within the function of the second kind.

2. Preliminaries

Throughout this paper we adopt the following set notations: $\mathbb{N}_0 := \{0\} \cup \mathbb{N} = \{0, 1, 2, 3, \ldots\}$, and $\mathbb{C}$ represents the complex numbers. Jacobi functions (and their special cases such as Gegenbauer, associated Legendre and Ferrers functions) have representations given in terms of Gauss hypergeometric functions, which can be defined as infinite series of ratios of shifted factorials (Pochhammer symbols). The shifted factorial can be defined for $a \in \mathbb{C}$, $n \in \mathbb{N}_0$ by $(a)_n := (a)(a+1)\cdots(a+n-1)$ [**11**, (5.2.4), (5.2.5)]. For $a \in \mathbb{C}\setminus -\mathbb{N}_0$, it is related to a ratio of two gamma functions by [**11**, Chapter 5]

$$(a)_n = \frac{\Gamma(a+n)}{\Gamma(a)}, \tag{2.1}$$

which allows one to extend the definition to non-positive integer values of n. Some other properties of shifted factorial which we will use are (for $n, k \in \mathbb{N}_0$, $n \geqslant k$)

$$\Gamma(a-n) = \frac{(-1)^n \Gamma(a)}{(1-a)_n}, \tag{2.2}$$

and

$$(-n)_k = \frac{(-1)^k n!}{(n-k)!}. \tag{2.3}$$

One also has an expression for the generalized binomial coefficient for $z \in \mathbb{C}$, $n \in \mathbb{N}_0$ [**11**, (1.2.6)],

$$\binom{z}{n} = \frac{(-1)^n(-z)_n}{n!}. \tag{2.4}$$

Define the multisets $\mathbf{a} := \{a_1, \ldots, a_r\}$, $\mathbf{b} := \{b_1, \ldots, b_s\}$. We will adhere to the common notational product convention that when $a_l \in \mathbb{C}$, $l \in \mathbb{N}$, $r, k \in \mathbb{N}_0$,

$$(\mathbf{a})_k := (a_1, \ldots, a_r)_k := (a_1)_k (a_2)_k \cdots (a_r)_k, \tag{2.5}$$

$$\Gamma(\mathbf{a}) := \Gamma(a_1, \ldots, a_r) := \Gamma(a_1) \cdots \Gamma(a_r). \tag{2.6}$$

Also, in the multiset notation $\mathbf{a} + t := \{a_1 + t, \ldots, a_r + t\}$.

For any expression of the form $(z^2-1)^\alpha$, we fix the branch of the power functions so that

$$(z^2-1)^\alpha := (z+1)^\alpha (z-1)^\alpha,$$

for any fixed $\alpha \in \mathbb{C}$ and $z \in \mathbb{C} \setminus \{-1, 1\}$. The generalized hypergeometric function [**11**, Chapter 16] is defined as the infinite series [**11**, (16.2.1)]

$$ {}_rF_s(\mathbf{a}; \mathbf{b}; z) := {}_rF_s\left({\mathbf{a} \atop \mathbf{b}}; z\right) := \sum_{k=0}^{\infty} \frac{(\mathbf{a})_k}{(\mathbf{b})_k} \frac{z^k}{k!}, \tag{2.7}$$

where $|z| < 1$, $b_j \notin -\mathbb{N}_0$, for $j \in \{1, \ldots, s\}$; and elsewhere by analytic continuation. The Olver-normalized (scaled or regularized) generalized hypergeometric series ${}_r\boldsymbol{F}_s(\mathbf{a}; \mathbf{b}; z)$ is defined by

$$ {}_r\boldsymbol{F}_s(\mathbf{a}; \mathbf{b}; z) := {}_r\boldsymbol{F}_s\left({\mathbf{a} \atop \mathbf{b}}; z\right) := \frac{1}{\Gamma(\mathbf{b})} {}_rF_s\left({\mathbf{a} \atop \mathbf{b}}; z\right) = \sum_{k=0}^{\infty} \frac{(a_1, \ldots, a_r)_k}{\Gamma(\mathbf{b}+k)} \frac{z^k}{k!}, \tag{2.8}$$

which is entire for all $a_l, b_j \in \mathbb{C}$, $l \in \{1, \ldots, r\}$, $j \in \{1, \ldots, s\}$. Both the generalized and Olver-normalized generalized hypergeometric series, if nonterminating, are entire if $r \leqslant s$, convergent for $|z| < 1$ if $r = s+1$, and divergent if $r \geqslant s+1$.

The special case of the generalized hypergeometric function with $r = 2$, $s = 1$ is referred to as the Gauss hypergeometric function [**11**, Chapter 15], or simply the hypergeometric function. It has many interesting properties, including linear transformations which were discovered by Euler and Pfaff. Euler's linear transformation is [**11**, (15.8.1)]

$$ {}_2\boldsymbol{F}_1\left({a, b \atop c}; z\right) = (1-z)^{c-a-b} {}_2\boldsymbol{F}_1\left({c-a, c-b \atop c}; z\right), \tag{2.9}$$

and Pfaff's linear transformation is [**11**, (15.8.1)]

(2.10)

$$ {}_2\boldsymbol{F}_1\left({a, b \atop c}; z\right) = (1-z)^{-a} {}_2\boldsymbol{F}_1\left({a, c-b \atop c}; \frac{z}{z-1}\right) = (1-z)^{-b} {}_2\boldsymbol{F}_1\left({b, c-a \atop c}; \frac{z}{z-1}\right).$$

2.1. The Gegenbauer and associated Legendre functions. The Gauss hypergeometric functions which satisfy quadratic transformations are Gegenbauer and associated Legendre functions of the first and second kind. As we will see, these functions correspond to Jacobi functions of the first and second kind with parameters satisfying certain relations. We now describe some of the properties of these functions, which have a deep and long history.

Let $n \in \mathbb{N}_0$. The Gegenbauer (ultraspherical) polynomial, which is an important specialization of the Jacobi polynomial for symmetric parameters values, is given in terms of a terminating Gauss hypergeometric series by [**11**, (18.7.1)]

$$C_n^\mu(z) = \frac{(2\mu)_n}{(\mu+\frac{1}{2})_n} P_n^{(\mu-\frac{1}{2},\mu-\frac{1}{2})}(z) = \frac{(2\mu)_n}{n!} {}_2F_1\left(\begin{matrix}-n, 2\mu+n\\ \mu+\frac{1}{2}\end{matrix}; \frac{1-z}{2}\right). \tag{2.11}$$

Note that the ultraspherical polynomials satisfy the parity relation [**11**]

$$C_n^\mu(-z) = (-1)^n C_n^\mu(z). \tag{2.12}$$

Gegenbauer functions, which generalize ultraspherical polynomials in having arbitrary degrees $n = \lambda \in \mathbb{C}$, are solutions $w = w(z) = w_\lambda^\mu(z)$ of the Gegenbauer differential equation [**11**, Table 18.8.1]

$$(z^2-1)\frac{\mathrm{d}^2 w(z)}{\mathrm{d}z^2} + (2\lambda+1)\,z\frac{\mathrm{d}w(z)}{\mathrm{d}z} - \lambda(\lambda+2\mu)w(z) = 0. \tag{2.13}$$

There are two linearly independent solutions of this second-order ordinary differential equation which are referred to as Gegenbauer functions of the first and second kind: C_λ^μ, D_λ^μ. A differential equation closely connected to the Gegenbauer differential equation (2.13) is the associated Legendre differential equation, which is [**11**, (14.2.2)]

$$(1-z^2)\frac{\mathrm{d}^2 w(z)}{\mathrm{d}z^2} - 2z\frac{\mathrm{d}w(z)}{\mathrm{d}z} + \left(\nu(\nu+1) - \frac{\mu^2}{1-z^2}\right) w(z) = 0. \tag{2.14}$$

Two linearly independent solutions of this equation are referred to as associated Legendre functions of the first and second kind: P_ν^μ, Q_ν^μ. In the following subsection we present the definitions of these important functions, which are Gauss hypergeometric functions that satisfy a quadratic transformation.

2.1.1. *Hypergeometric representations of the Gegenbauer and associated Legendre functions.* The Gegenbauer function of the first kind is defined by [**11**, (15.9.15)]

$$C_\lambda^\mu(z) := \frac{\sqrt{\pi}\,\Gamma(\lambda+2\mu)}{2^{2\mu-1}\Gamma(\mu)\Gamma(\lambda+1)}\, {}_2\boldsymbol{F}_1\left(\begin{matrix}-\lambda, 2\mu+\lambda\\ \mu+\frac{1}{2}\end{matrix}; \frac{1-z}{2}\right), \tag{2.15}$$

where $\lambda+2\mu \notin -\mathbb{N}_0$. It is a clear extension of the Gegenbauer polynomial, with the index allowed to be a complex number as well as a non-negative integer. Gegenbauer functions of the second kind, which will be useful to us in comparing to Jacobi functions of the second kind, are the following. They have two hypergeometric representations: for $\lambda+2\mu\notin-\mathbb{N}_0$ [**14**, (2.3)]

(2.16)
$$D^\mu_\lambda(z):=\frac{\mathrm{e}^{i\pi\mu}\Gamma(\lambda+2\mu)}{\Gamma(\mu)(2z)^{\lambda+2\mu}}\,{}_2\boldsymbol{F}_1\left(\begin{matrix}\frac12\lambda+\mu,\frac12\lambda+\mu+\frac12\\ \lambda+\mu+1\end{matrix};\frac{1}{z^2}\right)$$

(2.17)
$$=\frac{\mathrm{e}^{i\pi\mu}2^\lambda\Gamma(\lambda+\mu+\frac12)\Gamma(\lambda+2\mu)}{\sqrt{\pi}\,\Gamma(\mu)(z-1)^{\lambda+\mu+\frac12}(z+1)^{\mu-\frac12}}\,{}_2\boldsymbol{F}_1\left(\begin{matrix}\lambda+1,\lambda+\mu+\frac12\\ 2\lambda+2\mu+1\end{matrix};\frac{2}{1-z}\right),$$

and in the second representation $\lambda+\mu+\frac12\notin-\mathbb{N}_0$. The equality of these representations follows from a quadratic transformation of the Gauss hypergeometric function from Group 3 to Group 1 in [**11**, Table 15.8.1].

The associated Legendre function of the first kind is defined as [**11**, (14.3.6) and §14.21(i)]

(2.18)
$$P^\mu_\nu(z):=\left(\frac{z+1}{z-1}\right)^{\frac12\mu}{}_2\boldsymbol{F}_1\left(\begin{matrix}-\nu,\nu+1\\ 1-\mu\end{matrix};\frac{1-z}{2}\right),$$

for $|1-z|<2$, and elsewhere in z by analytic continuation. The associated Legendre function of the second kind $Q^\mu_\nu:\mathbb{C}\setminus(-\infty,1]\to\mathbb{C}$, $\nu+\mu\notin-\mathbb{N}$, has the two single Gauss hypergeometric function representations (see [**11**, (14.3.7) and §14.21], [**44**, entry 24, p. 161])

(2.19)
$$Q^\mu_\nu(z):=\frac{\sqrt{\pi}\,\mathrm{e}^{i\pi\mu}\Gamma(\nu+\mu+1)(z^2-1)^{\frac12\mu}}{2^{\nu+1}z^{\nu+\mu+1}}\,{}_2\boldsymbol{F}_1\left(\begin{matrix}\frac{\nu+\mu+1}{2},\frac{\nu+\mu+2}{2}\\ \nu+\frac32\end{matrix};\frac{1}{z^2}\right)$$

(2.20)
$$=\frac{2^\nu\mathrm{e}^{i\pi\mu}\Gamma(\nu+1)\Gamma(\nu+\mu+1)(z+1)^{\frac12\mu}}{(z-1)^{\frac12\mu+\nu+1}}\,{}_2\boldsymbol{F}_1\left(\begin{matrix}\nu+1,\nu+\mu+1\\ 2\nu+2\end{matrix};\frac{2}{1-z}\right),$$

where for the second representation, $\nu\notin-\mathbb{N}$. These representations are convergent as a Gauss hypergeometric series for $|z|>1$, respectively $|z-1|>2$, and elsewhere in $z\in\mathbb{C}\setminus(-\infty,1]$ by analytic continuation of the Gauss hypergeometric function.

REMARK 2.1. The associated Legendre functions of the first and second kind are related to the Gegenbauer functions of the first and second kind by [**11**, (14.3.22)]

(2.21)
$$P^\mu_\nu(z)=\frac{\Gamma(\frac12-\mu)\Gamma(\nu+\mu+1)}{2^\mu\sqrt{\pi}\,\Gamma(\nu-\mu+1)(z^2-1)^{\frac12\mu}}C^{\frac12-\mu}_{\nu+\mu}(z),$$

(2.22)
$$Q^\mu_\nu(z)=\frac{\mathrm{e}^{2\pi i(\mu-\frac14)}\sqrt{\pi}\,\Gamma(\frac12-\mu)\Gamma(\nu+\mu+1)}{2^\mu\Gamma(\nu-\mu+1)(z^2-1)^{\frac12\mu}}D^{\frac12-\mu}_{\nu+\mu}(z),$$

which are valid for $\mu\in\mathbb{C}\setminus\{\frac12,\frac32,\ldots\}$, $\nu+\mu\in\mathbb{C}\setminus-\mathbb{N}$. Equivalently, the inverse relationships are

(2.23)
$$C^\mu_\lambda(z)=\frac{\sqrt{\pi}\,\Gamma(\lambda+2\mu)}{2^{\mu-\frac12}\Gamma(\mu)\Gamma(\lambda+1)(z^2-1)^{\frac{\mu}{2}-\frac14}}P^{\frac12-\mu}_{\lambda+\mu-\frac12}(z),$$

(2.24)
$$D^\mu_\lambda(z)=\frac{\mathrm{e}^{2\pi i(\mu-\frac14)}\Gamma(\lambda+2\mu)}{\sqrt{\pi}\,2^{\mu-\frac12}\Gamma(\mu)\Gamma(\lambda+1)(z^2-1)^{\frac12\mu-\frac14}}Q^{\frac12-\mu}_{\lambda+\mu-\frac12}(z),$$

which are valid for $\lambda+2\mu\in\mathbb{C}\setminus-\mathbb{N}_0$.

REMARK 2.2. By comparing the Gauss hypergeometric representations of the various functions, one may express ${}_2\boldsymbol{F}_1(a,a+\frac{1}{2};c;z)$ in terms of associated Legendre functions of the first and second kind P_ν^μ, Q_ν^μ, and the Gegenbauer functions of the first and second kind C_λ^μ, D_λ^μ, by using the following very useful formulas. Let $z\in\mathbb{C}\setminus[1,\infty)$. Then

$$\begin{aligned}{}_2\boldsymbol{F}_1\left(\begin{matrix}a,a+\frac{1}{2}\\ c\end{matrix};z\right)&=2^{c-1}z^{\frac{1}{2}(1-c)}(1-z)^{\frac{1}{2}c-a-\frac{1}{2}}P_{2a-c}^{1-c}\left(\frac{1}{\sqrt{1-z}}\right)\\ &=\frac{2^{2c-2}\Gamma(c-\frac{1}{2})\Gamma(2(a-c+1))}{\sqrt{\pi}\,\Gamma(2a)(1-z)^a}C_{2a-2c+1}^{c-\frac{1}{2}}\left(\frac{1}{\sqrt{1-z}}\right),\end{aligned}\tag{2.25}$$

where $2c\notin\{1,-1,-3,\ldots\}$, $2a-2c\notin\{-2,-3,\ldots\}$, and

$$\begin{aligned}{}_2\boldsymbol{F}_1\left(\begin{matrix}a,a+\frac{1}{2}\\ c\end{matrix};z\right)&=\frac{e^{i\pi(c-2a-\frac{1}{2})}2^{c-\frac{1}{2}}(1-z)^{\frac{1}{2}c-a-\frac{1}{4}}}{\sqrt{\pi}\,\Gamma(2a)z^{\frac{1}{2}c-\frac{1}{4}}}Q_{c-\frac{3}{2}}^{2a-c+\frac{1}{2}}\left(\frac{1}{\sqrt{z}}\right)\\ &=\frac{e^{i\pi(2a-c)}2^{2c-2a-1}\Gamma(c-2a)(1-z)^{c-2a-\frac{1}{2}}}{\Gamma(2c-2a-1)z^{c-a-\frac{1}{2}}}D_{2a-1}^{c-2a}\left(\frac{1}{\sqrt{z}}\right),\end{aligned}\tag{2.26}$$

where $c, c-2a\notin-\mathbb{N}_0$.

2.1.2. *The Gegenbauer functions on the cut* $(-1,1)$ *and the Ferrers functions.* We will consider Jacobi functions of the second kind on-the-cut in Section 2.2.3. As we will see, for certain combinations of the parameters which we will describe below, the Jacobi functions of the first and second kind on-the-cut are related to the the Gegenbauer functions of the first and second kind on-the-cut and the associated Legendre functions of the first and second kind on-the-cut (Ferrers functions).

The Gegenbauer functions of the first and second kind on-the-cut are defined in terms of the Gegenbauer functions immediately above and below the segment $(-1,1)$ in the complex plane. These definition are [**12**, (3.3), (3.4)]

$$\mathsf{C}_\lambda^\mu(x):=D_\lambda^\mu(x+i0)+e^{-2\pi i\mu}D_\lambda^\mu(x-i0)=C_\lambda^\mu(x\pm i0),\quad x\in(-1,1],\tag{2.27}$$

$$\mathsf{D}_\lambda^\mu(x):=-iD_\lambda^\mu(x+i0)+ie^{-2\pi i\mu}D_\lambda^\mu(x-i0),\qquad x\in(-1,1).\tag{2.28}$$

Note that $\mathsf{C}_\lambda^\mu(x)$ and D_λ^μ are real for real values of λ and μ.

The Ferrers functions of the first and second kind are defined as [**11**, (14.23.1), (14.23.2)]

$$\begin{aligned}\mathsf{P}_\nu^\mu(x)&:=e^{\pm i\pi\mu}P_\nu^\mu(x\pm i0)\\ &=\frac{ie^{-i\pi\mu}}{\pi}\left(e^{-\frac{1}{2}i\pi\mu}Q_\nu^\mu(x+i0)-e^{\frac{1}{2}i\pi\mu}Q_\nu^\mu(x-i0)\right),\end{aligned}\tag{2.29}$$

$$\mathsf{Q}_\nu^\mu(x):=\frac{e^{-i\pi\mu}}{2}\left(e^{-\frac{1}{2}i\pi\mu}Q_\nu^\mu(x+i0)+e^{\frac{1}{2}i\pi\mu}Q_\nu^\mu(x-i0)\right).\tag{2.30}$$

Using the above definition one can readily obtain a single hypergeometric representation of the Gegenbauer function of the first kind on-the-cut, namely

$$\mathsf{C}_\lambda^\mu(x)=\frac{\sqrt{\pi}\,\Gamma(2\mu+\lambda)}{2^{2\mu-1}\Gamma(\mu)\Gamma(\lambda+1)}\,{}_2\boldsymbol{F}_1\left(\begin{matrix}-\lambda,2\mu+\lambda\\ \mu+\frac{1}{2}\end{matrix};\frac{1-x}{2}\right),\tag{2.31}$$

which is identical to the Gegenbauer function of the first kind defined by (2.15), because this function analytically continues to the segment $(-1,1)$; see (2.27). For the Gegenbauer function of the second kind on-the-cut, one can readily obtain a double hypergeometric representation by using the definition (2.28), using the interrelation between the Gegenbauer function of the second kind and the Legendre function of the second kind, and then comparing to the Ferrers function of the second kind through its definition (2.30).

However, first we will give hypergeometric representations of the Ferrers function of the first and second kind, which are easily found in the literature. The first author recently co-authored a paper with Park and Volkmer where all double hypergeometric representations of the Ferrers function of the second kind were computed [**10**]. Using (2.29) one can derive hypergeometric representations of the Ferrers function of the first kind (associated Legendre function of the first kind on-the-cut) $\mathsf{P}_\nu^\mu\colon (-1,1)\to\mathbb{C}$. For instance, one has a single hypergeometric representation [**11**, (14.3.1)]

$$\mathsf{P}_\nu^\mu(x) = \left(\frac{1+x}{1-x}\right)^{\frac12\mu} {}_2\boldsymbol{F}_1\left(\begin{matrix}-\nu,\nu+1\\1-\mu\end{matrix};\frac{1-x}{2}\right). \tag{2.32}$$

Let $\nu\in\mathbb{C}$, $\mu\in\mathbb{C}\setminus\mathbb{Z}$, $\nu+\mu\notin-\mathbb{N}$; then a double hypergeometric representation of the Ferrers function of the second kind is [**11**, (14.3.2)]

$$\begin{aligned}\mathsf{Q}_\nu^\mu(x) = \frac{\pi}{2\sin(\pi\mu)}\Bigg(&\cos(\pi\mu)\left(\frac{1+x}{1-x}\right)^{\frac12\mu} {}_2\boldsymbol{F}_1\left(\begin{matrix}-\nu,\nu+1\\1-\mu\end{matrix};\frac{1-x}{2}\right)\\ &-\frac{\Gamma(\nu+\mu+1)}{\Gamma(\nu-\mu+1)}\left(\frac{1-x}{1+x}\right)^{\frac12\mu} {}_2\boldsymbol{F}_1\left(\begin{matrix}-\nu,\nu+1\\1+\mu\end{matrix};\frac{1-x}{2}\right)\Bigg).\end{aligned} \tag{2.33}$$

LEMMA 2.3. *Let $x\in\mathbb{C}\setminus((-\infty,-1]\cup[1,\infty))$, $\lambda,\nu,\mu\in\mathbb{C}$. Then*

$$\mathsf{D}_\lambda^\mu(x) = \frac{\Gamma(\lambda+2\mu)}{2^{\mu-\frac32}\sqrt{\pi}\,\Gamma(\mu)\Gamma(\lambda+1)(1-x^2)^{\frac12\mu-\frac14}}\mathsf{Q}_{\lambda+\mu-\frac12}^{\frac12-\mu}(x) \tag{2.34}$$

when $\lambda+2\mu\notin-\mathbb{N}_0$, and

$$\mathsf{Q}_\nu^\mu(x) = \frac{\sqrt{\pi}\,\Gamma(\frac12-\mu)\Gamma(\nu+\mu+1)}{2^{\mu+1}\Gamma(\nu-\mu+1)(1-x^2)^{\frac12\mu}}\mathsf{D}_{\nu+\mu}^{\frac12-\mu}(x) \tag{2.35}$$

when $\mu\notin\{\frac12,\frac32,\ldots\}$ and $\nu+\mu\notin-\mathbb{N}$.

PROOF. Start with the definition (2.28) and use the interrelation between the Gegenbauer function of the second kind on-the-cut and the Ferrers function of the second kind, given by (2.24). Applying this relation to the double hypergeometric representation completes the proof. □

THEOREM 2.4. *Let $x\in\mathbb{C}\setminus((-\infty,-1]\cup[1,\infty))$, and $\lambda,\mu\in\mathbb{C}$, such that $\lambda+2\mu\notin-\mathbb{N}_0$. Then*

$$\begin{aligned}\mathsf{D}_\lambda^\mu(x) = &\frac{\sqrt{\pi}}{\cos(\pi\mu)2^{\mu-\frac12}\Gamma(\mu)}\\ &\times\left(\frac{\sin(\pi\mu)\Gamma(\lambda+2\mu)}{\Gamma(\lambda+1)(1+x)^{\mu-\frac12}}\,{}_2\boldsymbol{F}_1\left(\begin{matrix}\lambda+\mu+\frac12,\frac12-\lambda-\mu\\\frac12+\mu\end{matrix};\frac{1-x}{2}\right)\right.\\ &\left.-\frac{1}{(1-x)^{\mu-\frac12}}\,{}_2\boldsymbol{F}_1\left(\begin{matrix}\lambda+\mu+\frac12,\frac12-\lambda-\mu\\\frac32-\mu\end{matrix};\frac{1-x}{2}\right)\right).\end{aligned} \tag{2.36}$$

PROOF. Start with the definition (2.28) and use the interrelation between the Gegenbauer function of the second kind and the Legendre function of the second kind, given by (2.24). Then comparing with the double hypergeometric representation (2.33) completes the proof. □

Note that we also have an interrelation between the Ferrers function of the first kind and the Gegenbauer function of the first kind on-the-cut [**11**, (14.3.21)]:

$$\mathsf{P}_\nu^\mu(x) = \frac{\Gamma(\frac{1}{2}-\mu)\Gamma(\nu+\mu+1)}{2^\mu\sqrt{\pi}\,\Gamma(\nu-\mu+1)(1-x^2)^{\frac{1}{2}\mu}}\mathsf{C}_{\nu+\mu}^{\frac{1}{2}-\mu}(x), \tag{2.37}$$

where $\mu \notin \{\frac{1}{2}, \frac{3}{2}, \ldots\}$, $\nu+\mu \notin -\mathbb{N}$; or equivalently

$$\mathsf{C}_\lambda^\mu(x) = \frac{\sqrt{\pi}\,\Gamma(\lambda+2\mu)}{2^{\mu-\frac{1}{2}}\Gamma(\mu)\Gamma(\lambda+1)(1-x^2)^{\frac{1}{2}\mu-\frac{1}{4}}}\mathsf{P}_{\lambda+\mu-\frac{1}{2}}^{\frac{1}{2}-\mu}(x). \tag{2.38}$$

Finally we should add that the Legendre polynomial (the associated Legendre function of the first kind P_ν^μ and the Ferrers function of the first kind P_ν^μ with $\mu = 0$ and $\nu = n \in \mathbb{Z}$) is given by [**11**, (18.7.9)]

$$P_n(x) := P_n^0(x) = \mathsf{P}_n^0(x) = C_n^{\frac{1}{2}}(x) = P_n^{(0,0)}(x),$$

which vanishes for n negative.

2.2. Brief introduction to Jacobi functions of the first and second kind. Now we will discuss fundamental properties and special values and limits for the Jacobi functions. Jacobi functions are complex solutions $w = w(z) = w_\gamma^{(\alpha,\beta)}(z)$ of the Jacobi differential equation [**11**, Table 18.8.1]

$$(1-z^2)\frac{\mathrm{d}^2w}{\mathrm{d}z^2} + (\beta-\alpha-z(\alpha+\beta+2))\frac{\mathrm{d}w}{\mathrm{d}z} + \gamma(\alpha+\beta+\gamma+1)w = 0 \tag{2.39}$$

which is a second-order linear homogeneous differential equation. Solutions of this equation satisfy the three-term recurrence relation [**15**, (10.8.11), p. 169]

$$B_\gamma^{(\alpha,\beta)} w_\gamma^{(\alpha,\beta)}(z) + A_\gamma^{(\alpha,\beta)}(z) w_{\gamma+1}^{(\alpha,\beta)}(z) + w_{\gamma+2}^{(\alpha,\beta)}(z) = 0, \tag{2.40}$$

where

$$A_\gamma^{(\alpha,\beta)}(z) = -\frac{(\alpha+\beta+2\gamma+3)\left(\alpha^2-\beta^2+(\alpha+\beta+2\gamma+2)(\alpha+\beta+2\gamma+4)z\right)}{2(\gamma+2)(\alpha+\beta+\gamma+2)(\alpha+\beta+2\gamma+2)}, \tag{2.41}$$

$$B_\gamma^{(\alpha,\beta)} = \frac{(\alpha+\gamma+1)(\beta+\gamma+1)(\alpha+\beta+2\gamma+4)}{(\gamma+2)(\alpha+\beta+\gamma+2)(\alpha+\beta+2\gamma+2)}. \tag{2.42}$$

This three-term recurrence relation is very useful for deriving various solutions of (2.39) when solutions are known for values which have integer separations.

2.2.1. *The Jacobi function of the first kind.* The Jacobi function of the first kind is a generalization of the Jacobi polynomial (as given by (1.2)) when the degree is no longer restricted to be an integer. In the following material we derive properties of the Jacobi function of the first kind. In the following result we present the four single Gauss hypergeometric function representations of the Jacobi function of the first kind.

THEOREM 2.5. *Let $\alpha, \beta, \gamma \in \mathbb{C}$ such that $\alpha+\gamma \notin -\mathbb{N}$. Then, the Jacobi function of the first kind $P_\gamma^{(\alpha,\beta)} : \mathbb{C} \setminus (-\infty, -1] \to \mathbb{C}$ can be defined by*

(2.43)
$$P_\gamma^{(\alpha,\beta)}(z) := \frac{\Gamma(\alpha+\gamma+1)}{\Gamma(\gamma+1)} {}_2\boldsymbol{F}_1\left(\begin{matrix}-\gamma, \alpha+\beta+\gamma+1\\ \alpha+1\end{matrix}; \frac{1-z}{2}\right)$$

(2.44)
$$= \frac{\Gamma(\alpha+\gamma+1)}{\Gamma(\gamma+1)} \left(\frac{2}{z+1}\right)^\beta {}_2\boldsymbol{F}_1\left(\begin{matrix}-\beta-\gamma, \alpha+\gamma+1\\ \alpha+1\end{matrix}; \frac{1-z}{2}\right)$$

(2.45)
$$= \frac{\Gamma(\alpha+\gamma+1)}{\Gamma(\gamma+1)} \left(\frac{z+1}{2}\right)^\gamma {}_2\boldsymbol{F}_1\left(\begin{matrix}-\gamma, -\beta-\gamma\\ \alpha+1\end{matrix}; \frac{z-1}{z+1}\right)$$

(2.46)
$$= \frac{\Gamma(\alpha+\gamma+1)}{\Gamma(\gamma+1)} \left(\frac{2}{z+1}\right)^{\alpha+\beta+\gamma+1} {}_2\boldsymbol{F}_1\left(\begin{matrix}\alpha+\gamma+1, \alpha+\beta+\gamma+1\\ \alpha+1\end{matrix}; \frac{z-1}{z+1}\right).$$

PROOF. Start with (1.2) and replace the shifted factorial by a ratio of gamma functions using (2.1) and $n! = \Gamma(n+1)$, and substitute $n \mapsto \gamma \in \mathbb{C}$, $x \mapsto z$. Application of Euler's transformation (2.9) and Pfaff's transformation (2.10) provides the other three single hypergeometric representations. This completes the proof. □

There exist double Gauss hypergeometric representations of the Jacobi function of the first kind which can be obtained by using the linear transformation formulas for the Gauss hypergeometric function based on $z \mapsto (1-z^{-1})^{-1}$, $z \mapsto z^{-1}$, $z \mapsto (1-z)^{-1}$, $z \mapsto 1-z$, $z \mapsto 1-z^{-1}$ [**11**, (15.8.1)–(15.8.5)], respectively. However, these in general will involve a sum of two Gauss hypergeometric functions. We will not present the double hypergeometric function representations of the Jacobi function of the first kind here.

One has the following connection relation for the Jacobi function of the first kind.

COROLLARY 2.5.1. *Let $\gamma, \alpha, \beta \in \mathbb{C}$, $z \in \mathbb{C} \setminus (-\infty, -1]$, $\gamma \notin -\mathbb{N}$, $\beta+\gamma \notin \mathbb{N}_0$. Then*

$$P^{(\alpha,\beta)}_{-\gamma-\alpha-\beta-1}(z) = \frac{\Gamma(-\beta-\gamma)\Gamma(\gamma+1)}{\Gamma(-\gamma-\alpha-\beta)\Gamma(\alpha+\gamma+1)} P_\gamma^{(\alpha,\beta)}(z). \tag{2.47}$$

PROOF. This relation can be derived from (2.43) by making the replacement $\gamma \mapsto -\gamma-\alpha-\beta-1$, which leaves the parameters and argument of the hypergeometric function unchanged. Comparing the prefactors completes the proof. □

REMARK 2.6. One of the consequences of the definition of the Jacobi function of the first kind is the special value

$$P_\gamma^{(\alpha,\beta)}(1) = \frac{\Gamma(\alpha+\gamma+1)}{\Gamma(\alpha+1)\Gamma(\gamma+1)}, \tag{2.48}$$

where $\alpha+\gamma \notin -\mathbb{N}$. For $\gamma = n \in \mathbb{Z}$ one has

$$P_n^{(\alpha,\beta)}(1) = \frac{(\alpha+1)_n}{n!}, \qquad P_n^{(\alpha,\beta)}(-1) = (-1)^n \frac{(\beta+1)_n}{n!}, \tag{2.49}$$

which is consistent with (2.48) and the parity relation for Jacobi polynomials (see [**11**, Table 18.6.1]). From (2.43) we have

$$P_0^{(\alpha,\beta)}(z) = 1, \tag{2.50}$$

and $P_k^{(\alpha,\beta)}(z) = 0$ for all $k \in -\mathbb{N}$.

2.2.2. *The Jacobi function of the second kind.* The Jacobi function of the second kind $Q_\gamma^{(\alpha,\beta)}$, $\gamma \in \mathbb{C}$ is a generalization of the Jacobi function of the second kind $Q_n^{(\alpha,\beta)}$, $n \in \mathbb{N}_0$ (as given by [**15**, (10.8.18)]), where the degree is no longer restricted to be an integer. In the following material we derive properties of the Jacobi function of the second kind. Below we give the four single Gauss hypergeometric function representations of the Jacobi function of the second kind.

THEOREM 2.7. *Let $\gamma, \alpha, \beta, z \in \mathbb{C}$ such that $z \in \mathbb{C} \setminus [-1, 1]$, $\alpha + \gamma, \beta + \gamma \notin -\mathbb{N}$. Then, the Jacobi function of the second kind has the Gauss hypergeometric representations*

$$Q_\gamma^{(\alpha,\beta)}(z) := \frac{2^{\alpha+\beta+\gamma}\Gamma(\alpha+\gamma+1)\Gamma(\beta+\gamma+1)}{(z-1)^{\alpha+\gamma+1}(z+1)^\beta}\, {}_2\boldsymbol{F}_1\left(\begin{matrix}\gamma+1, \alpha+\gamma+1\\ \alpha+\beta+2\gamma+2\end{matrix}; \frac{2}{1-z}\right) \tag{2.51}$$

$$= \frac{2^{\alpha+\beta+\gamma}\Gamma(\alpha+\gamma+1)\Gamma(\beta+\gamma+1)}{(z-1)^{\alpha+\beta+\gamma+1}}\, {}_2\boldsymbol{F}_1\left(\begin{matrix}\beta+\gamma+1, \alpha+\beta+\gamma+1\\ \alpha+\beta+2\gamma+2\end{matrix}; \frac{2}{1-z}\right) \tag{2.52}$$

$$= \frac{2^{\alpha+\beta+\gamma}\Gamma(\alpha+\gamma+1)\Gamma(\beta+\gamma+1)}{(z-1)^{\alpha}(z+1)^{\beta+\gamma+1}}\, {}_2\boldsymbol{F}_1\left(\begin{matrix}\gamma+1, \beta+\gamma+1\\ \alpha+\beta+2\gamma+2\end{matrix}; \frac{2}{1+z}\right) \tag{2.53}$$

$$= \frac{2^{\alpha+\beta+\gamma}\Gamma(\alpha+\gamma+1)\Gamma(\beta+\gamma+1)}{(z+1)^{\alpha+\beta+\gamma+1}}\, {}_2\boldsymbol{F}_1\left(\begin{matrix}\alpha+\gamma+1, \alpha+\beta+\gamma+1\\ \alpha+\beta+2\gamma+2\end{matrix}; \frac{2}{1+z}\right). \tag{2.54}$$

PROOF. Start with [**15**, (10.8.18)] and let $n \mapsto \gamma \in \mathbb{C}$ and $x \mapsto z$. Application of Pfaff's ($z \mapsto z/(z-1)$) and Euler's ($z \mapsto z$) transformations, (2.9) and (2.10), provides the other three representations. This completes the proof. □

One has the following connection relation between Jacobi functions of the first kind and Jacobi functions of the second kind.

COROLLARY 2.7.1. *Let $\gamma, \alpha, \beta \in \mathbb{C}$, $z \in \mathbb{C} \setminus (-\infty, 1]$, $\alpha + \gamma, \beta + \gamma \notin -\mathbb{N}$, $\alpha + \beta + 2\gamma \notin \mathbb{Z}$. Then*

$$\begin{aligned} P_\gamma^{(\alpha,\beta)}(z) = \frac{-2\sin(\pi(\beta+\gamma))}{\pi\sin(\pi(\alpha+\beta+2\gamma+1))} \Big(&\sin(\pi\gamma) Q_\gamma^{(\alpha,\beta)}(z) \\ &- \sin(\pi(\alpha+\gamma)) \frac{\Gamma(\alpha+\gamma+1)\Gamma(\beta+\gamma+1)}{\Gamma(\gamma+1)\Gamma(\alpha+\beta+\gamma+1)} Q_{-\alpha-\beta-\gamma-1}^{(\alpha,\beta)}(z)\Big). \end{aligned} \tag{2.55}$$

PROOF. This can be derived by starting with (2.43), applying the linear transformation based on $z \mapsto z^{-1}$ [**11**, (15.8.2)], and then comparing twice with Theorem 2.7. □

REMARK 2.8. Using (2.55) one can see that for $\gamma = n \in \mathbb{N}_0$, $Q^{(\alpha,\beta)}_{-\alpha-\beta-\gamma-1}$ is a Jacobi polynomial, namely
(2.56)

$$Q^{(\alpha,\beta)}_{-\alpha-\beta-1-n}(z) = \frac{\Gamma(-\alpha)\Gamma(-\beta)}{2\Gamma(-\alpha-\beta)} \frac{n!(\alpha+\beta+1)_n}{(\alpha+1)_n(\beta+1)_n} P_n^{(\alpha,\beta)}(z)$$
$$= -\frac{\pi}{2} \frac{\sin(\pi(\alpha+\beta))}{\sin(\pi\alpha)\sin(\pi\beta)} \frac{n!\,\Gamma(\alpha+\beta+1+n)}{\Gamma(\alpha+1+n)\Gamma(\beta+1+n)} P_n^{(\alpha,\beta)}(z).$$

REMARK 2.9. From Theorem 2.7 one can derive special values for $Q^{(\alpha,\beta)}_{-1}(z)$ and $Q^{(\alpha,\beta)}_0(z)$, namely

$$Q^{(\alpha,\beta)}_{-1}(z) = \frac{2^{\alpha+\beta-1}\Gamma(\alpha)\Gamma(\beta)}{\Gamma(\alpha+\beta)(z-1)^{\alpha}(z+1)^{\beta}}, \tag{2.57}$$

$$Q^{(\alpha,\beta)}_0(z) = \frac{2^{\alpha+\beta}\Gamma(\alpha+1)\Gamma(\beta+1)}{(z+1)^{\alpha+\beta+1}} {}_2\boldsymbol{F}_1\left(\begin{matrix}\alpha+1, \alpha+\beta+1\\ \alpha+\beta+2\end{matrix}; \frac{2}{1+z}\right). \tag{2.58}$$

Using the three-term recurrence relation (2.40) one can derive values of the Jacobi function of the second kind at all negative integer values of the parameter γ. For instance, one can derive

$$Q^{(\alpha,\beta)}_{-2}(z) = \frac{2^{\alpha+\beta-2}\Gamma(\alpha-1)\Gamma(\beta-1)}{\Gamma(\alpha+\beta-1)(z-1)^{\alpha}(z+1)^{\beta}}(\alpha-\beta+(\alpha+\beta-2)z), \tag{2.59}$$

and expressions for Jacobi functions of the second kind with further negative integer values of γ.

If one examines the Gauss hypergeometric representations presented in Theorem 2.7 one can see that they are not defined for certain values of γ, α, β, since one must avoid $\alpha+\gamma$ and $\beta+\gamma$ being a negative integer. In fact, these singularities are removable and one is able to compute the values of these Jacobi functions. The Jacobi function of the second kind is evaluated when the parameters α, β and the degree γ are non-negative integers in the following result, which was inspired by the work in [**53**].

THEOREM 2.10. *Let $n, a, b \in \mathbb{N}_0$, $z \in \mathbb{C} \setminus [-1,1]$. Then*
(2.60)

$$Q_n^{(a,b)}(z) = \frac{(-1)^{a+n}}{2^{n+1}} \sum_{\substack{k=0\\ k\neq n}}^{a+b+2n} \frac{(-2)^k}{(n-k)} \left((z+1)^{n-k} - (z-1)^{n-k}\right) P_k^{(a+n-k,b+n-k)}(z)$$
$$+ \frac{(-1)^a}{2} \log\left(\frac{z+1}{z-1}\right) P_n^{(a,b)}(z).$$

PROOF. Start with the integral representation for the Jacobi function of the second kind [**49**, (4.61.1)]

$$Q^{(\alpha,\beta)}_{\gamma}(z) = \frac{1}{2^{\gamma+1}(z-1)^{\alpha}(z+1)^{\beta}} \int_{-1}^{1} \frac{(1-t)^{\alpha+\gamma}(1+t)^{\beta+\gamma}}{(z-t)^{\gamma+1}}\, dt, \tag{2.61}$$

where $\Re(\alpha+\gamma)$, $\Re(\beta+\gamma)>-1$ [**53**, (2.5)], and identify $(\gamma,\alpha,\beta)=(n,a,b)\in\mathbb{N}_0^3$. Then consider

$$\begin{aligned}\mu_{n,k}^{(a,b)}(z)&:=\frac{\mathrm{d}^k}{\mathrm{d}z^k}(1-z)^{n+a}(1+z)^{n+b}\\&=(-1)^k2^kk!(1-z)^{a+n-k}(1+z)^{b+n-k}P_k^{(a+n-k,b+n-k)}(z),\end{aligned}$$

where we have used the Rodrigues-type formula for Jacobi polynomials (see [**11**, Table 18.5.1]). It is easy to show that

$$(1-t)^{n+a}(1+t)^{n+b}=\sum_{k=0}^{2n+a+b}\mu_{n,k}^{(a,b)}(z)\frac{(t-z)^k}{k!},\tag{2.62}$$

and the right-hand side is valid for all $z\in\mathbb{C}$. Now start with (2.61) and insert (2.62) into the integrand and perform the integration over $t\in(-1,1)$, using

$$\int_{-1}^{1}(z-t)^{k-n-1}\,\mathrm{d}t=\begin{cases}\dfrac{(z+1)^{k-n}-(z-1)^{k-n}}{k-n} & \text{if } k\neq n,\\[2ex] \log\left(\dfrac{z+1}{z-1}\right) & \text{if } k=n,\end{cases}$$

which completes the proof. □

By using (2.53) we find that if $|z|\sim 1+\epsilon$, then as $\epsilon\to 0^+$ one has the behavior of the Jacobi function of the second kind near the singularity at $z=1$, as follows:

$$Q_\gamma^{(\alpha,\beta)}(1+\epsilon)\sim\frac{2^{\alpha-1}\Gamma(\alpha)\Gamma(\beta+\gamma+1)}{\Gamma(\alpha+\beta+\gamma+1)\epsilon^\alpha},\tag{2.63}$$

where $\Re\alpha>0$, $\beta+\gamma\notin-\mathbb{N}$. By using (2.53) we see that as $|z|\to\infty$ one has

$$Q_\gamma^{(\alpha,\beta)}(z)\sim\frac{2^{\alpha+\beta+\gamma}\Gamma(\alpha+\gamma+1)\Gamma(\beta+\gamma+1)}{\Gamma(\alpha+\beta+2\gamma+2)z^{\alpha+\beta+\gamma+1}},\tag{2.64}$$

where $\alpha+\gamma+1,\beta+\gamma\notin-\mathbb{N}$.

2.2.3. *Jacobi functions of the first and second kind on-the-cut.* We now refer to the real segment $(-1,1)$ as the cut and the Jacobi functions of the first and second kind on-the-cut as $\mathsf{P}_\gamma^{(\alpha,\beta)},\mathsf{Q}_\gamma^{(\alpha,\beta)}$. The natural definitions of these Jacobi functions are due to Durand and can be found in [**12**, (2.3), (2.4)] (see also [**4**]). They are

$$\begin{aligned}\mathsf{P}_\gamma^{(\alpha,\beta)}(x)&:=\frac{i}{\pi}\left(\mathrm{e}^{i\pi\alpha}Q_\gamma^{(\alpha,\beta)}(x+i0)-\mathrm{e}^{-i\pi\alpha}Q_\gamma^{(\alpha,\beta)}(x-i0)\right)\\&=P_\gamma^{(\alpha,\beta)}(x\pm i0)\end{aligned}\tag{2.65}$$

and

$$\mathsf{Q}_\gamma^{(\alpha,\beta)}(x):=\frac{1}{2}\left(\mathrm{e}^{i\pi\alpha}Q_\gamma^{(\alpha,\beta)}(x+i0)+\mathrm{e}^{-i\pi\alpha}Q_\gamma^{(\alpha,\beta)}(x-i0)\right).\tag{2.66}$$

Note that the Jacobi function of the first kind on-the-cut (2.65) is simply an analytic continuation of the Jacobi function of the first kind (see Theorem 2.5) since the complex-valued function is continuous across the real interval $(-1,1]$. On the other hand, the Jacobi function of the second kind on-the-cut is not an analytic continuation of the Jacobi function of the second kind (see Theorem 2.7). This is because $Q_\gamma^{(\alpha,\beta)}$ is not continuous across the real interval $(-1,1)$. Hence, an 'average' as shown in (2.66) must be taken of the function values with infinitesimal positive and negative arguments, to define it. Originally, in Szegő's book [**49**, §4.62.9] (see also [**15**, (10.8.22)]) the definition of the Jacobi function of the second kind on-the-cut was $\mathsf{Q}_\gamma^{(\alpha,\beta)}(x) := \frac{1}{2}(Q_\gamma^{(\alpha,\beta)}(x+i0) + Q_\gamma^{(\alpha,\beta)}(x-i0))$, but as is pointed out by Durand [**12**], Szegő's definition destroys the analogy between $\mathsf{P}_\gamma^{(\alpha,\beta)}(\cos\theta)$, $\mathsf{Q}_\gamma^{(\alpha,\beta)}(\cos\theta)$ and the trigonometric functions. Hence with the updated Durand definitions for the Jacobi functions of the first and second kind on-the-cut, namely (2.65) and (2.66), one has the asymptotics as $n \to \infty$ [**12**, p. 77]

$$Q_n^{(\alpha,\beta)}(\cos\theta \pm i0) \sim \frac{1}{2}\left(\frac{\pi}{n}\right)^{\frac{1}{2}} \left(\sin(\tfrac{1}{2}\theta)\right)^{-\alpha-\frac{1}{2}} \left(\cos(\tfrac{1}{2}\theta)\right)^{-\beta-\frac{1}{2}} \mathrm{e}^{\mp iN\theta \mp i\frac{\pi}{2}(\alpha+\frac{1}{2})},$$

where $N := n + \frac{1}{2}\alpha + \frac{1}{2}\beta + \frac{1}{2}$.

There are many double hypergeometric representations of the Jacobi function of the second kind on-the-cut $\mathsf{Q}_\gamma^{(\alpha,\beta)} : \mathbb{C}\setminus((-\infty,-1]\cup[1,\infty)) \to \mathbb{C}$. These representations follow by applying the definition (2.66) to Theorem 2.7, which provides Gauss hypergeometric representations for the Jacobi function of the second kind. Applying (2.66) causes the argument of each Gauss hypergeometric function to be just above or below the ray $(1,\infty)$, on which it is known that the Gauss hypergeometric function is discontinuous. The values above and below this ray may be transformed into values in a region where the function is continuous, by utilizing the transformations which one can find in [**10**, Appendix B]. Each of these transformations expresses a Gauss hypergeometric function with argument $x \pm i0$ as a sum of two such functions, with a common argument which is one of x^{-1}, $1-x$, $1-x^{-1}$ or $(1-x)^{-1}$. Eight Gauss hypergeometric function representations of the Jacobi function of the second kind on-the-cut can be obtained by starting with (2.51)–(2.54), applying the transformation [**10**, Theorem B.1] $z \mapsto z^{-1}$, and utilizing the Euler (2.9) or the Pfaff (2.10) transformations as needed. There are certainly more Gauss hypergeometric representations that can be obtained for the Jacobi function of the second kind on-the-cut by applying [**10**, Theorems B.2–B.4], but the derivation of these representations must be left to a later publication. We will give two of these here for $\gamma,\alpha,\beta \in \mathbb{C}$ with $\alpha,\beta \notin \mathbb{Z}$, $\alpha+\gamma, \beta+\gamma \notin -\mathbb{N}$, namely

(2.67)

$$\begin{aligned}\mathsf{Q}_\gamma^{(\alpha,\beta)}(x) &= \frac{\pi}{2\sin(\pi\alpha)}\Bigg(-\cos(\pi\alpha)\frac{\Gamma(\alpha+\gamma+1)}{\Gamma(\gamma+1)}\,{}_2\boldsymbol{F}_1\left(\begin{matrix}-\gamma,\alpha+\beta+\gamma+1\\1+\alpha\end{matrix};\frac{1-x}{2}\right)\\&+\frac{\Gamma(\beta+\gamma+1)}{\Gamma(\alpha+\beta+\gamma+1)}\left(\frac{2}{1-x}\right)^{\alpha}\left(\frac{2}{1+x}\right)^{\beta}{}_2\boldsymbol{F}_1\left(\begin{matrix}-\alpha-\beta-\gamma,\gamma+1\\1-\alpha\end{matrix};\frac{1-x}{2}\right)\Bigg),\\&= \frac{\pi}{2^{\gamma+1}\sin(\pi\alpha)}\Bigg(-\cos(\pi\alpha)\frac{\Gamma(\alpha+\gamma+1)}{\Gamma(\gamma+1)}(1+x)^{\gamma}\,{}_2\boldsymbol{F}_1\left(\begin{matrix}-\gamma,-\beta-\gamma\\1+\alpha\end{matrix};\frac{x-1}{x+1}\right)\\&+\frac{\Gamma(\beta+\gamma+1)}{\Gamma(\alpha+\beta+\gamma+1)}\frac{(1+x)^{\alpha+\gamma}}{(1-x)^{\alpha}}\,{}_2\boldsymbol{F}_1\left(\begin{matrix}-\alpha-\beta-\gamma,-\alpha-\gamma\\1-\alpha\end{matrix};\frac{x-1}{x+1}\right)\Bigg).\end{aligned}$$

Just as we were able to compute the values of the Jacobi function of the second kind with non-negative integer parameters and degree, the same evaluation can be accomplished for the Jacobi function of the second kind on-the-cut, which we present now.

THEOREM 2.11. *Let $n,a,b\in\mathbb{N}_0$, $x\in\mathbb{C}\setminus((-\infty,-1]\cup[1,\infty))$. Then*
(2.68)
$$\mathsf{Q}_n^{(a,b)}(x)=\frac{(-1)^n}{2^{n+1}}\sum_{\substack{k=0\\k\neq n}}^{a+b+2n}\frac{(-2)^k}{(n-k)}\left((1+x)^{n-k}-(x-1)^{n-k}\right)P_k^{(a+n-k,b+n-k)}(x)$$
$$+\frac{1}{2}\log\left(\frac{1+x}{1-x}\right)P_n^{(a,b)}(x).$$

PROOF. Start with Theorem 2.10 and use the definition (2.66), which completes the proof. □

Note that by setting $a=b$ in the above result we can obtain an interesting finite sum expression for the Ferrers functions of the second kind with non-negative integer degree and order, as a sum over ultraspherical polynomials.

COROLLARY 2.11.1. *Let $n,a\in\mathbb{N}_0$, Let $n,a\in\mathbb{N}_0$, $x\in\mathbb{C}\setminus((-\infty,-1]\cup[1,\infty))$. Then*
(2.69)
$$\mathsf{Q}_n^a(x)=\frac{(-1)^a(1-x^2)^{\frac{1}{2}a}}{2\sqrt{\pi}}\Bigg((-1)^{n+a}2^n(n+a)!$$
$$\times\sum_{\substack{k=0\\k\neq n-a}}^{2n}\frac{(-1)^k\Gamma(n-k+\frac{1}{2})}{2^k(2n-k)!(n-a-k)}\left((1+x)^{n-a-k}-(x-1)^{n-a-k}\right)C_k^{n-k+\frac{1}{2}}(x)$$
$$+2^a\Gamma(a+\tfrac{1}{2})\log\left(\frac{1+x}{1-x}\right)C_{n-a}^{a+\frac{1}{2}}(x)\Bigg).$$

PROOF. Start with (2.11) and set $a=b$. Then utilizing (2.91) below with (2.71) completes the proof. □

By using (2.67) with $x=1-\epsilon$ one has that

$$\mathsf{Q}_\gamma^{(\alpha,\beta)}(1-\epsilon)\sim\frac{2^{\alpha-1}\Gamma(\alpha)\Gamma(\beta+\gamma+1)}{\Gamma(\alpha+\beta+\gamma+1)\epsilon^\alpha}\quad\text{as }\epsilon\to0^+,\tag{2.70}$$

where $\beta+\gamma+1\notin-\mathbb{N}_0$ and $\Re\alpha>0$.

2.3. Specializations to Gegenbauer, associated Legendre, and Ferrers functions. Here we discuss some limiting cases where the Jacobi functions reduce to more elementary functions such as Gegenbauer, associated Legendre, and Ferrers functions.

The following identities involve symmetric and antisymmetric Jacobi functions of the first kind. The relation between the symmetric Jacobi function of the first kind and the Gegenbauer function of the first kind for $z\in\mathbb{C}\setminus(-\infty,-1]$ is

$$P_\gamma^{(\alpha,\alpha)}(z)=\frac{\Gamma(2\alpha+1)\Gamma(\alpha+\gamma+1)}{\Gamma(\alpha+1)\Gamma(2\alpha+\gamma+1)}C_\gamma^{\alpha+\frac{1}{2}}(z).\tag{2.71}$$

This follows by starting with (2.43) and comparing it to the Gauss hypergeometric representation of the Gegenbauer function of the first kind on the right-hand side, using (2.15).

REMARK 2.12. The relation between the symmetric Jacobi function of the first kind and the Ferrers function of the first kind is

$$P_\gamma^{(\alpha,\alpha)}(x) = \frac{2^\alpha \Gamma(\alpha+\gamma+1)}{\Gamma(\gamma+1)(1-x^2)^{\frac{1}{2}\alpha}} \mathsf{P}_{\alpha+\gamma}^{-\alpha}(x), \tag{2.72}$$

where $x \in \mathbb{C} \setminus ((-\infty,-1] \cup [1,\infty))$, and the relation between the symmetric Jacobi function of the first kind and the associated Legendre function of the first kind is

$$P_\gamma^{(\alpha,\alpha)}(z) = \frac{2^\alpha \Gamma(\alpha+\gamma+1)}{\Gamma(\gamma+1)(z^2-1)^{\frac{1}{2}\alpha}} P_{\alpha+\gamma}^{-\alpha}(z), \tag{2.73}$$

where $z \in \mathbb{C} \setminus (-\infty, 1]$. These are easily obtained through [**11**, (18.7.2)] and [**11**, (14.3.21), (14.3.22)].

REMARK 2.13. The relation between the antisymmetric Jacobi function of the first kind on-the-cut and the Ferrers function of the first kind and the Gegenbauer function of the first kind on-the-cut is

$$\begin{aligned} \mathsf{P}_\gamma^{(\alpha,-\alpha)}(x) &= \frac{\Gamma(\alpha+\gamma+1)}{\Gamma(\gamma+1)} \left(\frac{1+x}{1-x}\right)^{\frac{1}{2}\alpha} \mathsf{P}_\gamma^{-\alpha}(x) \\ &= \frac{\Gamma(2\alpha+1)\Gamma(\gamma-\alpha+1)}{2^\alpha \Gamma(\gamma+1)\Gamma(\alpha+1)} (1+x)^\alpha \mathsf{C}_{\gamma-\alpha}^{\alpha+\frac{1}{2}}(x), \end{aligned} \tag{2.74}$$

where $x \in \mathbb{C} \setminus ((-\infty,-1] \cup [1,\infty))$, and the relation between the antisymmetric Jacobi function of the first kind and the associated Legendre and Gegenbauer function of the first kind is

$$\begin{aligned} P_\gamma^{(\alpha,-\alpha)}(z) &= \frac{\Gamma(\alpha+\gamma+1)}{\Gamma(\gamma+1)} \left(\frac{z+1}{z-1}\right)^{\frac{1}{2}\alpha} P_\gamma^{-\alpha}(z) \\ &= \frac{\Gamma(2\alpha+1)\Gamma(\gamma-\alpha+1)}{2^\alpha \Gamma(\gamma+1)\Gamma(\alpha+1)} (z+1)^\alpha C_{\gamma-\alpha}^{\alpha+\frac{1}{2}}(z), \end{aligned} \tag{2.75}$$

where $z \in \mathbb{C} \setminus (-\infty,-1]$. These are obtained by comparing (2.43) with (2.32) and (2.18).

REMARK 2.14. One has the following quadratic transformations of the symmetric Jacobi functions of the first kind, which can be found in [**49**, Theorem 4.1]. Let $z \in \mathbb{C} \setminus (-\infty,-1]$, $\gamma, \alpha \in \mathbb{C}$, $\alpha+\gamma \notin -\mathbb{N}$. Then

$$P_{2\gamma}^{(\alpha,\alpha)}(z) = \frac{\sqrt{\pi}\,\Gamma(\alpha+2\gamma+1)}{2^{2\gamma}\Gamma(\gamma+\frac{1}{2})\Gamma(\alpha+\gamma+1)} P_\gamma^{(\alpha,-\frac{1}{2})}(2z^2-1), \tag{2.76}$$

where $\alpha+2\gamma \notin -\mathbb{N}$, $\gamma \notin -\mathbb{N}+\frac{1}{2}$, and

$$P_{2\gamma+1}^{(\alpha,\alpha)}(z) = \frac{\sqrt{\pi}\,\Gamma(\alpha+2\gamma+2)z}{2^{2\gamma+1}\Gamma(\gamma+\frac{3}{2})\Gamma(\alpha+\gamma+1)} P_\gamma^{(\alpha,\frac{1}{2})}(2z^2-1), \tag{2.77}$$

where $\alpha+2\gamma+1 \notin -\mathbb{N}$, $\gamma \notin -\mathbb{N}-\frac{1}{2}$. The restrictions on the parameters come directly by applying the restrictions on the parameters in Theorem 2.5 to the Jacobi functions of the first kind on both sides of the relations.

Below we present some identities which involve symmetric and antisymmetric Jacobi functions of the second kind.

THEOREM 2.15. *Two equivalent relations between the symmetric Jacobi function of the second kind and the associated Legendre function of the second kind are*

$$Q_\gamma^{(\alpha,\alpha)}(z) = \frac{2^\alpha e^{i\pi\alpha}\Gamma(\alpha+\gamma+1)}{\Gamma(\gamma+1)(z^2-1)^{\frac{1}{2}\alpha}} Q_{\alpha+\gamma}^{-\alpha}(z), \tag{2.78}$$

$$Q_\gamma^{(\alpha,\alpha)}(z) = \frac{2^\alpha e^{-i\pi\alpha}\Gamma(\alpha+\gamma+1)}{\Gamma(2\alpha+\gamma+1)(z^2-1)^{\frac{1}{2}\alpha}} Q_{\alpha+\gamma}^{\alpha}(z), \tag{2.79}$$

where $\alpha+\gamma \notin -\mathbb{N}$. Also, two equivalent relations between antisymmetric Jacobi functions of the second kind and the associated Legendre function of the second kind are

$$Q_\gamma^{(\alpha,-\alpha)}(z) = \frac{e^{-i\pi\alpha}\Gamma(\gamma-\alpha+1)}{\Gamma(\gamma+1)} \left(\frac{z+1}{z-1}\right)^{\frac{1}{2}\alpha} Q_\gamma^\alpha(z), \tag{2.80}$$

$$Q_\gamma^{(-\alpha,\alpha)}(z) = \frac{e^{-i\pi\alpha}\Gamma(\gamma-\alpha+1)}{\Gamma(\gamma+1)} \left(\frac{z-1}{z+1}\right)^{\frac{1}{2}\alpha} Q_\gamma^\alpha(z), \tag{2.81}$$

where $\gamma-\alpha \notin -\mathbb{N}$.

PROOF. By comparing (2.51) and (2.54) with (2.20) and by using the Legendre duplication formula [**11**, (5.5.5)] one can obtain all these formulas in a straightforward way. □

See [**8**, Section 3, (A.14)] for an interesting application of the symmetric relation for associated Legendre functions of the second kind.

REMARK 2.16. Observe that by identifying (2.78) and (2.79), for $z \in \mathbb{C}\setminus[-1,1]$ one has

$$Q_\gamma^{(\alpha,\alpha)}(z) = \frac{2^{2\alpha}}{(z^2-1)^\alpha} \begin{cases} Q_{\gamma+2\alpha}^{(-\alpha,-\alpha)}(z) & \text{if } z \in \mathbb{C}\setminus[-1,1] \\ & \text{such that } \Re z \geqslant 0, \\ e^{2\pi i\alpha} Q_{\gamma+2\alpha}^{(-\alpha,-\alpha)}(z) & \text{if } z \in \mathbb{C}\setminus[-1,1] \\ & \text{such that } \Re z < 0 \text{ and } \Im z < 0, \\ e^{-2\pi i\alpha} Q_{\gamma+2\alpha}^{(-\alpha,-\alpha)}(z) & \text{if } z \in \mathbb{C}\setminus[-1,1] \\ & \text{such that } \Re z < 0 \text{ and } \Im z \geqslant 0, \end{cases}$$

where the principal branches of complex powers are taken.

THEOREM 2.17. *Let $\alpha, \gamma \in \mathbb{C}$, $z \in \mathbb{C}\setminus[-1,1]$, $\alpha+\gamma \notin -\mathbb{N}$. Then the relations between the symmetric and antisymmetric Jacobi functions of the second kind to the Gegenbauer function of the second kind are*

$$Q_\gamma^{(\alpha,\alpha)}(z) = e^{-i\pi(\alpha+\frac{1}{2})}\sqrt{\pi}\, 2^{2\alpha} \frac{\Gamma(\alpha+\frac{1}{2})\Gamma(\alpha+\gamma+1)}{\Gamma(2\alpha+\gamma+1)} D_\gamma^{\alpha+\frac{1}{2}}(z), \tag{2.82}$$

where $\alpha \in \mathbb{C}\setminus\{-\frac{1}{2}, -\frac{3}{2}, -\frac{5}{2}, \ldots\}$, and

$$Q_\gamma^{(\alpha,-\alpha)}(z) = e^{i\pi(\alpha-\frac{1}{2})} 2^{2\gamma-\alpha+1} \frac{\Gamma(\alpha+\gamma+1)\Gamma(\frac{1}{2}-\alpha)\Gamma(\gamma+\frac{3}{2})}{\Gamma(2\gamma+2)(z-1)^\alpha} D_{\alpha+\gamma}^{\frac{1}{2}-\alpha}(z), \tag{2.83}$$

where $\alpha \in \mathbb{C}\setminus\{\frac{1}{2}, \frac{3}{2}, \frac{5}{2}, \ldots\}$, $\gamma \in \mathbb{C}\setminus\{-\frac{3}{2}, -\frac{5}{2}, -\frac{7}{2}, \ldots\}$.

PROOF. Start with the definition of the Jacobi function of the second kind (2.51) and take $\beta = \alpha$. Then comparing (2.17) using Euler's ($z \mapsto z$) transformation (2.9) produces (2.82). In order to produce (2.83), start with (2.51) and take $\beta = -\alpha$. Then compare (2.17) using Euler's transformation. This completes the proof. □

One has the following quadratic transformations of symmetric Jacobi functions of the second kind.

THEOREM 2.18. *Let $z \in \mathbb{C} \setminus [-1,1]$, $\gamma, \alpha \in \mathbb{C}$, $\alpha + \gamma \notin -\mathbb{N}$. Then*

$$Q_{2\gamma}^{(\alpha,\alpha)}(z) = \frac{\sqrt{\pi}\,\Gamma(\alpha+2\gamma+1)}{2^{2\gamma}\Gamma(\gamma+\frac{1}{2})\Gamma(\alpha+\gamma+1)} Q_{\gamma}^{(\alpha,-\frac{1}{2})}(2z^2-1), \tag{2.84}$$

where $\alpha + 2\gamma \notin -\mathbb{N}$, $\gamma \notin -\mathbb{N} + \frac{1}{2}$, and

$$Q_{2\gamma+1}^{(\alpha,\alpha)}(z) = \frac{\sqrt{\pi}\,\Gamma(\alpha+2\gamma+2)z}{2^{2\gamma+1}\Gamma(\gamma+\frac{3}{2})\Gamma(\alpha+\gamma+1)} Q_{\gamma}^{(\alpha,\frac{1}{2})}(2z^2-1), \tag{2.85}$$

where $\alpha + 2\gamma + 1 \notin -\mathbb{N}$, $\gamma \notin -\mathbb{N} - \frac{1}{2}$.

PROOF. Starting with the left-hand sides of (2.84), (2.85) and using the Gauss hypergeometric representation (2.51) yields ${}_2F_1$'s with parameters $(a, b; c)$ satisfying $c = 2b$. Then for both equations one uses a quadratic transformation of the Gauss hypergeometric function [**11**, (15.8.14)]. This transforms the ${}_2F_1$'s to forms which are recognizable as the right-hand sides through (2.54) and (2.51), respectively. This completes the proof. The restrictions on the parameters come directly by applying the restrictions on the parameters in Theorem 2.7 to the Jacobi functions of the second kind on both sides of the relations. □

There is also an interesting alternative quadratic transformation of the Jacobi function of the second kind with $\alpha = \pm\frac{1}{2}$. Note that there does not seem to be a corresponding transformation formula for the Jacobi function of the first kind since in this case the left-hand side would be a sum of two Gauss hypergeometric functions.

THEOREM 2.19. *Let $x \in \mathbb{C}\setminus((-\infty,-1] \cup [1,\infty))$, $\beta, \gamma \in \mathbb{C}$ such that $\beta+\gamma+\frac{1}{2} \notin -\mathbb{N}_0$. Then*

$$C_{2\gamma+1}^{\beta}(x) = \frac{2^{2\gamma+2}\Gamma(\beta+\gamma+\frac{1}{2})(1-x^2)^{-\beta-\gamma-\frac{1}{2}}}{\Gamma(-\gamma-\frac{1}{2})\Gamma(2\gamma+2)\Gamma(\beta)} Q_{-\gamma-1}^{(-\frac{1}{2},\beta+2\gamma+1)}\left(\frac{1+x^2}{1-x^2}\right), \tag{2.86}$$

$$C_{2\gamma}^{\beta}(x) = \frac{2^{2\gamma+1}\Gamma(\beta+\gamma+\frac{1}{2})x(1-x^2)^{-\beta-\gamma-\frac{1}{2}}}{\Gamma(-\gamma+\frac{1}{2})\Gamma(2\gamma+1)\Gamma(\beta)} Q_{-\gamma-1}^{(\frac{1}{2},\beta+2\gamma)}\left(\frac{1+x^2}{1-x^2}\right). \tag{2.87}$$

PROOF. These results are easily verified by starting with (2.51), (2.53), substituting the related values in the Jacobi function of the second kind, comparing with associated Legendre functions of the first kind with argument $\sqrt{(z-1)/(z+1)}$, and utilizing a quadratic transformation of the Gauss hypergeometric function which relates the two. This completes the proof. □

REMARK 2.20. Note that in Theorem 2.19, if the argument of the Jacobi function of the second kind has modulus greater than unity then the argument of the Gegenbauer function of the first kind has modulus less than unity.

COROLLARY 2.20.1. *Let $z,\beta,\gamma\in\mathbb{C}$ such that $z\in\mathbb{C}\setminus[-1,1]$. Then*

(2.88)
$$Q_\gamma^{(\frac12,\beta)}(z)=\frac{2^{\beta+3\gamma+\frac52}\Gamma(-2\gamma-1)\Gamma(\gamma+\frac32)\Gamma(\beta+2\gamma+2)}{\Gamma(\beta+\gamma+\frac32)(z-1)^{\frac12}(z+1)^{\beta+\gamma+1}}C_{-2\gamma-2}^{\beta+2\gamma+2}\left(\sqrt{\frac{z-1}{z+1}}\right),$$

where $-2\gamma-1,\gamma+\frac32,\beta+2\gamma+2\notin-\mathbb{N}_0$, and

(2.89)
$$Q_\gamma^{(-\frac12,\beta)}(z)=\frac{2^{\beta+3\gamma+\frac12}\Gamma(-2\gamma)\Gamma(\gamma+\frac12)\Gamma(\beta+2\gamma+1)}{\Gamma(\beta+\gamma+\frac12)(z+1)^{\beta+\gamma+\frac12}}C_{-2\gamma-1}^{\beta+2\gamma+1}\left(\sqrt{\frac{z-1}{z+1}}\right),$$

where $-2\gamma,\gamma+\frac12,\beta+2\gamma+1\notin-\mathbb{N}_0$.

PROOF. Inverting Theorem 2.19 completes the proof. □

Note that the above results imply the following corollary.

COROLLARY 2.20.2. *Let $z,\beta,\gamma\in\mathbb{C}$ such that $z\in\mathbb{C}\setminus[-1,1]$, $\gamma+\frac32,\beta+\gamma+1\notin-\mathbb{N}_0$. Then*

$$Q_\gamma^{(\frac12,\beta)}(z)=\frac{\Gamma(\gamma+\frac32)\Gamma(\beta+\gamma+1)}{\Gamma(\gamma+1)\Gamma(\beta+\gamma+\frac32)}\left(\frac{2}{z-1}\right)^{\frac12}Q_{\gamma+\frac12}^{(-\frac12,\beta)}(z).\tag{2.90}$$

PROOF. Equating the two right-hand sides in Theorem 2.19 completes the proof. □

THEOREM 2.21. *Let $x\in\mathbb{C}\setminus((-\infty,-1]\cup[1,\infty))$. Then the relation between the symmetric and antisymmetric Jacobi functions of the second kind on-the-cut and the Ferrers function of the second kind are*

$$\mathsf{Q}_\gamma^{(\alpha,\alpha)}(x)=\frac{2^\alpha\Gamma(\alpha+\gamma+1)}{\Gamma(\gamma+1)(1-x^2)^{\frac12\alpha}}\mathsf{Q}_{\gamma+\alpha}^{-\alpha}(x),\tag{2.91}$$

$$\mathsf{Q}_\gamma^{(\alpha,-\alpha)}(x)=\frac{\Gamma(\alpha+\gamma+1)}{\Gamma(\gamma+1)}\left(\frac{1+x}{1-x}\right)^{\frac12\alpha}\mathsf{Q}_\gamma^{-\alpha}(x),\tag{2.92}$$

where $\alpha+\gamma\notin-\mathbb{N}$, and

$$\mathsf{Q}_\gamma^{(-\alpha,\alpha)}(x)=\frac{\Gamma(\gamma-\alpha+1)}{\Gamma(\gamma+1)}\left(\frac{1-x}{1+x}\right)^{\frac12\alpha}\mathsf{Q}_\gamma^{\alpha}(x),\tag{2.93}$$

where $\gamma-\alpha\notin-\mathbb{N}$.

PROOF. The results follow by taking into account (2.67). It is the case that (cf. [**10**, Theorem 3.2])

(2.94)
$$\mathsf{Q}_\nu^\mu(x)=\frac{\pi}{2\sin(\pi\mu)}\left(\cos(\pi(\nu+\mu))\frac{\Gamma(\nu+\mu+1)}{\Gamma(\nu-\mu+1)}\left(\frac{1+x}{1-x}\right)^{\frac12\mu}{}_2\boldsymbol{F}_1\left(\begin{matrix}-\nu,\nu+1\\1+\mu\end{matrix};\frac{1+x}{2}\right)\right.$$
$$\left.-\cos(\pi\nu)\left(\frac{1-x}{1+x}\right)^{\frac12\mu}{}_2\boldsymbol{F}_1\left(\begin{matrix}-\nu,\nu+1\\1-\mu\end{matrix};\frac{1+x}{2}\right)\right),$$

where $\nu\in\mathbb{C}$, $\mu\in\mathbb{C}\setminus\mathbb{Z}$, such that $\nu+\mu\notin-\mathbb{N}$. The formula (2.91) is obtained by taking $\beta=\alpha$ and then comparing (2.67) with (2.94). The other identities follow by an analogous method, taking $\beta=-\alpha$. This completes the proof. □

3. Addition theorems for the Jacobi function of the first kind

The Flensted-Jensen–Koornwinder addition theorem for Jacobi functions of the first kind is an extension of the Koornwinder addition theorem for Jacobi polynomials to the case when the degree is allowed to be a complex number. This addition theorem has two separate contexts and some interesting special cases. We will refer to the contexts as the hyperbolic and trigonometric contexts. The hyperbolic context arises when the Jacobi function is analytically continued in the complex plane from the ray $[1, \infty)$. The trigonometric context arises when the argument of the Jacobi function is analytically continued from the real segment $(-1, 1)$. First we will present the addition theorem for the Jacobi function of the first kind in the hyperbolic context. As we will see, the Jacobi function in the trigonometric context can be obtained from the Jacobi function in the hyperbolic context (and vice versa). We now present the most general form of the addition theorem for Jacobi functions of the first kind in the hyperbolic and trigonometric contexts.

REMARK 3.1. The addition theorem for Jacobi polynomials was originally derived using group-theoretical methods as mentioned in [**35**], using the representation theory for $\mathrm{SU}(\alpha+2)/\mathrm{U}(\alpha+1)$ and $\mathrm{U}(\alpha+2)/\mathrm{U}(\alpha+1)$ combined with standard methods discussed, for example, in [**50**]. The Lie group-theoretic setting (and in much more detail for $\mathrm{SO}(p)\times\mathrm{SO}(q)$) is carefully described in [**31**] and is as follows. Let X be a compact symmetric group of rank one which is two-point homogeneous for an isometry group U. Normalize the metric so that the lengths of the closed geodesics is 2π. The choice of a base point in X gives an identification of X with U/K. The function space on X decomposes multiplicity-free as $\oplus\mathscr{H}^n$, and if $d(x, y)$ denotes the geodesic distance between points $x, y \in X$, then the elementary spherical function on $\mathscr{H}^n$ is the Jacobi polynomial $P_n^{(\alpha,\beta)}(\cos d(x, y))$ with the parameters α, β taking suitable 'group values,' depending on the choice of X. This observation was first made by Élie Cartan in 1929 [**5**]. For an orthonormal basis $s_k(x)$ of $\mathscr{H}^n$ it is clear that the real point-pair function $(x, y) \mapsto \sum_k s_k(x)s_k(y)$ only depends on $d(x, y)$, which in turn implies that

$$\sum_k s_k(x)\overline{s_k(y)} = c_n P_n^{(\alpha,\beta)}(\cos d(x, y)), \tag{3.1}$$

for suitable constants c_n. This identity is the Lie group-theoretic meaning of the addition formula for Jacobi polynomials. Picking a suitable basis for $\mathscr{H}^n$, this leads to an explicit summation formula for the Jacobi polynomials with an argument depending on θ_1, θ_2, ϕ, as a sum. Koornwinder rederived the addition theorem analytically in [**33**] without relying on group theory, and Flensted-Jensen and Koornwinder extended it to general n (i.e., λ) and $\alpha > \beta > -1/2$ in [**17**]. In fact, it should be pointed out that there is no group-theoretic argument which will yield addition theorems such as for Jacobi functions for general values of the parameters. Group-theoretic arguments rely on the completeness and orthogonality of the (unitary) representations of the group in question. However, the case of general parameters may be obtained by appropriate analytic continuation in the group parameters.

THEOREM 3.2. *Let* $\gamma, \alpha, \beta \in \mathbb{C}$, $\gamma \notin \mathbb{Z}$, $\alpha, \beta \notin -\mathbb{N}_0$, $z_1, z_2 \in \mathbb{C} \setminus (-\infty, -1]$, $x_1, x_2 \in \mathbb{C} \setminus ((-\infty, -1] \cup [1, \infty))$, $x, w \in \mathbb{C}$, *and let*

$$\begin{aligned} Z^\pm &:= Z^\pm(z_1, z_2, w, x) \\ &= 2z_1^2 z_2^2 + 2w^2(z_1^2 - 1)(z_2^2 - 1) \pm 4z_1 z_2 w x (z_1^2 - 1)^{\frac{1}{2}}(z_2^2 - 1)^{\frac{1}{2}} - 1 \end{aligned} \tag{3.2}$$

and

$$\begin{aligned} \mathsf{X}^\pm &:= \mathsf{X}^\pm(x_1, x_2, w, x) \\ &= 2x_1^2 x_2^2 + 2w^2(1 - x_1^2)(1 - x_2^2) \pm 4x_1 x_2 w x (1 - x_1^2)^{\frac{1}{2}}(1 - x_2^2)^{\frac{1}{2}} - 1 \end{aligned} \tag{3.3}$$

with the complex variables $\gamma, \alpha, \beta, z_1, z_2, x_1, x_2, x, w$ *required to be in some yet to be determined neighborhood of the real line. Then*

$$\begin{aligned} P_\gamma^{(\alpha,\beta)}(Z^\pm) = &\frac{\Gamma(\alpha+1)\Gamma(\gamma+1)}{\Gamma(\alpha+\gamma+1)} \sum_{k=0}^{\infty} \frac{(\alpha+1)_k(\alpha+\beta+\gamma+1)_k}{(\alpha+k)(\beta+1)_k(-\gamma)_k} \\ &\times \sum_{l=0}^{k} (\mp 1)^{k-l} \frac{(\alpha+k+l)(-\beta-\gamma)_l}{(\alpha+\gamma+1)_l} (z_1 z_2)^{k-l} \left((z_1^2 - 1)(z_2^2 - 1)\right)^{\frac{k+l}{2}} \\ &\times P_{\gamma-k}^{(\alpha+k+l,\beta+k-l)}(2z_1^2 - 1) P_{\gamma-k}^{(\alpha+k+l,\beta+k-l)}(2z_2^2 - 1) \\ &\times w^{k-l} P_l^{(\alpha-\beta-1,\beta+k-l)}(2w^2 - 1) \frac{\beta+k-l}{\beta} C_{k-l}^{\beta}(x) \end{aligned} \tag{3.4}$$

and

$$\begin{aligned} \mathsf{P}_\gamma^{(\alpha,\beta)}(\mathsf{X}^\pm) = &\frac{\Gamma(\alpha+1)\Gamma(\gamma+1)}{\Gamma(\alpha+\gamma+1)} \sum_{k=0}^{\infty} \frac{(\alpha+1)_k(\alpha+\beta+\gamma+1)_k}{(\alpha+k)(\beta+1)_k(-\gamma)_k} \\ &\times \sum_{l=0}^{k} (\mp 1)^{k-l} \frac{(\alpha+k+l)(-\beta-\gamma)_l}{(\alpha+\gamma+1)_l} (x_1 x_2)^{k-l} \left((1 - x_1^2)(1 - x_2^2)\right)^{\frac{k+l}{2}} \\ &\times \mathsf{P}_{\gamma-k}^{(\alpha+k+l,\beta+k-l)}(2x_1^2 - 1) \mathsf{P}_{\gamma-k}^{(\alpha+k+l,\beta+k-l)}(2x_2^2 - 1) \\ &\times w^{k-l} P_l^{(\alpha-\beta-1,\beta+k-l)}(2w^2 - 1) \frac{\beta+k-l}{\beta} C_{k-l}^{\beta}(x). \end{aligned} \tag{3.5}$$

PROOF. Start with the form of the Flensted-Jensen–Koornwinder addition theorem in [**17**, Theorem 2.1] (see also [**41**, (24)]). Define the Flensted-Jensen–Koornwinder–Jacobi function of the first kind [**17**, (2.1)] (Flensted-Jensen and Koornwinder refer to this function as the Jacobi function of the first kind)

$$\varphi_\lambda^{(\alpha,\beta)}(t) := {}_2F_1\left(\begin{matrix} \frac{1}{2}(\alpha+\beta+1+i\lambda), \frac{1}{2}(\alpha+\beta+1-i\lambda) \\ \alpha+1 \end{matrix} ; -\sinh^2 t \right), \tag{3.6}$$

and express it in terms of the Jacobi function of the first kind using

$$\varphi_\lambda^{(\alpha,\beta)}(t) = \frac{\Gamma(\alpha+1)\Gamma(-\frac{1}{2}(\alpha+\beta-1+i\lambda))}{\Gamma(\frac{1}{2}(\alpha-\beta+1-i\lambda))} P_{-\frac{1}{2}(\alpha+\beta+1+i\lambda)}^{(\alpha,\beta)}(\cosh(2t)), \tag{3.7}$$

which follows by comparing the Gauss hypergeometric representations of the functions. Replacing $\lambda = i(\alpha+\beta+2\gamma+1)$ and setting $z_1 = \cosh t_1$, $z_2 = \cosh t_2$, and $w = \cos\psi$ produces the form of the addition theorem (3.4). Then analytically continuing (3.4) to $\mathsf{X}^\pm \in (-1, 1)$ using (2.65) produces (3.5). This completes the proof. $\square$

In the limit as $\gamma \to n \in \mathbb{N}_0$, then Koornwinder's addition theorem for the Jacobi function of the first kind becomes the addition theorem for Jacobi polynomials and the double infinite sum becomes truncated as in (1.6).

REMARK 3.3. It is worth mentioning that in the definitions of $Z^\pm$ (3.2) and $\mathsf{X}^\pm$ (3.3), the influence of the ± 1 factor on the addition theorems in Theorem 3.2 and elsewhere in this paper is simply due to the influence of the parity relation (2.12) for ultraspherical polynomials, based upon the reflection map $x \mapsto -x$.

REMARK 3.4. Note that there are various ways of expressing the variables $Z^\pm$ and $\mathsf{X}^\pm$ defined by (3.2) and (3.3), which are useful in different applications. For instance, we may also write

(3.8)

$$\begin{aligned} Z^\pm &= 2z_1^2z_2^2(1-x^2) - 1 + 2(z_1^2-1)(z_2^2-1)\left(w \pm \frac{xz_1z_2}{\sqrt{(z_1^2-1)(z_2^2-1)}}\right)^2 \\ &= 2(z_1^2-1)(z_2^2-1)\left(\frac{2z_1^2z_2^2(1-x^2)-1}{2(z_1^2-1)(z_2^2-1)} + \left(w \pm \frac{xz_1z_2}{\sqrt{(z_1^2-1)(z_2^2-1)}}\right)^2\right) \end{aligned}$$

and

(3.9)

$$\begin{aligned} \mathsf{X}^\pm &= 2x_1^2x_2^2(1-x^2) - 1 + 2(1-x_1^2)(1-x_2^2)\left(w \pm \frac{xx_1x_2}{\sqrt{(1-x_1^2)(1-x_2^2)}}\right)^2 \\ &= 2(1-x_1^2)(1-x_2^2)\left(\frac{2x_1^2x_2^2(1-x^2)-1}{2(1-x_1^2)(1-x_2^2)} + \left(w \pm \frac{xx_1x_2}{\sqrt{(1-x_1^2)(1-x_2^2)}}\right)^2\right). \end{aligned}$$

First we will develop some tools which will help us prove the correct form of the double summation addition theorem for the Jacobi function of the second kind. Consider the orthogonality of the ultraspherical polynomials and the Jacobi polynomials with the argument $2w^2-1$.

LEMMA 3.5. *Let $m,n,p \in \mathbb{N}_0$, $\mu \in (-\frac{1}{2},\infty)\setminus\{0\}$, $\alpha > \beta > -1$. Then the ultraspherical and Jacobi polynomials satisfy the orthogonality relations*

$$\int_0^\pi C_m^\mu(\cos\phi)C_n^\mu(\cos\phi)(\sin\phi)^{2\mu}\,\mathrm{d}\phi = \frac{\pi\,\Gamma(2\mu+n)}{2^{2\mu-1}(\mu+n)n!\,\Gamma(\mu)^2}\delta_{m,n}, \tag{3.10}$$

(3.11)

$$\begin{aligned} &\int_0^1 P_m^{(\alpha-\beta-1,\beta+p)}(2w^2-1)P_n^{(\alpha-\beta-1,\beta+p)}(2w^2-1)w^{2\beta+2p+1}(1-w^2)^{\alpha-\beta-1}\,\mathrm{d}w \\ &\quad = \frac{\Gamma(\alpha-\beta+n)\Gamma(\beta+1+p+n)}{2(\alpha+p+2n)\Gamma(\alpha+p+n)n!}\delta_{m,n}. \end{aligned}$$

PROOF. These orthogonality relations follow easily from [**30**, (9.8.20), (9.8.2)] upon making the straightforward substitutions. □

3.1. The parabolic biangle orthogonal polynomial system. Define a system of 2-variable orthogonal polynomials, which are sometimes referred to as parabolic biangle polynomials, by [**40**]

$$\mathcal{P}_{k,l}^{(\alpha,\beta)}(w,\phi) := w^{k-l}P_l^{(\alpha-\beta-1,\beta+k-l)}(2w^2-1)C_{k-l}^\beta(\cos\phi), \tag{3.12}$$

where $k, l \in \mathbb{N}_0$ such that $l \leqslant k$. These 2-variable polynomials are orthogonal over $(w, \phi) \in (0,1) \times (0, \pi)$ with orthogonality measure

$$dm^{(\alpha,\beta)}(w,\phi) := (1-w^2)^{\alpha-\beta-1} w^{2\beta+1} (\sin\phi)^{2\beta}\, dw\, d\phi. \tag{3.13}$$

The orthogonal polynomial system $\mathcal{P}_{k,l}^{(\alpha,\beta)}$ is deeply connected to the addition theorem for Jacobi functions of the first and second kind. Using the orthogonality relations in Lemma 3.5 we can derive the orthogonality relation for the 2-variable parabolic biangle polynomials.

LEMMA 3.6. *Let $k, l, k', l' \in \mathbb{N}_0$ such that $l \leqslant k$, $l' \leqslant k'$, $\alpha > \beta > -1$. Then the 2-variable parabolic biangle polynomials satisfy the orthogonality relation*

$$\begin{aligned} &\int_0^1 \int_0^\pi \mathcal{P}_{k,l}^{(\alpha,\beta)}(w,\phi) \mathcal{P}_{k',l'}^{(\alpha,\beta)}(w,\phi)\, dm^{(\alpha,\beta)}(w,\phi) \\ &\quad = \frac{\pi\, \Gamma(\beta+1+k)\Gamma(2\beta+k-l)\Gamma(\alpha-\beta+l)}{2^{2\beta}\Gamma(\beta)^2(\alpha+k+l)(\beta+k-l)\Gamma(\alpha+k)(k-l)!\,l!} \delta_{k,k'}\delta_{l,l'}. \end{aligned} \tag{3.14}$$

PROOF. Starting with the definition of the 2-variable parabolic biangle polynomials (3.12), integrating over $(w, \phi) \in (0,1) \times (0, \pi)$ with measure (3.13), and using the orthogonality relations in Lemma 3.5 completes the proof. □

The following result is a Jacobi function of the first kind generalization of [**33**, (4.10)] for Jacobi polynomials.

THEOREM 3.7. *Let $k, l \in \mathbb{N}_0$ with $l \leqslant k$, $\gamma, \alpha, \beta \in \mathbb{C}$, $z_1, z_2 \in \mathbb{C}\setminus(-\infty,-1]$, $Z^{\pm}$ defined in (3.2), such that $x = \cos\phi$ and the complex variables $\gamma, \alpha, \beta, z_1, z_2$ are in some yet to be determined neighborhood of the real line. Then*

$$\begin{aligned} &\int_0^1 \int_0^\pi P_\gamma^{(\alpha,\beta)}(Z^\pm)\, w^{k-l} P_l^{(\alpha-\beta-1,\beta+k-l)}(2w^2-1) C_{k-l}^\beta(\cos\phi)\, dm^{(\alpha,\beta)}(w,\phi) \\ &\quad = (\mp 1)^{k+l}\, \mathsf{A}_{k,l}^{(\alpha,\beta,\gamma)} (z_1 z_2)^{k-l} ((z_1^2-1)(z_2^2-1))^{\frac{1}{2}(k+l)} \\ &\qquad \times P_{\gamma-k}^{(\alpha+k+l,\beta+k-l)}(2z_1^2-1) P_{\gamma-k}^{(\alpha+k+l,\beta+k-l)}(2z_2^2-1), \end{aligned} \tag{3.15}$$

where

$$\mathsf{A}_{k,l}^{(\alpha,\beta,\gamma)} := \frac{\pi\Gamma(\gamma+1)(\alpha+\beta+\gamma+1)_k \Gamma(2\beta+k-l)\Gamma(\alpha-\beta+l)(-\beta-\gamma)_l}{2^{2\beta}\Gamma(\beta)(-\gamma)_k\,(k-l)!\,l!\,\Gamma(\alpha+\gamma+1+l)}. \tag{3.16}$$

PROOF. Start with the addition theorem for the Jacobi function of the first kind (3.4) and consider the (k,l)-th term in the double series. It involves a product of two Jacobi functions of the first kind with degree $\gamma-k$ and parameters $(\alpha+k+l, \beta+k-l)$. Replace in (3.4) the summation indices k, l by k', l', multiply both sides of (3.4) by $\mathcal{P}_{k,l}^{(\alpha,\beta)}(w,\phi)\, dm^{(\alpha,\beta)}(w,\phi)$, and integrate both sides over $(w,\phi) \in (0,1)\times(0,\pi)$ using (3.10), (3.11). This completes the proof. □

We will return to the parabolic biangle polynomials in Section 4.

3.2. Special cases of the addition theorem for the Jacobi function of the first kind. In the case when $z_1, z_2, x_1, x_2, w, x = \cos\phi$ are real numbers then the argument of the Jacobi function of the first kind in the addition theorem takes a simpler form convenient form and was proved in Flensted-Jensen–Koornwinder [**17**].

REMARK 3.8. In the case where the variables z_1, z_2, x_1, x_2, x, w are real then one may write $Z^\pm$ and $\mathsf{X}^\pm$ as follows:

$$Z^\pm = 2\left|z_1z_2 \pm \mathrm{e}^{i\phi}\, w\,(z_1^2-1)^{\frac12}(z_2^2-1)^{\frac12}\right|^2 - 1, \tag{3.17}$$

$$\mathsf{X}^\pm = 2\left|x_1x_2 \pm \mathrm{e}^{i\phi}\, w\,(1-x_1^2)^{\frac12}(1-x_2^2)^{\frac12}\right|^2 - 1. \tag{3.18}$$

We now give a result which appears to be identical to Theorem 3.2, but it must be emphasized that it is only in the real case that we are able to write $Z^\pm$, $\mathsf{X}^\pm$ using (3.17), (3.18). Otherwise one must use (3.2), (3.3).

THEOREM 3.9. *Let $\gamma,\alpha,\beta\in\mathbb{C}$, $\gamma\notin\mathbb{Z}$, $\alpha,\beta\notin-\mathbb{N}_0$, $z_1,z_2\in(1,\infty)$, $x_1,x_2\in(-1,1)$, $w\in\mathbb{R}$, $\phi\in[0,\pi]$, and $Z^\pm$, $\mathsf{X}^\pm$ is defined as in* (3.17), (3.18) *respectively. Then*

$$\begin{aligned} P_\gamma^{(\alpha,\beta)}(Z^\pm) &= \frac{\Gamma(\alpha+1)\Gamma(\gamma+1)}{\Gamma(\alpha+\gamma+1)}\sum_{k=0}^\infty \frac{(\alpha+1)_k(\alpha+\beta+\gamma+1)_k}{(\alpha+k)(\beta+1)_k(-\gamma)_k} \\ &\quad\times\sum_{l=0}^k(\mp1)^{k-l}\frac{(\alpha+k+l)(-\beta-\gamma)_l}{(\alpha+\gamma+1)_l}(z_1z_2)^{k-l}\left((z_1^2-1)(z_2^2-1)\right)^{\frac{k+l}{2}} \\ &\qquad\times P_{\gamma-k}^{(\alpha+k+l,\beta+k-l)}(2z_1^2-1)P_{\gamma-k}^{(\alpha+k+l,\beta+k-l)}(2z_2^2-1) \\ &\qquad\times w^{k-l}P_l^{(\alpha-\beta-1,\beta+k-l)}(2w^2-1)\frac{\beta+k-l}{\beta}C_{k-l}^\beta(\cos\phi), \end{aligned} \tag{3.19}$$

and

$$\begin{aligned} \mathsf{P}_\gamma^{(\alpha,\beta)}(\mathsf{X}^\pm) &= \frac{\Gamma(\alpha+1)\Gamma(\gamma+1)}{\Gamma(\alpha+\gamma+1)}\sum_{k=0}^\infty \frac{(\alpha+1)_k(\alpha+\beta+\gamma+1)_k}{(\alpha+k)(\beta+1)_k(-\gamma)_k} \\ &\quad\times\sum_{l=0}^k(\mp1)^{k-l}\frac{(\alpha+k+l)(-\beta-\gamma)_l}{(\alpha+\gamma+1)_l}(x_1x_2)^{k-l}\left((1-x_1^2)(1-x_2^2)\right)^{\frac{k+l}{2}} \\ &\qquad\times \mathsf{P}_{\gamma-k}^{(\alpha+k+l,\beta+k-l)}(2x_1^2-1)\mathsf{P}_{\gamma-k}^{(\alpha+k+l,\beta+k-l)}(2x_2^2-1) \\ &\qquad\times w^{k-l}P_l^{(\alpha-\beta-1,\beta+k-l)}(2w^2-1)\frac{\beta+k-l}{\beta}C_{k-l}^\beta(\cos\phi). \end{aligned} \tag{3.20}$$

PROOF. Starting with Theorem 3.2 and restricting such that the variables $z_1, z_2, x_1, x_2, w, x=\cos\phi$ are real completes the proof. ◻

As in Theorem 3.2, in the limit when $\gamma\to n\in\mathbb{N}_0$, Koornwinder's addition theorem for the Jacobi function of the first kind becomes the addition theorem for Jacobi polynomials, and the double infinite sum becomes truncated as in (1.6). Note that for all results below, the double or single infinite sums on the right-hand side will always be truncated when the left-hand side is a polynomial (Jacobi polynomial or ultraspherical polynomial). We will not mention this again. Next we have a specialization of Theorem 3.9 when $w=1$.

COROLLARY 3.9.1. *Let $\gamma,\alpha,\beta\in\mathbb{C}$, $\gamma\notin\mathbb{Z}$, $\alpha,\beta\notin-\mathbb{N}_0$, $r_1,r_2\in[0,\infty)$, $\theta_1,\theta_2\in[0,\frac{\pi}{2}]$, $\phi\in[0,\pi]$, and*

$$Z^\pm := 2\left|\cosh r_1\cosh r_2 \pm \mathrm{e}^{i\phi}\sinh r_1\sinh r_2\right|^2 - 1 = \cosh(2r_1)\cosh(2r_2)\pm\sinh(2r_1)\sinh(2r_2)\cos\phi, \tag{3.21}$$

and

$$\mathsf{X}^\pm := 2\left|\cos\theta_1\cos\theta_2 \pm \mathrm{e}^{i\phi}\sin\theta_1\sin\theta_2\right|^2 - 1 = \cos(2\theta_1)\cos(2\theta_2)\pm\sin(2\theta_1)\sin(2\theta_2)\cos\phi. \tag{3.22}$$

Then

$$\begin{aligned} P_\gamma^{(\alpha,\beta)}(Z^\pm) &= \frac{\Gamma(\alpha+1)\Gamma(\gamma+1)}{\Gamma(\alpha+\gamma+1)} \\ &\times\sum_{k=0}^\infty \frac{(\alpha)_k(\frac{\alpha}{2}+1)_k(-\beta-\gamma)_k(\alpha+\beta+\gamma+1)_k}{(\frac{\alpha}{2})_k(\beta+1)_k(-\gamma)_k(\alpha+\gamma+1)_k}(\sinh r_1\sinh r_2)^{2k} \\ &\times\sum_{l=0}^k \frac{(\mp1)^l(\alpha-\beta)_{k-l}(-\alpha-\gamma-k)_l(-\alpha-2k+1)_l}{(k-l)!(-\alpha-2k)_l(\beta+\gamma+1)_l}(\coth r_1\coth r_2)^l \\ &\times P_{\gamma-k}^{(\alpha+2k-l,\beta+l)}(\cosh(2r_1))P_{\gamma-k}^{(\alpha+2k-l,\beta+l)}(\cosh(2r_2))\frac{\beta+l}{\beta}C_l^\beta(\cos\phi) \end{aligned} \tag{3.23}$$

and

$$\begin{aligned} \mathsf{P}_\gamma^{(\alpha,\beta)}(\mathsf{X}^\pm) &= \frac{\Gamma(\alpha+1)\Gamma(\gamma+1)}{\Gamma(\alpha+\gamma+1)} \\ &\times\sum_{k=0}^\infty \frac{(\alpha)_k(\frac{\alpha}{2}+1)_k(-\beta-\gamma)_k(\alpha+\beta+\gamma+1)_k}{(\frac{\alpha}{2})_k(\beta+1)_k(-\gamma)_k(\alpha+\gamma+1)_k}(\sin\theta_1\sin\theta_2)^{2k} \\ &\times\sum_{l=0}^k \frac{(\mp1)^l(\alpha-\beta)_{k-l}(-\alpha-\gamma-k)_l(-\alpha-2k+1)_l}{(k-l)!(-\alpha-2k)_l(\beta+\gamma+1)_l}(\cot\theta_1\cot\theta_2)^l \\ &\times P_{\gamma-k}^{(\alpha+2k-l,\beta+l)}(\cos(2\theta_1))P_{\gamma-k}^{(\alpha+2k-l,\beta+l)}(\cos(2\theta_2))\frac{\beta+l}{\beta}C_l^\beta(\cos\phi). \end{aligned} \tag{3.24}$$

PROOF. Start with Theorem 3.9 and let $w=1$ using (2.48); substituting $l\mapsto l'=k-l$ followed by relabeling $l'\mapsto l$ completes the proof. □

By letting $\alpha=\beta$ in Corollary 3.9.1 we can relate the above result to associated Legendre and Gegenbauer functions of the first kind. This is mentioned in [**35**], namely that Koornwinder's addition theorem for Jacobi polynomials generalizes Gegenbauer's addition theorem (1.4). Similarly, the extension to the Flensted-Jensen–Koornwinder addition theorem for Jacobi functions of the first kind generalizes the addition theorem for Gegenbauer functions of the first kind. First we define the variables

$$\mathcal{Z}^\pm := \mathcal{Z}^\pm(r_1,r_2,\phi) := \cosh r_1\cosh r_2\pm\sinh r_1\sinh r_2\cos\phi, \tag{3.25}$$

$$\mathcal{X}^\pm := \mathcal{X}^\pm(\theta_1,\theta_2,\phi) := \cos\theta_1\cos\theta_2\pm\sin\theta_1\sin\theta_2\cos\phi. \tag{3.26}$$

COROLLARY 3.9.2. *Let $\gamma, \alpha \in \mathbb{C}$, $2\alpha \neq 1, 0, -1, \ldots$, $\gamma \notin -\mathbb{N}$, $r_1, r_2 \in [0, \infty)$, $\theta_1, \theta_2 \in [0, \frac{\pi}{2}]$, $\phi \in [0, \pi]$, and $\mathcal{Z}^{\pm}$, $\mathcal{X}^{\pm}$ defined by* (3.25), (3.26), *respectively. Then*

$$C_\gamma^\alpha(\mathcal{Z}^\pm) = \frac{\Gamma(2\alpha)\Gamma(\gamma+1)}{\Gamma(2\alpha+\gamma)} \sum_{k=0}^{\infty} \frac{(\mp 1)^k\, 2^{2k} (\alpha)_k (\alpha)_k}{(-\gamma)_k (2\alpha+\gamma)_k} (\sinh(2r_1)\sinh(2r_2))^k \times C_{\gamma-k}^{\alpha+k}(\cosh(2r_1)) C_{\gamma-k}^{\alpha+k}(\cosh(2r_2)) \frac{\alpha - \frac{1}{2} + k}{\alpha - \frac{1}{2}} C_k^{\alpha-\frac{1}{2}}(\cos\phi) \tag{3.27}$$

and

$$C_\gamma^\alpha(\mathcal{X}^\pm) = \frac{\Gamma(2\alpha)\Gamma(\gamma+1)}{\Gamma(2\alpha+\gamma)} \sum_{k=0}^{\infty} \frac{(\mp 1)^k\, 2^{2k} (\alpha)_k (\alpha)_k}{(-\gamma)_k (2\alpha+\gamma)_k} (\sin(2\theta_1)\sin(2\theta_2))^k \times C_{\gamma-k}^{\alpha+k}(\cos(2\theta_1)) C_{\gamma-k}^{\alpha+k}(\cos(2\theta_2)) \frac{\alpha - \frac{1}{2} + k}{\alpha - \frac{1}{2}} C_k^{\alpha-\frac{1}{2}}(\cos\phi), \tag{3.28}$$

or equivalently

$$\frac{P_\gamma^{-\alpha}(\mathcal{Z}^\pm)}{(1-\mathcal{Z}^{\pm 2})^{\frac{1}{2}\alpha}} = \frac{2^\alpha \Gamma(\alpha+1)}{(\sinh(2\theta_1)\sinh(2\theta_1))^\alpha} \sum_{k=0}^{\infty} (\pm 1)^k (\alpha-\gamma)_k (\alpha+\gamma+1)_k \times P_\gamma^{-\alpha-k}(\cosh(2r_1)) P_\gamma^{-\alpha-k}(\cosh(2r_2)) \frac{\alpha+k}{\alpha} C_k^\alpha(\cos\phi) \tag{3.29}$$

and

$$\frac{\mathsf{P}_\gamma^{-\alpha}(\mathcal{X}^\pm)}{(1-\mathcal{X}^{\pm 2})^{\frac{1}{2}\alpha}} = \frac{2^\alpha \Gamma(\alpha+1)}{(\sin(2\theta_1)\sin(2\theta_1))^\alpha} \sum_{k=0}^{\infty} (\pm 1)^k (\alpha-\gamma)_k (\alpha+\gamma+1)_k \times \mathsf{P}_\gamma^{-\alpha-k}(\cos(2\theta_1)) \mathsf{P}_\gamma^{-\alpha-k}(\cos(2\theta_2)) \frac{\alpha+k}{\alpha} C_k^\alpha(\cos\phi). \tag{3.30}$$

PROOF. Start with Corollary 3.9.1 and let $\alpha = \beta$ using (2.71), (2.72) respectively for the Jacobi functions of the first kind on the left-hand side and those on the right-hand side. Then mapping $(2z_1^2-1, 2z_2^2-1) \mapsto (z_1, z_2)$, $(2x_1^2-1, 2x_2^2-1) \mapsto (x_1, x_2)$, where $z_1 = \cosh r_1$, $z_2 = \cosh r_2$, $x_1 = \cos\theta_1$, $x_2 = \cos\theta_2$, and simplifying using (2.71)–(2.73), completes the proof.

Another way to prove this result is to take $\beta = -\frac{1}{2}$, $w = \cos\psi = 1$, $\gamma \to 2\gamma$ in (3.19) and use the quadratic transformation (2.76). After using (2.71), this produces the left-hand side of (3.27) with degree 4γ and order $\alpha + \frac{1}{2}$. Because we set $w = 1$, the sum over l only survives for $l = 0, 1$. By taking $4\gamma \mapsto \gamma$ and expressing the contribution due to each of these terms, one can identify Gegenbauer's addition theorem through repeated application of (2.71) on the right-hand side; and using the fact that

$$\sum_{k=0}^{\infty} (f_{2k} + f_{2k+1}) = \sum_{k=0}^{\infty} f_k, \tag{3.31}$$

for any sequence $\{f\}_{k\in\mathbb{N}_0}$, one arrives at (3.27). The proof of (3.28) is similar. □

4. Addition theorems for the Jacobi function of the second kind

Now we present double summation addition theorems for the Jacobi functions of the second kind in the hyperbolic and trigonometric contexts.

REMARK 4.1. There is no direct group-theoretical derivation of an addition formula for the Jacobi functions of the second kind. However, those functions satisfy the same differential recurrence relations as the functions of the first kind, and the actions of those operators and the Jacobi differential equation (2.39) on functions of either kind for general parameters give realizations of the Lie algebras of the groups considered. One would therefore expect the Jacobi functions of the second kind to satisfy addition formulas with the same structure as those for the functions of the first kind. This is known to hold, for example, for the Gegenbauer functions of the first and second kind [**14**, §8].

4.1. The hyperbolic context for the addition theorem for the Jacobi function of the second kind.

Define

$$z_{\lessgtr} := \begin{matrix}\min\\ \max\end{matrix}\{z_1, z_2\}, \tag{4.1}$$

when $z_1, z_2 \in (1,\infty)$, and in the case when $z_1, z_2 \in \mathbb{C}$, one takes without loss of generality $z_1 = z_>$ to lie on an ellipse with foci at ± 1, and $z_2 = z_<$ to be in the interior of that ellipse.

THEOREM 4.2. *Let $\gamma, \alpha, \beta \in \mathbb{C}$, $z_1, z_2 \in \mathbb{C}\setminus(-\infty,-1]$, with $x, w \in \mathbb{C}$, $Z^\pm$ defined by* (3.2)*, such that the complex variables $\gamma, \alpha, \beta, z_1, z_2, x, w$ are in some yet to be determined neighborhood of the real line. Then*

$$\begin{aligned} Q_\gamma^{(\alpha,\beta)}(Z^\pm) = {}& \frac{\Gamma(\alpha+1)\Gamma(\gamma+1)}{\Gamma(\alpha+\gamma+1)} \sum_{k=0}^\infty \frac{(\alpha+1)_k(\gamma+1)_k}{(\alpha+k)(\beta+1)_k(1-\gamma)_k} \\ & \times \sum_{l=0}^k (\pm 1)^{k-l}(\alpha+k+l)(z_1z_2)^{k-l}\left((z_1^2-1)(z_2^2-1)\right)^{\frac{k+l}{2}} \\ & \times P_{\gamma-k}^{(\alpha+k+l,\beta+k-l)}(2z_<^2-1)\,Q_{\gamma-k}^{(\alpha+k+l,\beta+k-l)}(2z_>^2-1) \\ & \times w^{k-l}P_l^{(\alpha-\beta-1,\beta+k-l)}(2w^2-1)\frac{\beta+k-l}{\beta}C_{k-l}^\beta(x). \end{aligned} \tag{4.2}$$

PROOF. Start with the addition formula (3.4) for the Jacobi functions of the first kind. The series was shown to converge by Flensted-Jensen and Koornwinder [**17**, Theorem 2.1]. Now apply the connection relation (2.55), which relates the Jacobi function of the first kind to two Jacobi functions of the second kind, to both sides of (3.4): to the function of the first kind on the left and to the function of the first kind on the right with argument $2z_2^2-1$, assuming without loss of generality that $z_2 = z_>$. This yields an equation in which the asymptotic behavior of the two terms on the left as $z_2 \to \infty$ matches the term-by-term asymptotic behavior of the two corresponding series on the right, suggesting that they should be identified and that the series converge separately. To exploit this, project the left and right-hand sides by the parabolic biangle polynomials, as in Theorem 3.7, with the same application of the connection formula to the Jacobi functions of the first and second kind. This results in the equation

(4.3)

$$\begin{aligned}
&\mathsf{B}_\gamma^{(\alpha,\beta)} \int_0^1 \int_0^\pi Q_\gamma^{(\alpha,\beta)}(Z^\pm)\, w^{k-l} P_l^{(\alpha-\beta-1,\beta+k-l)}(2w^2-1) C_{k-l}^\beta(\cos\phi)\, \mathrm{d}m^{(\alpha,\beta)}(w,\phi) \\
&\quad + \mathsf{C}_{\gamma,k,l}^{(\alpha,\beta)} (z_1 z_2)^{k-l} ((z_1^2-1)(z_2^2-1))^{\frac{1}{2}(k+l)} \\
&\quad\quad \times P_{\gamma-k}^{(\alpha+k+l,\alpha+k-l)}(2z_<^2-1) Q_{\gamma-k}^{(\alpha+k+l,\alpha+k-l)}(2z_>^2-1) \\
&= \mathsf{D}_\gamma^{(\alpha,\beta)} \int_0^1 \int_0^\pi Q_{-\alpha-\beta-\gamma-1}^{(\alpha,\beta)}(Z^\pm) w^{k-l} P_l^{(\alpha-\beta-1,\beta+k-l)}(2w^2-1) C_{k-l}^\beta(\cos\phi)\, \mathrm{d}m^{(\alpha,\beta)}(w,\phi) \\
&\quad + \mathsf{E}_{\gamma,k,l}^{(\alpha,\beta)} (z_1 z_2)^{k-l} ((z_1^2-1)(z_2^2-1))^{\frac{1}{2}(k+l)} \\
&\quad\quad \times P_{-\alpha-\beta-\gamma-k-1}^{(\alpha+k+l,\alpha+k-l)}(2z_<^2-1) Q_{-\alpha-\beta-\gamma-k-1}^{(\alpha+k+l,\alpha+k-l)}(2z_>^2-1),
\end{aligned}$$

where

$$\mathsf{B}_\gamma^{(\alpha,\beta)} := \frac{-2\sin(\pi\gamma)\sin(\pi(\beta+\gamma))}{\pi\sin(\pi(\alpha+\beta+2\gamma+1))}, \tag{4.4}$$

$$\begin{aligned}
\mathsf{C}_{\gamma,k,l}^{(\alpha,\beta)} &:= \frac{\sin(\pi\gamma)\sin(\pi(\beta+\gamma))}{2^{2\beta-1}\sin(\pi(\alpha+\beta+2\gamma+1))} \\
&\quad \times \frac{\Gamma(\gamma+1)(\alpha+\beta+\gamma+1)_k \Gamma(2\beta+k-l)\Gamma(\alpha-\beta+l)(-\beta-\gamma)_l}{\Gamma(\beta)(-\gamma)_k (k-l)!\, l!\, \Gamma(\alpha+\gamma+1+l)},
\end{aligned} \tag{4.5}$$

and

(4.6)

$$\mathsf{D}_\gamma^{(\alpha,\beta)} := \frac{-2\sin(\pi(\alpha+\gamma))\sin(\pi(\beta+\gamma))\Gamma(\alpha+\gamma+1)\Gamma(\beta+\gamma+1)}{\pi\sin(\pi(\alpha+\beta+2\gamma+1))\Gamma(\gamma+1)\Gamma(\alpha+\beta+\gamma+1)},$$

(4.7)

$$\mathsf{E}_{\gamma,k,l}^{(\alpha,\beta)} := \frac{\sin(\pi(\alpha+\gamma))\sin(\pi(\beta+\gamma))\Gamma(\beta+\gamma+1)\Gamma(2\beta+k-l)\Gamma(\alpha-\beta+l)}{2^{2\beta-1}\sin(\pi(\alpha+\beta+2\gamma+1))\Gamma(\beta)\Gamma(\alpha+\beta+\gamma+1)(k-l)!\, l!}.$$

Now consider the asymptotics of all four terms of (4.3) as $z_2 \to \infty$. The asymptotic behavior of $Z^\pm$ as $z_2 \to \infty$ is $Z^\pm \sim z_2^2$. The behavior of the Jacobi function of the second kind as $|z| \to \infty$, by (2.64), is

$$Q_\gamma^{(\alpha,\beta)}(z) \sim \frac{1}{z^{\alpha+\beta+\gamma+1}}.$$

Therefore, one has the following asymptotic behaviors as $\zeta \to \infty$, where $\zeta = z_>$ with $z_<$ fixed:

$$Q_\gamma^{(\alpha,\beta)}(Z^\pm) \sim (Z^\pm)^{-\gamma-\alpha-\beta-1} \sim \zeta^{-2\gamma-2\alpha-2\beta-2}, \tag{4.8}$$

$$Q_{-\alpha-\beta-\gamma-1}^{(\alpha,\beta)}(Z^\pm) \sim (Z^\pm)^\gamma \sim \zeta^{2\gamma}, \tag{4.9}$$

and

$$Q_{\gamma-k}^{(\alpha+k+l,\beta+k-l)}(\zeta) \sim \zeta^{-\gamma-\alpha-\beta-k-1}, \tag{4.10}$$

$$Q_{-\alpha-\beta-\gamma-1-k}^{(\alpha+k+l,\beta+k-l)}(\zeta) \sim \zeta^{\gamma-k}. \tag{4.11}$$

The equation (4.3) with leading order asymptotic contribution as $z_2 \to \infty$ taken out can be written in terms of two analytic functions

$$\mathsf{f}(z_2^{-1}) := \mathsf{f}_{\gamma,k,l}^{(\alpha,\beta)}(z_1, z_2^{-1}),$$
$$\mathsf{g}(z_2^{-1}) := \mathsf{g}_{\gamma,k,l}^{(\alpha,\beta)}(z_1, z_2^{-1}),$$

as

$$z_2^{-2(\gamma+\alpha+\beta+1)}\mathsf{f}(z_2^{-1}) = z_2^{2\gamma}\mathsf{g}(z_2^{-1}). \tag{4.12}$$

For $4\Re\gamma \notin -2(\alpha+\beta+1)+\mathbb{Z}$, the only way the equation can be true is if f and g vanish identically. The case of general γ then follows by analytic continuation in γ. Therefore we have now verified separately all terms in the double series expansion of the Jacobi function of the second kind appearing in (4.2), and the corresponding series for $Q^{(\alpha,\beta)}_{-\alpha-\beta-\gamma-1}(Z^{\pm})$. The two series do not mix, have different asymptotic behaviors for $z_2 = z_> \to \infty$, and must converge separately given the overall convergence proved by Flensted-Jensen and Koornwinder [**17**, Theorem 2.1]. This completes the proof. □

REMARK 4.3. If one applies (2.56) to the Jacobi functions of the second kind on the left-hand side and right-hand side of (4.2), then it becomes the addition theorem for Jacobi polynomials in the hyperbolic context.

COROLLARY 4.3.1. *Let $k,l \in \mathbb{N}_0$ with $l \leqslant k$, $\gamma,\alpha,\beta \in \mathbb{C}$, $z_1, z_2 \in \mathbb{C}\setminus(-\infty,1]$, with $Z^{\pm}$ defined by* (3.2)*, such that $x=\cos\phi$ and the complex variables $\gamma,\alpha,\beta,z_1,z_2$ are in some yet to be determined neighborhood of the real line. Then*

$$\begin{aligned}\int_0^1\int_0^\pi & Q_\gamma^{(\alpha,\beta)}(Z^{\pm})w^{k-l}P_l^{(\alpha-\beta-1,\beta+k-l)}(2w^2-1)C_{k-l}^{\beta}(\cos\phi)\,\mathrm{d}m^{(\alpha,\beta)}(w,\phi),\\ &= (\pm1)^{k+l}\,\mathsf{A}_{k,l}^{(\alpha,\beta,\gamma)}(z_1z_2)^{k-l}((z_1^2-1)(z_2^2-1))^{\frac12(k+l)}\\ &\quad\times P_{\gamma-k}^{(\alpha+k+l,\beta+k-l)}(2z_<^2-1)Q_{\gamma-k}^{(\alpha+k+l,\beta+k-l)}(2z_>^2-1),\end{aligned} \tag{4.13}$$

where $\mathsf{A}_{k,l}^{(\alpha,\beta,\gamma)}$ is defined in (3.16).

PROOF. This follows directly from (4.12) since in the proof of Theorem 4.2, we showed that $\mathsf{f}=0$. The result $\mathsf{g}=0$ is equivalent to this result under the transformation $\gamma \mapsto -\gamma-\alpha-\beta-1$. □

REMARK 4.4. The integral representations Theorem 3.7 and Corollary 4.3.1 are equivalent to the double summation addition theorems for the Jacobi function of the first kind (3.4) and second kind, Theorem 4.2.

REMARK 4.5. One has the following well-known product representations which are the $k=l=0$ cases of the integral representations in Theorem 3.7 and Corollary 4.3.1. Let $x=\cos\phi$, $\gamma,\alpha,\beta\in\mathbb{C}$, $z_1,z_2\in\mathbb{C}\setminus(-\infty,-1]$, with $Z^{\pm}$ defined by (3.2), such that the complex variables $\gamma,\alpha,\beta,z_1,z_2$ are in some yet to be determined neighborhood of the real line. Then

$$P_\gamma^{(\alpha,\beta)}(2z_1^2-1)P_\gamma^{(\alpha,\beta)}(2z_2^2-1) = \frac{2\Gamma(\alpha+\gamma+1)}{\sqrt{\pi}\,\Gamma(\gamma+1)\Gamma(\beta+\frac12)\Gamma(\alpha-\beta)}\int_0^1\int_0^\pi P_\gamma^{(\alpha,\beta)}(Z^\pm)\,\mathrm{d}m^{(\alpha,\beta)}(w,\phi), \tag{4.14}$$

$$P_\gamma^{(\alpha,\beta)}(2z_<^2-1)Q_\gamma^{(\alpha,\beta)}(2z_>^2-1) = \frac{2\Gamma(\alpha+\gamma+1)}{\sqrt{\pi}\,\Gamma(\gamma+1)\Gamma(\beta+\frac12)\Gamma(\alpha-\beta)}\int_0^1\int_0^\pi Q_\gamma^{(\alpha,\beta)}(Z^\pm)\,\mathrm{d}m^{(\alpha,\beta)}(w,\phi). \tag{4.15}$$

For independent verification of these product representations, see [**16**, Theorem 4.1] for the product formula (4.14) and [**16**, p. 255] for the product formula (4.15).

REMARK 4.6. As pointed out by one of the referees, our proof of the addition theorem for the Jacobi function of the second kind was chosen so that it was the most economical, following as it does from the role of Jacobi functions as spherical functions on symmetric spaces [**35**], extended to general parameters as in [**17**]. However, this does not mean that our chosen method is the most insightful method of proof overall. In fact, Flensted-Jensen and Koornwinder (1979) [**17**] derived the addition theorem for the Jacobi function of the first kind from the product formula (4.14), which the same authors had obtained earlier in [**16**, Theorem 4.1]. Inspection of the proof of the product formula (4.14) in [**16**] shows that it really comes from the product formula (4.15) for Jacobi functions of the second kind [**16**, Proof of Theorem 4.1, p. 252]. In this proof, the authors expand the Jacobi function of the first kind using [**16**, (4.6)], which is simply the connection formula (2.55) which expresses the Jacobi function of the first kind as a linear combination of two Jacobi functions of the second kind with different degrees. So, as the referee pointed out, a more informative proof of the addition theorem for the Jacobi function of the second kind, Theorem 4.2, would be to derive it from the product formula (4.15) in the same way as Theorem 3.2 is derived in [**17**] from the product formula (4.14).

In the case when $z_1, z_2, w, x = \cos\phi$ are real numbers, the argument of the Jacobi function of the second kind in the addition theorem for the Jacobi function of the second kind takes a simpler and more convenient form. This is analogous to the Flensted-Jensen–Koornwinder addition theorem of the first kind, (3.19). We present this result now.

THEOREM 4.7. *Let* $\gamma,\alpha,\beta,w\in\mathbb{R}$*, such that* $\gamma\notin\mathbb{Z}$*,* $\alpha\notin-\mathbb{N}$*,* $\beta\in(-\frac12,\infty)\setminus\{0\}$*,* $\alpha+\gamma,\beta+\gamma\notin-\mathbb{N}$*,* $\phi\in[0,\pi]$*,* $z_1,z_2\in(1,\infty)$*, and* $Z^\pm$*,* $z_\lessgtr$*, defined by* (3.17), (4.1)*, respectively. Then*

$$\begin{aligned} Q_\gamma^{(\alpha,\beta)}(Z^\pm) = \frac{\Gamma(\alpha+1)\Gamma(\gamma+1)}{\Gamma(\alpha+\gamma+1)}&\sum_{k=0}^\infty \frac{(\alpha+1)_k(\alpha+\beta+\gamma+1)_k}{(\alpha+k)(\beta+1)_k(-\gamma)_k} \\ &\times\sum_{l=0}^k(\pm1)^{k+l}\frac{(\alpha+k+l)(-\beta-\gamma)_l}{(\alpha+\gamma+1)_l}(z_1z_2)^{k-l}\left((z_1^2-1)(z_2^2-1)\right)^{\frac{k+l}{2}} \\ &\quad\times P_{\gamma-k}^{(\alpha+k+l,\beta+k-l)}(2z_<^2-1)Q_{\gamma-k}^{(\alpha+k+l,\beta+k-l)}(2z_>^2-1) \\ &\quad\times w^{k-l}P_l^{(\alpha-\beta-1,\beta+k-l)}(2w^2-1)\frac{\beta+k-l}{\beta}C_{k-l}^\beta(\cos\phi). \end{aligned} \tag{4.16}$$

PROOF. This follows from Theorem 4.2 by setting the variables $\gamma, \alpha, \beta, z_1, z_2, w$, $x = \cos\phi$ to real numbers. $\square$

4.2. The trigonometric context of the addition theorem for the Jacobi function of the second kind. In the trigonometric context for the addition theorem for Jacobi functions of the second kind, one must use the Jacobi function of the second kind on-the-cut $\mathsf{Q}_\gamma^{(\alpha,\beta)}$ (2.66), which is defined in Section 2.2.3 and has a hypergeometric representation given by (2.67). Note that this representation is not unique and there are many other double Gauss hypergeometric representations of this function. For more about this see the discussion immediately above (2.67). Define

$$x_{\lessgtr} := \min_{\max}\{x_1, x_2\}, \tag{4.17}$$

where $x_1, x_2 \in (-1,1)$, and in the case where $x_1, x_2 \in \mathbb{C}$, then if one takes without loss of generality $x_1 = x_>$ to lie on an ellipse with foci at ± 1, then $x_2 = x_<$ must be chosen to be in the interior of that ellipse.

THEOREM 4.8. *Let $k, l \in \mathbb{N}_0$, $l \leqslant k$, $\gamma, \alpha, \beta \in \mathbb{C}$, $x_1, x_2 \in \mathbb{C}\setminus((-\infty,-1]\cup[1,\infty))$, $\alpha \notin \mathbb{Z}$, $\alpha+\gamma, \beta+\gamma \notin -\mathbb{N}$, with $\mathsf{X}^\pm$, $x_\lessgtr$ defined by* (3.3), (4.17) *respectively, such that the complex variables $\gamma, \alpha, \beta, x_1, x_2$ are in some yet to be determined neighborhood of the real line. Then*

$$\begin{aligned}\int_0^1\int_0^\pi &\mathsf{Q}_\gamma^{(\alpha,\beta)}(X^\pm)\, w^{k-l} P_l^{(\alpha-\beta-1,\beta+k-l)}(2w^2-1) C_{k-l}^\beta(\cos\phi)\, dm^{(\alpha,\beta)}(w,\phi)\\ &= (\mp 1)^{k+l}\mathsf{A}_{k,l}^{(\alpha,\beta,\gamma)}(x_1x_2)^{k-l}((1-x_1^2)(1-x_2^2))^{\frac12(k+l)}\\ &\quad\times \mathsf{Q}_{\gamma-k}^{(\alpha+k+l,\beta+k-l)}(2x_<^2-1)\mathsf{P}_{\gamma-k}^{(\alpha+k+l,\beta+k-l)}(2x_>^2-1),\end{aligned} \tag{4.18}$$

where $\mathsf{A}_{k,l}^{(\alpha,\beta,\gamma)}$ is defined in (3.16).

PROOF. Starting with Corollary 4.3.1 and applying the definition (2.66) completes the proof. $\square$

COROLLARY 4.8.1. *Let $\gamma, \alpha, \beta \in \mathbb{C}$, $x_1, x_2 \in \mathbb{C}\setminus((-\infty,-1]\cup[1,\infty))$, $\alpha \notin \mathbb{Z}$, $\alpha+\gamma, \beta+\gamma \notin -\mathbb{N}$, with $\mathsf{X}^\pm$, $x_\lessgtr$ defined by* (3.3), (4.17) *respectively, such that the complex variables $\gamma, \alpha, \beta, x_1, x_2$ are in some yet to be determined neighborhood of the real line. Then*

$$\begin{aligned}\int_0^1\int_0^\pi &\mathsf{Q}_\gamma^{(\alpha,\beta)}(X^\pm)\, dm^{(\alpha,\beta)}(w,\phi)\\ &= \frac{\sqrt{\pi}\,\Gamma(\gamma+1)\Gamma(\beta+\frac12)\Gamma(\alpha-\beta)}{2\Gamma(\alpha+\gamma+1)}\mathsf{Q}_\gamma^{(\alpha,\beta)}(2x_<^2-1)\mathsf{P}_\gamma^{(\alpha,\beta)}(2x_>^2-1).\end{aligned} \tag{4.19}$$

PROOF. Starting with Theorem 4.8 and setting $k = l = 0$ completes the proof. $\square$

THEOREM 4.9. *Let $\gamma, \alpha, \beta \in \mathbb{C}$, $x_1, x_2 \in \mathbb{C}\setminus((-\infty,-1]\cup[1,\infty))$, $\alpha \notin \mathbb{Z}$, $\alpha+\gamma, \beta+\gamma \notin -\mathbb{N}$, $x, w \in \mathbb{C}$ with $\mathsf{X}^\pm$, $x_\lessgtr$ defined by* (3.3), (4.17) *respectively, such that the complex variables $\gamma, \alpha, \beta, x_1, x_2, x, w$ are in some yet to be determined*

neighborhood of the real line. Then
(4.20)

$$\mathsf{Q}_\gamma^{(\alpha,\beta)}(\mathsf{X}^\pm) = \frac{\Gamma(\alpha+1)\Gamma(\gamma+1)}{\Gamma(\alpha+\gamma+1)} \sum_{k=0}^{\infty} \frac{(\alpha+1)_k(\alpha+\beta+\gamma+1)_k}{(\alpha+k)(\beta+1)_k(-\gamma)_k}$$
$$\times \sum_{l=0}^{k} (\mp 1)^{k-l} \frac{(\alpha+k+l)(-\beta-\gamma)_l}{(\alpha+\gamma+1)_l} (x_1x_2)^{k-l} \left((1-x_1^2)(1-x_2^2)\right)^{\frac{k+l}{2}}$$
$$\times \mathsf{Q}_{\gamma-k}^{(\alpha+k+l,\beta+k-l)}(2x_<^2-1)\mathsf{P}_{\gamma-k}^{(\alpha+k+l,\beta+k-l)}(2x_>^2-1)$$
$$\times w^{k-l} P_l^{(\alpha-\beta-1,\beta+k-l)}(2w^2-1)\frac{\beta+k-l}{\beta} C_{k-l}^\beta(x).$$

PROOF. The result follows by starting with the addition theorem for Jacobi functions of the second kind (4.16) and applying the definition (2.66) completes the proof. □

In the case when $z_1, z_2, w, x = \cos\phi$ are real numbers, then the argument of the Jacobi function of the second kind on-the-cut in the addition theorem for the Jacobi function of the second kind takes a simpler and more convenient form. This is analogous to the addition theorem (3.19). We present this result now.

COROLLARY 4.9.1. *Let $\gamma, \alpha, \beta, w \in \mathbb{R}$, such that $\gamma \notin \mathbb{Z}$, $\alpha, \alpha+\gamma, \beta+\gamma \notin -\mathbb{N}$, $x_1, x_2 \in (-1,1)$, $\phi \in [0,\pi]$, with $\mathsf{X}^\pm$, $x_\lessgtr$ defined by* (3.18), (4.17) *respectively. Then*
(4.21)

$$\mathsf{Q}_\gamma^{(\alpha,\beta)}(\mathsf{X}^\pm) = \frac{\Gamma(\alpha+1)\Gamma(\gamma+1)}{\Gamma(\alpha+\gamma+1)} \sum_{k=0}^{\infty} \frac{(\alpha+1)_k(\alpha+\beta+\gamma+1)_k}{(\alpha+k)(\beta+1)_k(-\gamma)_k}$$
$$\times \sum_{l=0}^{k} (\mp 1)^{k-l} \frac{(\alpha+k+l)(-\beta-\gamma)_l}{(\alpha+\gamma+1)_l} (x_1x_2)^{k-l} \left((1-x_1^2)(1-x_2^2)\right)^{\frac{k+l}{2}}$$
$$\times \mathsf{Q}_{\gamma-k}^{(\alpha+k+l,\beta+k-l)}(2x_<^2-1)\mathsf{P}_{\gamma-k}^{(\alpha+k+l,\beta+k-l)}(2x_>^2-1)$$
$$\times w^{k-l} P_l^{(\alpha-\beta-1,\beta+k-l)}(2w^2-1)\frac{\beta+k-l}{\beta} C_{k-l}^\beta(\cos\phi).$$

PROOF. This follows from Theorem 4.7 by setting the complex variables to be real. □

4.3. Olver-normalized Jacobi functions and their addition theorems. Koornwinder's addition theorem (1.6) for Jacobi polynomials of degree $n \in \mathbb{N}_0$ involves a terminating sum. One can attribute this to the infinite sum appearing in the generalizations (3.19), (3.20), first recognizing that Jacobi polynomials $P_{\gamma-k}^{(\alpha+k+l,\beta+k-l)}$ vanish for both $\gamma = n \in \mathbb{N}_0$ and $k \geqslant n+1$. Considering the limit as $\gamma \to n$ for all values of $k \in \mathbb{N}_0$ in Koornwinder's addition theorem, one sees that the factor $1/(-\gamma)_k$ blows up for $k \geqslant n+1$. On the other hand, this factor is multiplied in the limit by the Jacobi function of the first kind prefactor containing $1/\Gamma(\gamma-k+1)$, and considering the residues of the gamma function, one sees that the product will be finite, namely

$$\lim_{\gamma\to n} \frac{1}{(-\gamma)_k\Gamma(\gamma-k+1)} = \lim_{\gamma\to n} \frac{\Gamma(-\gamma)}{\Gamma(-\gamma+k)\Gamma(\gamma-k+1)} = \frac{(-1)^k}{n!}, \tag{4.22}$$

for $k \geqslant n+1$. When this finite factor is multiplied by the second Jacobi polynomial $P_{\gamma-k}^{(\alpha+\beta+k+l,\beta+k-l)}$, which vanishes for $k \geqslant n+1$, the resulting expression vanishes for all these k values, which yields a terminating sum over $k \in \{0,\ldots,n\}$.

Unlike Koornwinder's addition theorem for Jacobi polynomials, the addition theorem for the Jacobi functions of the second kind (see §4) involves a nonterminating sum. One can see this by examining (4.20), first recognizing that the Jacobi polynomials $P_{\gamma-k}^{(\alpha+k+l,\beta+k-l)}$ vanish for $\gamma = n \in \mathbb{N}_0$ and $k \geqslant n+1$. Considering the limit as $\gamma \to n$ for all values of $k \in \mathbb{N}_0$ in Koornwinder's addition theorem, one sees that the factor $1/(-\gamma)_k$ blows up for $k \geqslant n+1$. On the other hand, this factor in the limit is multiplied by the Jacobi function of the first kind prefactor containing $1/\Gamma(\gamma-k+1)$, and considering the residues of the gamma function, one sees that the product will be finite, as shown in (4.22), for $k \geqslant n+1$. This finite factor is then multiplied by the Jacobi function of the second kind $Q_{\gamma-k}^{(\alpha+\beta+k+l,\beta+k-l)}(2z_>^2-1)$, which does not vanish for $k \geqslant n+1$ for $\alpha, \beta \notin \mathbb{Z}$, unlike the case for Jacobi polynomials.

4.4. Olver-normalized Jacobi functions. In (2.8), we introduced Olver's normalization of the Gauss [**11**, (15.2.2)] and generalized hypergeometric function (see also [**11**, (16.2.5)]), which results in these functions being entire functions of all their parameters, including the denominator parameters. Olver applied this concept of a special normalization to the associated Legendre function of the second kind [**11**, (14.3.10)] (see also [**46**, pp. 170 and 178]). We now demonstrate how to apply it to the Jacobi functions of the first and second kind.

In the above description, instead of carefully determining the limits of the relevant functions when there are removable singularities due to the appearance of various gamma function prefactors, an alternative is to use appropriately defined Olver-normalized Jacobi functions and recast the addition theorems correspondingly. The benefit of using Olver-normalized definitions of the Jacobi functions is that one avoids complications due to gamma functions with removable singularities. Typical examples occur in often appearing examples when one has degrees γ and parameters α, β given by integers. In these cases, if the standard definitions such as those which appear in Theorems 2.5, 2.7 are used, the functions that appear are not defined and careful limits must be taken. However, if one adopts carefully chosen Olver-normalized definitions in which only Olver-normalized Gauss hypergeometric functions appear, then these functions will be entire in all their parameters. As we will see, by using these definitions, we arrive at formulas for the addition theorems which are elegant and highly useful! First we give our new choice of the Olver normalization, and then relate the Olver-normalized definitions to the usual definitions. Our definitions of Olver-normalized Jacobi functions of the first and second kind in the hyperbolic and trigonometric contexts are given by

$$\boldsymbol{P}_\gamma^{(\alpha,\beta)}(z) := {}_2\boldsymbol{F}_1\left(\begin{matrix}-\gamma, \alpha+\beta+\gamma+1\\ \alpha+1\end{matrix}; \frac{1-z}{2}\right), \tag{4.23}$$

$$\boldsymbol{Q}_\gamma^{(\alpha,\beta)}(z) := \frac{2^{\alpha+\beta+\gamma}}{(z-1)^{\alpha+\gamma+1}(z+1)^\beta}\, {}_2\boldsymbol{F}_1\left(\begin{matrix}\gamma+1, \alpha+\gamma+1\\ \alpha+\beta+2\gamma+2\end{matrix}; \frac{2}{1-z}\right), \tag{4.24}$$

and

$$\mathsf{P}_\gamma^{(\alpha,\beta)}(x) := {}_2\boldsymbol{F}_1\left(\begin{matrix}-\gamma, \alpha+\beta+\gamma+1\\ \alpha+1\end{matrix}; \frac{1-x}{2}\right), \tag{4.25}$$

(4.26)

$$\mathbf{Q}_\gamma^{(\alpha,\beta)}(x) := \tfrac{1}{2}\Gamma(\alpha+1)\left(\frac{1+x}{2}\right)^\gamma \times \left(\frac{\cos(\pi\alpha)\Gamma(\alpha+\gamma+1)}{\Gamma(\gamma+1)}\,{}_2\boldsymbol{F}_1\left(\begin{matrix}-\gamma,-\beta-\gamma\\1+\alpha\end{matrix};\frac{x-1}{x+1}\right) - \frac{\Gamma(\beta+\gamma+1)}{\Gamma(\alpha+\beta+\gamma+1)}\left(\frac{1+x}{1-x}\right)^\alpha {}_2\boldsymbol{F}_1\left(\begin{matrix}-\alpha-\gamma,-\alpha-\beta-\gamma\\1-\alpha\end{matrix};\frac{x-1}{x+1}\right)\right).$$

Therefore one has connection relations between the Jacobi functions of the first and second kinds and their Olver-normalized counterparts, namely

$$P_\gamma^{(\alpha,\beta)}(z) = \frac{\Gamma(\alpha+\gamma+1)}{\Gamma(\gamma+1)}\boldsymbol{P}_\gamma^{(\alpha,\beta)}(z), \tag{4.27}$$

$$Q_\gamma^{(\alpha,\beta)}(z) = \Gamma(\alpha+\gamma+1)\Gamma(\beta+\gamma+1)\,\boldsymbol{Q}_\gamma^{(\alpha,\beta)}(z), \tag{4.28}$$

$$\mathsf{P}_\gamma^{(\alpha,\beta)}(x) = \frac{\Gamma(\alpha+\gamma+1)}{\Gamma(\gamma+1)}\mathbf{P}_\gamma^{(\alpha,\beta)}(x). \tag{4.29}$$

Note that

$$\mathbf{P}_\gamma^{(\alpha,\beta)}(x) = \boldsymbol{P}_\gamma^{(\alpha,\beta)}(x \pm i0), \tag{4.30}$$

as in (2.65). Furthermore in the special case $\gamma = 0$ one has

$$\boldsymbol{Q}_0^{(\alpha,\beta)}(z) := \frac{2^{\alpha+\beta}}{(z-1)^{\alpha+1}(z+1)^\beta}\,{}_2\boldsymbol{F}_1\left(\begin{matrix}1,\alpha+1\\\alpha+\beta+2\end{matrix};\frac{2}{1-z}\right). \tag{4.31}$$

REMARK 4.10. As of the date of publication of this manuscript, we have been unable to find an Olver-normalized version of the Jacobi function of the second kind on-the-cut $\mathbf{Q}_\gamma^{(\alpha,\beta)}$. However we have found a special normalization of this function which works well when $\gamma = 0$ and the β parameter takes an integer value, which is of particular importance because this case appears in a very important application (see Section 5 below). Let $b \in \mathbb{N}_0$. Define

$$\mathcal{Q}_{-k}^{(\alpha+k+l,b+k-l)}(x) := \lim_{\gamma\to0,\beta\to b}(-\beta-\gamma)_l \mathsf{Q}_{\gamma-k}^{(\alpha+k+l,\beta+k-l)}(x), \tag{4.32}$$

which is a well-defined function for all α, b, x, k, l in its domain.

4.5. Addition theorems for the Olver-normalized Jacobi functions. Now that we have introduced the Olver-normalized Jacobi functions of the first and second kind in the hyperbolic and trigonometric contexts, we are in a position to perform the straightforward derivation of the corresponding addition theorems for these functions.

THEOREM 4.11. *Let $\gamma, \alpha, \beta \in \mathbb{C}$, $z_1, z_2 \in \mathbb{C}\setminus(-\infty,-1]$, $x_1, x_2 \in \mathbb{C}\setminus((-\infty,-1]\cup[1,\infty))$, $x, w \in \mathbb{C}$, with $Z^\pm$, $\mathsf{X}^\pm$ defined by (3.2), (3.3), respectively, such that the complex variables $\gamma, \alpha, \beta, z_1, z_2, x_1, x_2, x, w$ are in some yet to be determined* *neighborhood of the real line. Then*

(4.33)

$$P_\gamma^{(\alpha,\beta)}(Z^\pm)=\frac{\Gamma(\alpha+1)\Gamma(\alpha+\gamma+1)}{\Gamma(\gamma+1)}\sum_{k=0}^{\infty}\frac{(\alpha+1)_k(\alpha+\beta+\gamma+1)_k(-\gamma)_k}{(\alpha+k)(\beta+1)_k}$$
$$\times\sum_{l=0}^{k}(\mp1)^{k-l}(\alpha+k+l)(\alpha+\gamma+1)_l(-\beta-\gamma)_l(z_1z_2)^{k-l}$$
$$\times\left((z_1^2-1)(z_2^2-1)\right)^{\frac{k+l}{2}}\boldsymbol{P}_{\gamma-k}^{(\alpha+k+l,\beta+k-l)}(2z_1^2-1)\boldsymbol{P}_{\gamma-k}^{(\alpha+k+l,\beta+k-l)}(2z_2^2-1)$$
$$\times w^{k-l}P_l^{(\alpha-\beta-1,\beta+k-l)}(2w^2-1)\frac{\beta+k-l}{\beta}C_{k-l}^{\beta}(x),$$

(4.34)

$$Q_\gamma^{(\alpha,\beta)}(Z^\pm)=\Gamma(\alpha+1)\Gamma(\alpha+\gamma+1)\Gamma(\beta+\gamma+1)\sum_{k=0}^{\infty}\frac{(\alpha+1)_k(\alpha+\beta+\gamma+1)_k}{(\alpha+k)(\beta+1)_k}$$
$$\times\sum_{l=0}^{k}(\mp1)^{k-l}(\alpha+k+l)(\alpha+\gamma+1)_l(z_1z_2)^{k-l}\left((z_1^2-1)(z_2^2-1)\right)^{\frac{k+l}{2}}$$
$$\times\boldsymbol{Q}_{\gamma-k}^{(\alpha+k+l,\beta+k-l)}(2z_>^2-1)\boldsymbol{P}_{\gamma-k}^{(\alpha+k+l,\beta+k-l)}(2z_<^2-1)$$
$$\times w^{k-l}P_l^{(\alpha-\beta-1,\beta+k-l)}(2w^2-1)\frac{\beta+k-l}{\beta}C_{k-l}^{\beta}(x),$$

and

(4.35)

$$\mathsf{P}_\gamma^{(\alpha,\beta)}(\mathsf{X}^\pm)=\frac{\Gamma(\alpha+1)\Gamma(\alpha+\gamma+1)}{\Gamma(\gamma+1)}\sum_{k=0}^{\infty}\frac{(\alpha+1)_k(\alpha+\beta+\gamma+1)_k(-\gamma)_k}{(\alpha+k)(\beta+1)_k}$$
$$\times\sum_{l=0}^{k}(\mp1)^{k-l}(\alpha+k+l)(\alpha+\gamma+1)_l(-\beta-\gamma)_l(x_1x_2)^{k-l}$$
$$\times\left((1-x_1^2)(1-x_2^2)\right)^{\frac{k+l}{2}}\boldsymbol{\mathsf{P}}_{\gamma-k}^{(\alpha+k+l,\beta+k-l)}(2x_1^2-1)\boldsymbol{\mathsf{P}}_{\gamma-k}^{(\alpha+k+l,\beta+k-l)}(2x_2^2-1)$$
$$\times w^{k-l}P_l^{(\alpha-\beta-1,\beta+k-l)}(2w^2-1)\frac{\beta+k-l}{\beta}C_{k-l}^{\beta}(x),$$

(4.36)

$$\mathsf{Q}_\gamma^{(\alpha,\beta)}(\mathsf{X}^\pm)=\Gamma(\alpha+1)\sum_{k=0}^{\infty}\frac{(-1)^k(\alpha+\beta+\gamma+1)_k(\alpha+1)_k}{(\alpha+k)(\beta+1)_k}$$
$$\times\sum_{l=0}^{k}(\mp1)^{k-l}(\alpha+k+l)(-\beta-\gamma)_l(x_1x_2)^{k-l}\left((1-x_1^2)(1-x_2^2)\right)^{\frac{k+l}{2}}$$
$$\times\boldsymbol{\mathsf{Q}}_{\gamma-k}^{(\alpha+k+l,\beta+k-l)}(2x_<^2-1)\boldsymbol{\mathsf{P}}_{\gamma-k}^{(\alpha+k+l,\beta+k-l)}(2x_>^2-1)$$
$$\times w^{k-l}P_l^{(\alpha-\beta-1,\beta+k-l)}(2w^2-1)\frac{\beta+k-l}{\beta}C_{k-l}^{\beta}(x).$$

PROOF. Substituting (4.27)–(4.29) into (3.4), (3.5), (4.2), (4.20) as necessary completes the proof. □

There are corresponding expansions that are sometimes useful in which the l sum is reversed, obtained by making the replacements $l'=k-l$ and then $l'\mapsto l$ in Theorem 4.2. These are as follows.

COROLLARY 4.11.1. *Let* $\gamma,\alpha,\beta\in\mathbb{C}$, $z_1,z_2\in\mathbb{C}\setminus(-\infty,-1]$, $x_1,x_2\in\mathbb{C}\setminus((-\infty,-1]\cup[1,\infty))$, $x,w\in\mathbb{C}$, *with* $Z^\pm$, $\mathsf{X}^\pm$ *defined by* (3.2), (3.3), *respectively, such* *that the complex variables $\gamma, \alpha, \beta, z_1, z_2, x_1, x_2, x, w$ are in some yet to be determined neighborhood of the real line. Then*

(4.37)

$$P_\gamma^{(\alpha,\beta)}(Z^\pm) = x\frac{\Gamma(\alpha+1)\Gamma(\alpha+\gamma+1)}{\Gamma(\gamma+1)}\sum_{k=0}^{\infty}\frac{(\alpha+1)_k(\alpha+\beta+\gamma+1)_k(-\gamma)_k}{(\alpha+k)(\beta+1)_k}$$
$$\times\sum_{l=0}^{k}(\mp1)^l(\alpha+2k-l)(\alpha+\gamma+1)_{k-l}(-\beta-\gamma)_{k-l}(z_1z_2)^l((z_1^2-1)(z_2^2-1))^{\frac{2k-l}{2}}$$
$$\times \boldsymbol{P}_{\gamma-k}^{(\alpha+2k-l,\beta+l)}(2z_1^2-1)\boldsymbol{P}_{\gamma-k}^{(\alpha+2k-l,\beta+l)}(2z_2^2-1)$$
$$\times w^l P_{k-l}^{(\alpha-\beta-1,\beta+l)}(2w^2-1)\frac{\beta+l}{\beta}C_l^\beta(x),$$

(4.38)

$$Q_\gamma^{(\alpha,\beta)}(Z^\pm) = \Gamma(\alpha+1)\Gamma(\alpha+\gamma+1)\Gamma(\beta+\gamma+1)\sum_{k=0}^{\infty}\frac{(\alpha+1)_k(\alpha+\beta+\gamma+1)_k}{(\alpha+k)(\beta+1)_k}$$
$$\times\sum_{l=0}^{k}(\mp1)^l(\alpha+2k-l)(\alpha+\gamma+1)_{k-l}(z_1z_2)^l((z_1^2-1)(z_2^2-1))^{\frac{2k-l}{2}}$$
$$\times \boldsymbol{Q}_{\gamma-k}^{(\alpha+2k-l,\beta+l)}(2z_>^2-1)\boldsymbol{P}_{\gamma-k}^{(\alpha+2k-l,\beta+l)}(2z_<^2-1)$$
$$\times w^l P_l^{(\alpha-\beta-1,\beta+l)}(2w^2-1)\frac{\beta+l}{\beta}C_l^\beta(x),$$

and

(4.39)

$$\mathsf{P}_\gamma^{(\alpha,\beta)}(\mathsf{X}^\pm) = \frac{\Gamma(\alpha+1)\Gamma(\alpha+\gamma+1)}{\Gamma(\gamma+1)}\sum_{k=0}^{\infty}\frac{(\alpha+1)_k(\alpha+\beta+\gamma+1)_k(-\gamma)_k}{(\alpha+k)(\beta+1)_k}$$
$$\times\sum_{l=0}^{k}(\mp1)^l(\alpha+2k-l)(\alpha+\gamma+1)_{k-l}(-\beta-\gamma)_{k-l}(x_1x_2)^l$$
$$\times((x_1^2-1)(x_2^2-1))^{\frac{2k-l}{2}}\mathsf{P}_{\gamma-k}^{(\alpha+2k-l,\beta+l)}(2x_1^2-1)\mathsf{P}_{\gamma-k}^{(\alpha+2k-l,\beta+l)}(2x_2^2-1)$$
$$\times w^l P_{k-l}^{(\alpha-\beta-1,\beta+l)}(2w^2-1)\frac{\beta+l}{\beta}C_l^\beta(x),$$

(4.40)

$$\mathsf{Q}_\gamma^{(\alpha,\beta)}(\mathsf{X}^\pm) = \Gamma(\alpha+1)\sum_{k=0}^{\infty}(-1)^k\frac{(\alpha+1)_k(\alpha+\beta+\gamma+1)_k}{(\alpha+k)(\beta+1)_k}$$
$$\times\sum_{l=0}^{k}(\mp1)^l(\alpha+2k-l)(-\beta-\gamma)_{k-l}(x_1x_2)^l((x_1^2-1)(x_2^2-1))^{\frac{2k-l}{2}}$$
$$\times \mathsf{Q}_{\gamma-k}^{(\alpha+2k-l,\beta+l)}(2x_<^2-1)\mathsf{P}_{\gamma-k}^{(\alpha+2k-l,\beta+l)}(2x_>^2-1)$$
$$\times w^l P_{k-l}^{(\alpha-\beta-1,\beta+l)}(2w^2-1)\frac{\beta+l}{\beta}C_l^\beta(x).$$

PROOF. Making the replacement $l \mapsto k-l$ in Theorem 4.11 completes the proof. □

REMARK 4.12. It should be noted that removing the factors $\Gamma(\alpha+\gamma+1)/\Gamma(\gamma+1)$ from the P series and $\Gamma(\alpha+\gamma+1)\Gamma(\beta+\gamma+1)$ from the Q series would convert the P and Q on the left-hand sides to $\boldsymbol{P}$ and $\boldsymbol{Q}$, so that all expansions would be in terms of Olver-normalized functions.

By examining the expansion of the Jacobi function of the first kind, one can see that the $(-\gamma)_k$ shifted factorial in the above alternative expansion is moved from the denominator to the numerator, which is more natural for Jacobi polynomials where the sum is terminating. One can see the benefit: all the functions involved in the expansions are well-defined for all values of the parameters, including integer values. These expansions are extremely useful for expansions of fundamental solutions on rank-one symmetric spaces, where the degrees and parameters are given by integers. One no longer has any difficulties with various functions not being defined for certain parameter values. This is completely resolved. One example is that in the integer context of the Jacobi function of the second kind, the functions appear with degree equal to $\gamma - k$ for all $k \in \mathbb{N}_0$. They quickly become undefined for negative values of the degree. However, since the Olver-normalized Jacobi functions are entire functions, there is no longer any problem here. These alternative expansions are highly desirable!

5. Eigenfunction expansions of a fundamental solution of the Laplace–Beltrami operator on non-compact and compact symmetric spaces of rank one

As an application of the addition theorem for the Jacobi function of the second kind, we now give an introduction to the motivation for the material which has been presented in the previous sections. It is a study of the global analysis of the Laplace–Beltrami operator and the solutions of inhomogeneous elliptic equations: Poisson's equation on the Riemannian symmetric spaces of rank one.

Let $d = \dim_{\mathbb{R}} \mathbb{K}$, where $\mathbb{K}$ is equal to the real numbers $\mathbb{R}$, the complex numbers $\mathbb{C}$, the quaternions $\mathbb{H}$, or the octonions $\mathbb{O}$. For $d = 1$, namely the real case, the corresponding spaces are Riemannian manifolds of constant curvature, including Euclidean space $\mathbb{R}^n$, real hyperbolic geometry $\mathbb{R}\mathbf{H}^n_R$ (noncompact), and real hyperspherical geometry $\mathbb{R}\mathbf{S}^n_R$ (compact), in various models. For $d = 2, 4, 8$, it is well known that there are corresponding isotropic Riemannian manifolds of both noncompact and compact type which are referred to as rank-one symmetric spaces (see for instance [**27**]). They include the complex hyperbolic space $\mathbb{C}\mathbf{H}^n_R$, the quaternionic hyperbolic space $\mathbb{H}\mathbf{H}^n_H$, the octonionic hyperbolic plane $\mathbb{O}\mathbf{H}^2_R$, the complex projective space $\mathbb{C}\mathbf{P}^n_R$, the quaternionic projective space $\mathbb{H}\mathbf{P}^n_R$, and the octonionic projective (Cayley) plane $\mathbb{O}\mathbf{P}^2_R$. In each of the preceding, $R > 0$ is the radius of curvature. The complex, quaternionic, and octonionic rank-one symmetric spaces have respective real dimensions $2n, 4n, 16$. For a description of the Riemannian manifolds given by the rank-one symmetric spaces, see for instance [**25–27**] and the references therein.

Riemannian symmetric spaces, compact and non-compact, come in infinite series (four corresponding to complex simple groups and seven to real simple groups), together with a finite class of exceptional spaces (see [**25**, p. 516, 518]). Each symmetric space comes with a commutative algebra of invariant differential operators and correspondingly a class of eigenfunctions, its spherical functions. In the case of the rank-one symmetric spaces, these are just hypergeometric (Jacobi) functions with specified parameters. This has been used as motivation for several generalizations of the classical Gauss hypergeometric functions that are beyond the scope of this paper, for example the Heckman–Opdam functions on root systems [**22, 23, 47, 48**] and the work of Macdonald [**42, 43**].

Due to the isotropy of the symmetric spaces of rank one, a fundamental solution of the Laplace–Beltrami operator on any of these manifolds can be obtained by solving an inhomogeneous *ordinary* differential equation, given in terms of the geodesic distance. Laplace's equation is satisfied on the manifold when the Laplace–Beltrami operator acts on an unknown function and the result is zero. In geodesic polar coordinates the Laplace–Beltrami operator is given on the rank-one noncompact (hyperbolic) symmetric spaces by

$$\Delta = \frac{1}{R^2}\left\{\frac{\partial^2}{\partial r^2} + [d(n-1)\coth r + 2(d-1)\coth(2r)]\frac{\partial}{\partial r} + \frac{1}{\sinh^2 r}\Delta_{K/M}\right\} \tag{5.1}$$

$$= \frac{1}{R^2}\left\{\frac{\partial^2}{\partial r^2} + [(dn-1)\coth r + (d-1)\tanh r]\frac{\partial}{\partial r} + \frac{1}{\sinh^2 r}\Delta_{K/M}\right\} \tag{5.2}$$

$$=: \frac{1}{R^2}\left(\Delta_r + \frac{1}{\sinh^2 r}\Delta_{K/M}\right), \tag{5.3}$$

and on the rank-one compact (projective) spaces by

$$\Delta = \frac{1}{R^2}\left\{\frac{\partial^2}{\partial \theta^2} + [d(n-1)\cot\theta + 2(d-1)\cot(2\theta)]\frac{\partial}{\partial \theta} + \frac{1}{\sin^2\theta}\Delta_{K/M}\right\} \tag{5.4}$$

$$= \frac{1}{R^2}\left\{\frac{\partial^2}{\partial \theta^2} + [(dn-1)\cot\theta + (d-1)\tan\theta]\frac{\partial}{\partial \theta} + \frac{1}{\sin^2\theta}\Delta_{K/M}\right\} \tag{5.5}$$

$$=: \frac{1}{R^2}\left(\Delta_\theta + \frac{1}{\sin^2\theta}\Delta_{K/M}\right), \tag{5.6}$$

where r and θ are the geodesic distance coordinates on the noncompact and compact spaces respectively (see [**25**, Lemma 21]). For a spherically symmetric solution such as a fundamental solution, the contribution from $\Delta_{K/M}$ vanishes and the corresponding homogeneous equation becomes relatively simple, namely the radial form of Laplace's equation, which is

$$\Delta_r\, u(r) = 0 \quad \text{or} \quad \Delta_\theta\, v(\theta) = 0. \tag{5.7}$$

The solutions of this second-order ordinary differential equation are given by Jacobi/hypergeometric functions (see [**27**, p. 484]). It can be easily verified that 'radial' solutions, homogeneous or fundamental, are of the form

$$u(r) = a\,P_0^{(\alpha,\beta)}(\cosh(2r)) + b\,Q_0^{(\alpha,\beta)}(\cosh(2r)), \tag{5.8}$$

$$v(\theta) = c\,\mathsf{P}_0^{(\alpha,\beta)}(\cos(2\theta)) + d\,\mathsf{Q}_0^{(\alpha,\beta)}(\cos(2\theta)), \tag{5.9}$$

where for the complex, quaternionic, and octonionic rank-one symmetric spaces one has $\alpha = n-1, 2n-1, 7$ and $\beta = 0, 1, 3$ respectively [**18**, Table 1, p. 265]. Furthermore, a fundamental solution, which by definition is not a homogeneous solution, must be singular at the origin (i.e., at $r = 0$ or $\theta = 0$) and locally match to a Euclidean fundamental solution. This implies that any fundamental solution must be irregular at the origin. Therefore fundamental solutions must correspond to Jacobi functions of the second kind. For a fundamental solution, it is the case that $a = c = 0$; and we must determine b and d, which will depend on n and R.

REMARK 5.1. Note that the general solution as a function of the geodesic distance includes contributions from both the function of the first kind and the function of the second kind. However, the function of the first kind with $\gamma = 0$ simply contributes a constant a or c, since $P_0^{(\alpha,\beta)}(z) = 1$ as stated in (2.50) (and the same for the function on-the-cut). On the other hand, in the case of non-spherically symmetric solutions there will be a contribution from the function of the first kind because then the contribution of the $\Delta_{K/M}$ term in (5.3) or (5.6) will be non-zero.

Let $\mathbf{x}, \mathbf{x}' \in \mathbb{R}^s$. Then a Euclidean fundamental solution of Laplace's equation is (see for instance [**21**, p. 202])

$$\mathcal{G}^s(\mathbf{x}, \mathbf{x}') = \begin{cases} \dfrac{\Gamma(s/2)}{2\pi^{s/2}(s-2)} \|\mathbf{x} - \mathbf{x}'\|^{2-s} & \text{if } s = 1 \text{ or } s \geqslant 3, \\ \dfrac{1}{2\pi} \log \|\mathbf{x} - \mathbf{x}'\|^{-1} & \text{if } s = 2. \end{cases} \tag{5.10}$$

For a description of opposite antipodal fundamental solutions on the real hypersphere, see [**7**]. The above analysis leads to the following.

THEOREM 5.2. *A fundamental solution and an opposite antipodal fundamental solution, for the Laplace–Beltrami operators on the rank-one $n \geqslant 1$ noncompact and compact symmetric spaces respectively, given in terms of the respective geodesic distance coordinates $r \in [0, \infty)$ and $\theta \in [0, \pi/2]$, are*

$$\mathcal{G}^{\mathbf{CH}^n_R}(r) = \frac{(n-1)!}{2\pi^n R^{2n-2}} \mathcal{Q}_0^{(n-1,0)}(\cosh(2r)), \tag{5.11}$$

$$\mathcal{G}^{\mathbb{H}\mathbf{H}^n_R}(r) = \frac{(2n)!}{2\pi^{2n} R^{4n-2}} \mathcal{Q}_0^{(2n-1,1)}(\cosh(2r)), \tag{5.12}$$

$$\mathcal{G}^{\mathbb{O}\mathbf{H}^2_R}(r) = \frac{302\,400}{\pi^8 R^{14}} \mathcal{Q}_0^{(7,3)}(\cosh(2r)), \tag{5.13}$$

and

$$\mathcal{G}^{\mathbf{CP}^n_R}(\theta) = \frac{(n-1)!}{2\pi^n R^{2n-2}} \mathsf{Q}_0^{(n-1,0)}(\cos(2\theta)), \tag{5.14}$$

$$\mathcal{G}^{\mathbb{H}\mathbf{P}^n_R}(\theta) = \frac{(2n)!}{2\pi^{2n} R^{4n-2}} \mathsf{Q}_0^{(2n-1,1)}(\cos(2\theta)), \tag{5.15}$$

$$\mathcal{G}^{\mathbb{O}\mathbf{P}^2_R}(\theta) = \frac{302\,400}{\pi^8 R^{14}} \mathsf{Q}_0^{(7,3)}(\cos(2\theta)). \tag{5.16}$$

PROOF. The complex, quaternionic, and octonionic rank-one symmetric spaces all have even dimension: $s = 2n, 4n, 16$, respectively. It is easy to verify that the spherically symmetric solutions of Laplace's equation on these spaces are Jacobi functions of the first and second kind in the noncompact case, and Jacobi functions of the first and second kind on-the-cut in the compact case, having parameter $\gamma = 0$ and parameters $\alpha = n-1,\ 2n-1, 7$ and $\beta = 0, 1, 3$, respectively. Furthermore, each of the corresponding fundamental solutions must match to a Euclidean fundamental solution. Using (2.63) and (2.70), assuming that $\gamma = 0$, $\alpha = a$, $\beta = b$, $a \in \mathbb{N}$, $b \in \mathbb{N}_0$, one has that near the singularity at unity, the Jacobi function of the second kind and the Jacobi function of the second kind on-the-cut have the behaviors

$$\mathcal{Q}_0^{(a,b)}(1+\epsilon) \sim \mathsf{Q}_0^{(a,b)}(1-\epsilon) \sim \frac{2^{a-1}(a-1)!b!}{(a+b)!\,\epsilon^a} \qquad \text{as } \epsilon \to 0^+. \tag{5.17}$$

Referring to the geodesic distance coordinate of the hyperbolic manifolds as $r \in [0, \infty)$ and that of the compact manifolds as $\theta \in [0, \pi/2]$, one has

$$\cosh(2r) \sim \cosh(2\tfrac{\rho}{R}) \sim 1 + \tfrac{2\rho^2}{R^2} \qquad \text{as } r \to 0, \tag{5.18}$$

$$\cos(2\theta) \sim \cos(\tfrac{2\rho}{R}) \sim 1 - \tfrac{2\rho^2}{R^2} \qquad \text{as } \theta \to 0, \tag{5.19}$$

where ρ is the Euclidean geodesic distance. Matching locally to the Euclidean fundamental solution (5.10) in the $R \to \infty$ flat-space limit (see for instance [**9**, §2.4]), one can determine the constant of proportionality which multiplies the Jacobi functions of the second kind. This completes the proof. $\square$

Since fundamental solutions on the rank-one symmetric spaces all have $\gamma = 0$, we first present the expansions in this case. The $\gamma = 0$ case of any Jacobi function of the first kind equals unity. However, for the Jacobi functions of the second kind, the $\gamma = 0$ functions are quite rich, and expansions involving them are quite useful in that they allow one to produce separated eigenfunction expansions of a fundamental solution of Laplace's equation.

REMARK 5.3. The reader should be aware that the addition theorems presented below for the Jacobi functions of the second kind with $\gamma = 0$ are well-defined except in the case where the α and β parameters on the left-hand sides are non-negative integers. Then, special care must be taken (refer to Theorems 2.10, 2.11), though the functions at these parameter values may be obtained by taking an appropriate limit.

COROLLARY 5.3.1. *Let $\alpha, \beta \in \mathbb{C}$, $\beta \notin \mathbb{Z}$, $z_1, z_2 \in \mathbb{C} \setminus (-\infty, 1]$, $x_1, x_2 \in \mathbb{C} \setminus ((-\infty, -1] \cup [1, \infty))$, $x, w \in \mathbb{C}$, with $Z^\pm$, $\mathsf{X}^\pm$ defined by (3.2), (3.3), respectively, such that the complex variables $\alpha, \beta, z_1, z_2, x_1, x_2, x, w$ are in some yet to be determined neighborhood of the real line. Then*

$$
\begin{aligned}
Q_0^{(\alpha,\beta)}(Z^\pm) &= \Gamma(\alpha+1)\Gamma(\alpha+1)\Gamma(\beta+1)\sum_{k=0}^{\infty}\frac{(\alpha+1)_k(\alpha+\beta+1)_k}{(\alpha+k)(\beta+1)_k} \\
&\quad \times \sum_{l=0}^{k}(\mp 1)^{k-l}(\alpha+k+l)(\alpha+1)_l(z_1z_2)^{k-l}((z_1^2-1)(z_2^2-1))^{\frac{k+l}{2}} \\
&\quad \times \boldsymbol{Q}_{-k}^{(\alpha+k+l,\beta+k-l)}(2z_>^2-1)\boldsymbol{P}_{-k}^{(\alpha+k+l,\beta+k-l)}(2z_<^2-1) \\
&\quad \times w^{k-l}P_l^{(\alpha-\beta-1,\beta+k-l)}(2w^2-1)\frac{\beta+k-l}{\beta}C_{k-l}^{\beta}(x),
\end{aligned}
\tag{5.20}
$$

$$
\begin{aligned}
\mathsf{Q}_0^{(\alpha,\beta)}(\mathsf{X}^\pm) &= \Gamma(\alpha+1)\sum_{k=0}^{\infty}(-1)^k\frac{(\alpha+1)_k(\alpha+\beta+1)_k}{(\alpha+k)(\beta+1)_k} \\
&\quad \times \sum_{l=0}^{k}(\mp 1)^{k-l}(\alpha+k+l)(-\beta)_l(x_1x_2)^{k-l}((1-x_1^2)(1-x_2^2))^{\frac{k+l}{2}} \\
&\quad \times \mathsf{Q}_{-k}^{(\alpha+k+l,\beta+k-l)}(2x_<^2-1)\mathsf{P}_{-k}^{(\alpha+k+l,\beta+k-l)}(2x_>^2-1) \\
&\quad \times w^{k-l}P_l^{(\alpha-\beta-1,\beta+k-l)}(2w^2-1)\frac{\beta+k-l}{\beta}C_{k-l}^{\beta}(x).
\end{aligned}
\tag{5.21}
$$

PROOF. Substituting $\gamma = 0$ in Theorem 4.11 for the Jacobi functions of the second kind completes the proof. $\square$

Next we give examples of the expansions for complex and quaternionic hyperbolic spaces, where $\beta = 0, 1$, respectively. First we treat the complex case, which corresponds to complex hyperbolic and projective space. In order to do this we start with Theorem 4.11 and take the limit as $\beta \to 0$ using [**2**, (6.4.13)]

$$
\lim_{\mu\to 0}\frac{n+\mu}{\mu}C_n^{\mu}(x) = \epsilon_n T_n(x),
\tag{5.22}
$$

where $\epsilon_n := 2 - \delta_{n,0}$ is the Neumann factor commonly appearing in Fourier series.

COROLLARY 5.3.2. *Let $\alpha \in \mathbb{C}$, $z_1, z_2 \in \mathbb{C} \setminus (-\infty, 1]$, $x_1, x_2 \in \mathbb{C} \setminus ((-\infty, -1] \cup [1, \infty))$, $x, w \in \mathbb{C}$, with $Z^\pm$, $\mathsf{X}^\pm$ defined by (3.2), (3.3), respectively, such that the complex variables $\alpha, z_1, z_2, x_1, x_2, x, w$ are in some yet to be determined neighborhood of the real line. Then*

$$\begin{aligned} Q_0^{(\alpha,0)}(Z^\pm) = \Gamma(\alpha+1)\Gamma(\alpha+1) \sum_{k=0}^{\infty} \frac{(\alpha+1)_k(\alpha+1)_k}{(\alpha+k)k!} \\ \times \sum_{l=0}^{k} (\mp 1)^{k-l}(\alpha+k+l)(\alpha+1)_l(z_1z_2)^{k-l}((z_1^2-1)(z_2^2-1))^{\frac{k+l}{2}} \\ \times \boldsymbol{Q}_{-k}^{(\alpha+k+l,k-l)}(2z_>^2-1)\boldsymbol{P}_{-k}^{(\alpha+k+l,k-l)}(2z_<^2-1) \\ \times w^{k-l}P_l^{(\alpha-1,k-l)}(2w^2-1)\epsilon_{k-l}T_{k-l}(x), \end{aligned} \tag{5.23}$$

$$\begin{aligned} \mathsf{Q}_0^{(\alpha,0)}(\mathsf{X}^\pm) = \Gamma(\alpha+1) \sum_{k=0}^{\infty} (-1)^k \frac{(\alpha+1)_k(\alpha+1)_k}{(\alpha+k)k!} \\ \times \sum_{l=0}^{k} (\mp 1)^{k-l}(\alpha+k+l)(x_1x_2)^{k-l}((1-x_1^2)(1-x_2^2))^{\frac{k+l}{2}} \\ \times \mathcal{Q}_{-k}^{(\alpha+k+l,k-l)}(2x_<^2-1)\mathsf{P}_{-k}^{(\alpha+k+l,k-l)}(2x_>^2-1) \\ \times w^{k-l}P_l^{(\alpha-1,k-l)}(2w^2-1)\epsilon_{k-l}T_{k-l}(x). \end{aligned} \tag{5.24}$$

PROOF. Taking the limit as $\beta \to 0$ in Theorem 4.11 using (5.22) completes the proof. □

Now we treat the case of the quaternionic hyperbolic and projective spaces, which correspond to $\beta = 1$.

COROLLARY 5.3.3. *Let $\alpha \in \mathbb{C}$, $z_1, z_2 \in \mathbb{C} \setminus (-\infty, 1]$, $x_1, x_2 \in \mathbb{C} \setminus ((-\infty, -1] \cup [1, \infty))$, $x, w \in \mathbb{C}$, with $Z^\pm$, $\mathsf{X}^\pm$ defined by (3.2), (3.3), respectively, such that the complex variables $\alpha, z_1, z_2, x_1, x_2, x, w$ are in some yet to be determined neighborhood of the real line. Then*

$$\begin{aligned} Q_0^{(\alpha,1)}(Z^\pm) = \Gamma(\alpha+1)\Gamma(\alpha+1) \sum_{k=0}^{\infty} \frac{(\alpha+1)_k(\alpha+2)_k}{(\alpha+k)(2)_k} \\ \times \sum_{l=0}^{k} (\mp 1)^{k-l}(1+k-l)(\alpha+k+l)(\alpha+1)_l(z_1z_2)^{k-l}((z_1^2-1)(z_2^2-1))^{\frac{k+l}{2}} \\ \times \boldsymbol{Q}_{-k}^{(\alpha+k+l,1+k-l)}(2z_>^2-1)\boldsymbol{P}_{-k}^{(\alpha+k+l,1+k-l)}(2z_<^2-1) \\ \times w^{k-l}P_l^{(\alpha-2,1+k-l)}(2w^2-1)U_{k-l}(x), \end{aligned} \tag{5.25}$$

$$\begin{aligned} \mathsf{Q}_0^{(\alpha,1)}(\mathsf{X}^\pm) = \Gamma(\alpha+1) \sum_{k=0}^{\infty} (-1)^k \frac{(\alpha+1)_k(\alpha+2)_k}{(\alpha+k)(2)_k} \\ \times \sum_{l=0}^{k} (\mp 1)^{k-l}(1+k-l)(\alpha+k+l)(x_1x_2)^{k-l}((1-x_1^2)(1-x_2^2))^{\frac{k+l}{2}} \\ \times \mathcal{Q}_{-k}^{(\alpha+k+l,1+k-l)}(2x_<^2-1)\mathsf{P}_{-k}^{(\alpha+k+l,1+k-l)}(2x_>^2-1) \\ \times w^{k-l}P_l^{(\alpha-2,1+k-l)}(2w^2-1)U_{k-l}(x). \end{aligned} \tag{5.26}$$

PROOF. Take the limit as $\beta \to 1$ in Theorem 4.11 using [**11**, (18.7.4)], which connects the Chebyshev polynomial of the second kind to the Gegenbauer polynomial with parameter equal to unity, namely $C_n^1(x) = U_n(x)$. This completes the proof. $\square$

Now we treat the case of the octonionic hyperbolic and projective spaces, which correspond to $\beta = 3$.

COROLLARY 5.3.4. *Let $\alpha \in \mathbb{C}$, $z_1, z_2 \in \mathbb{C} \setminus (-\infty, 1]$, $x_1, x_2 \in \mathbb{C} \setminus ((-\infty, -1] \cup [1, \infty))$, $x, w \in \mathbb{C}$, with $Z^\pm$, $\mathsf{X}^\pm$ defined by (3.2), (3.3), respectively, such that the complex variables $\alpha, z_1, z_2, x_1, x_2, x, w$ are in some yet to be determined neighborhood of the real line. Then*

(5.27)

$$\begin{aligned}Q_0^{(\alpha,3)}(Z^\pm) &= 2\Gamma(\alpha+1)\Gamma(\alpha+1)\sum_{k=0}^{\infty}\frac{(\alpha+1)_k(\alpha+4)_k}{(\alpha+k)(4)_k}\\ &\quad\times\sum_{l=0}^{k}(\mp1)^{k-l}(3+k-l)(\alpha+k+l)(\alpha+1)_l(z_1z_2)^{k-l}((z_1^2-1)(z_2^2-1))^{\frac{k+l}{2}}\\ &\quad\times \boldsymbol{Q}_{-k}^{(\alpha+k+l,3+k-l)}(2z_>^2-1)\boldsymbol{P}_{-k}^{(\alpha+k+l,3+k-l)}(2z_<^2-1)\\ &\quad\times w^{k-l}P_l^{(\alpha-4,3+k-l)}(2w^2-1)C_{k-l}^3(x),\end{aligned}$$

(5.28)

$$\begin{aligned}\mathsf{Q}_0^{(\alpha,3)}(\mathsf{X}^\pm) &= \frac{1}{3}\Gamma(\alpha+1)\sum_{k=0}^{\infty}(-1)^k\frac{(\alpha+1)_k(\alpha+4)_k}{(\alpha+k)(4)_k}\\ &\quad\times\sum_{l=0}^{k}(\mp1)^{k-l}(3+k-l)(\alpha+k+l)(x_1x_2)^{k-l}((1-x_1^2)(1-x_2^2))^{\frac{k+l}{2}}\\ &\quad\times \mathcal{Q}_{-k}^{(\alpha+k+l,3+k-l)}(2x_<^2-1)\mathsf{P}_{-k}^{(\alpha+k+l,3+k-l)}(2x_>^2-1)\\ &\quad\times w^{k-l}P_l^{(\alpha-4,3+k-l)}(2w^2-1)C_{k-l}^3(x).\end{aligned}$$

PROOF. Setting $\beta = 3$ in Theorem 4.11 completes the proof. $\square$

The above calculations look almost trivial in that they are simply substitutions of the values $\beta = 0, 1, 3$ and $\gamma = 0$ into the addition theorems of Theorem 4.11. However, it should be understood that ordinarily these computations would be extremely difficult, particularly if one were to use the standard normalizations of the Jacobi functions. With standard normalizations, the Jacobi functions of the second kind at these parameter values, and in fact at arbitrary integer values of α, β and the degree γ, are not even defined. It is only because of the particular normalization that we have chosen that evaluations at these parameter values become quite easy. We will further take advantage of these expansions in later publications.

References

[1] M. Allé. Über die Eigenschaften derjenigen Gattung von Functionen, welche in der Entwicklung von $(1-2qx+q^2)^{-\frac{m}{2}}$ nach aufsteigenden Potenzen von q auftreten, und über die Entwicklung des Ausdruckes $\{1-2q[\cos\theta\cos\theta'+\sin\theta\sin\theta'\cos(\psi-\psi')]+q^2\}^{-\frac{m}{2}}$. *Sitzungsberichte der mathematisch-naturwissenschaftlichen Classe der kaiserlichen Akademie der Wissenschaften Wien*, 51:429–458, 1865.

[2] G. E. Andrews, R. Askey, and R. Roy, *Special functions*, Encyclopedia of Mathematics and its Applications, vol. 71, Cambridge University Press, Cambridge, 1999, DOI 10.1017/CBO9781107325937. MR1688958

[3] R. Askey, *Jacobi polynomials. I. New proofs of Koornwinder's Laplace type integral representation and Bateman's bilinear sum*, SIAM J. Math. Anal. **5** (1974), 119–124, DOI 10.1137/0505013. MR385197

[4] R. Askey, T. H. Koornwinder, and M. Rahman, *An integral of products of ultraspherical functions and a q-extension*, J. London Math. Soc. (2) **33** (1986), no. 1, 133–148, DOI 10.1112/jlms/s2-33.1.133. MR829394

[5] E. Cartan. Sur la détermination d'un système orthogonal complet dans un espace de Riemann symétrique clos. *Rendiconti del Circolo Matematico di Palermo*, 53:217–252, 1929.

[6] É. Cartan, *Leçons sur la géométrie projective complexe* (French), Gauthier-Villars, Paris, 1950. 2d ed. MR41456

[7] H. S. Cohl, *Fundamental solution of Laplace's equation in hyperspherical geometry*, SIGMA Symmetry Integrability Geom. Methods Appl. **7** (2011), Paper 108, 14, DOI 10.3842/SIGMA.2011.108. MR2861168

[8] H. S. Cohl, *Fourier, Gegenbauer and Jacobi expansions for a power-law fundamental solution of the polyharmonic equation and polyspherical addition theorems*, SIGMA Symmetry Integrability Geom. Methods Appl. **9** (2013), Paper 042, 26, DOI 10.3842/SIGMA.2013.042. MR3116178

[9] H. S. Cohl and R. M. Palmer, *Fourier and Gegenbauer expansions for a fundamental solution of Laplace's equation in hyperspherical geometry*, SIGMA Symmetry Integrability Geom. Methods Appl. **11** (2015), Paper 015, 23, DOI 10.3842/SIGMA.2015.015. MR3313691

[10] H. S. Cohl, J. Park, and H. Volkmer, *Gauss hypergeometric representations of the Ferrers function of the second kind*, SIGMA Symmetry Integrability Geom. Methods Appl. **17** (2021), Paper No. 053, 33, DOI 10.3842/SIGMA.2021.053. MR4261851

[11] *NIST Digital Library of Mathematical Functions.* `https://dlmf.nist.gov/`, Release 1.2.3 of 2024-12-15. F. W. J. Olver, A. B. Olde Daalhuis, D. W. Lozier, B. I. Schneider, R. F. Boisvert, C. W. Clark, B. R. Miller, B. V. Saunders, H. S. Cohl, and M. A. McClain, eds.

[12] L. Durand, *Product formulas and Nicholson-type integrals for Jacobi functions. I. Summary of results*, SIAM J. Math. Anal. **9** (1978), no. 1, 76–86, DOI 10.1137/0509007. MR466666

[13] L. Durand, *Addition formulas for Jacobi, Gegenbauer, Laguerre, and hyperbolic Bessel functions of the second kind*, SIAM J. Math. Anal. **10** (1979), no. 2, 425–437, DOI 10.1137/0510039. MR523856

[14] L. Durand, P. M. Fishbane, and L. M. Simmons Jr., *Expansion formulas and addition theorems for Gegenbauer functions*, J. Mathematical Phys. **17** (1976), no. 11, 1933–1948, DOI 10.1063/1.522831. MR422726

[15] A. Erdélyi, W. Magnus, F. Oberhettinger, and F. G. Tricomi. *Higher Transcendental Functions. Vol. II.* Robert E. Krieger Publishing Co. Inc., Melbourne, Fla., 1981.

[16] M. Flensted-Jensen and T. Koornwinder, *The convolution structure for Jacobi function expansions*, Ark. Mat. **11** (1973), 245–262, DOI 10.1007/BF02388521. MR340938

[17] M. Flensted-Jensen and T. H. Koornwinder, *Jacobi functions: the addition formula and the positivity of the dual convolution structure*, Ark. Mat. **17** (1979), no. 1, 139–151, DOI 10.1007/BF02385463. MR543509

[18] M. Flensted-Jensen and T. H. Koornwinder, *Positive definite spherical functions on a non-compact, rank one symmetric space*, Analyse harmonique sur les groupes de Lie (Sém., Nancy-Strasbourg 1976–1978), II, Lecture Notes in Math., vol. 739, Springer, Berlin, 1979, pp. 249–282. MR560841

[19] L. Gegenbauer. Über einige bestimmte Integrale. *Sitzungsberichte der Kaiserlichen Akademie der Wissenschaften. Mathematische-Naturwissenschaftliche Classe.*, 70:433–443, 1874.

[20] L. Gegenbauer. Das Additionstheorem der Functionen $C_n^\nu(x)$. *Sitzungsberichte der Kaiserlichen Akademie der Wissenschaften. Mathematische-Naturwissenschaftliche Classe.*, 102:942–950, 1893.

[21] I. M. Gel'fand and G. E. Shilov, *Generalized functions. Vol. I: Properties and operations*, Academic Press, New York-London, 1964. Translated by Eugene Saletan. MR166596

[22] G. J. Heckman, *Root systems and hypergeometric functions. II*, Compositio Math. **64** (1987), no. 3, 353–373. MR918417

[23] G. J. Heckman and E. M. Opdam, *Root systems and hypergeometric functions. I*, Compositio Math. **64** (1987), no. 3, 329–352. MR918416

[24] E. Heine. *Handbuch der Kugelfunctionen, Theorie und Anwendungen (volume 1)*. Druck und Verlag von G. Reimer, Berlin, 1878.

[25] S. Helgason, *Differential operators on homogeneous spaces*, Acta Math. **102** (1959), 239–299, DOI 10.1007/BF02564248. MR117681

[26] S. Helgason, *Differential geometry, Lie groups, and symmetric spaces*, Pure and Applied Mathematics, vol. 80, Academic Press, Inc. [Harcourt Brace Jovanovich, Publishers], New York-London, 1978. MR514561

[27] S. Helgason, *Groups and geometric analysis: Integral geometry, invariant differential operators, and spherical functions*, Pure and Applied Mathematics, vol. 113, Academic Press, Inc., Orlando, FL, 1984. MR754767

[28] M. E. H. Ismail (ed.), *Encyclopedia of special functions: the Askey-Bateman project. Vol. 1. Univariate orthogonal polynomials*, Cambridge University Press, Cambridge, 2020. with assistance by Walter Van Assche, DOI 10.1017/9780511979156. MR4421370

[29] C. G. J. Jacobi, *Untersuchungen über die Differentialgleichung der hypergeometrischen Reihe* (German), J. Reine Angew. Math. **56** (1859), 149–165, DOI 10.1515/crll.1859.56.149. MR1579090

[30] R. Koekoek, P. A. Lesky, and R. F. Swarttouw, *Hypergeometric orthogonal polynomials and their q-analogues*, Springer Monographs in Mathematics, Springer-Verlag, Berlin, 2010. With a foreword by Tom H. Koornwinder, DOI 10.1007/978-3-642-05014-5. MR2656096

[31] T. Koornwinder, *The addition formula for Jacobi polynomials and spherical harmonics*, SIAM J. Appl. Math. **25** (1973), 236–246, DOI 10.1137/0125027. MR346212

[32] T. Koornwinder, *Jacobi polynomials. II. An analytic proof of the product formula*, SIAM J. Math. Anal. **5** (1974), 125–137, DOI 10.1137/0505014. MR385198

[33] T. Koornwinder, *Jacobi polynomials. III. An analytic proof of the addition formula*, SIAM J. Math. Anal. **6** (1975), 533–543, DOI 10.1137/0506048. MR447659

[34] T. Koornwinder, *Yet another proof of the addition formula for Jacobi polynomials*, J. Math. Anal. Appl. **61** (1977), no. 1, 136–141, DOI 10.1016/0022-247X(77)90149-4. MR457816

[35] T. H. Koornwinder, *The addition formula for Jacobi polynomials. I. Summary of results*, Indag. Math. **34** (1972), 188–191. Nederl. Akad. Wetensch. Proc. Ser. A **75**. MR308476

[36] T. H. Koornwinder, *The addition formula for Jacobi polynomials. I. Summary of results*, Indag. Math. **34** (1972), 188–191. Nederl. Akad. Wetensch. Proc. Ser. A **75**. MR308476

[37] T. H. Koornwinder. The addition formula for Jacobi polynomials, II : the Laplace type integral representation and the product formula. *Stichting Mathematisch Centrum, Afdeling Toegepaste Wiskunde. TW: 133/72* `http://persistent-identifier.org/?identifier=urn:nbn:nl:ui:18-7722`, page 34, April 1972.

[38] T. H. Koornwinder. The addition formula for Jacobi polynomials, III : Completion of the proof. *Stichting Mathematisch Centrum, Afdeling Toegepaste Wiskunde. TW: 135/72*, `http://persistent-identifier.org/?identifier=urn:nbn:nl:ui:18-12598`, page 14, December 1972.

[39] T. H. Koornwinder, *Dual addition formulas associated with dual product formulas*, Frontiers in orthogonal polynomials and q-series, Contemp. Math. Appl. Monogr. Expo. Lect. Notes, vol. 1, World Sci. Publ., Hackensack, NJ, 2018, pp. 373–392. MR3791629

[40] T. H. Koornwinder and A. L. Schwartz, *Product formulas and associated hypergroups for orthogonal polynomials on the simplex and on a parabolic biangle*, Constr. Approx. **13** (1997), no. 4, 537–567, DOI 10.1007/s003659900058. MR1466065

[41] Z. Li and L. Peng, *Some representations of translations of the product of two functions for Hankel transforms and Jacobi transforms*, Constr. Approx. **26** (2007), no. 1, 115–125, DOI 10.1007/s00365-006-0668-x. MR2310691

[42] I. G. Macdonald, *Affine Hecke algebras and orthogonal polynomials*, Cambridge Tracts in Mathematics, vol. 157, Cambridge University Press, Cambridge, 2003, DOI 10.1017/CBO9780511542824. MR1976581

[43] I. G. Macdonald. Hypergeometric Functions, I., arXiv:1309.4568, page 63, 2013.

[44] W. Magnus, F. Oberhettinger, and R. P. Soni, *Formulas and theorems for the special functions of mathematical physics*, Third enlarged edition, Die Grundlehren der mathematischen Wissenschaften, Band 52, Springer-Verlag New York, Inc., New York, 1966. MR232968

[45] W. Miller Jr., *Symmetry and separation of variables*, Encyclopedia of Mathematics and its Applications, Vol. 4, Addison-Wesley Publishing Co., Reading, Mass.-London-Amsterdam, 1977. With a foreword by Richard Askey. MR460751

[46] F. W. J. Olver, *Asymptotics and special functions*, AKP Classics, A K Peters, Ltd., Wellesley, MA, 1997. Reprint of the 1974 original [Academic Press, New York; MR0435697 (55 #8655)]. MR1429619

[47] E. M. Opdam, *Root systems and hypergeometric functions. III*, Compositio Math. **67** (1988), no. 1, 21–49. MR949270

[48] E. M. Opdam, *Root systems and hypergeometric functions. IV*, Compositio Math. **67** (1988), no. 2, 191–209. MR951750

[49] G. Szegő, *Orthogonal polynomials*, 4th ed., American Mathematical Society Colloquium Publications, Vol. XXIII, American Mathematical Society, Providence, RI, 1975. MR372517

[50] N. Ja. Vilenkin, *Special functions and the theory of group representations*, Translations of Mathematical Monographs, Vol. 22, American Mathematical Society, Providence, RI, 1968. Translated from the Russian by V. N. Singh. MR229863

[51] N. Ja. Vilenkin and R. L. Šapiro, *Irreducible representations of the group* $\mathrm{SU}(n)$ *of class I relative to* $\mathrm{SU}(n-1)$ (Russian), Izv. Vysš. Učebn. Zaved. Matematika **1967** (1967), no. 7(62), 9–20. MR219662

[52] R. L. Šapiro, *Special functions related to representations of the group* $\mathrm{SU}(n)$, *of class I with respect to* $\mathrm{SU}(n-1)$ $(n \geqq 3)$ (Russian), Izv. Vysš. Učebn. Zaved. Matematika **1968** (1968), no. 4(71), 97–107. MR230955

[53] J. Wimp, P. McCabe, and J. N. L. Connor, *Computation of Jacobi functions of the second kind for use in nearside-farside scattering theory*, J. Comput. Appl. Math. **82** (1997), no. 1-2, 447–464, DOI 10.1016/S0377-0427(97)00083-6. 7th ICCAM 96 Congress (Leuven). MR1473558

Applied and Computational Mathematics Division, National Institute of Standards and Technology, Gaithersburg, Maryland 20899-8910
Email address: howard.cohl@nist.gov

Dept. of Quantitative Methods, Universidad Loyola Andalucía, E-41704 Seville, Spain
Email address: rscosa@gmail.com

Dept. of Physics, University of Wisconsin-Madison, Madison, WI 53706, USA, 415 Pearl Court, Aspen, CO 81611, USA
Email address: ldurandiii@comcast.net

Applied and Computational Mathematics Division, National Institute of Standards and Technology, Gaithersburg, MD 20899-8910, USA
Email address: camilo.montoya@nist.gov

Dept. of Mathematics, Louisiana State University, Baton Rouge, LA 70803-4918, USA
Email address: olafsson@lsu.edu

Contemporary Mathematics
Volume **818**, 2025
https://doi.org/10.1090/conm/818/16368

On bounds for ratios of contiguous hypergeometric functions

Javier Segura

ABSTRACT. We review recent results on analytical properties (monotonicity and bounds) for ratios of contiguous functions of hypergeometric type. The cases of parabolic cylinder functions and modified Bessel functions have been discussed with considerable detail in the literature, and we give a brief account of these results, completing some aspects in the case of parabolic cylinder functions. Different techniques for obtaining these bounds are considered. They are all based on simple qualitative descriptions of the solutions of associated ODEs (mainly Riccati equations, but not only Riccati). In spite of their simplicity, they provide the most accurate global bounds known so far. We also provide examples of application of these ideas to the more general cases of the Kummer confluent function and the Gauss hypergeometric function. The function ratios described in this paper are important functions appearing in a large number of applications, in which simple approximations are very often required.

1. Introduction

Many special functions, and in particular those of hypergeometric type, satisfy first order differential systems of the form

$$
\begin{aligned}
y_n' &= a_n(x)y_n(x) + d_n(x)y_{n-1}(x),\\
y_{n-1}' &= b_n(x)y_{n-1}(x) + e_n(x)y_n(x).
\end{aligned}
$$

This is the case for the Gauss hypergeometric functions $y_n = {}_2\mathrm{F}_1(a+\epsilon_1 n, b+\epsilon_2 n; c+\epsilon_3 n; x)$, $n \in \mathbb{N}$, for any ϵ_1, ϵ_2, $\epsilon_3 \in \mathbb{Z}$ and, as a consequence, for the confluent hypergeometric (Kummer) function $y_n = {}_1\mathrm{F}_1(a+\epsilon_1 n; c+\epsilon_3 n; x)$. The functions $y_n(x)$ and $y_{n-1}(x)$ are said to be contiguous functions.

In **[15]** it is discussed how to obtain bounds for the ratio of contiguous functions, $h_n(x) = y_n(x)/y_{n-1}(x)$, from the qualitative study of the solutions of the Riccati equation satisfied by this ratio:

$$
h_n'(x) = d_n(x) - (b_n(x) - a_n(x))h_n(x) - e_n h_n(x)^2,
$$

2020 *Mathematics Subject Classification.* Primary 33C15, 33C05; Secondary 33C10, 26D07, 41A99.

Key words and phrases. Confluent and Gauss hypergeometric functions, Weber parabolic cylinder functions, modified Bessel functions, bounds.

The author acknowledges support from Ministerio de Ciencia e Innovación, project PID2021-127252NB-I00 with funds from MCIN/AEI/10.13039/501100011033/ FEDER, UE. The author thanks the two anonymous reviewers for many useful comments and suggestions.

or the analogous equation for the reciprocal ratio $y_{n-1}(x)/y_n(x)$. This can be combined with the application of the three-term recurrence relation

$$e_{n+1}y_{n+1}(x) + (b_{n+1}(x) - a_n(x))y_n(x) - d_n(x)y_{n-1}(x) = 0.$$

These methods originating from the analysis of the Riccati equation have been carried out with particular detail for the case of modified Bessel functions, which is a sub-case of the confluent hypergeometric family with $\epsilon_1 = 1$, $\epsilon_3 = 2$ (see [**11**, 1039.5-6]). This is an important set of functions with an uncountable number of applications. It was recently proved in [**19**] that these methods suffice to characterize the best possible bounds of the form $(\alpha + \sqrt{\beta^2 + x^2})/x$ for the ratios of modified Bessel functions $I_{\nu-1}(x)/I_\nu(x)$ and $K_{\nu+1}(x)/K_\nu(x)$. An earlier application of these methods to the general confluent case can be found in [**16**], and a later one in [**13**].

Modifications of these techniques were considered in [**12**] and [**17**] for modified Bessel functions. In the first reference, bounds with improved accuracy are obtained by iterating the process of obtaining bounds from the Riccati equation. In the second, an analysis of the solutions of an equation of the type $\phi'(x) = P(x, f(\phi(x)))$ is considered, with $P(x, y)$ a third degree polynomial in y, f a simple algebraic function, and $\phi(x)$ the double ratio $\phi(x) = h_n(x)/h_{n+1}(x)$. Bounds with improved accuracy are obtained, at the cost of not so simple expressions compared to those in [**19**]. Double ratios were previously considered in [**18**] for parabolic cylinder functions, and very sharp bounds for the ratios of PCFs (sharp in three different limits) were obtained as a consequence.

In this paper, we summarize the basic ideas and techniques presented in [**12**, **17**–**19**], and we collect the most significant results for parabolic cylinder functions and modified Bessel functions. These ideas should be applicable to the more general case of hypergeometric functions, and we apply some of them to the Kummer function and the Gauss hypergeometric functions. The analysis of Kummer and Gauss functions cases will be far from complete; we expect that an exhaustive analysis like the one presented in [**19**] for modified Bessel functions can be carried out, of which the present work can be seen as only the starting point (following the first step in [**16**] for the Kummer function). In addition, we believe that it should also be possible to consider the extended methods of [**12**, **17**, **18**] in a more general setting.

2. Qualitative analysis and bounds

It appears that most of the results on bounds of ratios of consecutive hypergeometric functions known so far, if not all, can be obtained by a qualitative analysis of some first order differential equation satisfied by these or related ratios, combined with the use of the recurrence relation.

The most ubiquitous result is probably that concerning the bounds for solutions of Riccati equations, which has been used to obtain a good number of sharp inequalities for parabolic cylinder functions [**18**], modified Bessel functions [**4**, **12**, **14**, **17**, **20**, **24**] and confluent hypergeometric functions [**13**, **16**]. Many of these and related results have also been obtained using alternative methods, particularly for the case of modified Bessel functions (see for instance [**1**, **2**, **9**, **23**]) and also for confluent [**5**, **6**, **21**] and Gauss hypergeometric functions [**7**].

The approach based on the qualitative analysis of first order ODEs, together with the application of the recurrence relation, seems to be sufficient for obtaining all the known bounds. This was explicitly proved in [**19**] for the case of bounds of the type $(\alpha + \sqrt{\beta^2 + \gamma^2 x^2})/x$ for the modified Bessel function ratios. In this paper, we focus on these methods based on the qualitative analysis of ODEs.

Our analysis will start from the construction of bounds from the analysis of Riccati equations, based on the following result (see [**12**, Theorem 1]).

THEOREM 2.1. *Let $h(x)$ be a solution of $h'(x) = a(x) + b(x)h(x) + c(x)h(x)^2$ with $a(x)c(x) < 0$ and $a(x)$, $b(x)$, $c(x)$ continuous in $[a,b]$. Let $\lambda(x)$ be the positive solution of $a(x) + b(x)\lambda(x) + c(x)\lambda(x)^2 = 0$, then the following hold:*

1. *If $c(x) < 0$, $h(a^+) > 0$, $h'(a^+)\lambda'(a^+) > 0$ then $h(x) < \lambda(x)$ if $\lambda'(x) > 0$ and $h(x) > \lambda(x)$ if $\lambda'(x) < 0$.*
2. *If $c(x) > 0$, $h(b^-) > 0$, $h'(b^-)\lambda'(b^-) > 0$ then $h(x) < \lambda(x)$ if $\lambda'(x) < 0$ and $h(x) > \lambda(x)$ if $\lambda'(x) > 0$.*

In all these cases $\lambda'(x)h'(x) > 0$ for all $x \in (a,b)$, and $h(x)$ is a bounded function in any compact subset of (a,b).

PROOF. Let us consider the case $c(x) < 0$, $h(a^+) > 0$ and $\lambda(x)$ increasing. Then $h'(a^+) > 0$, and because $c(x) < 0$ this is only possible if $h(a^+) < \lambda(a^+)$ because $h'(x) = a(x) + b(x)h(x) + c(x)h(x)^2$ with $a(x)c(x) < 0$. Now, $h(x)$ must be increasing unless a value x_0 is reached such that $h(x_0) = \lambda(x_0)$, which implies that $h'(x_0) = 0$. However this cannot occur because the graph of $h(x)$ lies below that of $\lambda(x)$ at the left end of the interval, and $\lambda(x)$ is increasing, which means that if $h(x_0) = \lambda(x_0)$ then necessarily $h'(x_0) < 0$, in contradiction with the fact that $h'(x_0) = 0$. Because such x_0 does not exist we have $0 < h(x) < \lambda(x)$ for all $x \in (a,b)$ and then $h'(x) > 0$.

The rest of the cases follow similarly. □

REMARK 2.2. The condition $a(x)c(x) < 0$ is not essential, but it simplifies the analysis because the characteristic equation $a(x) + b(x)\lambda(x) + c(x)\lambda(x)^2 = 0$ has one negative and one positive real solution. For a more general situation, see [**15**]. On the other hand, the fact that only the positive root has been considered is not a restriction, because it is always possible to consider the Riccati equation for $-h(x)$ instead of that for $h(x)$.

It is also possible to extract bounds for other types of first order ODEs different from Riccati equations. In particular, in [**17, 18**] an analysis of the solutions of equations $\phi(x) = P(x, f(\phi(x)))$, with $P(x,y)$ a third degree polynomial in y, f a simple algebraic function and $\phi(x)$ the double ratio $\phi(x) = h_n(x)/h_{n+1}(x)$, is considered. The qualitative analysis is similar, but more involved given that we have three nullclines to be considered. Later we give more details on these methods when we describe the parabolic cylinder functions and the modified Bessel functions.

As commented before, most of the bounds that are available are related to a nullcline of a first order ODE (as in Theorem 2.1). In other cases, the use of an ODE satisfied by these ratios or related functions is also helpful, and one can check if a given function is a bound for one of the solutions by considering the next result.

THEOREM 2.3. *Let $P(x,y)$ be continuous in $(a,b) \times \mathbb{R}$ and $\phi(x)$ be a solution of the ODE $\phi'(x) = P(x, \phi(x))$ which is bounded in any compact subset of (a,b).*

Let $\lambda(x)$ be differentiable in (a,b). Denoting $\delta(x) = \lambda(x) - \phi(x)$ and $\Delta(x) = \lambda'(x) - P(x,\lambda(x))$ we have that

(1) *If $\delta(a^+)\Delta(x) > 0$ in (a,b) then $\delta(x)\Delta(x) > 0$ in (a,b).*
(2) *If $\delta(b^-)\Delta(x) < 0$ in (a,b) then $\delta(x)\Delta(x) < 0$ in (a,b).*

PROOF. Assume, for instance, that $\delta(a^+) > 0$ and $\Delta(x) > 0$ in (a,b); the rest of the cases follow similarly. With this we can prove that no $x_0 \in (a,b)$ exists such that $\delta(x) = 0$, and because $\delta(a^+) > 0$ this implies that $\delta(x) > 0$ in (a,b). For proving this, we suppose that x_0 is the smallest value in (a,b) for which $\delta(x_0) = 0$ and we arrive at a contradiction, which proves that such x_0 does not exist. We observe that $\delta(x) > 0$ in (a,x_0) because $\phi(x)$ is a bounded and continuous solution, and x_0 is the smallest value on (a,b) for which $\delta(x_0) = 0$.

Because $\delta'(x) = \lambda'(x) - \phi'(x) = \lambda'(x) - P(x,\phi(x))$, we have $\delta'(x_0) = \lambda'(x_0) - P(x_0,\phi(x_0)) = \Delta(x_0)$, where the last equality holds because $\delta(x_0) = 0$ and then $\phi(x_0) = \lambda(x_0)$. But because $\Delta(x_0) > 0$, $\delta'(x_0) > 0$, which contradicts the fact that $\delta(x) > 0$ in (a,x_0). □

This is a useful result for proving that $\lambda(x)$ is a bound for a given solution of the ODE $\phi'(x) = P(x,\phi(x))$. The only information required of the solution $\phi(x)$ is its behavior either at $x = a^+$ or $x = b^-$, and the assumption on continuity and boundedness. Observe that in Theorem 2.1 we did not need to require boundedness, and that boundedness was a consequence.

All the bounds we will describe in this paper are a consequence of either an analysis of nullclines as in Theorem 2.1 (maybe combined with the use of a recurrence relation), some variants for other types of first order ODEs (as done in [**17, 18**]) or, in some cases where the bounds are not directly related to nullclines, a consequence of Theorem 2.3.

With respect to the use of recurrence relations, we recall that the families of functions we will consider satisfy recurrence relations $y_{n+1}(x) + B_n(x)y_n(x) - C_n(x)y_{n-1}(x) = 0$. Denoting $h_n(x) = y_n(x)/y_{n-1}(x)$ and applying the recurrence in the backward direction,

$$h_n(x) = \frac{C_n(x)}{B_n(x) + h_{n+1}(x)},$$

which under certain conditions allows us to obtain a bound for $h_n(x)$ using a bound in the right hand side for $h_{n+1}(x)$. For instance, if $C_n(x)$ and $B_n(x)$ are positive and $U_n(x)$ is an upper bound for $h_n(x)$ then $C_n(x)/(B_n(x) + U_{n+1}(x))$ is a lower bound for $h_n(x)$. We refer to [**15**] for a discussion on the use of forward and backward recursion for obtaining bounds, and its relation to the existence of a minimal solution for the recurrence.

3. Parabolic cylinder functions

We present a brief account on bounds for the ratio

$$\Phi_n(x) = U(n-1,x)/U(n,x), \tag{3.1}$$

where $U(n,x)$ is the Weber parabolic cylinder function, which is a recessive solution as $x \to +\infty$ of the second order ODE

$$y''(x) - \left(\frac{x^2}{4} + n\right) y(x) = 0.$$

3.1. Sharp bounds from Riccati equations. In this subsection we summarize the main results of [**18**]. As an illustration of the ideas discussed in the previous section, we give some details for the analysis of the Riccati equation.

The PCF $U(n,x)$ satisfies the following difference–differential system (see 12.8.2 and 12.8.3 of [**22**]):

$$\begin{aligned} U'(n,x) &= \tfrac{x}{2}U(n,x) - U(n-1,x), \\ U'(n-1,x) &= -\tfrac{x}{2}U(n-1,x) - (n-1/2)U(n,x). \end{aligned} \tag{3.2}$$

This system is the only information required to obtain the bounds, together with the fact that the function $\Phi_n(x)$ is positive for all real x when $n > 1/2$ and increasing for large x , which is easy to check from the asymptotic expansions of $U(n,x)$ [**22**, 12.9.1]. Using that expansion we see that as $x \to +\infty$,[1]

$$\Phi_n(x) \sim x\left[1 + (n+1/2)x^{-2} - (n+1/2)(n+3/2)x^{-4} + \mathcal{O}(x^{-6})\right], \tag{3.3}$$

and the first term is enough to see that, indeed, $\Phi_n(x)$ is positive and increasing as $x \to +\infty$.

On the other hand, as $x \to -\infty$,[2]

$$\Phi_n(x) \sim -\frac{n-1/2}{x}\left[1 - (n-3/2)x^{-2} + 2(n-3/2)(n-2)x^{-4} + \mathcal{O}(x^{-6})\right]. \tag{3.4}$$

(notice that $\Phi_n(x)$ becomes negative for $n < 1/2$ and $x \to -\infty$).

Using (3.2) we see that $\Phi_n(x)$ is one of the solutions of the Riccati equation

$$h'(x) = h(x)^2 - xh(x) - (n-1/2), \tag{3.5}$$

which for $n > 1/2$ has the positive and increasing characteristic root $\lambda(x) = \frac{1}{2}\left(x + \sqrt{x^2+4n-2}\right)$. And because of (3.3) the hypotheses of Theorem 2.1 are met, with $(a,b) = (-\infty,+\infty)$, $c(x) > 0$ and $\lambda'(x) > 0$, which implies that $\Phi_n(x) > \lambda(x)$ for all real x. In addition, by applying the recurrence relation in the backward and the forward directions two additional (upper) bounds are obtained for $\Phi_n(x)$. We summarize those results in the following theorem. The second bound in the next theorem is obtained from the first bound and by applying the backward recurrence

$$\Phi_n(x) = x + \frac{n + \dfrac{1}{2}}{\Phi_{n+1}(x)}, \tag{3.6}$$

and the third bound is a consequence of the forward recurrence

$$\Phi_n(x) = \frac{n - \dfrac{1}{2}}{-x + \Phi_{n-1}(x)}. \tag{3.7}$$

[1] We note an obvious erratum in [**18**, Eq. (8)]: the exponent of the third term should be -4 and not 4.

[2] Again, the exponent of the third term in [**18**, Eq. (9)] should be -4 and not 4.

THEOREM 3.1. *The following bounds hold for all real x:*

$$\frac{U(n-1,x)}{U(n,x)} > B^{(2,1)}(x) = \frac{1}{2}\left(x+\sqrt{x^2+4n-2}\right) \qquad \text{for } n>1/2,$$

$$\frac{U(n-1,x)}{U(n,x)} < B^{(1,2)}(x) = \frac{1}{2}\left(x+\sqrt{x^2+4n+2}\right) \qquad \text{for } n>-1/2,$$

$$\frac{U(n-1,x)}{U(n,x)} < B^{(3,0)}(x) = \frac{1}{2}\frac{n-1/2}{n-3/2}\left(x+\sqrt{x^2+4n-6}\right) \quad \text{for } n>3/2.$$

The notation $B^{(m,n)}(x)$ for the bounds is analogous to that used for modified Bessel functions in [**19**], and denotes the number of terms of the expansions of

$$B(\alpha,\beta,\gamma,x) = \alpha x + \sqrt{\beta^2x^2+\gamma^2}, \quad \alpha>0, \tag{3.8}$$

which coincide with those of (3.4) as $x\to-\infty$ (m) and (3.3) as $x\to+\infty$ (n).

Assuming $\beta>0$ we have

$$B(\alpha,\beta,\gamma,x) = (\alpha\pm\beta)x \pm \frac{\gamma^2}{2\beta x} \mp \frac{\gamma^4}{8\beta^3x^3} \pm \frac{\gamma^6}{16\beta^5x^5} + \mathcal{O}(x^{-7}), \qquad \text{as } x\to\pm\infty.$$

Since in all the three bounds in the previous theorem $\alpha=\beta$, the first term in the previous expansion as $x\to-\infty$ is zero, coinciding with the expansion (3.4); therefore $m\geqslant 1$ for all of them. Also, we can easily check that the first bound has an additional correct term as $x\to-\infty$ and another one as $x\to+\infty$. Similarly, we can check the accuracy for the other two bounds. We observe that with three parameters, we have $m+n\leqslant 3$. It appears that the bound with $m=0$, $n=3$ should be possible.

Indeed, it is possible to give a bound of the type (3.8) with higher accuracy as $x\to+\infty$, by choosing α, β and γ such that the first three terms in the expansions as $x\to+\infty$ of $\Phi_n(x)$ and $B(\alpha,\beta,\gamma,x)$ coincide. This gives the values $\alpha=\frac{1}{2}(n+5/2)/(n+3/2)$, $\beta=\frac{1}{2}(n+1/2)/(n+3/2)$ and $\gamma^2=(n+1/2)^2/(n+3/2)$, and one can prove that this is indeed a bound using Theorem 2.3, though unsharp as $x\to-\infty$.

THEOREM 3.2. *If $n>-1/2$ then for all real x*

$$\frac{U(n-1,x)}{U(n,x)} > B^{(0,3)}(x) = \frac{(n+5/2)x+(n+1/2)\sqrt{x^2+4n+6}}{2(n+3/2)}.$$

PROOF. We start by checking that, denoting

$$\lambda_n(x) = \frac{(n+5/2)x+(n+1/2)\sqrt{x^2+4n+6}}{2(n+3/2)},$$

we have

$$\lambda_n(x)-\Phi_n(x) = -\frac{(n+1/2)(n+3/2)}{x^5} + \mathcal{O}(x^{-7}), \quad x\to+\infty.$$

With this, and considering Theorem 2.3, we only have to prove that $\Delta(x)>0$ for real x, where $\Delta(x)=\lambda'(x)-P(x,\lambda(x))$, with $P(x,y)=y^2-xy-(n-1/2)$ (see (3.5)).

We have

$$\Delta(x) = \frac{n+1/2}{(2n+3)^2}\left(2x^2+2n+3-\frac{x(2x^2+6n+9)}{\sqrt{x^2+4n+6}}\right),$$

and we see that $\Delta(x) > 0$ if $n > -1/2$ and $x \leqslant 0$. After some elementary algebra we can also write

$$\Delta(x) = \frac{2(n+1/2)(2n+3)}{\sqrt{x^2+4n+6}\left((2x^2+2n+2)\sqrt{x^2+4n+6}+x(2x^2+6n+9))\right)},$$

and $\Delta(x) > 0$ if $n > -1/2$ and $x > 0$, which completes the proof. □

It is also possible to prove the previous bound in a more straightforward way starting from the second bound in Theorem 3.1 and applying the backward recurrence (3.6). We give the proof using Theorem 2.3 as an illustration of application of this theorem.

The bound in Theorem 3.2 is new, unlike the bounds in Theorem 3.1, which were already discussed in [**18**]. We observe that the bound in Theorem 3.2 becomes negative for $x < -(n+1/2)$, which is a clear indication of the unsharpness as $x \to -\infty$. In any case, it is the best bound of the form (3.8) as $x \to +\infty$, and it completes the set of best bounds in the same way that the set of best bounds for ratios of modified Bessel functions was completed in [**19**], as we will also describe later in this paper.

For the case of modified Bessel functions, uniparametric sets of bounds linking the bounds of type $B^{(0,3)}$ with those of type $B^{(2,1)}$ and the bounds $B^{(3,0)}$ with the $B^{(1,2)}$ bounds were given in [**19**], as we will later summarize in Theorem 4.1. It is an open question whether the same type of analysis is possible for parabolic cylinder functions. Similarly, it seems feasible that best bounds could be found which have an osculatory character, meaning that the graphs of the bounds and the ratio of parabolic cylinder functions would be tangent at a point.

Subsequent applications of the recurrence are possible to obtain further bounds, but the resulting bounds become more complicated and are no longer of the form (3.8). For instance, considering the last bound in Theorem 3.1 and applying a further step of forward recurrence (3.7), we get the following.

THEOREM 3.3. *Let $n > 5/2$, the following bound holds for real x:*

$$\frac{U(n-1,x)}{U(n,x)} > B^{(4,0)}(x) = \frac{2(n-1/2)(n-5/2)}{(n-3/2)\sqrt{x^2+4n-10}-(n-7/2)x}.$$

This bound has the same three first terms as the expansion (3.4), but it is unsharp as $x \to +\infty$.

We notice that all the bounds given so far for parabolic cylinder functions are a consequence of the first bound in Theorem 3.1, which was obtained from the Riccati equation, and the application of the recurrence relation. The backward recurrence improves the accuracy at $+\infty$ but worsens it at $-\infty$; the opposite occurs with the forward recurrence. The only bound which is sharp at $\pm\infty$ is the first bound in Theorem 3.1. For obtaining bounds with higher accuracy both as $x \to -\infty$ and $x \to +\infty$, a different approach should be considered.

3.2. Beyond the Riccati bounds. Riccati equations are not the only possibility of obtaining sharp bounds for ratios of parabolic cylinder functions and other functions of hypergeometric type. As we see next, it is possible to use other differential equations which can give even sharper bounds. The possibility considered in [**18**] is to analyze the first order differential equation satisfied by the double ratio $\Phi_n(x)/\Phi_{n+1}(x)$. The rationale behind this approach is the fact that the double ratio has a slower variation than the simple ratio, and this facilitates finding more accurate bounds. Similar ideas were also considered for modified Bessel functions, as we later describe.

In [**18**] it was shown that the function

$$W_n(x) = \left(n+\frac{1}{2}\right)\frac{\Phi_n(x)}{\Phi_{n+1}(x)}$$

satisfies the differential equation

$$W_n'(x) = 2\left(\phi_n(x)^2 - V_n(x)\phi_n(x) - \frac{x}{4}\right),\quad V_n(x) = \frac{x^2}{4}+n, \tag{3.9}$$

where

$$\phi_n(x) = \Phi_n(x) - \frac{x}{2} = \sqrt{\frac{x^2}{4}+W_n(x)}.$$

An analysis similar to that carried out for Riccati equations in 2.1 can be considered for this differential equation. The analysis is more involved because solutions of a third degree equation have to be considered and we need to prove the monotonicity of some functions related to the roots, particularly for the largest root [**18**, Lemma 5]. For details we refer to [**18**]; we just recall Lemma 6 of that reference, and its consequence for the bounds.

LEMMA 3.4. *Let $y(x)$ satisfy the differential equation*

$$y'(x) = 2\left(z(x)^3 - \left(\frac{x^2}{4}+n\right)z(x) - \frac{x}{4}\right),\quad n>1/2, \tag{3.10}$$

where

$$z(x) = +\sqrt{\frac{x^2}{4}+y(x)}. \tag{3.11}$$

If $y(x)$ is positive and increasing as $x\to+\infty$, then

$$z(x) > \lambda_n^+(x) = f_n(x)\cos\left(\frac{1}{3}\arccos\left(\frac{x}{f_n(x)^3}\right)\right),\quad f_n(x) = \sqrt{\frac{x^2+4n}{3}},$$

$$y(x) > \lambda_n^+(x)^2 - \frac{x^2}{4},$$

and $y'(x)>0$ for all real x.

From the expansion (3.3) it is easy to check that $W_n(x)$ satisfies the hypothesis for $y(x)$ in the previous theorem, from which lower bounds for $W_n(x)$ and $\Phi_n(x)$ are derived. In addition, the monotonicity of $W_n(x)$ is also proved.

The following result gives the trigonometric bound for $\Phi_n(x)$ that stems from the previous theorem plus an additional algebraic bound with similar accuracy that is obtained using similar ideas as in Theorem 2.3.

THEOREM 3.5. *The following holds for any real x and $n>1/2$:*

$$\frac{U(n-1,x)}{U(n,x)} - \frac{x}{2} > f_n(x)\cos\left(\frac{1}{3}\arccos\left(\frac{x}{f_n(x)^3}\right)\right) > \sqrt{\frac{x^2}{4}+g_n(x)},$$

where $f_n(x)=\sqrt{\dfrac{x^2+4n}{3}}$, $g_n(x) = (n+\frac{1}{2})\dfrac{x+\sqrt{x^2+4n-2}}{x+\sqrt{x^2+4n+2}}$.

Both the trigonometric and the algebraic bounds are very sharp as $x \to \pm\infty$; the first three terms in the expansion (3.3) are reproduced, and the two first terms in (3.4). With the notation $B^{(n,m)}$ used before, these are $B^{(3,3)}$ bounds (recall that we considered $n \geqslant 1$ if the bound is $\mathcal{O}(x^{-1})$ as $x \to -\infty$, and then two correct terms means $n = 3$). They are also very sharp as $n \to +\infty$, see [**18**].

The forward and backward recurrences can again be considered. Starting with the bounds in Theorem 3.5 we get the upper bound $B^{(2,4)}$ by using the backward recurrence, while the forward recurrence gives the bound $B^{(4,2)}$.

The monotonicity of the double ratio $W_n(x)$ was earlier proved in [**8**] with a more indirect probabilistic approach. The original motivation of [**18**] was to prove that property by a direct method, but very sharp bounds were also obtained as a consequence. We end this section by formulating a conjecture that generalizes the property of monotonicity of the double ratio $W_n(x)$.

CONJECTURE 3.6. *Let* $R_n^{[1]}(x) = U(n-1,x)/U(n,x)$, $n > 1/2$, *and define* $R_n^{[k+1]} = R_n^{[k]}(x)/R_{n+1}^{[k]}(x)$; *then the functions* $R_n^{[k]}(x)$ *are positive increasing functions of* x *with* $R_n^{[k+1]}(x) > R_n^{[k]}(x)$. $R_n^{[k]}(x) < 1$ *if* $k \geqslant 2$.

The ratios $R_n^{[k]}(x) < 1$ have a sigmoidal shape for $k \geqslant 2$. From (3.3) we see that $\lim_{x\to+\infty} R_n^{[k]}(x) = 1$ and numerical experiments seem to suggest that $\lim_{k\to+\infty} R_n^{[k]}(x) = 1$ for $n > 1/2$, but not close to $n = 1/2$ (observe that (3.4) indicates $R_{1/2}^{[k]}(-\infty) = 0$).

4. Modified Bessel functions

Modified Bessel functions are, without any doubt, the functions of hypergeometric type for which the analysis of the bounds and monotonicity properties for the ratios of these functions have been most deeply studied (see [**1, 2, 4, 9, 12, 14, 17, 19, 20, 23, 24**]. This is not surprising, given the huge amount of applications where these ratios appear (see, for example, the applications cited in [**14, 19**]). In most of the papers (with the exception of [**12, 17**]) the bounds are of the form

$$B(\alpha,\beta,\gamma,x) = \frac{\alpha + \sqrt{\beta^2 + \gamma^2 x^2}}{x}. \tag{4.1}$$

These bounds are widely used because they can be quite sharp, they are simple and it is easy to operate with them. In [**19**], the analysis of these type of bounds was concluded, and the best possible bounds of this form were characterized and classified.

As for the rest of cases discussed in this paper, the main piece of information in our analysis is the difference–differential system [**11**, 10.29.2]

$$\begin{aligned} \mathcal{I}'_\nu(x) &= \mathcal{I}_{\nu-1}(x) - \frac{\nu}{x}\mathcal{I}_\nu(x), \\ \mathcal{I}'_{\nu-1}(x) &= \mathcal{I}_\nu(x) + \frac{\nu-1}{x}\mathcal{I}_{\nu-1}(x) \end{aligned} \tag{4.2}$$

(where $\mathcal{I}_\nu(x)$ denotes $I_\nu(x)$, $e^{i\pi\nu}K_\nu(x)$ or any linear combination of them), together with the unique behavior of $I_\nu(x)$ as $x \to 0^+$ and of $K_\nu(x)$ as $x \to +\infty$.

In the next section, we briefly summarize the main results given in [**19**]. The techniques employed are similar to the ideas of Theorems 2.1 and 2.3. After this, we summarize other types of bounds with higher accuracy (but not so simple), both of algebraic and trigonometric type.

4.1. Best bounds of the type $B(\alpha,\beta,\gamma,x) = (\alpha + \sqrt{\beta^2+\gamma^2x^2})/x$. One of the main results proved in [**19**] is that if α, β and γ are chosen such that $B(\alpha,\beta,\gamma,x) = (\alpha + \sqrt{\beta^2+\gamma^2x^2})/x$ is a sharp approximation for $\Phi_\nu(x) = I_{\nu-1}(x)/I_\nu(x)$ as $x \to 0^+$ (respectively $x \to +\infty$) and the graphs of the functions $B(\alpha,\beta,\gamma,x)$ and $\Phi_\nu(x)$ are tangent at some $x = x_* > 0$, then $B(\alpha,\beta,\gamma,x)$ is an upper (respectively lower) bound for $\Phi_\nu(x)$; the same is true for the ratio $\Phi_\nu(x) = K_{\nu+1}(x)/K_\nu(x)$ but interchanging lower and upper bounds. This provides the best possible bounds of the form $B(\alpha,\beta,\gamma,x)$ around any chosen value x_*.

There is no explicit expression for all the coefficients α, β and γ for the best bounds, except in the limits $x_* \to +\infty$ and $x_* \to 0$, when they give the best possible bounds at $x = 0$ and/or $x = +\infty$. The best possible bounds at $x = 0$ and/or $x = +\infty$ are therefore explicitly known, as described in [**19**].

These best bounds at $x = 0$ and/or $x = +\infty$ are particular cases of the four parametric bounds given in [**19**] for $I_{\nu-1}(x)/I_\nu(x)$ and $K_{\nu+1}(x)/K_\nu(x)$. These are the most accurate known bounds of the type $B(\alpha,\beta,\gamma,x)$ with explicit formulas; they are close to the best bounds described above, and they contain as particular cases the best bounds at $x = 0, +\infty$. We next summarize these four parametric bounds (upper and lower bounds for both $I_{\nu-1}(x)/I_\nu(x)$ and $K_{\nu+1}(x)/K_\nu(x)$). The first of these four theorems was already given in [**4**] in a different form; the other three are given in [**19**].

THEOREM 4.1. *The following holds for $\lambda \in [0,1/2]$, $\nu \geqslant \frac{1}{2} - \lambda$ and $x > 0$:*

$$\frac{I_{\nu-1}(x)}{I_\nu(x)} > L_\nu^{(I)}(\lambda,x) = B(\alpha_\nu^{(I)}(\lambda), \beta_\nu^{(I)}(\lambda), 1, x), \tag{4.3}$$

where $\alpha_\nu^{(I)}(\lambda) = \nu - 1/2 - \lambda$, $\beta_\nu^{(I)}(\lambda) = \sqrt{2\lambda} + \sqrt{\nu^2 - (\lambda - \frac{1}{2})^2}$.

THEOREM 4.2. *The following holds for $\lambda \in [0,1/2]$, $\nu \geqslant \frac{1}{2} - \lambda$ and $x > 0$:*

$$\frac{K_{\nu+1}(x)}{K_\nu(x)} < U_\nu^{(K)}(\lambda,x) = B(\alpha_\nu^{(K)}(\lambda), \beta_\nu^{(K)}(\lambda), 1, x),$$

where

$$\alpha_\nu^{(K)}(\lambda) = \nu + 1/2 + \lambda,\ \beta_\nu^{(K)}(\lambda) = -\sqrt{2\lambda} + \sqrt{\nu^2 - (\lambda - \frac{1}{2})^2}.$$

THEOREM 4.3. *The following holds for $\nu \geqslant 0$, $\lambda \in [1/2, 2]$ and $x > 0$:*

$$\frac{I_{\nu-1}(x)}{I_\nu(x)} < U_\nu^{(I)}(\lambda,x) = B(\nu - \lambda, \nu + \lambda, \sqrt{c_\nu^{(I)}(\lambda)}, x), \tag{4.4}$$

where

$$c_\nu^{(I)}(\lambda) = \frac{\nu + \lambda}{\nu - \lambda + 2\sqrt{2\lambda} - 1}.$$

THEOREM 4.4. *The following holds for $\lambda \in [1/2, 2]$, $\nu \geqslant \lambda$ and $x > 0$:*

$$\frac{K_{\nu+1}(x)}{K_\nu(x)} > L_\nu^{(K)}(\lambda,x) = B(\nu + \lambda, \nu - \lambda, \sqrt{c_\nu^{(K)}(\lambda)}, x),$$

where

$$c_\nu^{(K)}(\lambda) = \frac{\nu - \lambda}{\nu + \lambda - 2\sqrt{2\lambda} + 1}.$$

(n,m)	α	β	γ	Range	Bound
$(2,1)$	$\nu-1$	$\nu+1$	1	$\nu\geqslant 0$	$L_\nu^{(I)}(\frac{1}{2},x)$
$(0,3)$	$\nu-\frac{1}{2}$	$\sqrt{\nu^2-\frac{1}{4}}$	1	$\nu\geqslant\frac{1}{2}$	$L_\nu^{(I)}(0,x)$
$(1,2)$	$\nu-\frac{1}{2}$	$\nu+\frac{1}{2}$	1	$\nu\geqslant 0$	$U_\nu^{(I)}(\frac{1}{2},x)$
$(3,0)$	$\nu-2$	$\nu+2$	$\sqrt{(\nu+2)/(\nu+1)}$	$\nu\geqslant 0$	$U_\nu^{(I)}(2,x)$
$(2,1)$	$\nu+1$	$\nu-1$	1	$\nu\in\mathbb{R}$	$U_\nu^{(K)}(\frac{1}{2},x)$
$(0,3)$	$\nu+\frac{1}{2}$	$\sqrt{\nu^2-\frac{1}{4}}$	1	$\nu>1/2$	$U_\nu^{(K)}(0,x)$
$(1,2)$	$\nu+\frac{1}{2}$	$\nu-\frac{1}{2}$	1	$\nu>1/2$	$L_\nu^{(K)}(\frac{1}{2},x)$
$(3,0)$	$\nu+2$	$\nu-2$	$\sqrt{(\nu-2)/(\nu-1)}$	$\nu\geqslant 2$	$L_\nu^{(K)}(2,x)$

TABLE 1. Bounds for the ratios $I_{\nu-1}(x)/I_\nu(x)$ and $K_{\nu+1}(x)/K_\nu(x)$ of the type $B(\alpha,\beta,\gamma,x)=(\alpha+\sqrt{\beta^2+\gamma^2x^2})/x$ classified according to their accuracies at $x=0$ (n) and $x=+\infty$ (m). The range of validity of the bounds is shown, and the relation with the parametric bounds of Theorems 4.1–4.4 is given in the last column (the lower bounds are denoted with an L and the upper bounds with a U).

As we did earlier in this paper, it is possible to find the bounds of the form $B(\alpha,\beta,\gamma,x)$ which are most accurate at $x=0$ or $x=+\infty$ by comparing the expansions at these points. We say that a bound has accuracy (n,m) if the first n terms of its expansion around $x=0$ are exact and the same happens with the first m terms at $x=+\infty$. In Table 1 we summarize such bounds for both $I_{\nu-1}/I_\nu(x)$ and $K_{\nu+1}/K_\nu(x)$; all those bounds are particular cases of the previous four theorems, and this relation is also given in the table.

For the ratio $I_{\nu-1}(x)/I_\nu(x)$, all the bounds appearing in Table 1 had been already described before, but they were first classified and ordered in [**19**]. In contrast, the set of best bounds at $x=0$ and/or $x=+\infty$ for $K_{\nu+1}(x)/K_\nu(x)$ was not complete, as the cases $(0,3)$ and $(3,0)$ had not been considered earlier and were first described in [**19**].

The bounds in Table 1 exhaust the best bounds at 0 and/or $+\infty$ of the form $B(\alpha,\beta,\gamma,x)$. However, of course, other forms may be available with higher accuracy. For instance, in [**19**] the following bounds are proved using arguments similar to those of Theorem 2.3.

THEOREM 4.5. *Let $\phi_{-,\nu}(x)=xI_{\nu-1}(x)/I_\nu(x)$ and $\phi_{+,\nu}(x)=xK_{\nu+1}(x)/K_\nu(x)$, then both functions satisfy the following properties for $\nu\geqslant 1/2$ and $x>0$:*

$$0<\phi'_{\pm,\nu}(x)\leqslant 1,$$

$$B_\nu^{(1,3)}(x)\equiv\nu+\sqrt{\nu^2+x(x-1)}<\phi_{\pm,\nu}(x)\leqslant\nu+\sqrt{\nu^2+x(x+1)}\equiv\hat{B}_\nu^{(1,3)}(x),$$

where the equality only takes place for $\phi_{+,\nu}(x)$ when $\nu=1/2$. The upper bound for $\phi_{+,\nu}(x)$ and the lower bound for $\phi_{-,\nu}(x)$ are of accuracy $(1,3)$.

Of course, further bounds are possible by application of the recurrence relation. For instance, using the backward recurrence in the case of first kind Bessel functions we get, starting from the lower bound of the previous theorem that, for all $x > 0$ and $\nu > 0$,

$$\frac{I_{\nu-1}(x)}{I_\nu(x)} < \frac{2\nu}{x} + \frac{x}{\nu+1+\sqrt{(\nu+1)^2+x(x-1)}},$$

and this bound has accuracy $(2,3)$.

4.2. Other bounds. It is possible to obtain bounds with higher accuracy using modified methods, as we will next describe. It is also possible, as described before, to improve the accuracy of the bounds at $x = 0$ by using the recurrence relation. In all these cases, the improvement in the accuracy of the bounds is accompanied by more involved expressions for them. We briefly describe the bounds obtained from the iteration of the Riccati equation in [**12**] and from the use of the ODE satisfied by double ratios in [**17**].

4.2.1. *Bounds from the iteration of the Riccati equation.* The idea is to start from a Riccati equation

$$h_0'(x) = A_0(x) + B_0(x)h_0(x) + C_0(x)h_0(x)^2,$$

and to consider the function $h_1(x) = h_0(x)/\phi_0(x)$, where $\phi_0(x) \equiv \beta_0(x)$ is a function of convenience. We choose $\phi_0(x)$ as one of the roots of $A_0(x) + B_0(x)\phi_0(x) + C_0(x)\phi_0(x)^2 = 0$, which is a bound for $h_0(x)$ if the conditions of Theorem 2.1 are met. The next step is to consider the Riccati equation for $h_1(x)$,

$$\begin{aligned} &h_1'(x) = A_1(x) + B_1(x)h_1(x) + C_1(x)h_1(x)^2,\\ &A_1(x) = \frac{A_0(x)}{\phi_0(x)}, \quad B_1(x) = B_0(x) - \frac{\phi_0'(x)}{\phi_0(x)}, \quad C_1(x) = \phi_0(x)C_0(x). \end{aligned} \tag{4.5}$$

If $\phi_1(x)$, one of the solutions of the characteristic equation $A_1(x) + B_1(x)\phi_1(x) + C_1(x)\phi_1(x)^2 = 0$, turns out to be a bound for $h_1(x)$, then $\beta_1(x) = \phi_1(x)\beta_0(x) = \phi_1(x)\phi_0(x)$ will be a bound for $h_1(x)$. For studying whether $h_1(x)$ is a bound, the same type of analysis as for the original Riccati equation is considered, based on Theorem 2.1.

This iteration of Riccati equations was introduced in [**12**], starting from the Riccati equations for $h(x) = x^{-\alpha}I_\nu(x)/I_{\nu-1}(x)$ and $h(x) = x^{-\alpha}K_{\nu-1}(x)/K_\nu(x)$ (it would be equivalent to consider the Riccati equations for $\Phi(x) = 1/h(x)$). We summarize these results and compare them with those more elementary (but also accurate) bounds in [**19**].

For first kind Bessel functions, and after one iteration, the best bounds that are obtained are for $\alpha = 0$ and $\alpha = 2$, and we have the following.

THEOREM 4.6. *Let*

$$B_\alpha(\nu,x) = \frac{\delta_\alpha(\nu,x) + \sqrt{\delta_\alpha(\nu,x)^2+x^2}}{x},$$

where

$$\delta_\alpha(\nu,x) = (\nu - 1/2) + \frac{\lambda}{2\sqrt{\lambda^2+x^2}}, \qquad \lambda = \nu + (\alpha-1)/2;$$

then

$$\frac{I_{\nu-1}(x)}{I_\nu(x)} > B_0(\nu,x), \quad \nu \geqslant 1/2$$

and

$$\frac{I_{\nu-1}(x)}{I_\nu(x)} < B_2(\nu, x), \quad \nu \geqslant 0.$$

The accuracy of the first bound is $(1,3)$ while the second bound has accuracy $(1,2)$. We may compare these bounds with the bounds of equal accuracy obtained in [**19**], particularly with the bound in Theorem 4.5 for the $(1,3)$ case, which is also a lower bound of the same accuracy, and with the third bound in Table 1. In both cases, comparing the expansions at $x = 0$ and $x = +\infty$ we conclude that the bounds of the previous theorem are generally better (though more complicated). Numerical tests show that indeed, the $(1,2)$ bound of the previous theorem appears to be better for all x and $\nu > 0$ than the bound in Table 1, and that the $(1,3)$ bound of the previous theorem is also superior when $\nu > 3/2$.

The bounds from the previous theorem can be improved by applying the recurrence relation in the backward direction. We refer to [**12**] for the explicit result. The bounds improve their accuracy at $x = 0$ by one unit with respect to the bounds in Theorem 4.6, and therefore they have accuracies $(2,3)$ and $(2,2)$. We also refer to [**12**, Theorem 9], for bounds of the same type for the second kind Bessel function.

4.2.2. *Trigonometric bounds from double ratios.* Similarly as described in Section 3.2, the analysis of the first order ODE satisfied by double ratios of modified Bessel functions can be used to obtain very sharp trigonometric bounds for ratios of modified Bessel functions, as was described in [**17**].

We start from the ratio, $\Phi_\nu(x) = \mathcal{I}_{\nu-1}(x)/\mathcal{I}_\nu(x)$, where $\mathcal{I}_\nu(x)$ can be any of the solutions of (4.2). Then, using (4.2) one can easily prove that the double ratio

$$W_\nu(x) = \frac{\Phi_\nu(x)}{\Phi_{\nu+1}(x)}$$

satisfies the first order ODE

$$W_\nu'(x) = -\frac{2}{x^3}\left(\psi_\nu(x)^3 + \psi_\nu(x)^2 - (\nu^2 + x^2)\psi_\nu(x) - \nu^2\right),$$

where

$$W_\nu(x) = \frac{1}{x^2}\left(\psi_\nu(x)^2 - \nu^2\right),$$

and we also have

$$\psi_\nu(x) = x\Phi_\nu(x) - \nu = x\frac{\mathcal{I}_\nu'(x)}{\mathcal{I}_\nu(x)}.$$

We refer to [**17**] for further details.

A qualitative analysis of the solution of the ODE for $W_\nu(x)$, together with the behavior of the solutions as $x \to 0$ and $\to +\infty$, provides information on the monotonicity of the ratios as well as trigonometric bounds for these simple or double ratios. In particular, for the first and second kind Bessel functions, the following result was proved in [**17**].

THEOREM 4.7. *For $x > 0$ and $\nu \geqslant 0$ the following holds:*

$$\begin{aligned}
&\frac{I_{\nu-1}(x)}{I_\nu(x)} < \frac{2g_\nu(x)}{3x}\cos\left(\frac{1}{3}\arccos\left(\frac{h_\nu(x)}{g_\nu(x)^3}\right)\right) + \frac{\nu - 1/3}{x},\\
&\frac{K_{\nu-1}(x)}{K_\nu(x)} < \frac{2g_\nu(x)}{3x}\cos\left(\frac{1}{3}\arccos\left(\frac{h_\nu(x)}{g_\nu(x)^3}\right) - \frac{\pi}{3}\right) - \frac{\nu - 1/3}{x},
\end{aligned} \tag{4.6}$$

where $g_\nu(x) = \sqrt{3(\nu^2 + x^2) + 1}$, $h_\nu(x) = 9\nu^2 - \frac{9}{2}x^2 - 1$.

The bound for the first kind function has accuracy $(3,2)$, and the accuracy for the second kind function is $(2,2)$ when $\nu > 1$. It is again possible to use the recurrence to improve the accuracy at $x = 0$.

We notice that from the bounds for ratios of modified Bessel functions, it is possible to derive other types of bounds, like for instance bounds for products of Bessel functions or bounds on Turánians. With respect to the product, we mention that it is easy to prove that (see, for example, [**17**, section 2.1])

$$\frac{I_{\nu-1}(x)}{I_\nu(x)} + \frac{K_{\nu-1}(x)}{K_\nu(x)} = \frac{1}{xI_\nu(x)K_\nu(x)},$$

and then, as a consequence, the following.

COROLLARY 4.8.

$$I_\nu(x)K_\nu(x) > \frac{\sqrt{3}}{2g_\nu(x)\sin\left(\dfrac{1}{3}\arccos\left(\dfrac{h_\nu(x)}{g_\nu(x)^3}\right)+\dfrac{\pi}{3}\right)} > \frac{1}{2\sqrt{x^2+\nu^2+\dfrac{1}{3}}},$$

where $g_\nu(x) = \sqrt{3(\nu^2+x^2)+1}$, $h_\nu(x) = 9\nu^2 - \frac{9}{2}x^2 - 1$.

We notice that in [**17**] it was conjectured that

$$I_\nu(x)K_\nu(x) > \frac{1}{2\sqrt{x^2+\nu^2+\dfrac{1}{5}}}.$$

We do not discuss possible applications of these or the other bounds for bounding Turánians. Bounds for Turánians are given, for example, in [**3, 14**].

5. Confluent hypergeometric functions

Modified Bessel functions and parabolic functions are particular cases of confluent hypergeometric functions. It is therefore natural to analyze whether similar techniques are applicable to more general cases, depending on more than one parameter. We start with confluent hypergeometric functions, and in the last section we consider the Gauss hypergeometric case. The goal is not to be exhaustive with the analysis, as done in the previous examples, but to illustrate that similar techniques are also fruitful in more general cases.

For confluent hypergeometric functions, a first example of application of the technique based on Riccati equation is given in the Appendix of [**16**]; similar ideas were later considered in [**13**]. Related results can also be obtained by an analysis of the log-concavity and log-convexity of series, as described in [**5, 6, 21**].

Confluent hypergeometric functions are the solutions of the ODE

$$xy''(x) + (b-x)y'(x) - ay(x) = 0. \tag{5.1}$$

We consider the regular solution at the origin, that is, the Kummer confluent hypergeometric function

$$M(a,b,x) = \sum_{k=0}^{\infty} \frac{(a)_n}{(b)_n n!} x^n. \tag{5.2}$$

In our analysis, we prefer to consider an alternative normalization:

$$m(a,b,x) = \frac{\Gamma(a)}{\Gamma(b)} M(a,b,x).$$

With this, we have that $m(a,b,x)$ satisfies the difference–differential relation

$$m'(a,b,x)=m(a+1,b+1,x), \tag{5.3}$$

which, considering the differential equation (5.1), leads to the recurrence relation

$$xm(a+2,b+2,x)+(b-x)m(a+1,b+1,x)-am(a,b,x)=0. \tag{5.4}$$

Denoting

$$h(a,b,x)=\frac{m'(a,b,x)}{m(a,b,x)}=\frac{m(a+1,b+1,x)}{m(a,b,x)},$$

taking the derivative and using the differential equation (5.1) we get

$$h'(a,b,x)=\frac{m''(a,b,x)}{m(a,b,x)}-h(a,b,x)^2=\frac{a}{x}+\left(1-\frac{b}{x}\right)h(a,b,x)-h(a,b,x)^2. \tag{5.5}$$

The characteristic roots of this Riccati equation, λ-solutions of $x\lambda^2+(b-x)\lambda-a=0$, are

$$\lambda_{\pm}(a,b,x)=\frac{x-b\pm\sqrt{(x-b)^2+4ax}}{2x}.$$

The relevant root will be the positive one, which from now on we denote by $\lambda(a,b,x)$. This root is increasing if $b>a$, decreasing if $a>b$ and constant if $a=b$. In addition, we see that as $x\to 0$

$$\lambda(a,b,x)=\frac{a}{b}\left[1+\frac{b-a}{b^2}x+\mathcal{O}(x^2)\right]. \tag{5.6}$$

On the other hand we have that

$$h(a,b,x)=\frac{m(a+1,b+1,x)}{m(a,b,x)}=\frac{a}{b}\left[1+\frac{b-a}{b(b+1)}x+\frac{(b-a)(b-2a)}{b^2(b+1)(b+2)}x^2+\dots\right], \tag{5.7}$$

and therefore $h(a,b,0^+)>0$ and $h'(a,b,0^+)$ has the sign of $b-a$ (the same monotonicity as $\lambda(a,b,x)$ close to $x=0$).

The information on the monotonicity of $\lambda(a,b,x)$ and the sign of $h(a,b,0^+)$ and $h'(a,b,0^+)$ is enough to prove the following result, which is a direct consequence of Theorem 2.3 and which was described earlier in [**16**, Thm. 3].

THEOREM 5.1. *Let us assume that $a,b>0$. Then, $h(a,b,x)=m(a+1,b+1,x)/m(a,b,x)$ is monotonic as a function of $x>0$, $h(a,b,x)$ is strictly increasing if $b>a$, constant if $b=a$ and strictly decreasing if $b<a$. The following inequalities hold for $x>0$:*

(1) *$h(a,b,x)<\lambda(a,b,x)$ if $b>a$.*
(2) *$h(a,b,x)=\lambda(a,b,x)=a/b$ if $b=a$.*
(3) *$h(a,b,x)>\lambda(a,b,x)$ if $b<a$.*

As in previous examples, the recurrence relation can be used to obtain further bounds. We write the recurrence (5.4) as

$$h(a,b,x)=\frac{a}{b-x+xh(a+1,b+1,x)},$$

which is equivalent to applying the recurrence in the backward direction.

We denote

$$\begin{aligned}\tilde{\lambda}(a,b,x) &= \frac{a}{b-x+x\lambda(a+1,b+1,x)}\\ &= \frac{2a}{b-x-1+\sqrt{(x-b-1)^2+4(a+1)x}},\end{aligned}$$

which is positive for $a,b,x>0$.

THEOREM 5.2. *Let $a,b,x>0$, then*

$$\tilde{\lambda}(a,b,x) < h(a,b,x) < \lambda(a,b,x) \quad \textit{if } b>a.$$

The inequalities are reversed if $b<a$, and they become equalities if $a=b$.

Let us now write Theorem 5.2 in terms of the Kummer function.

THEOREM 5.3. *Let $a,b,x>0$. The Kummer function satisfies the inequalities*

$$b-x+\sqrt{(b-x)^2+4ax} < 2b\frac{M(a,b,x)}{M(a+1,b+1,x)} < b-x-1+\sqrt{(x-b-1)^2+4(a+1)x}$$

if $b>a$, and the inequalities are reversed if $b<a$. The inequalities turn to equalities if $a=b$.

We can, as before, measure the accuracy of the bounds by checking how many terms coincide in the expansions at $x=0$ and $x=+\infty$

Starting with the upper bound $\lambda(a,b,x)$, comparing (5.6) with (5.7) we see that the first term in the expansions at $x=0$ coincide. On the other hand, as $x\to+\infty$,

$$\lambda(a,b,x) = 1-\frac{b-a}{x}+\frac{a(b-a)}{x^2}+\mathcal{O}(x^{-3}), \tag{5.8}$$

and the first two terms coincide with the expansion of $h(a,b,x)$, which is

$$h(a,b,x) = 1-\frac{b-a}{x}+\frac{(a-1)(b-a)}{x^2}+\mathcal{O}(x^{-3}). \tag{5.9}$$

With this, we can say that the accuracy of the upper bound is $(1,2)$.

With respect to the accuracy of the lower bounds $\tilde{\lambda}(a,b,x)$ for $h(a,b,x)$, considering that

$$\tilde{\lambda}(a,b,x) = \frac{a}{b}+\frac{a(b-a)}{b^2(b+1)}x+\mathcal{O}(x^2)$$

and comparing with (5.7) we see that the first two terms coincide (the third term is not shown but it is different). On the other hand,

$$\tilde{\lambda}(a,b,x) = 1+\frac{a(a-b+1)-b}{ax}+\mathcal{O}(x^{-2})$$

and the first term coincides with (5.9). Therefore the accuracy for the lower bound $\tilde{\lambda}(a,b,x)$ is $(2,1)$.

5.1. Further bounds. As happened for the case of modified Bessel functions, the bounds that are obtained by the use of the Riccati equation and the application of the recurrence relation are of type $(1,2)$ and $(2,1)$. It is natural to ask if, as in the case of Bessel functions, we can obtain uniparametric bounds which continuously connect the $(1,2)$ to $(3,0)$ cases and the $(2,1)$ to the $(0,3)$. We will not answer this question here, but we advance one result in this direction, which is the obtention of a bound of the type $(0,3)$.

A candidate for such $(0,3)$ bound is $\lambda(a-1,b-1,x)$ because considering (5.8) and (5.9), we indeed observe that the first three terms coincide. Considering an additional term in the expansion, we see that as $x \to +\infty$

$$h(a,b,x) - \lambda(a-1,b-1,x) = \frac{(a-1)(b-a)}{x^3} + \mathcal{O}(x^{-4}) \tag{5.10}$$

if $a > 1$, and as $x \to 0$

$$h(a,b,x) - \lambda(a-1,b-1,x) = \frac{b-a}{(b-1)b} + \mathcal{O}(x) \tag{5.11}$$

if $b > 1$.

THEOREM 5.4. *p Let $x > 0$ and $a, b > 1$, then $B^{(0,3)}(x) \equiv \lambda(a-1,b-1,x) < h(a,b,x)$ if $a < b$; the inequality is reversed if $a > b$ and becomes an equality if $a = b$.*

PROOF. We exclude the trivial case $a = b$.
We define

$$\delta(x) = q(x) - \phi(x),$$

$$q(x) = 1/B^{(0,3)}(x) = \frac{b-1-x+\sqrt{(x-b+1)^2+4(a-1)x}}{2(a-1)},$$

$$\phi(x) = 1/h(a,b,x),$$

where $\phi(x)$ satisfies

$$\phi'(x) = 1 - \left(1 - \frac{b}{x}\right)\phi(x) - \frac{a}{x}\phi(x)^2.$$

Considering (5.10) and (5.11) we know that $\delta(0^+)$ has the same sign as $b-a$, which is the sign of $\delta(+\infty)$ only if $a > 1$. Therefore, we can only have a bound if $a > 1$. Next we prove that $\lambda(a-1,b-1,x)$ is always a bound in that case provided $b > 1$.

Now, for proving the result we apply Theorem 2.3. Because the sign of $\delta(0^+)$ is the same as the sign of $b-a$, all that remains to be proved is that $(b-a)\Delta(x) > 0$ for $x > 0$, with

$$\Delta(x) = q'(x) - 1 + \left(1 - \frac{b}{x}\right)q(x) + \frac{a}{x}q^2(x).$$

After some algebra, one can verify that, as a function of a, $\Delta(x)$ only vanishes at $a = b$ and that

$$\left.\frac{\partial\Delta}{\partial a}\right|_{a=b} = -\frac{b-1}{x(b-1+x)} < 0 \quad \text{if } b > 1.$$

Therefore $\Delta(x)$ has the same sign as $b-a$, which ends the proof. □

5.2. Bounds for other ratios of contiguous functions. So far, we have considered bounds for the ratios $m(a+1,b+1,x)/m(a,b,x)$, but we could also consider other ratios like, for instance, $m(a+1,b,x)/m(a,b,x)$ or $m(a+1,b+2,x)/m(a,b,x)$, this last case being related to modified Bessel functions, as we will see. The different cases can be related through recurrence relations.

For example, the relation [**10**, 13.3.4] can be written

$$m(a+1,b,x) - am(a,b,x) - xm(a+1,b+1,x) = 0, \tag{5.12}$$

and therefore

$$\frac{m(a+1,b,x)}{m(a,b,x)} = a + x\frac{m(a+1,b+1,x)}{m(a,b,x)}.$$

With this the bounds we have obtained so far translate easily to bounds for the ratios $m(a+1,b,x)/m(a,b,x)$ (related with the results of [**16**, Thm. 4]). The accuracy of the bounds is maintained because there are no subtractions. This is not always the case, and we will illustrate this with the recurrence related to the modified Bessel functions.

We consider now the relation [**10**, 13.3.4], which we write in the form

$$m(a+1,b,x) - (b+x)m(a+1,b+1,x) + x(b-a)m(a+1,b+2,x) = 0. \tag{5.13}$$

Combining Eqs. (5.12) and (5.13) to eliminate $m(a+1,b,x)$ we get

$$am(a,b,x) = bm(a+1,b+1,x) + x(a-b)m(a+1,b+2,x) \tag{5.14}$$

from which we have

$$\frac{m(a+1,b+2,x)}{m(a,b,x)} = \frac{1}{x(a-b)}\left(a - b\frac{m(a+1,b+1,x)}{m(a,b,x)}\right). \tag{5.15}$$

Using the bounds of Theorem 5.2 on the right-hand side of this equation we get bounds for $H(a,b,x) = \dfrac{m(a+1,b+2,x)}{m(a,b,x)}$. Unlike the case of Theorem 5.2 we do not get that the inequalities are reversed when going from the case $a < b$ to the case $a > b$ for the M function (notice the $a-b$ in the denominator of (5.15)).

We start with the upper bound for $h(a,b,x)$, $\lambda(a,b,x)$, and write

$$\lambda(a,b,x) = 1 + f, \quad f = \frac{\sqrt{(x-b)^2+4ax} - x - b}{2x},$$

which we can also write, denoting $\delta = a - b$

$$f = \frac{\sqrt{(x+b)^2+4\delta x} - x - b}{2x} = \frac{2\delta}{x+b+\sqrt{(x-b)^2+4ax}}.$$

With this we get the upper bound for $H(a,b,x)$:

$$\eta(a,b,x) \equiv \frac{1}{x(a-b)}\left(a - b\lambda(a,b,x)\right) = \frac{1}{x}\left(1 - \frac{2b}{x+b+\sqrt{(x-b)^2+4ax}}\right).$$

Proceeding similarly with the lower bound $\tilde{\lambda}(a,b,x)$:

$$\tilde{\lambda}(a,b,x) = \frac{2(b+\delta)}{2b+\Delta},$$

where

$$\Delta = \sqrt{(x+b+1)^2+4\delta x} - (x+b+1) = \frac{4\delta x}{x+b+1+\sqrt{(x+b+1)^2+4\delta x}}.$$

Now we can write

$$\tilde{\lambda}(a,b,x) = 1 + g,\ g = \delta\frac{1-p}{b+\delta p},\ p = \frac{2x}{x+b+1+\sqrt{(x+b+1)^2+4(a-b)x}},$$

and we get the lower bound for $H(a,b,x)$

$$\tilde{\eta}(a,b,x) = \frac{1}{x(a-b)}\left(a - b\tilde{\lambda}(a,b,x)\right) = \frac{ap}{x(ap+b(1-p))}.$$

Then we have the following result.

THEOREM 5.5. *For $a, b, x > 0$ the following holds:*

$$\tilde{\eta}(a,b,x) < \frac{m(a+1,b+2,x)}{m(a,b,x)} < \eta(a,b,x).$$

It is instructive to compare now these bounds with those obtained earlier for modified Bessel functions [**19**] and summarized earlier in this paper.

Considering the relation [**10**, 13.6.9]

$$M(\nu+1/2, 2\nu+1, 2z) = \Gamma(1+\nu)e^z(z/2)^{-\nu}I_\nu(z)$$

we have

$$\frac{I_\nu(z)}{I_{\nu-1}(z)} = 2z\frac{m(a+1,b+2,2z)}{m(a,b,2z)}, \quad a = \nu - 1/2, \quad b = 2\nu - 1,$$

and a straightforward computation shows that the upper bound $\eta(a,b,x)$ in this case corresponds to the bound $(0,2)$ in [**19**, Table 3.1].

With respect to the lower bound $\tilde{\eta}$, we obtain a bound of type $(1,1)$ which is not in [**19**, Table 3.1], namely

$$\frac{I_{\nu-1}(x)}{I_\nu(x)} < \frac{\nu + \sqrt{x^2+\nu^2+x}}{x}.$$

The bound $(1,1)$ of Table 3.1 of [**19**] is clearly better, because it does not have the last sumand inside the square root.

We observe that we started with bounds with accuracies $(1,2)$ and $(2,1)$ for bounding $m(a+1,b+1,x)/m(a,b,x)$ and we have ended with $(0,2)$ and $(1,1)$ bounds for $m(a+1,b+2,x)/m(a,b,x)$. We conclude that for the particular case of the recurrence (5.15), the relation with the case we have studied earlier is not convenient because a cancellation appears which reduces the accuracy at $x = 0$. For this case, and surely for others, an independent analysis is convenient.

6. Gauss hypergeometric function

We finally provide some new bounds for the ratios of Gauss hypergeometric functions, and discuss their relation with the bounds we have described so far for the confluent hypergeometric case.

As a previous result on bounds of ratios of Gauss hypergeometric functions, we can mention [**7**, example 3], where bounds for the ratio ${}_2F_1(a+1,b;c+1;x)/{}_2F_1(a,b;c;x)$ were established. We expect that the analysis based on the qualitative analysis of the Riccati equations can also be used to obtain those results, but we choose as an illustration of the Riccati methods the case of the bounds for the ratio ${}_2F_1(a+1,b+1;c+1;x)/{}_2F_1(a,b;c;x)$ and leave for a later analysis the cases of other ratios of Gauss hypergeometric functions.

We define $y(a,b,c,x) = \dfrac{\Gamma(a)\Gamma(b)}{\Gamma(c)}\,{}_2\mathrm{F}_1(a,b;c;x)$, which satisfies the Gauss differential equation

$$x(1-x)y''(a,b,c,x) + [c-(a+b+1)x]\,y'(a,b,c,x) - aby(a,b,c,x) = 0, \tag{6.1}$$

and the difference–differential relation

$$y'(a,b,c,x) = y(a+1,b+1,c+1).$$

Combining both we have the recurrence relation

$$x(1-x)y(a+2,b+2,c+2,x)+[c-(a+b+1)x]\,y(a+1,b+1,c+1,x) \\ -aby(a,b,c,x)=0.$$

Consider now the ratio

$$h(a,b,c,x)=y(a+1,b+1,c+1,x)/y(a,b,x)=y'(a,b,x)/y(a,b,x),$$

we can write the recurrence relation as

$$h(a,b,c,x)=\frac{ab}{c-(a+b+1)x+x(1-x)h(a+1,b+1,c+1,x)}. \tag{6.2}$$

For brevity, except when needed, we drop the parameters a,b,c from the notation of $h(a,b,c,x)$.

Differentiating and using (6.1) we have

$$h'(x)=-\frac{1}{x(1-x)}\left[x(1-x)h(x)^2+[c-(a+b+1)x]\,h(x)-ab\right].$$

We consider the positive root of the characteristic equation

$$\lambda(x)=\frac{(a+b+1)x-c+\sqrt{((a+b+1)x-c)^2+4abx(1-x)}}{2x(1-x)}.$$

As $x\to 0$ we have

$$h(x)=\frac{ab}{c}\left(1+\frac{c(a+b+1)-ab}{c(c+1)}x+\mathcal{O}(x^2)\right). \tag{6.3}$$

On the other hand, for $c>0$,

$$\lambda(x)=\frac{ab}{c}\left(1+\frac{c(a+b+1)-ab}{c^2}x+\mathcal{O}(x^2)\right). \tag{6.4}$$

With this and the following lemma we will have the basic ingredients for obtaining a first bound for the ratio $h(x)$.

LEMMA 6.1. *If $a,b,c>0$ with $c>ab/(a+b+1)$ then $\lambda(x)$ is increasing in $[0,1)$.*

PROOF. We write

$$\lambda(x)=\frac{2ab}{\phi(x)},\quad \phi(x)=c-(a+b+1)x+\sqrt{((a+b+1)x-c)^2+4abx(1-x)}$$

and we prove that $\phi(x)$ is monotonically decreasing in $[0,1)$ if $a,b,c>0$ with $c>ab/(a+b+1)$.

We have $\phi'(0)=2\left(\frac{ab}{c}-(a+b+1)\right)$ and then $\phi'(0)<0$ if $c>ab/(a+b+1)$. In addition, $\phi(0)=2c$ (recall that $c>0$), while

$$\phi(1)=\begin{cases}0, & a+b+1-c\geqslant 0\\ 2(c-a-b-1), & a+b+1-c<0.\end{cases}$$

Therefore, $\phi(0)>\phi(1)\geqslant 0$ if a, b, $c>0$.

On the other hand

$$\phi''(x)=\frac{4ab((a+b+1)c-ab-c^2)}{(((a+b+1)x-c)^2+4abx(1-x))^{3/2}},$$

and we observe that $\phi''(x)$ does not change sign in $[0,1)$ and it has the same sign as $f(c)=(a+b+1)c-ab-c^2$. We observe that the quadratic function $f(c)$ is such that $f(\pm\infty)=-\infty$, and has a maximum at $c=c_m=(a+b+1)/2>0$ where $f(c_m)=\frac{1}{4}\left((a-b)^2+2(a+b)+1\right)>0$; we observe that $f(a+b+1)=-ab<0$ and thefore $f(c)<0$ (and $\phi''(x)<0$ in $[0,1)$) if $c\geqslant a+b+1$.

If $\phi''(x)<0$ then necessarily $\phi'(x)<0$ in $(0,1)$ because $\phi'(0)<0$ and $\phi'(x)$ would decrease in $(0,1)$. This is the situation when $c\geqslant(a+b+1)$, because we have proved that $\phi''(x)<0$ in that case. If $c<a+b+1$ the same would be true provided $\phi''(x)<0$.

In the cases $c<a+b+1$ for which $\phi''(x)>0$, it still holds that $\phi'(x)<0$ in $[0,1)$ under the hypothesis of the theorem, as we prove now. We have $\phi'(1)<0$ and $\phi'(x)$ is increasing in $[0,1)$. Then, if there existed $x_0\in(0,1)$ such that $\phi'(x_0)=0$ (and $\phi(x_0)>0$ because $\phi(x)$ is positive in $(0,1)$) we would have $\phi'(x)>0$ in $(x_0,1)$ because $\phi'(x)$ is increasing; but this implies that $0<\phi(x_0)<\phi(1)$, in contradiction with the fact that $\phi(1)=0$. □

THEOREM 6.2. *Let $a,b,c>0$, $c>ab/(a+b+1)$ then $h(x)<\lambda(x)$ for all $x\in(0,1)$ and $h(x)$ is monotonically increasing in $(0,1)$.*

PROOF. From (6.3) we see that $h(0^+)>0$, $h'(0^+)>0$. Considering also the monotonicity of $\lambda(x)$, the result is an immediate consequence of Theorem 2.1. □

The recurrence relation can be used, similarly as was done before in other cases, to obtain further bounds. In particular, applying the recurrence (6.2) to the bound of Theorem 6.2, we obtain an additional bound. We give those two bounds in terms of the Gauss hypergeometric function in the next theorem.

THEOREM 6.3. *Suppose $a,b,c>0$, $c>ab/(a+b+1)$, and denote*

$$H(x)=2c\frac{{}_2F_1(a,b;c;x)}{{}_2F_1(a+1,b+1;c+1;x)},$$

$F(x)=4x(1-x)$ and $d=a+b+1$. The following bounds hold:

$$H(x)>c-dx+\sqrt{(dx-c)^2+abF(x)},$$

$$H(x)<c-1-(d-2)x+\sqrt{((d+2)x-(c+1))^2+(a+1)(b+1)F(x)}.$$

The validity of the upper bound can be extended to $c>\dfrac{ab-2}{a+b+3}$.

We observe that in the confluent limit we recover the bounds described before for the ratio of confluent hypergeometric functions. In the confluent limit we make the replacement $x\to x/b$ and take the limit $b\to\infty$. After this, in order to make the connection with the notation for the confluent case (we did not use the parameter c) we rename c as b. This means that in the previous theorem $2c\,{}_2F_1(a,b;c;x)/{}_2F_1(a+1,b+1;c+1;x)$ would be replaced by $2b\,M(a,b,x)/M(a+1,b+1,x)$ and in the bounds we must consider the replacements $c\to b$, $(d+m)x\to x$, $(b+m)F(x)\to 4x$, m being any constant value. With this, we recover Theorem 5.3 for the case $b>a$.

The case $a<b$ of Theorem 5.3, however, appears to be disconnected from Theorem 6.3.

6.1. Future work. There are many possibilities for exploring additional bounds for the confluent and Gauss hypergeometric functions. To begin with, and comparing with the best studied case (modified Bessel functions), there are a number of voids that need to be filled in order to have a result similar to that of

Table 1, where no gaps in the description of bounds exist and there exist uniparametric sets of bounds connecting the best upper and lower bounds. Also, it is not known whether other types of bounds, like those obtained from the iteration of the Riccati equation (as shown in the Bessel case) or from other type of differential equations (as in the sections on parabolic cylinder functions and Bessel functions) are feasible.

References

[1] D. E. Amos, *Computation of modified Bessel functions and their ratios*, Math. Comp. **28** (1974), 239–251, DOI 10.2307/2005830. MR333287

[2] Á. Baricz, *On a product of modified Bessel functions*, Proc. Amer. Math. Soc. **137** (2009), no. 1, 189–193, DOI 10.1090/S0002-9939-08-09571-3. MR2439440

[3] Á. Baricz, *Bounds for Turánians of modified Bessel functions*, Expo. Math. **33** (2015), no. 2, 223–251, DOI 10.1016/j.exmath.2014.07.001. MR3342624

[4] K. Hornik and B. Grün, *Amos-type bounds for modified Bessel function ratios*, J. Math. Anal. Appl. **408** (2013), no. 1, 91–101, DOI 10.1016/j.jmaa.2013.05.070. MR3079949

[5] S. I. Kalmykov and D. B. Karp, *Log-concavity for series in reciprocal gamma functions and applications*, Integral Transforms Spec. Funct. **24** (2013), no. 11, 859–872, DOI 10.1080/10652469.2013.764874. MR3171999

[6] S. I. Kalmykov and D. B. Karp, *Log-convexity and log-concavity for series in gamma ratios and applications*, J. Math. Anal. Appl. **406** (2013), no. 2, 400–418, DOI 10.1016/j.jmaa.2013.04.061. MR3062548

[7] S. I. Kalmykov and D. B. Karp, *Logarithmic concavity of series in gamma ratios*, Russian Math. (Iz. VUZ) **58** (2014), no. 6, 63–68, DOI 10.3103/S1066369X14060073. Translation of Izv. Vyssh. Uchebn. Zaved. Mat. **2014**, no. 6, 70–77. MR3298931

[8] T. Koch, *Universal bounds and monotonicity properties of ratios of Hermite and parabolic cylinder functions*, Proc. Amer. Math. Soc. **148** (2020), no. 5, 2149–2155, DOI 10.1090/proc/14896. MR4078099

[9] A. Laforgia and P. Natalini, *Some inequalities for modified Bessel functions*, J. Inequal. Appl., posted on 2010, Art. ID 253035, 10, DOI 10.1155/2010/253035. MR2592860

[10] A. B. Olde Daalhuis, *Confluent hypergeometric functions*, NIST handbook of mathematical functions, U.S. Dept. Commerce, Washington, DC, 2010, pp. 321–349. MR2655353

[11] F. W. J. Olver and L. C. Maximon, *Bessel functions*, NIST handbook of mathematical functions, U.S. Dept. Commerce, Washington, DC, 2010, pp. 215–286. MR2655350

[12] D. Ruiz-Antolín and J. Segura, *A new type of sharp bounds for ratios of modified Bessel functions*, J. Math. Anal. Appl. **443** (2016), no. 2, 1232–1246, DOI 10.1016/j.jmaa.2016.06.011. MR3514344

[13] L. Sablica and K. Hornik, *On bounds for Kummer's function ratio*, Math. Comp. **91** (2022), no. 334, 887–907, DOI 10.1090/mcom/3690. MR4379980

[14] J. Segura, *Bounds for ratios of modified Bessel functions and associated Turán-type inequalities*, J. Math. Anal. Appl. **374** (2011), no. 2, 516–528, DOI 10.1016/j.jmaa.2010.09.030. MR2729238

[15] J. Segura, *On bounds for solutions of monotonic first order difference-differential systems*, J. Inequal. Appl., 2012, 2012:65, 17, DOI 10.1186/1029-242X-2012-65. MR2915630

[16] J. Segura, *Sharp bounds for cumulative distribution functions*, J. Math. Anal. Appl. **436** (2016), no. 2, 748–763, DOI 10.1016/j.jmaa.2015.12.024. MR3446977

[17] J. Segura, *Monotonicity properties for ratios and products of modified Bessel functions and sharp trigonometric bounds*, Results Math. **76** (2021), no. 4, Paper No. 221, 22, DOI 10.1007/s00025-021-01531-1. MR4328474

[18] J. Segura, *Uniform (very) sharp bounds for ratios of parabolic cylinder functions*, Stud. Appl. Math. **147** (2021), no. 2, 816–833, DOI 10.1111/sapm.12401. MR4313183

[19] J. Segura, *Simple bounds with best possible accuracy for ratios of modified Bessel functions*, J. Math. Anal. Appl. **526** (2023), no. 1, Paper No. 127211, 26, DOI 10.1016/j.jmaa.2023.127211. MR4562060

[20] H. C. Simpson and S. J. Spector, *Some monotonicity results for ratios of modified Bessel functions*, Quart. Appl. Math. **42** (1984), no. 1, 95–98, DOI 10.1090/qam/736509. MR736509

[21] S. Sra and D. Karp, *The multivariate Watson distribution: maximum-likelihood estimation and other aspects*, J. Multivariate Anal. **114** (2013), 256–269, DOI 10.1016/j.jmva.2012.08.010. MR2993885
[22] N. M. Temme, *Parabolic cylinder functions*, NIST handbook of mathematical functions, U.S. Dept. Commerce, Washington, DC, 2010, pp. 303–319. MR2655352
[23] Z.-H. Yang and S.-Z. Zheng, *Monotonicity of the ratio of modified Bessel functions of the first kind with applications*, J. Inequal. Appl., 2018, Paper No. 57, 21, DOI 10.1186/s13660-018-1648-4. MR3773091
[24] L. Yuan and J. D. Kalbfleisch, *On the Bessel distribution and related problems*, Ann. Inst. Statist. Math. **52** (2000), no. 3, 438–447, DOI 10.1023/A:1004152916478. MR1794244

Departamento de Matemáticas, Estadística y Computación. Universidad de Cantabria. 39005-Santander.

Email address: javier.segura@unican.es

Contemporary Mathematics
Volume **818**, 2025
https://doi.org/10.1090/conm/818/16369

Expansion of the Ferrers function of the first kind in a trigonometric series

Hans Volkmer

ABSTRACT. This note derives a trigonometric expansion of the Ferrers functions $\mathsf{P}_\nu^\mu(x)$ of the first kind. Conditions ensuring convergence of the expansion are discussed.

Consider the Ferrers function Q_ν^μ of the second kind [**5**, (14.13.2)], also known as the associated Legendre function of the second kind on the cut. It is an analytic function defined on the domain $D_1 := \mathbb{C} \setminus ((-\infty, -1] \cup [1, \infty))$. In [**1**, §6] we obtained the following representation of Q_ν^μ in terms of two hypergeometric functions ${}_2F_1\left({a,b \atop c}; w\right)$ with the same set of parameters a, b, c but different arguments w. Let $x \in D_1$ and $\nu, \mu \in \mathbb{C}$ such that $\nu + \mu \notin -\mathbb{N}$. Then

(1)

$$\mathsf{Q}_\nu^\mu(x) = \sqrt{\pi}\, 2^{\mu-1}(1-x^2)^{\frac{1}{2}\mu} \frac{\Gamma(\nu+\mu+1)}{\Gamma(\nu+\frac{3}{2})} \times \left(u^{\nu+\mu+1} {}_2F_1\left({\mu+\frac{1}{2}, \nu+\mu+1 \atop \nu+\frac{3}{2}}; \frac{u}{v}\right) + v^{\nu+\mu+1} {}_2F_1\left({\mu+\frac{1}{2}, \nu+\mu+1 \atop \nu+\frac{3}{2}}; \frac{v}{u}\right)\right),$$

where $u = x + i\sqrt{1-x^2}$, $v = x - i\sqrt{1-x^2} = \frac{1}{u}$. All powers are defined by their principal values. If we set $x = \cos\theta$, $\Re\theta \in (0, \pi)$, then (1) gives us a representation of $\mathsf{Q}_\nu^\mu(\cos\theta)$ with $\frac{u}{v} = e^{2i\theta}$, $\frac{v}{u} = e^{-2i\theta}$. In the most interesting case $\theta \in (0, \pi)$ we discussed the convergence properties of the resulting trigonometric expansion of $\mathsf{Q}_\nu^\mu(\cos\theta)$.

In this note we obtain analogous results for the Ferrers function P_ν^μ of the first kind. We start with a known representation of the associated Legendre function $P_\nu^\mu(z)$ of the first kind in terms of two hypergeometric functions with different sets of parameters a, b, c but with the same argument ζ. This representation [**2**, p. 129, Entry 27], [**3**, p. 158, Entry 14] is given by

$$P_\nu^\mu(z) = A_1\, {}_2F_1\left({a_1, b_1 \atop c_1}; \zeta\right) + A_2\, {}_2F_1\left({a_2, b_2 \atop c_2}; \zeta\right), \tag{2}$$

2020 *Mathematics Subject Classification.* Primary 33C05, 42A10.
Key words and phrases. Ferrers function of the first kind, trigonometric expansion.

where

$$A_1 = \frac{2^\mu}{\sqrt{\pi}} \frac{\Gamma(-\nu-\frac{1}{2})}{\Gamma(-\nu-\mu)}(z^2-1)^{\frac{1}{2}\mu}(z+(z^2-1)^{\frac{1}{2}})^{-\nu-\mu-1},$$
$$A_2 = \frac{2^\mu}{\sqrt{\pi}} \frac{\Gamma(\nu+\frac{1}{2})}{\Gamma(\nu-\mu+1)}(z^2-1)^{\frac{1}{2}\mu}(z+(z^2-1)^{\frac{1}{2}})^{\nu-\mu},$$

$$a_1 = \mu+\tfrac{1}{2},\ b_1 = \nu+\mu+1,\ c_1 = \nu+\tfrac{3}{2}, \quad a_2 = \mu+\tfrac{1}{2},\ b_2 = \mu-\nu,\ c_2 = \tfrac{1}{2}-\nu,$$

and

$$\zeta = \frac{z-(z^2-1)^{\frac{1}{2}}}{z+(z^2-1)^{\frac{1}{2}}}. \tag{3}$$

The transformation (3) can be traced back to [**4**]. The identity (2) holds for $z > 1$ and is then analytically continued to the domain $D_2 := \mathbb{C} \setminus (-\infty, 1]$. In order to obtain the right-hand side of (2) as an analytic function of $z \in D_2$, we write the power $(z^2-1)^\alpha$ for any $\alpha \in \mathbb{C}$ in the form

$$(z^2-1)^\alpha = (z-1)^\alpha (z+1)^\alpha$$

and use the principal value of the powers $(z-1)^\alpha$ and $(z+1)^\alpha$. If $z \in D_2$ then $|\zeta| < 1$, so that the hypergeometric functions in (2) are given by hypergeometric power series.

We now derive a representation of P_ν^μ which is quite similar to the representation (1).

THEOREM 1. *Let $x \in D_1$ and $\nu, \mu \in \mathbb{C}$ such that $\nu + \mu \notin -\mathbb{N}$. Then*

(4)
$$\mathsf{P}_\nu^\mu(x) = \frac{2^\mu}{i\sqrt{\pi}}(1-x^2)^{\frac{1}{2}\mu}\frac{\Gamma(\nu+\mu+1)}{\Gamma(\nu+\frac{3}{2})} \times \left(u^{\nu+\mu+1}{}_2F_1\left({\mu+\frac{1}{2}, \nu+\mu+1 \atop \nu+\frac{3}{2}} ; \frac{u}{v}\right) - v^{\nu+\mu+1}{}_2F_1\left({\mu+\frac{1}{2}, \nu+\mu+1 \atop \nu+\frac{3}{2}} ; \frac{v}{u}\right)\right),$$

where $u = x + i\sqrt{1-x^2}$, $v = x - i\sqrt{1-x^2} = \frac{1}{u}$.

PROOF. Both sides of (4) are analytic functions on D_1, so it is sufficient to prove it for $x \in (-1,1)$. If $x \in (-1,1)$ we use (2) and $\mathsf{P}_\nu^\mu(x) = e^{i\frac{1}{2}\pi\mu}P_\nu^\mu(x+i0)$ [**3**, p. 166] to obtain

(5)
$$\mathsf{P}_\nu^\mu(x) = \frac{2^\mu}{\sqrt{\pi}}(1-x^2)^{\frac{1}{2}\mu}\frac{\Gamma(\nu+\mu+1)}{\Gamma(\nu+\frac{3}{2})}\left(B_1 v^{\nu+\mu+1}{}_2F_1\left({a_1, b_1 \atop c_1}; \frac{v}{u}\right) + B_2 u^{\nu-\mu}{}_2F_1\left({a_2, b_2 \atop c_2}; \frac{v}{u}\right)\right),$$

where

$$B_1 = e^{i\pi\mu}\frac{\Gamma(-\nu-\frac{1}{2})\Gamma(\nu+\frac{3}{2})}{\Gamma(-\nu-\mu)\Gamma(\nu+\mu+1)} = e^{i\pi\mu}\frac{\sin(\nu+\mu)\pi}{\cos\nu\pi},$$
$$B_2 = e^{i\pi\mu}\frac{\Gamma(\nu+\frac{1}{2})\Gamma(\nu+\frac{3}{2})}{\Gamma(\nu-\mu+1)\Gamma(\nu+\mu+1)}.$$

We replace the second hypergeometric function in (5) by [**5**, (15.8.2)]

$${}_2F_1\left({a_2, b_2 \atop c_2}; \frac{v}{u}\right) = C_1 u^{2\mu+1}{}_2F_1\left({a_1, b_1 \atop c_1}; \frac{u}{v}\right) + C_2 u^{-2\nu-1}{}_2F_1\left({a_1, b_1 \atop c_1}; \frac{v}{u}\right),$$

where

$$C_1 = \frac{1}{i} e^{-i\pi\mu} \frac{\Gamma(\nu-\mu+1)\Gamma(\nu+\mu+1)}{\Gamma(\nu+\frac{1}{2})\Gamma(\nu+\frac{3}{2})},$$

$$C_2 = \frac{1}{i} e^{i\pi\nu} \frac{\Gamma(\nu-\mu+1)\Gamma(\nu+\mu+1)\Gamma(-\nu-\frac{1}{2})}{\Gamma(\nu+\frac{1}{2})\Gamma(\mu+\frac{1}{2})\Gamma(\frac{1}{2}-\mu)}.$$

Now we can verify that

$$B_2 C_1 = \frac{1}{i}, \qquad B_1 + B_2 C_2 = -\frac{1}{i},$$

so the desired identity (4) follows. □

If we set $x = \cos\theta$, $\Re\theta \in (0,\pi)$, then Theorem 1 gives

(6)

$$\mathsf{P}_\nu^\mu(\cos\theta) = \frac{2^\mu}{i\sqrt{\pi}} (\sin\theta)^\mu \frac{\Gamma(\nu+\mu+1)}{\Gamma(\nu+\frac{3}{2})} \times \left(e^{i(\nu+\mu+1)\theta} {}_2F_1\left({\mu+\frac{1}{2}, \nu+\mu+1 \atop \nu+\frac{3}{2}} ; e^{2i\theta} \right) - e^{-i(\nu+\mu+1)\theta} {}_2F_1\left({\mu+\frac{1}{2}, \nu+\mu+1 \atop \nu+\frac{3}{2}} ; e^{-2i\theta} \right) \right).$$

If $\theta \in (0,\pi)$ then the arguments $w = e^{\pm 2i\theta}$ of the hypergeometric function lie on the unit circle $|w| = 1$. Provided the hypergeometric series converges at $e^{\pm 2i\theta}$ we obtain [**5**, (14.13.1)]

$$(7)\quad \mathsf{P}_\nu^\mu(\cos\theta) = \frac{2^{\mu+1}}{\sqrt{\pi}} (\sin\theta)^\mu \sum_{k=0}^\infty \frac{\Gamma(\nu+\mu+k+1)}{\Gamma(\nu+k+\frac{3}{2})} \frac{(\mu+\frac{1}{2})_k}{k!} \sin((\nu+\mu+2k+1)\theta).$$

Regarding the convergence of the series in (7) we have the following result. The proofs are similar to those presented in [**1**, § 6]. We include them in order to make this note self-contained.

THEOREM 2. *Let $\theta \in (0,\pi)$, $\nu, \mu \in \mathbb{C}$ such that $\nu + \mu \notin -\mathbb{N}$.*

(a) *If $\Re\mu < 0$ then the series in (7) converges absolutely.*

(b) *If $0 \leqslant \Re\mu < \frac{1}{2}$ then the series in (7) converges, but, if $\theta \neq \frac{1}{2}\pi$, it does not converge absolutely.*

(c) *If $\Re\mu \geqslant \frac{1}{2}$ and $\theta \neq \frac{1}{2}\pi$, then the series in (7) diverges.*

PROOF. (a) It is known [**5**, §15.2(i)] that the Gauss hypergeometric series ${}_2F_1\left({a,b \atop c}; w\right)$ converges absolutely on the unit circle $|w| = 1$ if $\Re(c-a-b) > 0$. In our case $a = \mu + \frac{1}{2}$, $b = \nu+\mu+1$, $c = \nu + \frac{3}{2}$ so $c - a - b = -2\mu$. If $\Re\mu < 0$ it follows that the series in (7) is the sum of two absolutely convergent series and so is itself absolutely convergent.

(b) Suppose that $0 \leqslant \Re\mu < \frac{1}{2}$. If $-1 < \Re(c-a-b) \leqslant 0$ then the Gauss hypergeometric series converges conditionally at $|w| = 1$, $w \neq 1$ [**5**, §15.2(i)]. It follows that the series in (7) is the sum of two convergent series and so is itself convergent. However, in general, it is not true that the sum of two conditionally convergent series is conditionally convergent. We still have to show that the series in (7) does not converge absolutely if $\theta \neq \frac{1}{2}\pi$. According to [**5**, (5.11.12)],

$$\frac{\Gamma(a+z)}{\Gamma(b+z)} \sim z^{a-b}$$

as $z \to +\infty$. Therefore, as $k \to \infty$,

$$\frac{\Gamma(\nu+\mu+k+1)}{\Gamma(\nu+k+\frac{3}{2})}\frac{(\mu+\frac{1}{2})_k}{k!}$$
$$= \frac{\Gamma(\nu+\mu+k+1)}{\Gamma(\nu+k+\frac{3}{2})}\frac{\Gamma(\mu+\frac{1}{2}+k)}{\Gamma(k+1)\Gamma(\mu+\frac{1}{2})} \sim \frac{k^{\mu-\frac{1}{2}}k^{\mu-\frac{1}{2}}}{\Gamma(\mu+\frac{1}{2})} = \frac{k^{2\mu-1}}{\Gamma(\mu+\frac{1}{2})}.$$

Since $1/\Gamma(\mu+\frac{1}{2}) \neq 0$, there are positive constants κ and K such that

$$\left|\frac{\Gamma(\nu+\mu+k+1)}{\Gamma(\nu+k+\frac{3}{2})}\frac{(\mu+\frac{1}{2})_k}{k!}\right| \geqslant \frac{\kappa}{k}$$

for $k \geqslant K$. The second part of statement (b) now follows from Lemma 3.

(c) Suppose that $\Re\mu \geqslant \frac{1}{2}$ and the series in (7) converges. Then the terms of the series must converge to 0. It follows that $\sin((\nu+\mu+2k+1)\theta) \to 0$ as $k \to \infty$. By Lemma 4 this is impossible unless $\theta = \frac{1}{2}\pi$. Therefore, the series in (7) will diverge for $\theta \neq \frac{1}{2}\pi$. □

Lemma 3. *Let $a \in \mathbb{C}$ and $\theta \in (0,\pi)$. Then*

$$\sum_{k=1}^{\infty} \frac{1}{k}|\sin((a+2k)\theta)| = \infty$$

unless $\sin a = 0$ *and* $\theta = \frac{1}{2}\pi$.

Proof. If $\theta = \frac{1}{2}\pi$ then $|\sin((a+2k)\theta)|$ is independent of k and the assertion follows. Now suppose that $\theta \neq \frac{1}{2}\pi$. Since $|\sin z| \geqslant |\sin(\Re z)|$ for all $z \in \mathbb{C}$, it is enough to consider $a \in \mathbb{R}$. Then $|\sin((a+2k)\theta)| \geqslant \sin^2((a+2k)\theta)$. Therefore, it is enough to show that

$$\sum_{k=1}^{\infty} \frac{1}{k}\sin^2((a+2k)\theta) = \infty.$$

Now

$$\frac{1}{k}\sin^2((a+2k)\theta) = \frac{1}{2k} - \frac{1}{2k}\cos(2(a+2k)\theta).$$

The series $\sum_{k=1}^{\infty} 1/(2k)$ diverges. The series $\sum_{k=1}^{\infty}\cos(2(a+2k)\theta)/(2k)$ converges by the Dirichlet test provided the partial sums $\sum_{k=1}^{n}\cos(2(a+2k)\theta)$ form a bounded sequence. This is true for $\theta \in (0,\pi)$, $\theta \neq \frac{\pi}{2}$. □

Lemma 4. *Let $a, b \in \mathbb{C}$ and $\sin b \neq 0$. Then the sequence $\sin(a+bn)$ does not converge to 0 as $n \to \infty$.*

Proof. Suppose that $\sin(a+bn) \to 0$ as $n \to \infty$. We have

$$\sin(a+b(n+1)) = \cos(a+bn)\sin b + \sin(a+bn)\cos b.$$

If we let $n \to \infty$ and use $\sin b \neq 0$ we obtain that $\cos(a+bn) \to 0$. Since $\cos^2 x + \sin^2 x = 1$ this is a contradiction. □

Acknowledgment

The author thanks Howard Cohl for encouraging this work, and two referees whose suggestions led to several improvements of the paper.

References

[1] H. S. Cohl, J. Park, and H. Volkmer, *Gauss hypergeometric representations of the Ferrers function of the second kind*, SIGMA Symmetry Integrability Geom. Methods Appl. **17** (2021), Paper No. 053, 33, DOI 10.3842/SIGMA.2021.053. MR4261851

[2] H. van Haeringen and L. P. Kok, *Table errata: Higher transcendental functions, Vol. II [McGraw-Hill, New York, 1953; MR* **15**, *419] by A. Erdélyi, W. Magnus, F. Oberhettinger and F. G. Tricomi*, Math. Comp. **41** (1983), no. 164, 778. MR717722

[3] W. Magnus, F. Oberhettinger, and R. P. Soni, *Formulas and theorems for the special functions of mathematical physics*, Third enlarged edition, Die Grundlehren der mathematischen Wissenschaften, Band 52, Springer-Verlag New York, Inc., New York, 1966. MR232968

[4] R. Olbricht, Studien ueber die Kugel- und Cylinder-funktionen (Halle, 1887), Nova Acta Acad. Leop. LII (1888), 1–48.

[5] F. W. J. Olver, D. W. Lozier, R. F. Boisvert, and C. W. Clark (eds.), *NIST handbook of mathematical functions*, U.S. Department of Commerce, National Institute of Standards and Technology, Washington, DC; Cambridge University Press, Cambridge, 2010. MR2723248

Dept. of Mathematical Sciences, University of Wisconsin - Milwaukee
Email address: volkmer@uwm.edu

Contemporary Mathematics
Volume **818**, 2025
https://doi.org/10.1090/conm/818/16370

An infinite product of incomplete Beta function-type hypergeometric functions, and its probabilistic origins

N. S. Witte

ABSTRACT. Recently it has been shown that the α-Sun density $h(x)$ [*J. Math. Anal. Appl.* **527** (2023), 127371], which interpolates between the Fréchet density and that of the positive, stable distributions whose density is given by a Fox H-function, has a Mellin transform involving an infinite product of ratios of incomplete Beta functions. We develop systematic, but asymptotic, approximations for such products and consequently for the behaviour of the density as $x \to 0^+$, which complement the recent exact form for this by Simon [*Electron. Commun. Probab.* **28** (2023), 1–13.] The systematic expansion is an example of a Power Product Expansion, and in our case we derive bounds and estimates which show that it is not convergent and thus is only an asymptotic expansion.

1. The α-Sun density

In [**32**] the analytic properties of a density function $h(x; \alpha, \gamma)$, $x \in (0, \infty)$, $\gamma > 0$, $0 < \alpha < 1$, which arises from the domain of attraction problem for a statistic interpolating between the supremum and sum of random variables, were investigated. The parameter α controls the interpolation between these two cases, while γ parametrises the type of extreme value distribution from which the underlying random variables are drawn. For $\alpha = 0$, $\gamma > 0$, the density reduces to the Fréchet density, whereas for $\alpha = 1$, $0 < \gamma < 1$, it is a particular Fox H-function appropriate to positive, stable distributions with index γ. However, for intermediate α an entirely new function appears, which is not one of the extensions to the hypergeometric function considered to date.

In their study of this model—the α-Sun process—Greenwood and Hooghiemstra [**19**, Theorem 2, Eq. (2.4)] have shown that this density $h(x; \alpha, \gamma)$ satisfies the linear, homogeneous integral equation

$$h(x) = \frac{\gamma}{x} \int_0^x du \, \frac{h(u)}{(x - \alpha u)^\gamma}, \tag{1.1}$$

subject to the conditions $\alpha \in (0, 1)$, $\gamma \in (0, \infty)$, identifying the domain of attraction as above, with $h(x)$ a real, normalised probability density on $x \in (0, \infty)$. It was observed that the α-Sun process, for all α in $(0, 1)$, behaves in a way that is much more like the supremum of the input variables than their sum. Furthermore there is non-trivial limiting behaviour as $\alpha \to 1^-$.

2020 *Mathematics Subject Classification.* Primary 60E07, 60G70, 30A80, 33.20, 33A35.

The central result of [**32**], Eq. (3.41) in Prop. 3.9, gives for $h(x)$ a Mellin–Barnes integral representation, which is a standard way of defining many special functions, such as the Meijer-G and Fox-H functions. Under the conditions $0 \leqslant \alpha < 1$, $\gamma > 0$, $0 < x < \infty$, the density $h(x;\alpha,\gamma)$ for any $c < 1$ has the representation

(1.2)
$$h(x) = \frac{\gamma}{2\pi i x}\int_{c-i\infty}^{c+i\infty} dt\, x^{-\gamma t}\Gamma(1-t)(1-\alpha)^{-\gamma t}\prod_{j=1}^{\infty}\frac{{}_2F_1(\gamma,(j-t)\gamma;1+(j-t)\gamma;\alpha)}{{}_2F_1(\gamma,j\gamma;1+j\gamma;\alpha)},$$

where ${}_2F_1(a,b;c;z)$ is the Gauss hypergeometric function [**27**, §15]. In the general case the Mellin transform of the density $h(x)$ is defined by

$$H(s) := \int_0^\infty dx\, x^{s-1}h(x) \tag{1.3}$$

for the vertical strip in the complex s-plane, $1-\gamma < \Re(s) < 1+\gamma$. Prop. 3.1 of [**32**] states that for $\Re(s) < 1+\gamma$ the Mellin transform $H(s;\alpha,\gamma)$ satisfies the linear, homogeneous functional equation

$$H(s) = \frac{\gamma}{1+\gamma-s}\,{}_2F_1(\gamma,1+\gamma-s;2+\gamma-s;\alpha)H(s-\gamma). \tag{1.4}$$

Firstly with regard to the analytical character of ${}_2F_1(\gamma,1+\gamma-s;2+\gamma-s;\alpha)$ with respect to s we can make the following immediate observations: For $0 < |\alpha| < 1$ the ${}_2F_1$ in (1.4) is finite for all $\Re(s) < 1+\gamma$, whereas it has simple poles on the right-hand side of the complex s-plane at $s = 1+\gamma+k$, $k \in \mathbb{Z}_{\geqslant 0}$. For $|\alpha| = 1$ we require in addition that $\Re(\gamma) < 1$, and the poles are the same as for $|\alpha| < 1$. This ${}_2F_1$ function is one of a pair of solutions to the hypergeometric differential equation about $\alpha = 0$ that is bounded as $s \to -\infty$ (by $(1-\alpha)^{-\gamma}$). This distinguishes itself from the other member of the pair, which has the simple algebraic form $\alpha^{s-\gamma-1}$, and which diverges as $s \to -\infty$. This hypergeometric function can be linearly mapped into other forms via the Kummer or linear fractional transformations of the α-plane, and we record one version [**27**, 15.8.1] of these:

$${}_2F_1(\gamma,1+\gamma-s;2+\gamma-s;\alpha) = (1-\alpha)^{-\gamma}{}_2F_1(\gamma,1;2+\gamma-s;\tfrac{\alpha}{\alpha-1}).$$

A useful alternative integral representation of the ${}_2F_1$ is

$${}_2F_1(\gamma,1;2+\gamma-s;\tfrac{\alpha}{\alpha-1}) = 1-\alpha\gamma(1-\alpha)^{\gamma}\int_0^1 dt\, t^{1+\gamma-s}(1-\alpha t)^{-\gamma-1}. \tag{1.5}$$

The hypergeometric function can be identified in a number of ways. In one instance it is an incomplete Beta function (see [**27**, 8.17.7]),

$${}_2F_1(\gamma,1+\gamma-s;2+\gamma-s;\alpha) = (1+\gamma-s)\alpha^{s-\gamma-1}B_\alpha(1+\gamma-s,1-\gamma).$$

In particular a sequence of these ${}_2F_1$ will feature prominently.

Definition 1.1. Let $F_j := {}_2F_1(\gamma,j\gamma;1+j\gamma;\alpha)$ for $j \in \mathbb{N}_0$. The series expansion in α also generates a partial fraction form with respect to j which will be useful in the sequel,

$$F_j = \sum_{l=0}^{\infty}\alpha^l\frac{(\gamma)_l}{l!}\frac{j}{j+l\gamma^{-1}}. \tag{1.6}$$

Some basic properties of this sequence are the following: Let $0 < \alpha < 1$ and $\gamma > 0$. The hypergeometric factors F_j satisfy $F_0 = 1$, are monotonically increasing, $F_{j+1} > F_j$, and are bounded above, $F_j < (1-\alpha)^{-\gamma}$.

Simon [**29**, Thm. 2] has recently given the asymptotic form of the density as $x \to 0^+$,

$$h(x;\alpha,\gamma) \underset{x\to 0^+}{\sim} c(\alpha,\gamma)x^{-1-\gamma(1-\alpha)^{-1}}e^{-(1-\alpha)^{-\gamma}x^{-\gamma}}, \tag{1.7}$$

where the constant factor is

$$c = \gamma(1-\alpha)^{\frac{\gamma}{\alpha-1}}\exp\left(\frac{\alpha\psi(1)}{\alpha-1}\right)\prod_{k=1}^{\infty}\frac{\exp\left(\frac{\alpha}{(\alpha-1)k}\right)}{{}_2F_1\left(1,\gamma;1+k\gamma;\frac{\alpha}{\alpha-1}\right)}. \tag{1.8}$$

The modus operandi of our study is to develop alternative representations for the infinite products of hypergeometric functions, ideally as identities or alternatively as approximations in any sense, especially those which transparently exhibit the behaviour of the density as $\alpha \to 1^-$ and for all γ. These are the regions where the magnitudes of the products grow or shrink the fastest, and have an interesting interpretation for the underlying probability models. The infinite products are very slowly converging and are completely opaque regarding even the gross behaviour of the product, in these parameter regions. Simon's constant in the regime $x \to 0^+$ is no more helpful in this regard, although it is a significant step to know this result. In keeping with this aim we develop a product formula for the F_j factors themselves using a recursive formula in §2. In the following section §3 we situate our product formula within a class of generating functions known as power product expansions and generalised Lambert series, recalling a number of known results for recursive and convolution identities for relations between these objects in a general setting. Using these relations we derive bounds and estimates in §4 for our particular application, and demonstrate the asymptotic character of our product formula due to the super-exponential growth of the coefficients. Our final section §5 employs our product formula for the F_j in the product over j, switches the order, and employs an identity for products of rational roots-of-unity factors in terms of Gamma functions, to arrive at an asymptotic approximation for the density, which is given in Proposition 5.1. Some simple Gamma function product formulae are given for the first two successive approximations to the constant factor in (5.4) and (5.5), and these are numerically compared to Simon's exact formula.

2. The product formula

From Prop. 3.5 of [**32**] the consecutive product of hypergeometric functions F_k has the leading-order behaviour

$$\prod_{k=1}^{N}{}_2F_1(\gamma,k\gamma;1+k\gamma;\alpha) \underset{N\to\infty}{\sim} C(1-\alpha)^{-N\gamma}\left[N+1+\frac{1}{\gamma}\right]^{-\alpha/(1-\alpha)}, \tag{2.1}$$

where C is a constant independent of N. Another useful function turns out to be

$$G(t) := \frac{1}{\Gamma\left(\dfrac{1+\gamma-s}{\gamma}\right)}H(s), \qquad t := \frac{1-s}{\gamma}, \tag{2.2}$$

which eliminates the simple poles at $s = 1+k\gamma$, $k \in \mathbb{N}$, so that in the finite t-plane $|G(t)| \leqslant (1-\alpha)^{-\gamma}\left[\cosh(\pi\Im(t))\right]^{1/2}$. At the prescribed values $s = 1-k\gamma$, $k \in \mathbb{Z}_{\geqslant 0}$, we deduce

$$G_k := G(t=k) = \frac{1}{\prod_{l=1}^{k} {}_2F_1(\gamma, l\gamma; 1+l\gamma; \alpha)}, \tag{2.3}$$

which has the growth property given above.

Furthermore the finite product G_k defined by (2.3) was shown to be [**32**, Lemma 3.3] a function of complex k analytic in the half-plane $\Re(k) > 1+\gamma^{-1}$, having the infinite product form

$$G_k = (1-\alpha)^{\gamma k} \prod_{j=1}^{\infty} \frac{F_{j+k}}{F_j}, \tag{2.4}$$

and an interpolating function

$$G(-t) = (1-\alpha)^{-\gamma t} \prod_{j=1}^{\infty} \frac{F_{j-t}}{F_j}. \tag{2.5}$$

Our new result is really a large-j expansion for F_j, but not in the usual asymptotic sense of Poincaré with a sum of descending monomials; rather in the sense of a Padé approximant with only denominator factors.

Proposition 2.1. *Let $0 \leqslant \alpha < 1$, $\gamma > 0$. There is a sequence $\{f_m(\alpha,\gamma)\}_{m=1}^{\infty}$ such that for each $m \in \mathbb{N}$,*

$$\begin{aligned} &(1-\alpha)^{\gamma}\left(1+\frac{1}{j}f_1\right)\left(1+\frac{1}{j^2}f_2\right)\cdots\left(1+\frac{1}{j^m}f_m\right)F_j \\ &\qquad = 1 + \mathrm{O}(j^{-m-1}) \qquad \text{as } j \to \infty. \end{aligned} \tag{2.6}$$

Proof. The values of f_m with $m = 1,2,3,\ldots$ can be generated by hand or otherwise, by expanding successive products with F_j as a series expansion in large j; see (1.6). For example, to begin with one observes that the leading term $\mathrm{O}(j^{-1})$ is

$$\log\left[(1-\alpha)^{\gamma}{}_2F_1(\gamma, j\gamma; j\gamma+1; \alpha)\right] = -\frac{\alpha}{1-\alpha}j^{-1} + \mathrm{O}(j^{-2})$$

and that to eliminate this requires multiplication by an additional factor $(1+\frac{\alpha}{1-\alpha}j^{-1})$. Re-expanding one observes a new leading term of order $\mathrm{O}(j^{-2})$, i.e.,

$$\log\left[(1-\alpha)^{\gamma}\left(1+\frac{\alpha}{(1-\alpha)}j^{-1}\right){}_2F_1(\gamma, j\gamma; j\gamma+1; \alpha)\right] = \frac{\alpha}{\gamma(1-\alpha)^2}j^{-2} + \mathrm{O}(j^{-3}).$$

And so on. We will now show that this holds in general for all $m > 1$ by induction. We assume its veracity for m, i.e.,

$$\begin{aligned} &(1-\alpha)^{\gamma}\left(1+\frac{1}{j}f_1\right)\left(1+\frac{1}{j^2}f_2\right)\cdots\left(1+\frac{1}{j^m}f_m\right)F_j \\ &\qquad = 1 + \mathrm{O}(j^{-m-1}) = 1 + C_{m+1}j^{-m-1} + \mathrm{O}(j^{-m-2}), \end{aligned}$$

where the sub-leading terms contain decreasing powers of j which drop by one unit each term. It should be observed that C_{m+1} has no dependence on j, and by algebraic construction from terms arising from the expansion of the products on the left-hand side it can only have contributions where the sum of non-zero indices add up to $m+1$, that is to say, it can only have dependence on $f_1, \ldots, f_m$. This follows from the fact that F_j, by (1.6), has an analytic expansion about $j = \infty$. Now we compute

$$(1-\alpha)^{\gamma}\left(1+\frac{1}{j}f_1\right)\left(1+\frac{1}{j^2}f_2\right)\cdots\left(1+\frac{1}{j^m}f_m\right)\left(1+\frac{1}{j^{m+1}}f_{m+1}\right)F_j$$
$$= 1 + [f_{m+1} + C_{m+1}]\, j^{-m-1} + \mathrm{O}(j^{-m-2}).$$

As before, we choose $f_{m+1} = -C_{m+1}$, and the leading correction is now $\mathrm{O}(j^{-m-2})$, thus proving our claim. □

REMARK 2.1. We will find subsequently in Proposition 4.7 that the product (2.6) does not converge as $m \to \infty$, because the growth of f_m is super-exponential (see (4.23)), so the representation is asymptotic.

There is a parallel expansion for large γ, but in each factor both the numerator and denominator contain j, and this has disadvantages for our subsequent applications.

REMARK 2.2. The combination $(1-\alpha)^m f_m$ is a polynomial in α whose leading and trailing coefficients in respect of α are given by

(2.7)
$$(1-\alpha)^m\gamma^{m-1}f_m = (-1)^{m-1}\alpha\Big[\left((1+\gamma)^{m-1}-\gamma^{m-1}\right)\alpha^{m-2}+\dots$$
$$+\left(\tfrac{1}{2}(2^m-2m)+\tfrac{1}{4}(2^{m+1}-2m-3-(-1)^m)\gamma\right)\alpha+1\Big].$$

The first six f_n are

$$(1-\alpha)f_1 = \alpha, \tag{2.8}$$

$$(1-\alpha)^2\gamma f_2 = -\alpha, \tag{2.9}$$

$$(1-\alpha)^3\gamma^2 f_3 = \alpha[\alpha(2\gamma+1)+1], \tag{2.10}$$

$$(1-\alpha)^4\gamma^3 f_4 = -\alpha\left[\alpha^2\left(3\gamma^2+3\gamma+1\right)+\alpha(5\gamma+4)+1\right], \tag{2.11}$$

and

$$(1-\alpha)^5\gamma^4 f_5 = \alpha\left[\alpha^3\left(4\gamma^3+6\gamma^2+4\gamma+1\right)+\alpha^2\left(16\gamma^2+25\gamma+11\right)+\alpha(13\gamma+11)+1\right], \tag{2.12}$$

$$(1-\alpha)^6\gamma^5 f_6 = -\alpha\left[\alpha^4\left(5\gamma^4+10\gamma^3+10\gamma^2+5\gamma+1\right)+\alpha^3\left(33\gamma^3+83\gamma^2+78\gamma+26\right)+\alpha^2\left(65\gamma^2+129\gamma+66\right)+2\alpha(14\gamma+13)+1\right]. \tag{2.13}$$

Prior to continuing we require a preliminary identity that recasts an infinite sum as a finite sum.

LEMMA 2.1. *Let $m \in \mathbb{Z}_{\geqslant 0}$, $0 < \alpha < 1$, and $\gamma > 0$. Then*

$$\sum_{k=0}^{\infty}\alpha^k\frac{(\gamma)_k}{k!}k^m = (1-\alpha)^{-\gamma}\sum_{l=0}^{m}S(m,l)(\gamma)_l\left(\frac{\alpha}{1-\alpha}\right)^l, \tag{2.14}$$

where $S(m,i) = \left\{ {m \atop i} \right\}$ is the Stirling number of the second kind or the Stirling "partition" number in the notation of [**27**, §26.8(i)], *i.e., the number of partitions of $\{1,2,\dots,m\}$ into exactly i non-empty subsets.*

PROOF. The Stirling numbers of the second kind are the matrix elements of the transformation matrix giving the monomial basis in terms of the descending factorial basis (see [**27**, §26.8(i)]), which is expressed as

$$k^m = \sum_{l=0}^{m} S(m,l)(k-l+1)_l = \sum_{l=0}^{m} S(m,l)\frac{k!}{(k-l)!}.$$

Employing this in the left-hand side, and interchanging the summations as the k-sum is uniformly and absolutely convergent, we have

$$\begin{aligned}
\sum_{k=0}^{\infty} \alpha^k \frac{(\gamma)_k}{k!} k^m &= \sum_{l=0}^{m} S(m,l) \sum_{k=l}^{\infty} \alpha^k \frac{(\gamma)_k}{(k-l)!} \\
&= \sum_{l=0}^{m} S(m,l) \sum_{i=0}^{\infty} \alpha^{i+l} \frac{(\gamma)_{i+l}}{i!} \\
&= \sum_{l=0}^{m} S(m,l)\alpha^l (\gamma)_l \sum_{i=0}^{\infty} \alpha^i \frac{(\gamma+l)_i}{i!} && \text{using } (\gamma)_{i+l} = (\gamma)_l(\gamma+l)_i \\
&= \sum_{l=0}^{m} S(m,l)\alpha^l (\gamma)_l (1-\alpha)^{-\gamma-l} && \text{by the binomial theorem} \\
&= (1-\alpha)^{-\gamma} \sum_{l=0}^{m} S(m,l)(\gamma)_l \left(\frac{\alpha}{1-\alpha}\right)^l .
\end{aligned}$$

□

In the proof of Proposition 2.1 we indicated how successive f_n could be determined in principle, but did not give an explicit formula—our next result gives an explicit recursive means to compute the coefficients f_m.

PROPOSITION 2.2. *Let $0 \leqslant \alpha < 1$, $\gamma > 0$. The coefficients f_m, $m \geqslant 1$ satisfy the recursive formula*

(2.15)

$$\begin{aligned}
f_{m+1} &= (-1)^m \gamma^{-m-1} \sum_{i=0}^{m+1} S(m+1,i)(\gamma)_i \left(\frac{\alpha}{1-\alpha}\right)^i \\
&\quad + \sum_{n=0}^{m} (-1)^{n+1}\gamma^{-n} \sum_{i=0}^{n} S(n,i)(\gamma)_i \left(\frac{\alpha}{1-\alpha}\right)^i \times \sum_{\substack{m\geqslant l_s > \cdots > l_1 \geqslant 1 \\ \sum_{k=1}^{s} l_k = m+1-n}} f_{l_s}\cdots f_{l_1}.
\end{aligned}$$

Therefore $(1-\alpha)^{m+1}\gamma^m f_{m+1}$ is a polynomial in α, γ of degrees $m, m-1$ respectively. The initial coefficient f_1 is given by the first term of the right-hand side at $m = 0$, the second term having an empty summand.

PROOF. In order to determine f_{m+1} we only need to extract the coefficient of j^{-m-1} in

$$(1-\alpha)^{\gamma}\left(1+\frac{1}{j}f_1\right)\cdots\left(1+\frac{1}{j^m}f_m\right)F_j.$$

In addition we are going to split the sum for F_j, (1.6), using

$$\frac{j}{j+k\gamma^{-1}} = \sum_{n=0}^{m+1}(-1)^n\left(\frac{k}{j\gamma}\right)^n + (-1)^{m+2}\left(\frac{k}{j\gamma}\right)^{m+2}\frac{j}{j+k\gamma^{-1}}.$$

Thus we have

$$
\begin{aligned}
-f_{m+1} &= [j^{-m-1}](1-\alpha)^{\gamma}\sum_{k\geqslant 0}\alpha^k\frac{(\gamma)_k}{k!}\sum_{n=0}^{m+1}\left(1+\frac{1}{j}f_1\right)\cdots\left(1+\frac{1}{j^m}f_m\right)(-1)^n\left(\frac{k}{j\gamma}\right)^n \\
&= [j^{-m-1}](1-\alpha)^{\gamma}\sum_{k\geqslant 0}\alpha^k\frac{(\gamma)_k}{k!}\sum_{n=0}^{m}\left(1+\frac{1}{j}f_1\right)\cdots\left(1+\frac{1}{j^m}f_m\right)(-1)^n\left(\frac{k}{j\gamma}\right)^n \\
&\quad + (1-\alpha)^{\gamma}\sum_{k\geqslant 0}\alpha^k\frac{(\gamma)_k}{k!}(-1)^{m+1}\left(\frac{k}{\gamma}\right)^{m+1} \\
&= (-1)^{m+1}\gamma^{-m-1}(1-\alpha)^{\gamma}\sum_{k\geqslant 0}\alpha^k\frac{(\gamma)_k}{k!}k^{m+1} \\
&\quad + [j^{-m-1}](1-\alpha)^{\gamma}\sum_{k\geqslant 0}\alpha^k\frac{(\gamma)_k}{k!}\sum_{n=0}^{m}(-1)^n\left(\frac{k}{\gamma}\right)^n j^{-n}\sum_{m\geqslant l_s>\cdots>l_1\geqslant 1} j^{-l_1}f_{l_1}\cdots j^{-l_s}f_{l_s} \\
&= (-1)^{m+1}\gamma^{-m-1}(1-\alpha)^{\gamma}\sum_{k\geqslant 0}\alpha^k\frac{(\gamma)_k}{k!}k^{m+1} \\
&\quad + (1-\alpha)^{\gamma}\sum_{k\geqslant 0}\alpha^k\frac{(\gamma)_k}{k!}\sum_{n=0}^{m}(-1)^n\left(\frac{k}{\gamma}\right)^n\sum_{\substack{m\geqslant l_s>\cdots>l_1\geqslant 1\\ \sum_{i=1}^{s}l_i=m+1-n}} f_{l_1}\cdots f_{l_s}.
\end{aligned}
$$

Applying identity (2.14) to the k-sums appearing above then yields (2.15). □

Let us define a t-sequence which contains all the α, γ dependencies that occur in the recursive formula (2.15), by

$$
t_n(\alpha,\gamma) := (-1)^n\gamma^{-n}\sum_{i=0}^{n}S(n,i)(\gamma)_i\left(\frac{\alpha}{1-\alpha}\right)^i, \qquad n\geqslant 0. \tag{2.16}
$$

We note for standard conditions, i.e., $0<\alpha<1$, $\gamma>0$, and by the combinatorial interpretation of $S(n,i)$, that t_n oscillates with period 2 in n, being positive for even n. The first six t_n are: $t_{-1}=0$, $t_0=1$,

$$(1-\alpha)t_1 = -\alpha, \tag{2.17}$$

$$(1-\alpha)^2\gamma t_2 = \left[\alpha^2\gamma+\alpha\right], \tag{2.18}$$

$$(1-\alpha)^3\gamma^2 t_3 = -\left[\alpha^3\gamma^2+\alpha^2(3\gamma+1)+\alpha\right], \tag{2.19}$$

$$(1-\alpha)^4\gamma^3 t_4 = \left[\alpha^4\gamma^3+\alpha^3\left(6\gamma^2+4\gamma+1\right)+\alpha^2(7\gamma+4)+\alpha\right], \tag{2.20}$$

and

$$
\begin{aligned}
&(1-\alpha)^5\gamma^4 t_5 = \\
&\quad -\left[\alpha^5\gamma^4+\alpha^4\left(10\gamma^3+10\gamma^2+5\gamma+1\right)+\alpha^3\left(25\gamma^2+30\gamma+11\right)\right. \\
&\qquad \left.+\alpha^2(15\gamma+11)+\alpha\right],
\end{aligned} \tag{2.21}
$$

$$
\begin{aligned}
&(1-\alpha)^6\gamma^5 t_6 = \\
&\quad \left[\alpha^6\gamma^5+\alpha^5\left(15\gamma^4+20\gamma^3+15\gamma^2+6\gamma+1\right)\right. \\
&\quad +\alpha^4\left(65\gamma^3+120\gamma^2+91\gamma+26\right)+\alpha^3\left(90\gamma^2+146\gamma+66\right) \\
&\quad \left.+\alpha^2(31\gamma+26)+\alpha\right].
\end{aligned} \tag{2.22}
$$

Then the recursive formula (2.15) takes the generic form

$$f_{m+1} = -t_{m+1} - \sum_{n=0}^{m} t_n \sum_{\substack{m \geqslant l_s > \cdots > l_1 \geqslant 1 \\ \sum_{k=1}^{s} l_k = m+1-n}} f_{l_s} \cdots f_{l_1}. \tag{2.23}$$

A unified formula can be written for(2.23) by allowing for a zero index $l = 0$ on the f_l along with $f_0 := 1$, so that we have

$$\sum_{n=0}^{m+1} t_n \sum_{\substack{m+1 \geqslant l_s > \cdots > l_1 \geqslant 0 \\ n+\sum_{k=1}^{s} l_k = m+1}} f_{l_s} \cdots f_{l_1} = 0. \tag{2.24}$$

Now the t_{m+1} term corresponds to $n = m+1$ and the f_{m+1} term to $n = 0$ in the expanded inner summation. Consequently we have the expansion

$$(1-\alpha)^{\gamma} F_j = \sum_{k \geqslant 0} j^{-k} t_k(\alpha, \gamma) \tag{2.25}$$

and will defer the question of convergence or otherwise of this sum until §4.

3. Power product expansions and Lambert series

At this point it is advantageous to step back from the specific problem at hand and recognise a larger picture—what we have been doing is constructing a *power product expansion* (PPE) for a function that is given in terms of an analytic expansion about $x = 0$, i.e., in terms of an infinite product with a particular structure of its factors. For some background on PPEs one should consult [**2**], [**15**], [**16**], [**17**]. We define this via the relation

$$A(x) := 1 + \sum_{n \geqslant 1} a_n x^n = \prod_{n \geqslant 1} (1 + f_n x^n), \tag{3.1}$$

where the left-hand side encodes a sequence $\{a_n\}_{n \geqslant 1}$ and the right-hand side another sequence $\{f_n\}_{n \geqslant 1}$. These sums and products may be convergent or be purely formal as generating functions. On the right-hand side we see an obvious combinatorial interpretation in terms of partitions without repetitions, or strict partitions. In the traditional combinatorial approach there is a "simple" expression for the a-coefficient, such that for $m \geqslant 1$ one has

$$a_m = \sum_{\substack{m \geqslant l_s > \cdots > l_1 \geqslant 1 \\ \sum_{k=1}^{s} l_k = m}} f_{l_s} \cdots f_{l_1}. \tag{3.2}$$

The first six a_n in terms of the f_n are

$$a_1 = f_1, \tag{3.3}$$

$$a_2 = f_2, \tag{3.4}$$

$$a_3 = f_3 + f_2 f_1, \tag{3.5}$$

$$a_4 = f_4 + f_3 f_1, \tag{3.6}$$

$$a_5 = f_5 + f_4 f_1 + f_3 f_2, \tag{3.7}$$

$$a_6 = f_6 + f_5 f_1 + f_4 f_2 + f_3 f_2 f_1. \tag{3.8}$$

However, in contrast to many studies of additive combinatorics, we are not going to study examples where the f_n are given with simple forms and the problem is to determine the consequent properties of the a_n which have the combinatorial interpretation, but to do the reverse. Thus we will take the a_n as given in some simple explicit form and wish to construct the f_n in terms of the former, or study their properties. In this context the a_n are indeterminates, whether or not they have any explicit form. It appears that there is not an equivalent "simple" expression in this inverse situation. Inverting the relation (3.2) for the first members gives

$$f_1 = a_1, \tag{3.9}$$

$$f_2 = a_2, \tag{3.10}$$

$$f_3 = a_3 - a_2 a_1, \tag{3.11}$$

$$f_4 = a_4 - a_3 a_1 + a_2 a_1^2, \tag{3.12}$$

$$f_5 = a_5 - a_4 a_1 - a_3 a_2 + a_3 a_1^2 + a_2^2 a_1 - a_1^3 a_2, \tag{3.13}$$

$$f_6 = a_2 a_1^4 - a_3 a_1^3 - a_2^2 a_1^2 + a_4 a_1^2 + a_2 a_3 a_1 - a_5 a_1 - a_2 a_4 + a_6. \tag{3.14}$$

See Comtet [**8**], Chap. II, Supp. & Ex. 16, pp. 120–121; Eq. (10) of [**15**]; p. 1221 of [**16**]; pp. 94–95 of [**18**]. A related thread closely connected to this problem is the study of generalised Lambert series and their arithmetic properties, see [**6**], [**7**], [**10**], [**11**], [**13**]. On the applications of Möbius inversion in combinatorial analysis see [**5**]. In addition many of these studies have treated a generalised power product expansion (GPPE) [**17**] of the form

$$1 + \sum_{n\geqslant 1} a_n x^n = \prod_{n\geqslant 1} (1 + f_n x^n)^{r_n} .$$

Let us recall the generating function definition (3.1) and form the logarithmic derivative

$$x\frac{A'}{A} := \sum_{n=1}^{\infty} d_n x^n = \sum_{n=1}^{\infty} \frac{n f_n x^n}{1 + f_n x^n}, \tag{3.15}$$

which introduces the Lambert series (see Knopp [**22**, Chap. XII, §58 C]). Let us write the partition of n as a sum of multiplicities $\lambda(n) \equiv (1^{\lambda_1} 2^{\lambda_2} \dots) \vdash n = 1\,\lambda_1 + 2\,\lambda_2 + \dots + n\,\lambda_n$. (See p. 1 of [**3**] for an explanation of the notation.) Gingold and Knopfmacher [**16**, Lemma 2.1, Eq. (2.3)] deduced

$$f_n = \sum_{\lambda \vdash n} c(\lambda) a_1^{\lambda_1} \cdots a_n^{\lambda_n}$$

as the form of the f_n in terms of the a_n, without giving a general, closed formula for the coefficients $c(\lambda)$. They investigated the absolute sum of these coefficients,

$$B(n) := \sum_{\lambda(n)} |c(\lambda)|,$$

which is a sequence recorded in the OEIS [**1**] as `A220418`. They found some properties and in particular bounds for $B(n)$,

$$\frac{2^{n-1}}{n} \leqslant B(n) < \frac{2^n}{n}, \qquad n \geqslant 1,$$

along with the asymptotic growth estimate

$$B(n) \underset{n\to\infty}{=} \frac{2^n}{n} \left(1 + \mathrm{O}(n^{-1})\right).$$

3.1. t_n, a_n and d_n-coefficients. The t_n and a_n coefficients are related by the linear relation

$$-t_{m+1} = t_m a_1 + t_{m-1} a_2 + \ldots + t_1 a_m + t_0 a_{m+1}, \quad m \geqslant 0. \tag{3.16}$$

This follows from comparing (2.24) with (3.2), from which one can conclude that

$$\sum_{l=0}^{m+1} t_l a_{m+1-l} = 0,$$

taking $a_0 = 1$. The above relation can be formulated as a matrix equation with a lower triangular Toeplitz structure, whose solution for the a_n coefficients is given by

$$\begin{pmatrix} a_1 \\ a_2 \\ a_3 \\ a_4 \\ \vdots \end{pmatrix} = - \begin{pmatrix} t_0 & 0 & 0 & 0 & 0 & \cdots \\ t_1 & t_0 & 0 & 0 & 0 & \cdots \\ t_2 & t_1 & t_0 & 0 & 0 & \cdots \\ t_3 & t_2 & t_1 & t_0 & 0 & \cdots \\ \vdots & \vdots & \vdots & \vdots & \vdots & \end{pmatrix}^{-1} \begin{pmatrix} t_1 \\ t_2 \\ t_3 \\ t_4 \\ \vdots \end{pmatrix}. \tag{3.17}$$

Another formula for a_n, $n \geqslant 0$, can be found via geometrical generating functions:

$$a_n = -[x^n] \frac{\sum_{j=1}^{\infty} t_j x^j}{1 + \sum_{j=1}^{\infty} t_j x^j}. \tag{3.18}$$

PROPOSITION 3.1. *We have*

$$a_n = \sum_{\substack{m_l \geqslant 0 \\ \sum_{l \geqslant 1} l m_l = n}} (-1)^{\sum_{l \geqslant 1} m_l} \binom{\sum_{l \geqslant 1} m_l}{m_1, m_2, \ldots} \prod_{l \geqslant 1} t_l^{m_l}, \qquad n \geqslant 1, \tag{3.19}$$

or alternatively

$$A(x) = \left(1 + \sum_{l=1}^{\infty} t_l x^l \right)^{-1}. \tag{3.20}$$

PROOF. By recalling (3.1), relation (3.20) is derived directly from (3.18). Without loss of generality we can take b_0 to be unity in a series expansion of the reciprocal,

$$\left(1 + \sum_{n=1}^{\infty} b_n x^n \right)^{-1} = \sum_{n=0}^{\infty} x^n \sum_{\substack{m_l \geqslant 0 \\ \sum_{l=1} l m_l = n}} (-1)^{\sum_l m_l} \binom{\sum_{l \geqslant 1} m_l}{m_1, m_2, \ldots} \prod_{l \geqslant 1} b_l^{m_l}, \tag{3.21}$$

using the multinomial expansion (see [**8**, §3.5] and [**30**]). □

Explicit solutions for low coefficients are

$$a_1 = -t_1, \tag{3.22}$$

$$a_2 = -t_2 + t_1^2, \tag{3.23}$$

$$a_3 = -t_3 + 2t_2t_1 - t_1^3, \tag{3.24}$$

$$a_4 = -t_4 + 2t_3t_1 + t_2^2 - 3t_2t_1^2 + t_1^4, \tag{3.25}$$

$$a_5 = -t_5 + 2t_4t_1 + 2t_3t_2 - 3t_3t_1^2 - 3t_2^2t_1 + 4t_2t_1^3 - t_1^5, \tag{3.26}$$

(3.27)

$$a_6 = -t_6 + 2t_5t_1 + 2t_4t_2 - 3t_4t_1^2 + t_3^2 - 6t_3t_2t_1 + 4t_3t_1^3 - t_2^3 + 6t_2^2t_1^2 - 5t_2t_1^4 + t_1^6.$$

Note that the foregoing relations are completely reflexive, i.e., symmetrical under $t_n \leftrightarrow a_n$, as one can observe from identity (3.20).

Recalling (3.15), we can deduce a similar linear recurrence relation linking the d_n coefficients to the t_m coefficients.

PROPOSITION 3.2. *The d_n coefficients and the t_m coefficients satisfy the convolution relation*

$$-mt_m = \sum_{n=1}^{m} t_{m-n}d_n, \qquad m \geqslant 1, \tag{3.28}$$

which can be written as a recurrence relation for the d_n coefficients in the forward direction,

$$d_m = -d_{m-1}t_1 - \ldots - d_1t_{m-1} - mt_m, \quad d_0 = 0, \quad m \geqslant 1. \tag{3.29}$$

PROOF. It follows from (3.15) and (3.20) that

$$\sum_{n\geqslant 1} d_nx^n = x\frac{d}{dx}\log A(x) = -x\frac{d}{dx}\log\left(1 + \sum_{l\geqslant 1} t_lx^l\right),$$

whence (3.29). □

In addition the d_n coefficients are related to the a_n coefficients by the simple convolution relation

$$na_n = d_n + \sum_{k=1}^{n-1} d_ka_{n-k}, \qquad n \geqslant 1, \tag{3.30}$$

which follows from the definition.

3.2. f_n-coefficients. The first six f_n in terms of the t_n are:

$$f_1 = -t_1, \tag{3.31}$$

$$f_2 = t_1^2 - t_2, \tag{3.32}$$

$$f_3 = t_1t_2 - t_3, \tag{3.33}$$

$$f_4 = t_1^4 - 2t_2t_1^2 + t_3t_1 + t_2^2 - t_4, \tag{3.34}$$

$$f_5 = t_2t_1^3 - t_3t_1^2 - t_2^2t_1 + t_4t_1 + t_2t_3 - t_5, \tag{3.35}$$

$$f_6 = t_3t_1^3 + t_2^2t_1^2 - t_4t_1^2 - 3t_2t_3t_1 + t_5t_1 + t_3^2 + t_2t_4 - t_6. \tag{3.36}$$

A fundamental relation derived by many authors [**28**], [**11**], [**16**] relates the f_n and the d_n, with a number of these studies treating the more general setting of the GPPE.

PROPOSITION 3.3 ([**11**, Eq. 2.12], [**16**], [**28**]). *The f_n and the d_n satisfy the arithmetic convolution relation*

$$f_n = \frac{1}{n} d_n + \sum_{\substack{d|n \\ d>1}} \frac{1}{d} \left(-f_{n/d}\right)^d . \tag{3.37}$$

Note that (3.37) is not a Dirichlet convolution, so some of the powerful analytical number theory tools are not available here.

PROOF. The proof of this is well-known and we reproduce it here for the convenience of the reader. Starting with (3.15) one computes

$$\begin{aligned}
\frac{A'(x)}{A(x)} := \sum_{n=1}^{\infty} d_n x^{n-1} &= \sum_{n=1}^{\infty} \frac{n f_n x^{n-1}}{1+f_n x^n} \\
&= \sum_{n=1}^{\infty} n f_n x^{n-1} \sum_{m=0}^{\infty} (-1)^m f_n^m x^{mn} \\
&= \sum_{n\geqslant 1, m\geqslant 0} (-1)^m n f_n^{m+1} x^{(m+1)n-1} \\
&= -\sum_{l\geqslant 1} x^{l-1} \sum_{n|l, n\geqslant 1} n(-1)^{l/n} f_n^{l/n} \quad \text{with } l=(m+1)n \\
&= -\sum_{l\geqslant 1} x^{l-1} \sum_{d|l, d\geqslant 1} (-1)^d \frac{l}{d} f_{l/d}^d
\end{aligned}$$

and compares coefficients. Separating out the term with $d = 1$ gives (3.37). □

REMARK 3.1. Ritt [**28**], Feld [**11**] and subsequent authors have undertaken studies of the growth of the f_n coefficients and the consequent convergence of the associated power-product expansion, albeit in a more general structure, but with the assumption that $r = \limsup_{n\geqslant 1} |d_n|^{1/n}$ exists. However as we will see this does not apply here, and in our case the rate of growth is faster.

4. Bounds and growth estimates

We seek estimates and bounds on the growth of $f_m(\alpha,\gamma)$ as $m \to \infty$, for all $0 \leqslant \alpha < 1$ and bounded $\gamma > 0$. We will undertake this task in a step-by-step manner, firstly for t_n, then for d_n, and finally f_n. Let us begin by recalling the definition (2.16) of the t_n-sequence which contains all the α, γ dependencies that occur in the recursive formula (2.15), for $n \geqslant 0$.

REMARK 4.1. In the defining sum (2.16) we can write the Pochhammer symbol as a descending factorial $(-x)_i = (-1)^i x(x-1)\cdots(x-i+1)$, and employ the other defining characteristic of the second-kind Stirling numbers as the coefficients of the descending factorial expansion of monomials (see [**27**, Eq. 26.8.10]), that is

$$\sum_{i=1}^{n} S(n,i)(-1)^i(-x)_i = \sum_{i=1}^{n} S(n,i)x(x-1)\cdots(x-i+1) = x^n .$$

This allows us to deduce the limiting identity $t_n(\alpha,-\gamma) \underset{\alpha\to\infty}{\to} (-1)^n$, which is obvious from the explicit formulae (2.17)–(2.22).

The t_n coefficients possess an exponential generating function, which we will subsequently show to be convergent about $z = 0$ (see Proposition 4.4), namely

$$\sum_{n=0}^{\infty} \frac{z^n}{n!} t_n = (1-\alpha)^\gamma \left[1 - \alpha e^{-z/\gamma}\right]^{-\gamma}, \tag{4.1}$$

so that

$$t_n = (-1)^n (1-\alpha)^\gamma \gamma^{-n} \frac{d^n}{dz^n} \left[1 - \alpha e^z\right]^{-\gamma} \Big|_{z=0} \tag{4.2}$$

or alternatively

$$t_n = (-1)^n (1-\alpha)^\gamma \gamma^{-n} \frac{n!}{2\pi i} \oint_{C_0} \frac{dz}{z^{n+1}} \left[1 - \alpha e^z\right]^{-\gamma}, \tag{4.3}$$

where C_0 is a loop contour enclosing the origin and none of the branch points $z = z_m = -\log\alpha + 2\pi i m$, $m \in \mathbb{Z}$. The first result follows from the exponential generating function for the Stirling numbers of the second kind in [**27**, Eq. 26.8.12]. In addition these coefficients possess the purely formal geometrical generating function

$$\sum_{n=0}^{\infty} z^n t_n = (1-\alpha)^\gamma {}_2F_1(\gamma, \gamma z^{-1}; 1 + \gamma z^{-1}; \alpha) = (1-\alpha)^\gamma \sum_{l \geqslant 0} \alpha^l \frac{(\gamma)_l}{l!} \frac{\gamma}{\gamma + lz}, \tag{4.4}$$

which is evident from the sequence of poles at $z = -\gamma/m$, $m \in \mathbb{N}$, accumulating at $z = 0$. This latter result can be deduced from [**27**, Eq. 26.8.11], but such a deduction involves violating a technical condition known to apply here.

Next we give an identity which is the same as in Lemma 2.1, but we reverse the logic that was employed earlier.

PROPOSITION 4.1. *The t_n coefficients are given by a convergent expansion about $\alpha = 0$, namely*

$$t_n(\alpha, \gamma) = (-)^n \gamma^{-n} (1-\alpha)^\gamma \sum_{l=0}^{\infty} l^n \frac{(\gamma)_l}{l!} \alpha^l, \qquad |\alpha| < 1. \tag{4.5}$$

PROOF. We choose a circular loop centred on the origin for C_0, $z = re^{i\theta}$ with $0 < r < \log\alpha^{-1}$. Then we expand the radical factor of the integral (4.3) via the binomial theorem, arriving at

$$t_n = (-1)^n \gamma^{-n} (1-\alpha)^\gamma n! r^{-n} \sum_{l=0}^{\infty} \alpha^l \frac{(\gamma)_l}{l!} \int_{-\pi}^{\pi} \frac{d\theta}{2\pi} e^{-in\theta} \exp(lre^{i\theta}).$$

The angular integral is easily evaluated by series-expanding the exponential in the last factor in the integrand, and we find this integral is $(lr)^n/n!$. This facilitates the cancelling of all the r-dependence, and we have (4.5). The ratio test then gives us the radius of convergence, as stated. □

PROPOSITION 4.2. *The $t_n(\alpha, \gamma)$ coefficients satisfy the mixed difference equation*

$$t_n(\alpha, \gamma) = t_{n-1}(\alpha, \gamma) - \frac{1}{1-\alpha} \left(1 + \gamma^{-1}\right)^{n-1} t_{n-1}(\alpha, \gamma + 1), \tag{4.6}$$

or in terms of the re-defined coefficient $\tilde{t}_n(\alpha, \gamma) = \gamma^n t_n(\alpha, \gamma)$,

$$\tilde{t}_n(\alpha, \gamma) = \gamma \tilde{t}_{n-1}(\alpha, \gamma) - \frac{\gamma}{1-\alpha} \tilde{t}_{n-1}(\alpha, \gamma + 1). \tag{4.7}$$

PROOF. The Stirling numbers of the second kind satisfy the recurrence relation (see [**27**, Eq. 26.8.22])

$$S(n,i) = iS(n-1,i) + S(n-1,i-1), \quad 0 < i \leqslant n-1, \quad n > 1.$$

Employing this in (2.16) we have two terms on the right-hand side: in the first of these we write $i = \gamma + i - \gamma$ so that $i(\gamma)_i = (\gamma)_{i+1} - \gamma(\gamma)_i = \gamma(\gamma+1)_i - \gamma(\gamma)_i$; in the second we replace $i \mapsto i+1$ and use $(\gamma)_{i+1} = \gamma(\gamma+1)_i$ again. As a result we have three terms which are of the same form as the definition (2.16) with additional factors, and two of these coalesce. Simplifying we have (4.6). □

Central to our task will be knowledge of the growth of the t_n coefficients with respect to n, which will be obtained in two independent ways—via explicit upper and lower bounds for the magnitude of t_n, and secondly via asymptotic estimates of the value of t_n. Our bounds will in fact be sharp, and both will converge to our estimates.

One can find upper and lower bounds to the magnitude of the t_n coefficients with a very similar form.

PROPOSITION 4.3. *Let $0 < \gamma < 1$ and $0 < \alpha < 1$. The magnitudes of the t_n coefficients are bounded above through the formula*

$$|t_n| < (1-\alpha)^\gamma \frac{\gamma^{-n}}{\Gamma(\gamma)} \operatorname{Li}_{-n-\gamma+1}(\alpha), \tag{4.8}$$

where $\operatorname{Li}_s(z)$ is the polylogarithm as defined by [**27**, Eq. 25.12.10]. *The magnitudes of the t_n coefficients are bounded below through the formula*

$$|t_n| > (1-\alpha)^\gamma \frac{\gamma^{-n}}{\Gamma(\gamma)} \left[\operatorname{Li}_{-n-\gamma+1}(\alpha) - (1-\gamma)\operatorname{Li}_{-n-\gamma+2}(\alpha)\right]. \tag{4.9}$$

PROOF. Our strategy is to apply bounds for the ratio $(\gamma)_l/l!$ in formula (4.5). The simplest bound suitable for this purpose is Gautschi's inequality (see [**27**, Eq. 5.6.4] or the original [**14**]), which states, subject to $x > 0$ and $0 < s < 1$,

$$(x+1)^{s-1} < \frac{\Gamma(x+s)}{\Gamma(x+1)} < x^{s-1}.$$

For the upper bound we use

$$\frac{\Gamma(l+\gamma)}{\Gamma(l+1)} < l^{\gamma-1}$$

and the definition of the polylogarithm. For the lower bound we deduce a shifted identity derived from Gautschi's lower bound, namely

$$\frac{\Gamma(l+\gamma)}{\Gamma(l+1)} > (l+\gamma-1)l^{\gamma-2},$$

which introduces an additional, negative term. □

REMARK 4.2. For the asymptotic forms of the bounds in Proposition 4.3 we require polylogarithms with large, negative order, and the following series representation suffices for this. Let $|\mu| < 2\pi$ and $s \neq 1, 2, 3, \ldots$; then we have [**27**, Eq. 25.12.12]

$$\operatorname{Li}_s(e^\mu) = \Gamma(1-s)(-\mu)^{s-1} + \sum_{k=0}^{\infty} \frac{\zeta(s-k)}{k!}\mu^k,$$

where $\zeta(s)$ is the Riemann zeta function, analytically continued to $\Re(s) < 0$. Taking only the leading-order term of this yields both (4.10) and (4.11). If the large n asymptotics of the polylogarithm are employed then the above bounds reduce to

$$|t_n| < (1-\alpha)^\gamma \frac{(\gamma)_n}{\gamma^n} \left(\log\alpha^{-1}\right)^{-n-\gamma} \tag{4.10}$$

and

$$|t_n| > (1-\alpha)^\gamma \frac{(\gamma)_n}{\gamma^n} \left(\log\alpha^{-1}\right)^{-n-\gamma} \left[1 - \frac{(1-\gamma)}{n+\gamma-1}\log\alpha^{-1}\right]. \tag{4.11}$$

In addition to the proof of Proposition 4.3, there is an alternative approach to the upper bound.

PROPOSITION 4.4. *The magnitudes of the t_n coefficients are bounded above through the formula*

$$|t_n| \leqslant (1-\alpha)^\gamma n! \left(\gamma r_0\right)^{-n-\gamma} \left(n+\gamma r_0\right)^\gamma, \tag{4.12}$$

where r_0 is given by

$$r_0 + \frac{n}{\gamma} = W_0(\tfrac{n}{\alpha\gamma}e^{n/\gamma}) \tag{4.13}$$

with $W_0(\cdot)$ being the principal branch of the Lambert W-function [**27**, §4.13]. *If the asymptotic formula for the Lambert W_0-function is again employed in the above equation, we have*

$$\alpha e^{r_0} \underset{n\to\infty}{\sim} \frac{n}{n-\gamma\log\alpha(1-\gamma n^{-1}+\mathrm{O}(n^{-2}))}. \tag{4.14}$$

PROOF. Starting from (4.3) specialised to a circular loop of radius r, one can deduce the upper bound

$$|t_n| \leqslant (1-\alpha)^\gamma \gamma^{-n} n! \sup_{z=re^{i\theta},\, -\pi<\theta<\pi} r^{-n}\left|1-\alpha e^z\right|^{-\gamma}.$$

The right-hand side has a supremum at $\theta = 0$ of $r^{-n}(1-\alpha e^r)^{-\gamma}$. This supremum can be minimised with respect to r at $r = r_0$, given by

$$\alpha e^{r_0} = \frac{n}{n+\gamma r_0}$$

or by (4.13). Clearly $r_0 < \log\alpha^{-1}$ as $1-\alpha e^{r_0} = \alpha\gamma n^{-1} r_0 e^{r_0}$. Furthermore, due to the elementary inequality $(1+r)^{-1} > e^{-r}$ for $r > 0$, we can deduce the lower bound $n(n+\gamma)^{-1}\log\alpha^{-1} < r_0$. The large n asymptotics of r_0 follow from those of the W_0-Lambert function taken to third order[1] in [**27**, Eq. 4.13.1_1]. □

PROPOSITION 4.5. *The t_n coefficients possess the large-n asymptotic growth formula*

$$t_n \underset{n\to\infty}{\sim} (-1)^n \frac{(1-\alpha)^\gamma}{\sqrt{2\pi}} n!(n+1)^{-1/2} z_0^{-n-\gamma} \left(z_0 + \frac{n+1}{\gamma}\right)^\gamma \left(z_0 + \frac{n+1}{\gamma} + 1\right)^{-1/2}, \tag{4.15}$$

where z_0 is given by

$$z_0 + \tfrac{n+1}{\gamma} = W_0(\tfrac{n+1}{\alpha\gamma}e^{(n+1)/\gamma}) \tag{4.16}$$

[1]The author is indebted to a reviewer's insight that one has to take the expansion to third order to obtain the correct sub-leading terms. If the expansion is terminated prematurely one gets contributions such as $\log n$ terms which end up being cancelled at the next order.

with $W(\cdot)$ being the principal branch of the Lambert W-function [**27**, §4.13].

If the asymptotic formula for the Lambert W_0-function is employed for z_0 in the above equation (*see* [**27**, Eq. 4.13.1_1]), *one has*

$$t_n \underset{n\to\infty}{\sim} (-1)^n \frac{1}{\sqrt{2\pi}}(1-\alpha)^\gamma \gamma^{1/2} n!(n+1)^{\gamma-1}\left(\gamma \log \alpha^{-1}\right)^{-n-\gamma} \tag{4.17}$$

and the n-th root growth

$$|t_n|^{1/n} \underset{n\to\infty}{\sim} \frac{n}{e\gamma \log \alpha^{-1}}. \tag{4.18}$$

A variant of the above estimate which is simpler for our purposes is the formula

$$t_n \underset{n\to\infty}{\sim} (-1)^n (1-\alpha)^\gamma \frac{(\gamma)_n}{\gamma^n}\left(\log \alpha^{-1}\right)^{-n-\gamma}. \tag{4.19}$$

PROOF. We perform a steepest descent analysis of the integral (4.3). An excellent reference for this method is [**12**, Chap. 8]. The phase function is $f(z) = -(n+1)\log z - \gamma \log(1-\alpha e^z)$, with a unique critical point denoted by z_0, the solution of

$$\frac{n+1}{z_0} = \alpha\gamma \frac{e^{z_0}}{1-\alpha e^{z_0}},$$

which is solved by (4.16). Employing the second derivative

$$f_0'' = \frac{n+1}{z_0^2}\left[1 + \frac{n+1}{\gamma} + z_0\right], \tag{4.20}$$

we find the leading term to be given by (4.15). □

The growth of the d_n coefficients with respect to n is required and it turns out that this is controlled by the growth of the t_m coefficients: the last term on the right-hand side of (3.29), mt_m, dominates the other terms.

PROPOSITION 4.6. *As $m \to \infty$, d_m grows as*

$$-\frac{d_m}{mt_m} = 1 + \mathrm{O}(m^{-1}). \tag{4.21}$$

PROOF. Since $t_m \neq 0$ we can rewrite the convolution (3.29) as

$$\frac{d_m}{mt_m} = -1 - \sum_{j=1}^{m-1} \frac{j}{m}\frac{t_{m-j}t_j}{t_m}\frac{d_j}{jt_j},$$

motivating the definition $u_m := d_m/(mt_m)$. This allows us to deduce the inequality

$$|u_m| \leqslant 1 + \sum_{j=1}^{m-1} \frac{j}{m}\frac{|t_{m-j}||t_j|}{|t_m|}|u_j|.$$

This, in turn, facilitates the use of discrete analogues to the Gronwall–Bellman type inequalities ([**24**] and [**25**, Theorem 1.2.3], and given in their original formulation for the continuous setting by [**4**], [**20**], [**23**], and [**26**]). The result is the explicit inequality for the u_m coefficients

$$|u_m| \leqslant 1 + \sum_{j=1}^{m-1} \frac{j}{m}\frac{|t_{m-j}||t_j|}{|t_m|} \prod_{k=j+1}^{m-1}\left(1 + \frac{k}{m}\frac{|t_{m-k}||t_k|}{|t_m|}\right).$$

The relevant controlling aspect of the growth of u_m with $m \to \infty$ is computed using the bounds 4.8 and 4.9, to be

$$\begin{aligned}\frac{|t_{m-k}||t_k|}{|t_m|} &\leqslant (1-\alpha)^\gamma \left(\log\alpha^{-1}\right)^{-\gamma} \frac{(\gamma)_{m-k}(\gamma)_k}{(\gamma)_m}\left[1-\frac{1-\gamma}{m+\gamma-1}\log\alpha^{-1}\right]^{-1}\\ &= (1-\alpha)^\gamma \left(\log\alpha^{-1}\right)^{-\gamma}\frac{(\gamma)_k}{(m-1+\gamma)\cdots(m-k+\gamma)}\left[1+\mathrm{O}(m^{-1})\right]\\ &= \mathrm{O}(m^{-k})\end{aligned}$$

when $k=\mathrm{O}(1)$. As this ratio is symmetrical under $k \mapsto m-k$, the same conclusion applies when $k=\mathrm{O}(m)$. As is visible from the first line displayed above, the ratio $|t_{m-k}||t_k|/|t_m|$ is essentially a reciprocal binomial coefficient, and this decreases rapidly away from the $k=0,m$ endpoints to a minimum around $k=m/2$. This implies that the inner product is of order $1+\mathrm{O}(m^{-j-1})$ with $j \geqslant 1$, and consequently $|u_m| \leqslant 1+\mathrm{O}(m^{-1})$. □

From the previous result we conclude that $\limsup_{n\geqslant 1} n^{-1}|d_n|^{1/n} = \frac{1}{e\gamma\log\alpha^{-1}}$. Our final estimates concern the growth of the f_n coefficients, and we observe that this in turn is completely controlled by the growth of the d_n coefficients.

PROPOSITION 4.7. *As $n\to\infty$, f_n grows as*

$$\frac{nf_n}{d_n} = 1+\mathrm{O}(\kappa^n(\log n)^{\log 2}), \tag{4.22}$$

where $\kappa = 0.7514184\ldots$. Consequently we have

$$\lim_{n\to\infty}\frac{f_n}{t_n} = -1, \qquad \limsup_{n\geqslant 1} n^{-1}|f_n|^{1/n} = \frac{1}{e\gamma\log\alpha^{-1}}. \tag{4.23}$$

Therefore expansion (2.6) of Proposition 2.1 for our particular application is not convergent and is in fact asymptotic in the sense of large γ and small α.

PROOF. Our strategy is to adapt the method of **[11]** to our situation. Starting with (3.37) rephrased slightly differently, we deduce the inequality

$$n|f_n| \leqslant |d_n| + \sum_{\substack{p|n\\ 1\leqslant p<n}} p|f_p|^{n/p}, \tag{4.24}$$

the summation being over the proper divisors p of n. Let p_1 denote the divisor contributing the greatest term to the above sum. Then if we use $n/2$ as an upper bound for the number of proper divisors of n, we have

$$\frac{n|f_n|}{|d_n|} \leqslant 1+\tfrac{1}{2}np_1\frac{|f_{p_1}|^{n/p_1}}{|d_n|} = 1+\tfrac{1}{2}np_1^{1-n/p_1}\frac{|d_{p_1}|^{n/p_1}}{|d_n|}\left(\frac{p_1|f_{p_1}|}{|d_{p_1}|}\right)^{n/p_1}. \tag{4.25}$$

We know from Proposition 4.6 that

$$\frac{|d_{p_1}|^{n/p_1}}{|d_n|} \leqslant \frac{p_1^{n/p_1}}{n}\frac{|t_{p_1}|^{n/p_1}}{|t_n|}\left(1+\mathrm{O}(n^{-1},p_1^{-1})\right),$$

and from Proposition 4.3 that

$$\frac{|t_{p_1}|^{n/p_1}}{|t_n|} \leqslant \left[\sqrt{2\pi}\frac{e^{-\gamma}}{\Gamma(\gamma)}\left(\frac{1-\alpha}{\log\alpha^{-1}}\right)^\gamma\right]^{n/p_1-1}\left[\frac{(p_1+\gamma)^{n/p_1}}{n+\gamma}\right]^{\gamma-1/2}\left(\frac{p_1+\gamma}{n+\gamma}\right)^n.$$

If we make the simplifying definitions

$$C = \frac{\sqrt{2\pi}}{\Gamma(\gamma)}\left(\frac{1-\alpha}{\log\alpha^{-1}}\right)^{\gamma}, \qquad V_{p_1} := C^{1-p_1/n}\left[\frac{p_1+\gamma}{(n+\gamma)^{p_1/n}}\right]^{\gamma-1/2}\frac{p_1|f_{p_1}|}{|d_{p_1}|},$$

we deduce that the inequality (4.25) becomes

$$V_n \leqslant 1 + \tfrac{1}{2}p_1\left(\frac{p_1+\gamma}{n+\gamma}\right)^n V_{p_1}^{n/p_1}.$$

We will now apply this inequality recursively by constructing a sequence of proper divisors $p_0 := n, p_1, p_2, \dots, p_k, p_{k+1}, \dots$, where $p_{k+1}|p_k$, $p_{k+1} < p_k$, and p_{k+1} is the proper divisor with the largest contribution $p_{k+1}|f_{p_{k+1}}|^{p_k/p_{k+1}}$ to the second term on the right-hand side of (4.25), out of all such proper divisors. This sequence will terminate because of the strict inequality applying, all satisfy $p_k \geqslant 1$, and we denote the terminal divisor by $p_{m+1} = 1$. However the utility of this construction is that we do not need to identify these divisors because of the simple inequality $p_{k+1} \leqslant p_k/2$. Therefore starting from the top and descending, we can generate a sequence of upper bounds $p_k \leqslant 2^{-k}n$; whereas if we start from the terminating divisor and ascend, we generate a sequence of lower bounds $2^{m+1-k} \leqslant p_k$. We note that combining these bounds we deduce that $2^{m+1} \leqslant n$, and we find the upper bound $m+1 \leqslant \left\lceil \frac{\log n}{\log 2}\right\rceil$. To simplify matters further, we note that

$$\left(\frac{p_{k+1}+\gamma}{p_k+\gamma}\right)^{p_k} \leqslant e^{\gamma}\, 2^{-p_k}.$$

To see how this inequality arises take the even case ($p_k = 2l$) first:

$$\begin{aligned}
&p_{k+1}+\gamma \leqslant \tfrac{1}{2}p_k + \gamma = \tfrac{1}{2}(p_k+2\gamma)\\
\Longrightarrow\ &\frac{p_{k+1}+\gamma}{p_k+\gamma} \leqslant \tfrac{1}{2}\left[1+\gamma(p_k+\gamma)^{-1}\right] \leqslant \tfrac{1}{2}\left[1+\gamma p_k^{-1}\right]\\
\Longrightarrow\ &\left(\frac{p_{k+1}+\gamma}{p_k+\gamma}\right)^{p_k} \leqslant 2^{-p_k}\left[1+\gamma p_k^{-1}\right]^{p_k} \leqslant 2^{-p_k}e^{\gamma},
\end{aligned}$$

whereas for the odd case ($p_k = 2l+1$), $p_{k+1} \leqslant l + 1/2$ implies $p_{k+1} \leqslant l$, thus

$$\begin{aligned}
&p_{k+1}+\gamma \leqslant \tfrac{1}{2}p_k - \tfrac{1}{2} + \gamma = \tfrac{1}{2}\left(p_k+\gamma\right) + \tfrac{1}{2}(\gamma-1)\\
\Longrightarrow\ &\frac{p_{k+1}+\gamma}{p_k+\gamma} \leqslant \tfrac{1}{2}\left[1+(\gamma-1)(p_k+\gamma)^{-1}\right] \leqslant \tfrac{1}{2}\left[1+\gamma p_k^{-1}\right]
\end{aligned}$$

and the remaining steps follow the even case. Thus our final inequality becomes a nested recursion

$$V_{p_k} \leqslant 1 + \tfrac{1}{2}p_{k+1}2^{-p_k}V_{p_{k+1}}^2, \qquad k = 0, 1, \dots, m+1.$$

Iterating this we find $V_n \leqslant 1 + 2^{-g_0}x_{m+1}$, where the exponents are $g_k = k+1-m+2^{m+1-k}$ and the numerators are generated by the recurrence $x_n = 1 + x_{n-1}^2$ with $x_1 = 1$. This is the OEIS sequence [1] A003095, where x_n counts the number of binary trees of height less than or equal to n and has the asymptotic growth as $n \to \infty$ of $x_n \underset{n\to\infty}{\sim} c^{2^n}$ with $c = 1.2259024435287485386279474959130085213\dots$. In conclusion we find exponential convergence of V_n to unity, i.e.,

$$V_n \leqslant 1 + 2^{-2^{m+1}+m-1}c^{2^{m+2}} = 1 + \tfrac{1}{2}\left(\tfrac{1}{2}c^2\right)^n\left(\tfrac{1}{2}n\right)^{\log 2},$$

and (4.22) follows. □

5. Gamma function products

In our final section we employ the product expansion (2.6) as an asymptotic expansion, to derive some leading-order approximations for the α-Sun density. At this juncture it is important to clarify a distinction between the true and rigorous convergent nature of the j product in (1.2), and what we are proposing to do: each factor of F_j is replaced by an internal product of m factors via (2.6) up to a fixed cut-off, and the order of products is reversed in order to perform the j product exactly, so that one has a tractable result for the Mellin transform $H(s)$ and can compute the inverse Mellin transform explicitly. We will only carry this out for the first two orders, and compare the result with the recent related result of Simon [**29**].

To begin with we recall some classical results concerning evaluations of infinite products of degree-m monomial factors which evaluate as m-fold Gamma function products.

LEMMA 5.1. *Let the m-th root of unity be denoted by $\omega_m = e^{2\pi i/m}$. Then for $z \notin \mathbb{N}\, \omega_m^{-j}$ with $j = 0, \ldots, m-1$, $m \in \mathbb{N}$, $m \geqslant 2$, we have* [**21**]

$$\prod_{n=1}^{\infty} \frac{n^m}{n^m - z^m} = \prod_{j=0}^{m-1} \Gamma(1 - \omega_m^j z). \tag{5.1}$$

Furthermore let $|z| < 1$; then the above product can also be evaluated as [**31**]

$$\prod_{j=0}^{m-1} \Gamma(1 - \omega_m^j z) = \exp\left(\sum_{k=1}^{\infty} \frac{\zeta(mk)}{k} z^{mk} \right). \tag{5.2}$$

Our first technical result is the evaluation of infinite products resulting from the truncation of the m-expansion in (2.6).

COROLLARY 5.1. *Let $\omega_m^{(k+1/2)} = e^{2\pi i/m \cdot (k+\frac{1}{2})}$ for $m \in \mathbb{N}$, $k \in \mathbb{Z}_{\geqslant 0}$. Furthermore let $t \in \mathbb{C}$, excluding all $1 - f_m^{1/m} \omega_m^{(k+1/2)}$. Then*

$$\begin{aligned} &\prod_{j=1}^{\infty} \frac{1 + f_1 j^{-1}}{1 + f_1 (j-t)^{-1}} \cdots \frac{1 + f_m j^{-m}}{1 + f_m (j-t)^{-m}} \\ &\qquad = \frac{\Gamma(1 - t + f_1)}{\Gamma(1-t)\Gamma(1+f_1)} \cdots \prod_{k=0}^{m-1} \frac{\Gamma(1 - t - f_m^{1/m} \omega_m^{(k+1/2)})}{\Gamma(1-t)\Gamma(1 - f_m^{1/m} \omega_m^{(k+1/2)})}. \end{aligned} \tag{5.3}$$

PROOF. Using (5.1), we compute for $J \in \mathbb{N}$ and $m > 1$ the finite product

$$\begin{aligned} \prod_{j=1}^{J} \left(1 + \frac{f_m}{(j-t)^m} \right) &= \prod_{k=0}^{m-1} \frac{\Gamma(J + 1 - t - f_m^{1/m} \omega_m^{(k+1/2)})}{\Gamma(J+1-t)} \frac{\Gamma(1-t)}{\Gamma(1 - t - f_m^{1/m} \omega_m^{(k+1/2)})} \\ &\underset{J \to \infty}{\to} (J+1)^{-f_m^{1/m} \sum_{k=0}^{m-1} \omega_m^{(k+1/2)}} \prod_{k=0}^{m-1} \frac{\Gamma(1-t)}{\Gamma(1 - t - f_m^{1/m} \omega_m^{(k+1/2)})} \\ &= \prod_{k=0}^{m-1} \frac{\Gamma(1-t)}{\Gamma(1 - t - f_m^{1/m} \omega_m^{(k+1/2)})}. \end{aligned}$$

The first equality is derived in the same way as the proof of (5.1); that is, by performing a root-of-unity factorisation of each factor, which results in a m-fold product of polynomials all of which are linear in j, and for each of those linear factors the product over j can be evaluated as a ratio of Gamma functions. The final equality holds because of the sum identity $\sum_{k=0}^{m-1} \omega_m^{(k+1/2)} = 0$. For $m = 1$ one can use the fact that only ratios are actually required in $\prod_{j=1}^{J} \left(1 + \frac{f_1}{j}\right) / \left(1 + \frac{f_1}{(j-t)}\right)$, and the convergence as $J \to \infty$ follows, as this ratio is $1 + \mathrm{O}(j^{-2})$. □

Let us first write products and ratios of Gamma functions in the following concise notation: for arbitrary complex indeterminates $a_1, a_2, \ldots, b_1, b_2, \ldots$ subject to $a_k \neq -\mathbb{Z}_{\geqslant 0}$,

$$\Gamma\begin{pmatrix} a_1 & a_2 & \ldots \\ b_1 & b_2 & \ldots \end{pmatrix} = \frac{\Gamma(a_1)\Gamma(a_2)\cdots}{\Gamma(b_1)\Gamma(b_2)\cdots}.$$

A similar array notation has been employed for the parameters of the ${}_PF_Q$ hypergeometric functions.

PROPOSITION 5.1. *Let $0 \leqslant \alpha < 1$. At the first order $m = 1$ we compute the Mellin inversion and find the previously given result* [**32**, Eq. 3.61]

$$h(x) \underset{\gamma \gg 1 \gg \alpha}{\sim} \frac{\gamma (1-\alpha)^{-\gamma(1-\alpha)^{-1}}}{\Gamma(\frac{1}{1-\alpha})} x^{-1-\gamma/(1-\alpha)} e^{-(1-\alpha)^{-\gamma} x^{-\gamma}}, \qquad \textit{as } x \to 0^+. \tag{5.4}$$

At the second order we have

$$\begin{aligned} h(x) \underset{\gamma \gg 1 \gg \alpha}{\sim} & \frac{\gamma x^{-1}}{\Gamma\left(\frac{1}{1-\alpha}, 1 + \frac{\alpha^{1/2}\gamma^{-1/2}}{1-\alpha}, 1 - \frac{\alpha^{1/2}\gamma^{-1/2}}{1-\alpha}\right)} \\ & \times \Bigg\{ \Gamma\begin{pmatrix} \frac{[\alpha^{1/2}\gamma^{-1/2} - \alpha]}{1-\alpha} & -\frac{[\alpha^{1/2}\gamma^{-1/2} + \alpha]}{1-\alpha} \\ -\frac{\alpha}{1-\alpha} & -\frac{\alpha}{1-\alpha} \end{pmatrix} [(1-\alpha)x]^{-\gamma/(1-\alpha)} \\ & \qquad \times {}_2F_2\left(\begin{matrix} \frac{1}{1-\alpha}, & \frac{1}{1-\alpha} \\ \frac{[1 - \alpha^{1/2}\gamma^{-1/2}]}{1-\alpha}, & \frac{[1 + \alpha^{1/2}\gamma^{-1/2}]}{1-\alpha} \end{matrix} ; -(1-\alpha)^{-\gamma} x^{-\gamma} \right) \\ & + \Gamma\begin{pmatrix} \frac{[-\alpha^{1/2}\gamma^{-1/2} + \alpha]}{1-\alpha} & -2\frac{\alpha^{1/2}\gamma^{-1/2}}{1-\alpha} \\ -\frac{\alpha^{1/2}\gamma^{-1/2}}{1-\alpha} & -\frac{\alpha^{1/2}\gamma^{-1/2}}{1-\alpha} \end{pmatrix} [(1-\alpha)x]^{-\gamma[1 + (1-\alpha)^{-1}\alpha^{1/2}\gamma^{-1/2}]} \\ & \qquad \times {}_2F_2\left(\begin{matrix} 1 + \frac{\alpha^{1/2}\gamma^{-1/2}}{1-\alpha}, & 1 + \frac{\alpha^{1/2}\gamma^{-1/2}}{1-\alpha} \\ 1 - \frac{[\alpha - \alpha^{1/2}\gamma^{-1/2}]}{1-\alpha}, & 1 + 2\frac{\alpha^{1/2}\gamma^{-1/2}}{1-\alpha} \end{matrix} ; -(1-\alpha)^{-\gamma} x^{-\gamma} \right) \\ & + \Gamma\begin{pmatrix} \frac{[\alpha^{1/2}\gamma^{-1/2} + \alpha]}{1-\alpha} & 2\frac{\alpha^{1/2}\gamma^{-1/2}}{1-\alpha} \\ \frac{\alpha^{1/2}\gamma^{-1/2}}{1-\alpha} & \frac{\alpha^{1/2}\gamma^{-1/2}}{1-\alpha} \end{pmatrix} [(1-\alpha)x]^{-\gamma[1 - (1-\alpha)^{-1}\alpha^{1/2}\gamma^{-1/2}]} \\ & \qquad \times {}_2F_2\left(\begin{matrix} 1 - \frac{\alpha^{1/2}\gamma^{-1/2}}{1-\alpha}, & 1 - \frac{\alpha^{1/2}\gamma^{-1/2}}{1-\alpha} \\ 1 - \frac{[\alpha + \alpha^{1/2}\gamma^{-1/2}]}{1-\alpha}, & 1 - 2\frac{\alpha^{1/2}\gamma^{-1/2}}{1-\alpha} \end{matrix} ; -(1-\alpha)^{-\gamma} x^{-\gamma} \right) \Bigg\}. \end{aligned} \tag{5.5}$$

PROOF. We give details just for the first-order case, as this is typical of the general situation. Applying (5.3) to the specific case at hand we find that

$$\prod_{j=1}^{\infty} \frac{1 + f_1 j^{-1}}{1 + f_1 (j-t)^{-1}} = \frac{\Gamma(\frac{1}{1-\alpha} - t)}{\Gamma(1-t)\Gamma(\frac{1}{1-\alpha})}.$$

Replacing the factor $\prod_{j=1}^{\infty} F_{j-t}/F_j$ in (1.2) with the above expression, we have the Mellin–Barnes integral

$$h(x) \underset{\gamma\gg 1\gg\alpha}{\sim} \frac{\gamma}{2\pi i x}\int_{c-i\infty}^{c+i\infty} dt\; x^{-\gamma t}(1-\alpha)^{-\gamma t}\frac{\Gamma(\frac{1}{1-\alpha}-t)}{\Gamma(\frac{1}{1-\alpha})}$$

with $c < 1$. We note that the only singularities of the integrand are simple poles at $t = t_l = l + (1-\alpha)^{-1}$, $l \geqslant 0$.

However, because the integrand has an exponential decay as $t \to c \pm i\infty$ arising from the Gamma function factor $\Gamma((1-\alpha)^{-1} - t)$, one can fold the upper and lower arms of the contour over the positive-t axis, thus enclosing all of these poles. Computing the residues in the standard way [**9**] one has

$$h(x) \underset{\gamma\gg 1\gg\alpha}{\sim} \frac{\gamma x^{-1}}{\Gamma(\frac{1}{1-\alpha})}\sum_{l=0}^{\infty}\frac{(-1)^l}{l!}\,[(1-\alpha)x]^{-\gamma(l+(1-\alpha)^{-1})}$$

and the summation is trivial, leading to (5.4). When including the second-order factors as well, one has two new sequences of simple poles at $t = t_l = l + 1 \pm (1-\alpha)^{-1}\alpha^{1/2}\gamma^{-1/2}$, in addition to the first-order set. We assume that these three sequences do not overlap, as in the case of generic α, γ. After some computation we recognise the series definition of the ${}_2F_2$ hypergeometric function, and simplifying we arrive at (5.5). □

COROLLARY 5.2. *The small-x asymptotic form of the second-order approximation to the density* (5.5) *is*

$$\text{(5.6)}\quad h(x) \underset{x\to 0^+}{\sim} \frac{\gamma(1-\alpha)^{-\gamma/(1-\alpha)}}{\Gamma\left(\frac{1}{1-\alpha}, 1+\frac{\alpha^{1/2}\gamma^{-1/2}}{1-\alpha}, 1-\frac{\alpha^{1/2}\gamma^{-1/2}}{1-\alpha}\right)}\, x^{-1-\gamma/(1-\alpha)} e^{-(1-\alpha)^{-\gamma}x^{-\gamma}},$$

which has precisely the same x-dependence as (5.4) *and only differs in the constant factor.*

PROOF. Utilising in (5.5) the large-argument asymptotic leading-order terms for the hypergeometric function ([**27**, Eq. 16.11.7] with $\kappa = 1$, $\nu = a_1 + a_2 - b_1 - b_2$),

$$\frac{\Gamma(a_1)\Gamma(a_2)}{\Gamma(b_1)\Gamma(b_2)}\,{}_2F_2\left(\begin{matrix} a_1, & a_2 \\ b_1, & b_2\end{matrix}; z\right) \underset{z\to\infty}{\sim} z^{\nu}e^{z}$$
$$+ \frac{\Gamma(a_1)\Gamma(a_1-a_2)}{\Gamma(b_1-a_1)\Gamma(b_2-a_1)}\left(e^{\pm\pi i}z\right)^{-a_1} + \frac{\Gamma(a_2)\Gamma(a_2-a_1)}{\Gamma(b_1-a_2)\Gamma(b_2-a_2)}\left(e^{\pm\pi i}z\right)^{-a_2},$$

we find significant simplification occurs because all three terms end up with the same x-dependence, and there is an amalgamation of their coefficients into a single common factor, resulting in (5.6). □

In concluding we present clear numerical evidence of our approximations versus the recent result in [**29**, Thm. 2], which gives the asymptotic form of the density as $x \to 0^+$ in (1.7) and (1.8). Note that (5.4) is a global approximation to the density, as the constant factor is its true normalisation even though the x-dependent part is identical to the true asymptotic form as $x \to 0^+$, as given by the result in [**29**, Thm. 2]. Such approximations are then a large-γ and small-α asymptotic development, in the sense of $(1-\alpha)^{-1}$ being of $\mathrm{O}(1)$ magnitude. When such conditions are violated, as is apparent in the plots in Figures 1–4, the approximation can vanish and change sign. Another point to note from Figure 5 is that the constant $c(\alpha, \gamma)$ of (1.8) diverges to 0 or ∞ as $\alpha \to 1^-$ when $\gamma < 1$ or $\gamma \geqslant 1$ respectively.

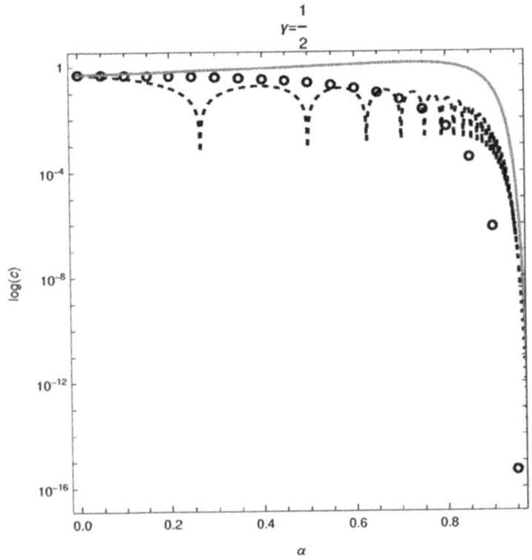

FIGURE 1. Plot of the constant factor $c = c(\alpha, \gamma)$ in the $x \to 0^+$ asymptotic form of $h(x)$, versus α, for $\gamma = \frac{1}{2}$. Open circles are the exact c given by (1.8), the full grey line is the first-order approximation (5.4), and the dashed black line is the absolute value of the second-order approximation (5.6).

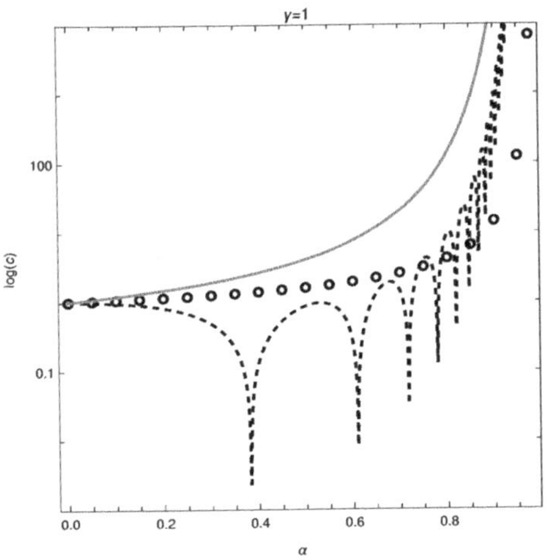

FIGURE 2. As per Figure 1, for $\gamma = 1$.

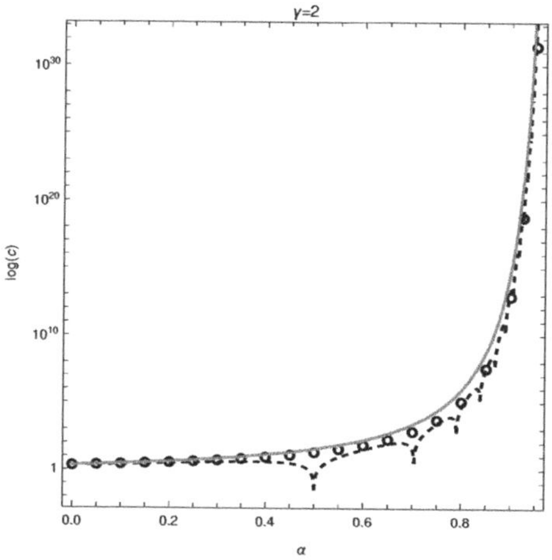

FIGURE 3. As per Figure 1, for $\gamma = 2$.

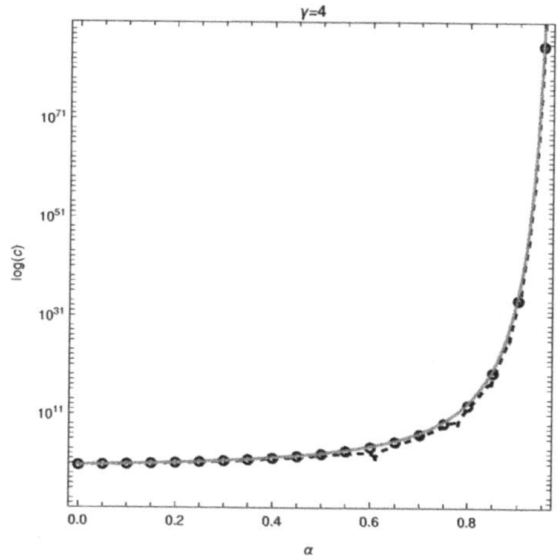

FIGURE 4. As per Figure 1, for $\gamma = 4$.

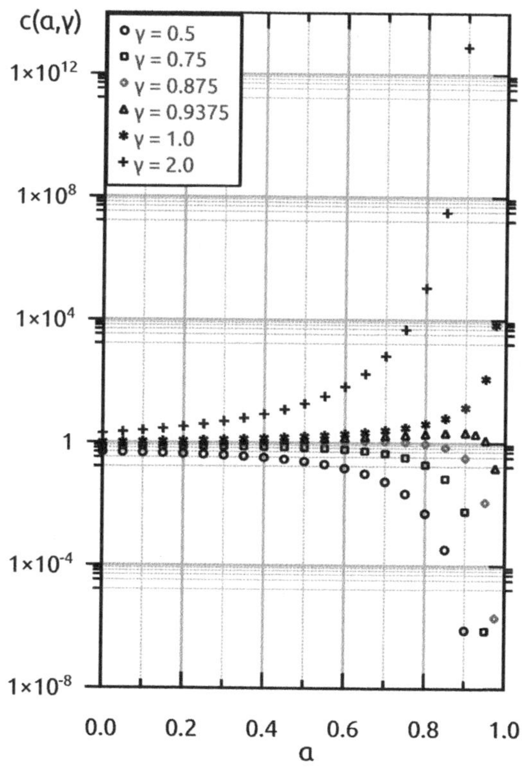

FIGURE 5. Plot of the constant factor (1.8) in the $x \to 0^+$ asymptotic form of $h(x)$ versus $\alpha \in [0, 1)$ for $\gamma = \frac{1}{2}, \dots, 2$.

Acknowledgments

The author thanks Howard Cohl for the invitation to present a preliminary version of the current work to the 2022 AMS fall sectional meeting *Hypergeometric Functions, q-Series and Generalizations* in Salt Lake City, Utah, and thanks Cindy Greenwood and Thomas Simon for much appreciated correspondence. He would also like express gratitude to the School of Mathematics and Statistics, University of Melbourne, for hosting his visit during the final stages of this study. Lastly the presentation and clarity of this manuscript has greatly benefited from the attention to detail and thoroughness of the referees.

References

[1] The On-Line Encyclopedia of Integer Sequences. `https://oeis.org`.

[2] G. Alkauskas, *A curious proof of Fermat's little theorem*, Amer. Math. Monthly **116** (2009), no. 4, 362–364, DOI 10.4169/193009709X470236. MR2503323

[3] G. E. Andrews, *The theory of partitions*, Cambridge Mathematical Library, Cambridge University Press, Cambridge, 1998. Reprint of the 1976 original. MR1634067

[4] R. Bellman, *The stability of solutions of linear differential equations*, Duke Math. J. **10** (1943), 643–647. MR9408

[5] E. A. Bender and J. R. Goldman, *On the applications of Möbius inversion in combinatorial analysis*, Amer. Math. Monthly **82** (1975), no. 8, 789–803, DOI 10.2307/2319793. MR376360

[6] S. Borofsky, *Expansion of analytic functions into infinite products*, Ann. of Math. (2) **32** (1931), no. 1, 23–36, DOI 10.2307/1968411. MR1502978

[7] S. Borofsky, *Expansion of analytic functions into infinite products*, Ann. of Math. (2) **32** (1931), no. 1, 23–36, DOI 10.2307/1968411. MR1502978

[8] L. Comtet, *Advanced combinatorics: The art of finite and infinite expansions*, Revised and enlarged edition, D. Reidel Publishing Co., Dordrecht, 1974. MR460128

[9] B. Davies, *Integral transforms and their applications*, 3rd ed., Texts in Applied Mathematics, vol. 41, Springer-Verlag, New York, 2002, DOI 10.1007/978-1-4684-9283-5. MR1867914

[10] W. C. Doyle, *A generalized Lambert series and its Moebius function*, Ann. of Math. (2) **40** (1939), no. 2, 353–359, DOI 10.2307/1968923. MR1503462

[11] J. M. Feld, *The expansion of analytic functions in generalized Lambert series*, Ann. of Math. (2) **33** (1932), no. 1, 139–142, DOI 10.2307/1968106. MR1503039

[12] P. Flajolet and R. Sedgewick, *Analytic combinatorics*, Cambridge University Press, Cambridge, 2009, DOI 10.1017/CBO9780511801655. MR2483235

[13] M. C. Garvin, *A Generalized Lambert Series*, Amer. J. Math. **58** (1936), no. 3, 507–513, DOI 10.2307/2370967. MR1507173

[14] W. Gautschi, *Some elementary inequalities relating to the gamma and incomplete gamma function*, J. Math. and Phys. **38** (1959/60), 77–81, DOI 10.1002/sapm195938177. MR103289

[15] H. Gingold, H. W. Gould, and M. E. Mays, *Power product expansions*, Utilitas Math. **34** (1988), 143–161. MR982255

[16] H. Gingold and A. Knopfmacher, *Analytic properties of power product expansions*, Canad. J. Math. **47** (1995), no. 6, 1219–1239, DOI 10.4153/CJM-1995-062-9. MR1370515

[17] H. Gingold and J. Quaintance, *Approximations of analytic functions via generalized power product expansions*, J. Approx. Theory **188** (2014), 19–38, DOI 10.1016/j.jat.2014.08.006. MR3274227

[18] H. Gingold and J. Quaintance, *Different representations for power product generating functions*, Util. Math. **116** (2020), 91–118. MR4243169

[19] P. E. Greenwood and G. Hooghiemstra, *On the domain of attraction of an operator between supremum and sum*, Probab. Theory Related Fields **89** (1991), no. 2, 201–210, DOI 10.1007/BF01366906. MR1110538

[20] T. H. Gronwall, *Note on the derivatives with respect to a parameter of the solutions of a system of differential equations*, Ann. of Math. (2) **20** (1919), no. 4, 292–296, DOI 10.2307/1967124. MR1502565

[21] E. R. Hansen. *A Table of Series and Products*. Prentice-Hall, 1975.

[22] K. Knopp, *Theorie und Anwendung der Unendlichen Reihen* (German), Springer-Verlag, Berlin-Heidelberg, 1947. 4th ed. MR28430

[23] Ch. E. Mikeladze, *De la résolution numérique des équations intégrales*, Bull. Acad. Sci. URSS, 1935(2):255–300.

[24] B. G. Pachpatte, *On the discrete generalizations of Gronwall's inequality*, J. Indian Math. Soc. (N.S.) **37** (1973), 147–156 (1974). MR361496

[25] B. G. Pachpatte, *Inequalities for finite difference equations*, Monographs and Textbooks in Pure and Applied Mathematics, vol. 247, Marcel Dekker, Inc., New York, 2002. MR1867845

[26] G. Peano, *Sull'integrabilità delle equazioni differenziali di primo ordine*, Atti della R. Accademia delle Scienze di Torino, 21:677–685, 1886.

[27] NIST Digital Library of Mathematical Functions. `http://dlmf.nist.gov/` (release 1.1.10 of 2023-06-15).

[28] J. F. Ritt, *Representation of analytic functions as infinite products*, Math. Z. **32** (1930), no. 1, 1–3, DOI 10.1007/BF01194617. MR1545149

[29] T. Simon, *A note on the α-sun distribution*, Electron. Commun. Probab. **28** (2023), Paper No. 19, 13. MR4596534

[30] W. Wang and T. Wang, *General identities on Bell polynomials*, Comput. Math. Appl. **58** (2009), no. 1, 104–118, DOI 10.1016/j.camwa.2009.03.093. MR2535972

[31] X. Wang, *A new series representation involving root of unity for the values of Riemann zeta function at integer arguments*, `arXiv:2010.07112` (October 2020).

[32] N. S. Witte and P. E. Greenwood, *On the density arising from the domain of attraction of an operator interpolating between sum and supremum: the α-Sun operator*, J. Math. Anal. Appl. **527** (2023), no. 1, Paper No. 127371, 31, DOI 10.1016/j.jmaa.2023.127371. MR4589206

School of Mathematics and Statistics, Victoria University of Wellington, PO Box 600, Wellington 6140, New Zealand

Email address: `n.s.witte@protonmail.com`

Generalizations and Extensions

Contemporary Mathematics
Volume **818**, 2025
https://doi.org/10.1090/conm/818/16372

The quadratic reciprocity law and the Gauss–Schering lemma

Heng Huat Chan and Teoh Guan Chua

ABSTRACT. In this article, we present a proof of the Gauss–Schering Lemma using the Quadratic Reciprocity Laws for the Jacobi symbol.

1. Introduction

DEFINITION 1.1 (Legendre symbol). Let p be an odd prime. The Legendre symbol $\left(\frac{a}{p}\right)_L$ is defined to be

$$\left(\frac{a}{p}\right)_L = \begin{cases} 0 & \text{if } (a,p) \neq 1, \\ 1 & \text{if } x^2 \equiv a \pmod{p} \text{ is solvable}, \\ -1 & \text{otherwise.} \end{cases}$$

The main part of the famous Gauss's Law of Quadratic Reciprocity is the following.

THEOREM 1.2 (Law of Quadratic Reciprocity). *If p and q are distinct odd primes then*

$$\left(\frac{p}{q}\right)_L \left(\frac{q}{p}\right)_L = (-1)^{(p-1)(q-1)/4}. \tag{1.1}$$

There are many proofs of (1.1) and several of these proofs involve a lemma, now known as Gauss's Lemma. Before we state Gauss's Lemma, we introduce some notations.

Let

$$\begin{aligned} r_n(s) &= \text{ the least non-negative residue of } s \text{ modulo } n, \\ H_n &= \{j \mid 1 \leqslant j \leqslant (n-1)/2\}, \\ S_{a,n} &= \{r_n(aj) \mid j \in H_n\}, \end{aligned}$$

and

$$\mu(a,n) = |\{s \in S_{a,n} \mid s \notin H_n\}|.$$

Gauss's Lemma gives a relation between the Legendre symbol and $\mu(a,p)$.

2020 *Mathematics Subject Classification.* Primary 11A15.

LEMMA 1.3 (Gauss's Lemma). *Let p be an odd prime number and a be any positive integer such that $(a,p)=1$. Then*

$$\left(\frac{a}{p}\right)_L = (-1)^{\mu(a,p)}.$$

A usual proof of Lemma 1.3 uses Euler's criterion which states that (see for example [**5**, p. 101, Corollary 2.38])

$$\left(\frac{a}{p}\right)_L \equiv a^{(p-1)/2} \pmod{p}. \tag{1.2}$$

REMARK 1.4. Another proof of Gauss's Lemma which involves the "transfer map" in group theory can be found in [**10**, p. 91, Section 7.3].

The Legendre symbol has a natural extension to composite odd positive integers. It is now known as the Jacobi symbol.

DEFINITION 1.5 (Jacobi symbol). Let a be any integer. The Jacobi symbol $\left(\frac{a}{b}\right)_J$ is defined for odd positive integer b by

$$\left(\frac{a}{b}\right)_J = \prod_{j=1}^{k} \left(\frac{a}{p_j}\right)_L^{\alpha_j}$$

if $b = \prod_{j=1}^k p_j^{\alpha_j}$.

Surprisingly, the Jacobi symbol satisfies a relation similar to (1.1).

THEOREM 1.6 (Law of Quadratic Reciprocity for the Jacobi symbol). *If a and b are odd positive integers satisfying $(a,b)=1$, then*

$$\left(\frac{a}{b}\right)_J \left(\frac{b}{a}\right)_J = (-1)^{(a-1)(b-1)/4}. \tag{1.3}$$

REMARK 1.7. The proof of (1.3) follows from (1.1) and requires only the simple congruence

$$\frac{a-1}{2} + \frac{b-1}{2} \equiv \frac{ab-1}{2} \pmod{2},$$

where a and b are odd positive integers. See [**5**, p. 145] for more details.

Since the discovery of (1.3), there are proofs of (1.3) which are independent of (1.1). Two of the earliest published works on proving (1.3) directly are due to M. Jenkins [**3**] and E. Schering [**8**] (see also [**4**] for a modern treatment of Schering's work). Both Jenkins and Schering used a lemma analogous to Lemma 1.3 but Schering's version is an exact analogue of Lemma 1.3 with the prime p replaced by a composite odd positive integer b. As such, the following lemma is sometimes known as Gauss–Schering Lemma (see [**2**]).

LEMMA 1.8 (Gauss–Schering Lemma). *Let b be a positive odd integer and a be any positive integer such that $(a,b)=1$. Then*

$$\left(\frac{a}{b}\right)_J = (-1)^{\mu(a,b)}.$$

The beauty of Lemma 1.8 is that it is exactly Lemma 1.3 with the prime p replaced by a composite odd positive integer b. The proofs of Lemma 1.3, however, cannot be adapted to prove Lemma 1.8. This is due to the fact that Euler's criterion (1.2) is false when the prime p is replaced by a composite odd integer n. Since (1.3) follows from (1.1), we would like to know if we could deduce Lemma 1.8 from (1.3). This will be discussed in the next section.

2. A proof of the Gauss–Schering Lemma

One of the proofs of (1.1) which is due to Eisenstein (see [**9**, p. 10]) uses the following trigonometric identity:

LEMMA 2.1. *Let m be an odd positive integer. Then*

$$\frac{\sin mx}{\sin x} = (2i)^{m-1} \prod_{j\in H_m} \left(\sin^2 x - \sin^2 \frac{2\pi j}{m}\right). \tag{2.1}$$

Lemma 2.1 can be proved using the identity

$$x^m - 1 = \prod_{j=1}^{m} (x - \zeta_m^j),$$

where $\zeta_m = e^{4\pi i/m}$ where m is odd. By writing the identity in terms of the sine function, one arrives at the identity

$$\sin mx = (2i)^{m-1} \sin x \prod_{j=1}^{(m-1)/2} \sin(x - 2\pi j/m) \sin(x + 2\pi j/m)$$

and Lemma 2.1 follows by using the identity

$$\sin(a+b)\sin(a-b) = (\sin a + \sin b)(\sin a - \sin b).$$

We now use Lemma 2.1 to show the following:

LEMMA 2.2. *Let a and b be odd positive integers such that $(a,b) = 1$. Then*

$$\prod_{s\in H_b} \frac{\sin(2as\pi/b)}{\sin(2s\pi/b)} = (-1)^{\mu(a,b)}.$$

PROOF. Let

$$S_1 = \{s | s \in S_{a,b} \cap H_b\}$$

and

$$S_2 = \{b - s | s \in S_{a,b} \text{ but } s \notin H_b\}.$$

Note that $|S_2| = \mu(a,b)$.

If $s \in S_1$ then $\sin(2as\pi/m) = \sin(2u\pi/m)$ for some $u \in H_b$. If $s \in S_2$ then $\sin(2as\pi/m) = -\sin(2u\pi/m)$ for some $u \in H_b$. Note that since $S_1 \cup S_2 = H_b$,

$$\prod_{u\in H_b} \sin(2au\pi/b) = (-1)^{\mu(a,b)} \prod_{u\in H_b} \sin(2u\pi/b),$$

and this completes the proof of the lemma. □

Reversing the roles of a and b, we find by Lemma 2.2 that

$$\prod_{t\in H_a} \sin(2bt\pi/a) = (-1)^{\mu(b,a)} \prod_{t\in H_a} \sin(2t\pi/a).$$

By Lemma 2.1, we conclude that

$$\begin{aligned}(-1)^{\mu(a,b)+\mu(b,a)} = (-1)^{\mu(a,b)-\mu(b,a)} &= \prod_{s\in H_a}\prod_{t\in H_b} \frac{\sin^2(2s\pi/b)-\sin^2(2\pi t/a)}{\sin^2(2t\pi/a)-\sin^2(2s\pi/b)} \\ &= (-1)^{(a-1)(b-1)/4}.\end{aligned}$$

By (1.3),

$$\left(\frac{a}{b}\right)_J \left(\frac{b}{a}\right)_J = (-1)^{(a-1)(b-1)/4} = (-1)^{\mu(a,b)+\mu(b,a)}. \tag{2.2}$$

We are now ready to prove Lemma 1.8. Let q be an odd prime and b be any positive odd integer such that $(b,q) = 1$. Then by Lemma 1.3, we find that

$$\left(\frac{b}{q}\right)_J = (-1)^{\mu(b,q)}.$$

By (2.2), we conclude that

$$\left(\frac{q}{b}\right)_J = (-1)^{\mu(q,b)}. \tag{2.3}$$

Next, suppose a is any odd positive integer. Since $(a,b) = 1$, by the Dirichlet Theorem on primes in arithmetic progression (see for example [**1**, Chapter 7]), there exists a prime Q of the form $a + bn$. By (2.3),

$$\left(\frac{a}{b}\right)_J = \left(\frac{Q}{b}\right)_J = (-1)^{\mu(Q,b)}.$$

But $\mu(Q,b) = \mu(a,b)$ since $S_{Q,b} = S_{a,b}$ Therefore,

$$\left(\frac{a}{b}\right)_J = (-1)^{\mu(a,b)}$$

and the proof of Lemma 1.8 is complete.

3. Two representations of the Jacobi symbol

In this section, we mention two other representations of the Jacobi symbol. These representations were established in the literature to provide direct proofs of (1.3). We will deduce these representations from Lemma 1.8.

THEOREM 3.1. *Let $\lfloor x \rfloor$ denote the largest integer less than x, where x is any real number. Define*

$$s(a,b) = \sum_{j=1}^{(b-1)/2} \left\lfloor \frac{aj}{b} \right\rfloor$$

and

$$t(a,b) = \sum_{j=1}^{(b-1)/2} \left\lfloor \frac{2aj}{b} \right\rfloor.$$

If a and b are odd positive integers such that $(a,b)=1$, then

$$\left(\frac{a}{b}\right)_J = (-1)^{s(a,b)} \tag{3.1}$$

$$= (-1)^{t(a,b)}. \tag{3.2}$$

PROOF. For each $j \in H_b$ and odd positive integer a, write

$$ja = q_j b + r_j \tag{3.3}$$

with $1 \leqslant r_j \leqslant b-1$. Rewrite ja as

$$ja = \begin{cases} q_j b + r_j & \text{if } r_j \in H_b, \\ q_j b + b - s_j & \text{if } r_j = b - s_j \notin H_b. \end{cases}$$

Note that $s_j \in H_b$ and that there are exactly $\mu(a,b)$ number of j such that $r_j \notin H_b$. Therefore,

$$\begin{aligned} a\sum_{j=1}^{(b-1)/2} j &\equiv b\sum_{j=1}^{(b-1)/2} q_j + b\mu(a,b) + \sum_{r_j\in H_b} r_j - \sum_{r_k\notin H_b} s_k \\ &\equiv \sum_{j=1}^{(b-1)/2} q_j + \mu(a,b) + \sum_{j=1}^{(b-1)/2} j \pmod 2, \end{aligned}$$

where in the last relation, we have used the fact that

$$\{r_j | r_j \in H_b\} \cup \{s_k | r_k = b - s_k \notin H_b\} = H_b$$

and

$$\sum_{r_j\in H_b} r_j - \sum_{r_k\notin H_b} s_k \equiv \sum_{r_j\in H_b} r_j + \sum_{r_k\notin H_b} s_k \pmod 2.$$

This implies that

$$\sum_{j=1}^{(b-1)/2} \left\lfloor \frac{aj}{b} \right\rfloor \equiv \mu(a,b) \pmod 2,$$

and hence

$$\left(\frac{a}{b}\right)_J = (-1)^{\mu(a,b)} = (-1)^{s(a,b)}. \tag{3.4}$$

This completes the proof of (3.1).

To prove (3.2), we return to (3.3) and write

$$2ja = \begin{cases} 2q_j b + 2r_j = \lfloor 2ja/b \rfloor b + 2r_j & \text{if } r_j \in H_b, \\ (2q_j+1)b + b - 2s_j = \lfloor 2ja/b \rfloor b + b - 2s_j & \text{if } r_j = b - s_j \notin H_b. \end{cases}$$

The second case holds since

$$b + 1 \leqslant 2r_j = 2b - 2s_j < 2b$$

implies that

$$2ja = (2q_j+1)b + b - 2s_j, 1 \leqslant b - 2s_j < b.$$

Hence

$$2a\sum_{j=1}^{(b-1)/2} j \equiv b\sum_{j=1}^{(b-1)/2} \left\lfloor \frac{2ja}{b} \right\rfloor + \sum_{r_j\in H_b} 2r_j + b\mu(a,b) + 2\sum_{r_j\notin H_b} s_j \pmod 2,$$

which implies that

$$\sum_{j=1}^{(b-1)/2} \left\lfloor \frac{2aj}{b} \right\rfloor \equiv \mu(a,b) \pmod 2$$

and the proof of (3.2) is complete. □

REMARK 3.2. The identity (3.2) appears in the work of H. Rademacher in his study of the Dedekind sum [**6**, p. 159]. Rademacher established (3.2) using the reciprocity relations satisfied by the Dedekind sum. He then proved (1.3) using (3.2).

There is another way of proving Lemma 1.8 without the use of (1.3) due to E. I. Zolotarev [**12**]. Zolotarev's proof took a different approach from Schering's proof. He defined

$$\left(\frac{a}{b}\right)_Z = \mathrm{sgn}(\lambda_a),$$

where λ_a is the permutation of $\mathbf{Z}/b\mathbf{Z}$ which sends $[x]_b$ to $[ax]_b$, and $\mathrm{sgn}(\sigma)$ is the "signum" of the permutation σ. Zolotarev then showed that for odd positive integers a and b satisfying $(a,b)=1$,

$$\left(\frac{a}{b}\right)_Z \left(\frac{b}{a}\right)_Z = (-1)^{(a-1)(b-1)/4}$$

and

$$\left(\frac{a}{b}\right)_Z = \left(\frac{a}{b}\right)_J.$$

For a modern treatment of Zolotarev's work, see [**7**] and [**11**].

References

[1] H. H. Chan, *Analytic number theory for undergraduates*, Monographs in Number Theory, vol. 3, World Scientific Publishing Co. Pte. Ltd., Hackensack, NJ, 2009, DOI 10.1142/7252. MR2523579

[2] R. Hill, *Metaplectic covers of* GL_n *and the Gauss-Schering lemma* (English, with English and French summaries), J. Théor. Nombres Bordeaux **13** (2001), no. 1, 189–199. 21st Journées Arithmétiques (Rome, 2001). MR1838080

[3] M. Jenkins, *Proof of an Arithmetical Theorem leading, by means of Gauss's Fourth Demonstration of Legendre's Law of Reciprocity, to the extension of that Law*, Proc. Lond. Math. Soc. **2** (1866/69), 29–32, DOI 10.1112/plms/s1-2.1.29. MR1576676

[4] A. Kuroki and S.-i. Katayama, *A variation of Takagi's proof for quadratic reciprocity laws of Jacobi symbols*, J. Math. Univ. Tokushima **43** (2009), 9–23. MR2656202

[5] I. Niven, H. S. Zuckerman, and H. L. Montgomery, *An introduction to the theory of numbers*, 5th ed., John Wiley & Sons, Inc., New York, 1991. MR1083765

[6] H. Rademacher, *Topics in analytic number theory*, Die Grundlehren der mathematischen Wissenschaften, Band 169, Springer-Verlag, New York-Heidelberg, 1973. MR364103

[7] G. Rousseau, *On the Jacobi symbol*, J. Number Theory **48** (1994), no. 1, 109–111, DOI 10.1006/jnth.1994.1057. MR1284879

[8] E. Schering, *Zur Theorie der Quadratischen Reste* (French), Acta Math. **1** (1882), no. 1, 153–170, DOI 10.1007/BF02391842. MR1554581

[9] J.-P. Serre, *A course in arithmetic*, Graduate Texts in Mathematics, No. 7, Springer-Verlag, New York-Heidelberg, 1973. Translated from the French. MR344216

[10] J.-P. Serre, *Finite groups: an introduction*, Surveys of Modern Mathematics, vol. 10, International Press, Somerville, MA; Higher Education Press, Beijing, 2016. MR3469786

[11] M. Szyjewski, *Zolotarev's proof of Gauss reciprocity and Jacobi symbols*, Serdica Math. J. **37** (2011), no. 3, 251–260. MR2951412

[12] E. I. Zolotarev. Nouvelle démonstration de la loi de réciprocité de Legendre. Nouv. Ann. Math (2) **11** (1872), 354–362.

Mathematical Research Center, Shandong University, No. 1 Building, 5 Hongjialou Road, Jinan, 250100, PR China
Email address: chanhh6789@sdu.edu.cn

Dept. of Mathematics, Ngee Ann Secondary School, Singapore, 529283, Singapore
Email address: teohguan.chua@gmail.com

Contemporary Mathematics
Volume **818**, 2025
https://doi.org/10.1090/conm/818/16373

Apéry-like sequences defined by four-term recurrence relations

Shaun Cooper

Dedicated to the memory of Richard Askey

ABSTRACT. The Apéry numbers may be defined by a cubic three-term recurrence relation, that is, a three-term relation where the coefficients are polynomials in the index of degree 3.

In this work, we first provide a systematic review of Apéry numbers and other related sequences that satisfy quadratic or cubic three-term recurrence relations, and show how they are interrelated and how they may be classified. This leads to sequences defined by cubic k-term recurrence relations. The cases corresponding to $k = 2$ in this framework lead to Ramanujan's theories of elliptic functions to alternative bases, while the cases corresponding to $k = 3$ correspond to the Apéry, Domb, Almkvist–Zudilin numbers and other sequences that are well-studied.

We conduct a detailed analysis for the case $k = 4$. Some of the sequences that arise are new. Of particular interest are ten sequences that are said to be self-starting in the sense that a single initial condition is enough to start the recurrence relation. Of additional interest are two sequences which take values in $\mathbb{Z}[i]$ and two others with values in $\mathbb{Z}[\sqrt{2}]$. Congruence properties and asymptotic expansions for the ten self-starting sequences are investigated and several conjectures are presented. For example, we conjecture that the integer-valued sequence defined by the recurrence relation

$$(n+1)^3T(n+1) = 2(2n+1)(5n^2+5n+2)T(n) - 8n(7n^2+1)T(n-1) + 22n(2n-1)(n-1)T(n-2)$$

and initial condition $T(0) = 1$ satisfies a Lucas congruence for every prime p. Moreover, the sequence is conjectured to satisfy the supercongruence

$$T(pn) \equiv T(n) \pmod{p^2} \quad \text{for all positive integers } n \tag{0.1}$$

if $p = 2$, 59, or 5581, and for no other primes $p < 10^4$.

1. Introduction

The Apéry numbers may be defined by the recurrence relation

$$(n+1)^3A(n+1) = (2n+1)(17n^2+17n+5)A(n) - n^3A(n-1) \tag{1.1}$$

2020 *Mathematics Subject Classification.* Primary 11F11, 33-02; Secondary 05A10, 11A07, 11B65, 11F20, 33C20.

Key words and phrases. Almkvist–Zudilin numbers, Apéry numbers, asymptotic expansion, Clausen's identity, diagonal coefficient, Domb numbers, genus zero, Lucas congruence, modular form, multivariate Taylor expansion, Ramanujan's alternative theories, self-starting sequence, sporadic sequence, supercongruence, theta function identity.

and initial condition $A(0) = 1$; the single initial condition is enough to start the recurrence, as can be seen by taking $n = 0$ in (1.1). The Apéry numbers are famous for having been introduced and used by R. Apéry to prove that $\zeta(3)$ is irrational, e.g., see [**4**] or [**41**]. An alternative formula for the Apéry numbers is the binomial sum

$$A(n) = \sum_{k=0}^{n} \binom{n}{k}^2 \binom{n+k}{k}^2. \tag{1.2}$$

This reveals that $A(n)$ is always an integer, a fact not obvious from the recurrence relation (1.1). The method of creative telescoping can be used to show that the $A(n)$ defined by the binomial sum (1.2) satisfy the recurrence relation (1.1). For example, using the computer algebra system Maple, and typing the commands:

```
> with(sumtools):
> sumrecursion(binomial(n,k)^2*binomial(n+k,n)^2, k, A(n));
```

produces the recurrence relation

$$(n-1)^3 A(n-2) - (2n-1)(17n^2 - 17n + 5)A(n-1) + n^3 A(n)$$

that is equal to zero. Hence, we deduce (1.1) from (1.2). Conversely, there does not seem to be an algorithm for determining (1.2) from (1.1) apart from for searching a database.

It is a remarkable result of F. Beukers [**7**] (see also [**16**, Theorem 6.38]) that the Apéry numbers are parameterised by modular forms according to the formula

$$\prod_{j=1}^{\infty} \frac{(1-q^{2j})^7(1-q^{3j})^7}{(1-q^j)^5(1-q^{6j})^5} = \sum_{n=0}^{\infty} A(n) \left(q \prod_{j=1}^{\infty} \frac{(1-q^j)^{12}(1-q^{6j})^{12}}{(1-q^{2j})^{12}(1-q^{3j})^{12}} \right)^n, \tag{1.3}$$

valid in a neighbourhood of $q = 0$.

We briefly mention four other properties of the Apéry numbers. First, it has been shown by I. Gessel [**23**] that the Apéry numbers $A(n)$ satisfy the Lucas congruences for every prime p. This means that for any positive integer n and for $T(n) = A(n)$, the congruence

$$T(n) \equiv T(n_0)T(n_1)\cdots T(n_r) \pmod{p} \tag{1.4}$$

holds, where $n = n_0 + n_1 p + \cdots + n_r p^r$ is the p-adic expansion of n in base p.

Second, the Apéry numbers have an asymptotic formula, stated without proof by A. van der Poorten [**41**] and attributed to H. Cohen, given by

$$A(n) \sim \frac{1}{2^{9/4}(\pi n)^{3/2}} (1+\sqrt{2})^{4n+2} \qquad \text{as } n \to \infty. \tag{1.5}$$

For a proof, see the work of M. Hirschhorn [**26**].

Third, it has been noted by various authors, e.g., [**8**], [**12**], [**30**], [**38**] that the Apéry numbers occur in several ways as diagonal coefficients of multivariate Taylor expansions. For example, A. Straub [**38**] showed that if

$$\frac{1}{(1-x_1-x_2)(1-x_3-x_4) - x_1x_2x_3x_4}$$
$$= \sum_{n_1=0}^{\infty}\sum_{n_2=0}^{\infty}\sum_{n_3=0}^{\infty}\sum_{n_4=0}^{\infty} c(n_1,n_2,n_3,n_4) x_1^{n_1}x_2^{n_2}x_3^{n_3}x_4^{n_4}$$

then

$$c(n_1,n_2,n_3,n_4) = \sum_{k=0}^{\min\{n_1,n_3\}} \binom{n_1}{k}\binom{n_3}{k}\binom{n_1+n_2-k}{n_1}\binom{n_3+n_4-k}{n_3}.$$

The diagonal coefficients are the coefficients $c(n,n,n,n)$ and are given by

$$c(n,n,n,n) = \sum_{k=0}^{n} \binom{n}{k}^2 \binom{2n-k}{n}^2.$$

Replacing k with $n-k$ gives

$$c(n,n,n,n) = \sum_{k=0}^{n} \binom{n}{n-k}^2 \binom{n+k}{n}^2 = \sum_{k=0}^{n} \binom{n}{k}^2 \binom{n+k}{n}^2.$$

Hence by (1.2), the diagonal terms are the Apéry numbers. Further results have been given by O. Gorodetsky [**24**].

Finally, it was indicated by T. Sato [**36**] that the Apéry numbers can be used to produce series for $1/\pi$, for example

$$\frac{1}{\pi} = \left(72\sqrt{15} - 160\sqrt{3}\right) \sum_{n=0}^{\infty} A(n) \left(\frac{1}{2} - \frac{3\sqrt{5}}{20} + n\right) \left(\frac{1-\sqrt{5}}{2}\right)^{12n}.$$

Such formulas have been investigated in detail by H. H. Chan and H. Verrill [**11**] and further in [**16**].

In view of the rich and varied properties of the Apéry numbers, there is interest in determining and classifying other sequences with similar properties. The goal of this work is to exhibit some new sequences that satisfy recurrence relations similar to (1.1). One significant difference is that some of the new sequences are not integer-valued instead but take values in $\mathbb{Z}[i]$ or $\mathbb{Z}[\sqrt{2}]$. The modular parameterisations are established, and Lucas type properties are conjectured. That leads to another significant difference in that some of the conjectured congruences are for certain primes.

This work is organised as follows. Sections 2, 3 and 4 are largely expository and aim to provide a summary of the known Apéry-like sequences along with their classification and interrelations. Thus, in Section 2 we review Zagier's six sporadic sequences and their modular parameterisations. These sequences are integer-valued and are defined by a three-term recurrence relation with quadratic coefficients. Because the parameterising modular forms have weight one, we will also refer to Zagier's sequences as the weight one examples.

In Section 3 we review the Apéry numbers along with other integer-valued sequences defined by three-term recurrence relations with cubic coefficients. This leads to a theorem of Almkvist, van Straten and Zudilin that places these sequences in one-to-one correspondence with Zagier's sequences by means of a Clausen type identity involving the square of the generating function for the weight one sequences. Because the parameterising modular forms have weight two, we will refer to these sequences as weight two examples. Key in understanding the connection between the weight one and weight two examples is the non-linear differential equation

$$\mathrm{D}^2 Z - \frac{1}{2Z}\left(\mathrm{D}Z\right)^2 = HZ, \tag{1.6}$$

where D is the differential operator

$$\mathrm{D} = BX\frac{\mathrm{d}}{\mathrm{d}X}$$

and where B^2 and H are polynomials in X. In Theorem 3.1 it will be shown that (1.6) can be used to write down a simple $(k+1)$-term recurrence relation for the coefficients $T(n)$ in the expansion

$$Z = \sum_{n=0}^{\infty} T(n) X^n, \tag{1.7}$$

where k is the maximum of the degrees of B^2 and H. It will be shown further in Section 10 how parameters in the asymptotic expansion for $T(n)$ can be obtained from the polynomials B^2 and H.

In Section 4 we analyse a modular parameterisation of Ramanujan in detail, and use (1.6) along with standard properties of theta functions to give a complete and direct proof without using the usual inversion formula for elliptic functions or Clausen's formula. Ramanujan's example is important because it provides a prototype for further modular parameterisations, and this leads to further Apéry-type sequences which are defined by recurrence relations involving two or more terms. A summary involving a large number of examples akin to Ramanujan's prototype is provided in Tables 3–4 and 5–6. We indicate how six of the examples in Tables 3–4 are related to Zagier's weight one sequences by another Clausen type theorem of Chan, Tanigawa, Yang and Zudilin.

The sequences in Tables 3–4 that satisfy four-term recurrence relations are studied in Sections 5–9. Sections 5, 6, 7 and 8 provide analyses of the sequences for levels 11, 13, 14 and 15, respectively, and the remaining sequences are studied in Section 9. The sequences for levels 11, 14, 15 and 24 stand out in that they are self-starting, by which we mean that the single initial condition $T(0) = 1$ is enough to start the recurrence relations. The self-starting sequences are conjectured to satisfy Lucas congruences. On the other hand, the non self-starting sequences appear not to satisfy Lucas congruences.

The level 13 sequence is analysed in detail in Section 6. It is neither self-starting nor integer valued but is shown to become integer-valued when multiplied by 4^n. No conjectural congruences have been found for this sequence.

In Section 7, four sequences are obtained corresponding to level 14, two of which are new. The new sequences take values in $\mathbb{Z}[\sqrt{2}]$ and are conjectured to satisfy Lucas congruences for certain primes p.

The level 15 sequences are studied on Section 8. Four sequences emerge, three of which are new, and two of which take values in $\mathbb{Z}[i]$ and are conjectured to satisfy Lucas congruences for certain primes p.

The sequences for levels 20, 21 and 24 are discussed in Section 9.

In Section 10 we outline how parameters in the asymptotic expansions of the coefficients $T(n)$ in (1.7) can be derived from the polynomials B^2 and H in (1.6).

Finally, in Section 11 we consider supercongruences satisfied by the self-starting sequences defined by four-term recurrence relations. We find that such congruences exist, but are weaker, i.e., they are modulo p^2 instead of modulo p^3. Moreover, such congruences seem to be fairly rare and involve only certain primes.

Throughout this work we shall make use of the notation

$$\eta_N = q^{N/24}\prod_{j=1}^{\infty}(1-q^{jN})$$

where N is any positive integer and $|q|<1$. We will occasionally use the compact notation for products from the theory of basic hypergeometric series such as

$$(a;q)_\infty = \prod_{j=0}^{\infty}(1-aq^j) \quad \text{and} \quad (a,b;q)_\infty = (a;q)_\infty(b;q)_\infty$$

so that, for example,

$$\eta_N = q^{N/24}(q^N;q^N)_\infty.$$

2. Zagier's sporadic sequences: the six weight one examples

An old result of Franel [**22**] is that the sequence defined by

$$t_c(n) = \sum_{k=0}^{n}\binom{n}{k}^3$$

satisfies the three-term recurrence relation

$$(2.1) \qquad (n+1)^2 t_c(n+1) = (7n^2+7n+2)t_c(n) + 8n^2 t_c(n-1), \quad n \geqslant 0.$$

Observe that the coefficients in (2.1) are polynomials in n of degree 2, whereas the coefficients in the recurrence relation (1.1) for the Apéry numbers in (1.1) have degree 3. Another example where the coefficients are polynomials of degree 2, also due to Apéry [**4**], is the sequence defined by

$$t_5(n) = \sum_{k=0}^{n}\binom{n}{k}^2\binom{n+k}{k}.$$

It satisfies the three-term recurrence relation

$$(2.2) \qquad (n+1)^2 t_5(n+1) = (11n^2+11n+3)t_5(n) + n^2 t_5(n-1), \quad n \geqslant 0.$$

Motivated by the common form of the relations (2.1) and (2.2), D. Zagier [**45**] was led to consider sequences of the form

$$(2.3) \qquad (n+1)^2 t(n+1) = (\alpha n^2+\alpha n+\beta)t(n) + \gamma n^2 t(n-1), \quad n \geqslant 0,$$

where α, β and γ are integers. He performed a computer search for other integral sequences that satisfy the three-term recurrence relation. The search yielded six sequences that are not either terminating (two successive terms that are zero), polynomial, or the product of a polynomial with a multinomial coefficient, corresponding to the parameter values $(\alpha,\beta,\gamma)\in S$ where

$$(2.4) \qquad S = \{(11,3,1), (-17,-6,-72), (10,3,-9), (7,2,8), (12,4,-32), (-9,-3,-27)\}.$$

They are now called Zagier's sporadic sequences. Remarkably, all six sequences can be expressed as sums of binomial coefficients, and they can be parameterised by modular forms. The details are summarised in Table 1. The levels of the modular forms are listed in the first column, and the letters (A), (B) and (C) in that column have no special meaning other than to distinguish the three different examples that occur for level 6. The functions z in Table 1 are modular forms of weight one, and for this reason we shall refer to the sequences as the weight one examples.

TABLE 1. Zagier's sporadic sequences, their modular parameterisations and entries in Sloane's OEIS [**37**]. The notation is $z = \sum_{n=0}^{\infty} t(n)x^n$, where $t(0) = 1$ and $(n+1)^2 t(n+1) = (\alpha n^2 + \alpha n + \beta)t(n) + \gamma n^2 t(n-1)$ for $n \geqslant 0$.

Level	(α, β, γ)	x	z	$t(n)$	OEIS
5	$(11,3,1)$	$q\dfrac{(q,q^4;q^5)_\infty^5}{(q^2,q^3;q^5)_\infty^5}$	$\dfrac{(q;q)_\infty^2}{(q,q^4;q^5)_\infty^5}$	$\sum_j \binom{n}{j}^2\binom{n+j}{j}$	A005258
6 (A)	$(-17,-6,-72)$	$\dfrac{\eta_2\eta_6^5}{\eta_1^5\eta_3}$	$\dfrac{\eta_1^6\eta_6}{\eta_2^3\eta_3^2}$	$\sum_{j,\ell}(-8)^{n-j}\binom{n}{j}\binom{j}{\ell}^3$	A093388
6 (B)	$(10,3,-9)$	$\dfrac{\eta_1^4\eta_6^8}{\eta_2^8\eta_3^4}$	$\dfrac{\eta_2^6\eta_3}{\eta_1^3\eta_6^2}$	$\sum_j \binom{n}{j}^2\binom{2j}{j}$	A002893
6 (C)	$(7,2,8)$	$\dfrac{\eta_1^3\eta_6^9}{\eta_2^3\eta_3^9}$	$\dfrac{\eta_2\eta_3^6}{\eta_1^2\eta_6^3}$	$\sum_j \binom{n}{j}^3$	A000172
8	$(-12,-4,-32)$	$\dfrac{\eta_2^2\eta_8^4}{\eta_1^4\eta_4^2}$	$\dfrac{\eta_1^4}{\eta_2^2}$	$(-1)^n\sum_j \binom{n}{j}\binom{2j}{j}\binom{2n-2j}{n-j}$	A081085
9	$(-9,-3,-27)$	$\dfrac{\eta_9^3}{\eta_1^3}$	$\dfrac{\eta_1^3}{\eta_3}$	$\sum_j(-3)^{n-3j}\binom{n}{j}\binom{n-j}{j}\binom{n-2j}{j}$	A006077

Each function x in Table 1 satisfies the differentiation formula

$$q\frac{\mathrm{d}x}{\mathrm{d}q} = z^2 x(1 - \alpha x - \gamma x^2) \tag{2.5}$$

and each z satisfies the second order linear differential equation with respect to x given by

$$\frac{\mathrm{d}}{\mathrm{d}x}\left(x(1 - \alpha x - \gamma x^2)\frac{\mathrm{d}z}{\mathrm{d}x}\right) = (\beta + \gamma x)z. \tag{2.6}$$

The proofs of (2.5) are case-by-case for each pair of functions x and z in Table 1. For example, see [**16**, p. 315] for the case of level 5; use [**16**, (6.29) and Theorem 6.22] for the three level 6 cases; use [**16**, Theorem 8.23 and (8.64)] to obtain the level 8 result; and use [**16**, Theorem 9.7 and (9.16)] in the case of level 9.

The differential equation (2.6) is equivalent to the recurrence relation (2.3). For proofs that each pair x and z in Table 1 satisfies the differential equation (2.6), see, for example, Theorems 5.29, 6.34, 8.28, 9.10 in [**16**].

3. Apéry numbers and the six weight two examples

It has already been noted that in the recurrence relation (1.1) for the Apéry numbers the coefficients are polynomials of degree 3, whereas the coefficients in the recurrence relation (2.3) for Zagier's examples, are polynomials of degree 2. Thus, the Apéry numbers are not an instance of one of Zagier's examples. It is however a remarkable fact that Zagier's examples are in one-to-one correspondence with another set of sequences which include the Apéry numbers as one of the examples. We shall now describe the correspondence.

Suppose

$$z = \sum_{n=0}^{\infty} t(n)x^n \tag{3.1}$$

where the sequence $\{t(n)\}$ satisfies the three-term recurrence relation

$$(n+1)^2 t(n+1) = (\alpha n^2 + \alpha n + \beta)t(n) + \gamma n^2 t(n-1) \tag{3.2}$$

and initial condition $t(0) = 1$. Equivalently, z satisfies the second order linear differential equation

$$\frac{\mathrm{d}}{\mathrm{d}x}\left(x(1-\alpha x-\gamma x^2)\frac{\mathrm{d}z}{\mathrm{d}x}\right) = (\beta+\gamma x)z$$

and initial condition $z(0) = 1$. Now make the change of variables

$$w = \frac{x}{1-\alpha x-\gamma x^2} \quad \text{and} \quad y = (1-\alpha x-\gamma x^2)z^2. \tag{3.3}$$

By the chain rule, the differentiation formula (2.5) in the new variables becomes

$$q\frac{\mathrm{d}w}{\mathrm{d}q} = y\,w\,A \tag{3.4}$$

where

$$A = \sqrt{(1+\alpha w)^2 + 4\gamma w^2}. \tag{3.5}$$

Moreover, applying the same change of variables to the differential equation (2.6) (e.g., see [**16**, Theorem 5.41] for details) shows that y satisfies the second order non-linear differential equation with respect to w given by

$$\mathrm{D}^2 y - \frac{1}{2y}(\mathrm{D}y)^2 = Hy \tag{3.6}$$

where D is the differential operator defined by

$$\mathrm{D} = A\,w\,\frac{\mathrm{d}}{\mathrm{d}w}$$

and A is as in (3.5), and H is given by

$$H = (2\beta-\alpha)w - \frac{(\alpha^2+4\gamma)w^2}{2}. \tag{3.7}$$

In order to solve (3.6) it is useful to prove the following slightly more general result.

THEOREM 3.1. *Suppose $Z = Z(X)$ is analytic in a neighbourhood of $X = 0$ with $Z(0) = Z_0 \neq 0$, and satisfies the second order non-linear differential equation*

$$\mathrm{D}^2 Z - \frac{1}{2Z}(\mathrm{D}Z)^2 = HZ, \tag{3.8}$$

where D *is the differential operator*

$$\mathrm{D} = BX\frac{\mathrm{d}}{\mathrm{d}X},$$

where

$$B = \sqrt{G} \quad \text{and} \quad G = 1 + \sum_{j=1}^{k} g_j X^j, \tag{3.9}$$

and

$$H = \sum_{j=1}^{k} h_j X^j. \tag{3.10}$$

Then the coefficients $T(n)$ in the power series expansion

$$Z = \sum_{n=0}^{\infty} T(n) X^n$$

satisfy the $(k+1)$-term recurrence relation

$$\begin{aligned}(n+1)^3 T(n+1) &+ (n+1)\sum_{j=1}^{k} g_j (n+1-\tfrac{j}{2})(n+1-j)T(n+1-j) \\ &= 2\sum_{j=1}^{k} h_j (n+1-\tfrac{j}{2})T(n+1-j)\end{aligned} \tag{3.11}$$

where g_j and h_j are the coefficients in the polynomials G and H in (3.9) *and* (3.10)*, respectively.*

PROOF. Apply the operator D to both sides of (3.8) to obtain

$$\begin{aligned}\mathrm{D}^3 Z &= \frac{1}{Z}(\mathrm{D}Z)(\mathrm{D}^2 Z) - \frac{1}{2Z^2}(\mathrm{D}Z)^3 + \mathrm{D}(HZ) \\ &= \frac{(\mathrm{D}Z)}{Z}\left\{\mathrm{D}^2 Z - \frac{1}{2Z}(\mathrm{D}Z)^2\right\} + H(\mathrm{D}Z) + (\mathrm{D}H)Z.\end{aligned}$$

The differential equation (3.8) can be used to simplify the term in braces, and the result is

$$\mathrm{D}^3 Z = \frac{(\mathrm{D}Z)}{Z} HZ + H(\mathrm{D}Z) + (\mathrm{D}H)Z = 2H(\mathrm{D}Z) + (\mathrm{D}H)Z.$$

It follows that Z satisfies the third order linear differential equation with respect to X given by

$$\mathrm{D}^3 Z - 2H(\mathrm{D}Z) - (\mathrm{D}H)Z = 0,$$

or equivalently, after dividing by B,

$$X\frac{\mathrm{d}}{\mathrm{d}X}(D^2 Z) - 2HX\frac{\mathrm{d}Z}{\mathrm{d}X} - X\frac{\mathrm{d}H}{\mathrm{d}X}Z = 0. \tag{3.12}$$

By direct calculation, and recalling from (3.9) that $G = B^2$, it is easy to check that

$$X\frac{\mathrm{d}}{\mathrm{d}X}(D^2 Z) = G(\Delta^3 Z) + \frac{3}{2}(\Delta G)(\Delta^2 Z) + \frac{1}{2}(\Delta^2 G)(\Delta Z)$$

where

$$\Delta = X\frac{\mathrm{d}}{\mathrm{d}X}.$$

Hence the differential equation (3.12) takes the form

$$G(\Delta^3 Z) + \frac{3}{2}(\Delta G)(\Delta^2 Z) + \frac{1}{2}(\Delta^2 G)(\Delta Z) = 2H(\Delta Z) + (\Delta H)Z.$$

Now substitute the expansions

$$Z = \sum_{j=0}^{\infty} T(j) X^{j+r}, \quad G = 1 + \sum_{j=1}^{k} g_j X^j \quad \text{and} \quad H = \sum_{j=1}^{k} h_j X^j$$

to obtain

$$\left(1+\sum_{j=1}^{k} g_j X^j\right)\left(\sum_{j=0}^{\infty}(j+r)^3 T(j) X^{j+r}\right)$$
$$+\frac{3}{2}\left(\sum_{j=1}^{k} j g_j X^j\right)\left(\sum_{j=0}^{\infty}(j+r)^2 T(j) X^{j+r}\right)+\frac{1}{2}\left(\sum_{j=1}^{k} j^2 g_j X^j\right)\left(\sum_{j=0}^{\infty}(j+r) T(j) X^{j+r}\right)$$
$$=2\left(\sum_{j=1}^{k} h_j X^j\right)\left(\sum_{j=0}^{\infty}(j+r) T(j) X^{j+r}\right)+\left(\sum_{j=1}^{k} j h_j X^j\right)\left(\sum_{j=0}^{\infty} T(j) X^{j+r}\right).$$

Extracting the coefficient of X^r gives the indicial equation $r^3 = 0$, from which it follows that there is a unique solution that is analytic at $X = 0$ and satisfies the initial condition $Z(0) = Z_0$. Finally, setting $r = 0$ in (3.13) and extracting the coefficient of x^{n+1} leads to

$$(n+1)^3 T(n+1)+\sum_{j=1}^{k} g_j (n+1-j)^3 T(n+1-j)$$
$$+\frac{3}{2}\sum_{j=1}^{k} j g_j (n+1-j)^2 T(n+1-j)+\frac{1}{2}\sum_{j=1}^{k} j^2 g_j (n+1-j) T(n+1-j)$$
$$=2\sum_{j=1}^{k} h_j (n+1-j) T(n+1-j)+\sum_{j=1}^{k} j h_j T(n+1-j),$$

where $T(n+1-j)$ is defined to be zero if $j > n+1$. This simplifies to give (3.11). □

Now we can return to solving (3.6) for the data given by (3.5) and (3.7). Application of Theorem 3.1 with

$$Z = y, \quad X = w, \quad B^2 = 1+2\alpha w+(\alpha^2+4\gamma)w^2 \text{ and } H = (2\beta-\alpha)w-\frac{(\alpha^2+4\gamma)w^2}{2}$$

gives

$$y = \sum_{n=0}^{\infty} s(n) w^n, \tag{3.13}$$

where the coefficients $s(n)$ satisfy the recurrence relation

$$(n+1)^3 s(n+1)+2\alpha(n+1)\left(n+\frac{1}{2}\right) n s(n)+(\alpha^2+4\gamma)(n+1)(n)(n-1)s(n-1)$$
$$=2(2\beta-\alpha)\left(n+\frac{1}{2}\right)s(n)-(\alpha^2+4\gamma) n s(n-1),$$

and this simplifies to

$$(n+1)^3 s(n+1) = -(2n+1)(\alpha n^2+\alpha n+\alpha-2\beta)s(n)-(\alpha^2+4\gamma)n^3 s(n-1). \tag{3.14}$$

On combining (3.1), (3.2), (3.3), (3.13) and (3.14), we deduce the following theorem of G. Almkvist, D. van Straten and W. Zudilin (see [**1**] or [**16**, p. 342]).

THEOREM 3.2. *Let $\{t(n)\}$ be the sequence that satisfies the initial condition*

$$t(0) = 1$$

TABLE 2. Apéry-like sequences, their modular parameterisations, and entries in Sloane's OEIS [**37**]. The Apéry numbers correspond to the sequence labelled 6(A). The notation is $y = \sum_{n=0}^{\infty} s(n)\, w^n$, where $s(0) = 1$ and

$$(n+1)^3 s(n+1) = -(2n+1)(\alpha n^2 + \alpha n + \alpha - 2\beta)s(n) - (\alpha^2 + 4\gamma)n^3 s(n-1).$$

Level	(α, β, γ)	w	y	$s(n)$	OEIS
5	$(11, 3, 1)$	$\frac{\eta_5^6}{\eta_1^6}$	$\frac{\eta_1^5}{\eta_5}$	$\sum_j (-1)^{j+n} \binom{n}{j}^3 \binom{4n-5j}{3n}$	A229111
6 (A)	$(-17, -6, -72)$	$\frac{\eta_1^{12}\eta_6^{12}}{\eta_2^{12}\eta_3^{12}}$	$\frac{\eta_2^7\eta_3^7}{\eta_1^5\eta_6^5}$	$\sum_j \binom{n}{j}^2 \binom{n+j}{j}^2$	A005259
6 (B)	$(10, 3, -9)$	$\frac{\eta_2^6\eta_6^6}{\eta_1^6\eta_3^6}$	$\frac{\eta_1^4\eta_3^4}{\eta_2^2\eta_6^2}$	$(-1)^n \sum_j \binom{n}{j}^2 \binom{2j}{j} \binom{2n-2j}{n-j}$	A002895
6 (C)	$(7, 2, 8)$	$\frac{\eta_3^4\eta_6^4}{\eta_1^4\eta_2^4}$	$\frac{\eta_1^3\eta_2^3}{\eta_3\eta_6}$	$\sum_j (-3)^{n-3j} \binom{n+j}{j} \binom{n}{j} \binom{n-j}{j} \binom{n-2j}{j}$	A125143
8	$(12, 4, -32)$	$\frac{\eta_1^8\eta_8^8}{\eta_2^8\eta_4^8}$	$\frac{\eta_2^6\eta_4^6}{\eta_1^4\eta_8^4}$	$\sum_j \binom{n}{j}^2 \binom{2j}{n}^2$	A290575
9	$(-9, -3, -27)$	$\frac{\eta_1^6\eta_9^6}{\eta_3^{12}}$	$\frac{\eta_3^{10}}{\eta_1^3\eta_9^3}$	$\sum_{j,\ell} \binom{n}{j}^2 \binom{n}{\ell} \binom{j}{\ell} \binom{j+\ell}{n}$	A290576

and the quadratic three-term recurrence relation

$$(n+1)^2 t(n+1) = (\alpha n^2 + \alpha n + \beta)t(n) + \gamma n^2 t(n-1)$$

for $n \geqslant 0$. Then in a neighborhood of $x = 0$, the following identity holds:

$$\left(\sum_{n=0}^{\infty} t(n) x^{n+\frac{1}{2}}\right)^2 = \sum_{n=0}^{\infty} s(n) \left(\frac{x}{1 - \alpha x - \gamma x^2}\right)^{n+1}, \tag{3.15}$$

where $\{s(n)\}$ satisfies the cubic three-term relation

$$(n+1)^3 s(n+1) = -(2n+1)(\alpha n^2 + \alpha n + \alpha - 2\beta)s(n) - (\alpha^2 + 4\gamma)n^3 s(n-1) \tag{3.16}$$

and initial condition $s(0) = 1$.

In particular, taking $(\alpha, \beta, \gamma) = (-17, -6, -72)$ in the recurrence (3.16) gives the recurrence (1.1) for the Apéry numbers, and since the initial conditions are the same, we have $s(n) = A(n)$. Hence, this instance of the theorem gives

$$\left(\sum_{n=0}^{\infty} t(n) x^{n+\frac{1}{2}}\right)^2 = \sum_{n=0}^{\infty} s(n) \left(\frac{x}{(1+8x)(1+9x)}\right)^{n+1}$$

where $s(n)$ are the Apéry numbers and $t(n)$ are the numbers in the sequence labelled 6 (A) in Table 1.

The six sequences $\{s(n)\}$ that correspond to the parameter values $(\alpha, \beta, \gamma) \in S$, where S is given by (2.4), are all integer valued and have expressions as sums of terms involving binomial coefficients and the details are summarised in Table 2. Explicit formulas for the respective modular forms w and y are also given in Table 2. Once again, proofs are case-by-case and can be found, for example, in Chapters 5, 6, 8 and 9 of [**16**].

The functions z in Table 1 are modular forms of weight one, while the functions y in Table 2 are modular forms of weight two. For that reason we shall refer to the sequences $t(n)$ and $s(n)$ as the weight one and weight two examples, respectively.

4. Jacobi's parameterisation, Ramanujan's alternative theories, and other levels

Perhaps the oldest modular parameterisation is due to Jacobi and involves the complete elliptic integral of the first kind. Ramanujan took the square of Jacobi's result to get [**34**, (25)]

$$\left(\frac{2K}{\pi}\right)^2 = 1+\left(\frac{1}{2}\right)^3 (2kk')^2+\left(\frac{1\cdot 3}{2\cdot 4}\right)^3 (2kk')^4+\cdots = \sum_{n=0}^{\infty}\binom{2n}{n}^3\left(\frac{kk'}{4}\right)^{2n}. \tag{4.1}$$

He then utilised q-expansions for K, k and k' that are originally due to Jacobi. A complete proof of the modular parameterisation of (4.1), e.g., see [**16**, Th. 4.30], is normally quite long and involves first establishing Jacobi's results and then invoking Clausen's identity for the square of a ${}_2F_1$ hypergeometric function. In view of the fundamental importance of (4.1) in Ramanujan's work, especially in deriving series for $1/\pi$, it is instructive and useful to have a direct proof. We shall give such a proof that uses only well-known properties of theta functions that can be found in [**5**, Ch. 16 and 17] or [**16**, Ch. 3]. We will write Z for $(2K/\pi)^2$ and X for $(kk'/4)^2$. Here is a detailed formulation of Ramanujan's result (4.1).

THEOREM 4.1. *Let w, X and Z be defined by*

$$w = \frac{\eta_4^8}{\eta_1^8}, \quad \frac{1}{X} = \frac{1}{w} + 32 + 256w \quad \text{and} \quad Z = \frac{\eta_1^2\eta_4^2}{X^{5/12}}.$$

Then:

(a) *Another formula for X is*

$$X = \frac{\eta_1^{24}\eta_4^{24}}{\eta_2^{48}}.$$

Hence,

$$Z = \frac{\eta_2^{20}}{\eta_1^8\eta_4^8} \quad \text{and} \quad Z^2X = \frac{\eta_1^8\eta_4^8}{\eta_2^8}.$$

(b) *The differentiation formula*

$$q\frac{\mathrm{d}X}{\mathrm{d}q} = Z\,X\,B$$

holds, where $B = \sqrt{1-64X}$.

(c) *The differential equation*

$$\mathrm{D}^2 Z - \frac{1}{2Z}\left(\mathrm{D}Z\right)^2 = HZ$$

holds, where

$$\mathrm{D} = B\,X\,\frac{\mathrm{d}}{\mathrm{d}X} \quad \textit{and} \quad H = 8X,$$

and B is as above.

(d) *In a neighbourhood of $q=0$, the following power series expansion holds:*

$$Z = \sum_{n=0}^{\infty} \binom{2n}{n}^3 X^n.$$

PROOF OF THEOREM. (a) By the definition of X we have

$$\frac{1}{X} = \left(\frac{1}{w^{1/2}} + 16w^{1/2}\right)^2 = \left(\frac{\eta_1^4}{\eta_4^4} + 16\frac{\eta_4^4}{\eta_1^4}\right)^2 = \frac{\eta_2^8}{\eta_1^8\eta_4^8}\left(\frac{\eta_1^8}{\eta_2^4} + 16\frac{\eta_4^8}{\eta_2^4}\right)^2.$$

The terms in parentheses can be expressed in terms of theta functions, e.g., see [**16**, Th. 3.5], to yield

$$\frac{1}{X} = \frac{\eta_2^8}{\eta_1^8\eta_4^8}\left(\varphi^4(-q) + 16q\psi^4(q)\right)^2,$$

where

$$\varphi(q) = \sum_{j=-\infty}^{\infty} q^{j^2} \quad \text{and} \quad \psi(q) = \sum_{j=0}^{\infty} q^{j(j+1)/2}.$$

By an identity of Jacobi, e.g., [**16**, Th. 3.7] this simplifies to

$$\frac{1}{X} = \frac{\eta_2^8}{\eta_1^8\eta_4^8} \times \varphi^8(q),$$

and on applying [**16**, Th. 3.5] again to convert the theta function to an eta-quotient we obtain

$$\frac{1}{X} = \frac{\eta_2^8}{\eta_1^8\eta_4^8} \times \left(\frac{\eta_2^5}{\eta_1^2\eta_4^2}\right)^8 = \frac{\eta_2^{48}}{\eta_1^{24}\eta_4^{24}}.$$

The result for X follows on taking reciprocals, while expressions for Z and Z^2X are obtained by trivially by using the result for X in the definition of Z. This completes the proof of (a).

Now will prove (b). The strategy will be to first compute the derivative of w and then use the chain rule to calculate the derivative of X. We have

$$\begin{aligned} q\frac{\mathrm{d}}{\mathrm{d}q}\log w &= q\frac{\mathrm{d}}{\mathrm{d}q}\log\left(q\prod_{j=1}^{\infty}\frac{(1-q^{4j})^8}{(1-q^j)^8}\right) \\ &= 1 - 32\sum_{j=1}^{\infty}\frac{jq^{4j}}{1-q^{4j}} + 8\sum_{j=1}^{\infty}\frac{jq^j}{1-q^j} \\ &= \frac{4P(q^4) - P(q)}{3}, \end{aligned} \tag{4.2}$$

where

$$P(q) = 1 - 24\sum_{j=1}^{\infty}\frac{jq^j}{1-q^j}.$$

We now apply the sum of four squares formula [**16**, Th. 3.26] followed by [**16**, Th. 3.5] to convert the resulting theta function to a product, to obtain

$$q\frac{\mathrm{d}}{\mathrm{d}q}\log w = \varphi^4(q) = \frac{\eta_2^{20}}{\eta_1^8\eta_4^8}.$$

By Part (a) this simplifies further to

$$q\frac{\mathrm{d}}{\mathrm{d}q}\log w = Z. \tag{4.3}$$

For future reference, we note that (4.2) and (4.3) imply

$$Z = \frac{4P(q^4) - P(q)}{3}. \tag{4.4}$$

Next, applying the operator $q\frac{\mathrm{d}}{\mathrm{d}q}$ to the relation

$$\frac{1}{X} = \frac{1}{w} + 32 + 256w$$

we obtain

$$-\frac{1}{X^2}q\frac{\mathrm{d}X}{\mathrm{d}q} = \left(\frac{-1}{w^2} + 256\right)q\frac{\mathrm{d}w}{\mathrm{d}q},$$

and so

$$q\frac{\mathrm{d}X}{\mathrm{d}q} = X^2\left(\frac{1}{w} - 256w\right)q\frac{\mathrm{d}}{\mathrm{d}q}\log w.$$

By (4.3) and by further calculation, this becomes

$$\begin{aligned}
q\frac{\mathrm{d}X}{\mathrm{d}q} &= X^2 \times \sqrt{\left(\frac{1}{w} - 256w\right)^2} \times Z \\
&= ZX^2 \times \sqrt{\left(\frac{1}{w} + 256w\right)^2 - 1024} \\
&= ZX^2 \times \sqrt{\left(\frac{1}{X} - 32\right)^2 - 1024} \qquad \text{by the definition of } X \\
&= ZX^2 \times \sqrt{\frac{1}{X^2} - \frac{64}{X}} \\
&= ZXB.
\end{aligned}$$

Now we prove (c). First, by the chain rule and part (b) of the theorem, the differential operator D is related to differentiation with respect to q by

$$\mathrm{D} = BX\frac{\mathrm{d}}{\mathrm{d}X} = BX\,q\frac{\mathrm{d}}{\mathrm{d}q}\Big/\left(q\frac{\mathrm{d}X}{\mathrm{d}q}\right) = \frac{1}{Z}q\frac{\mathrm{d}}{\mathrm{d}q}. \tag{4.5}$$

Next, by the definition of Z we have

$$Z^{12} = \frac{\eta_1^{24}\eta_4^{24}}{X^5}.$$

Take the logarithm and apply $q\frac{\mathrm{d}}{\mathrm{d}q}$ to obtain

$$\frac{12}{Z}q\frac{\mathrm{d}Z}{\mathrm{d}q} = P(q) + 4P(q^4) - 5q\frac{\mathrm{d}}{\mathrm{d}q}\log X.$$

Now use the result of part (b) and (4.5) to deduce

$$12DZ + 5BZ = 4P(q^4) + P(q). \tag{4.6}$$

Next, apply the differential operator D to (4.6) and make use of (4.5) to obtain

$$12\mathrm{D}^2 Z + 5\mathrm{D}(BZ) = \frac{1}{Z} q\frac{\mathrm{d}}{\mathrm{d}q}\left(4P(q^4) + P(q)\right).$$

It remains to calculate the derivative on the right hand side. On using the formula [**16**, p. 29]

$$q\frac{\mathrm{d}}{\mathrm{d}q}P(q) = \frac{P^2(q) - Q(q)}{12} \quad \text{where } Q(q) = 1 + 240\sum_{j=1}^{\infty}\frac{j^3 q^j}{1-q^j}$$

we obtain

$$12\mathrm{D}^2 Z + 5\mathrm{D}(BZ) = \frac{1}{12Z}\left(16P^2(q^4) - 16Q(q^4) + P^2(q) - Q(q)\right). \tag{4.7}$$

To handle the $16Q(q^4)$ and $Q(q)$ terms, use the sum of eight squares and sum of eight triangular numbers formulas, e.g., see [**16**, Th. 3.39], to deduce

$$16Q(q^4) + Q(q) = 16\varphi^8(q) + \varphi^8(-q) + 256q^2\psi^8(q^2),$$

which we rewrite as

$$16Q(q^4) + Q(q) = 16\varphi^8(q) + \left(\varphi^4(-q) + 16q\psi^4(q^2)\right)^2 - 32q\varphi^4(-q)\psi^4(q^2).$$

By Jacobi's identity [**16**, Th. 3.7] this simplifies to

$$16Q(q^4) + Q(q) = 17\varphi^8(q) - 32q\varphi^4(-q)\psi^4(q^2).$$

Now use [**16**, Th. 3.5] to convert the theta functions to products, then use part (a) of the theorem to deduce

$$16Q(q^4) + Q(q) = 17\frac{\eta_2^{40}}{\eta_1^{16}\eta_4^{16}} - 32\frac{\eta_1^8\eta_4^8}{\eta_2^8} = 17Z^2 - 32Z^2X.$$

Hence, (4.7) simplifies to

$$12\mathrm{D}^2 Z + 5\mathrm{D}(BZ) = \frac{1}{12Z}\left(16P^2(q^4) + P^2(q) + Z^2(32X - 17)\right). \tag{4.8}$$

In order to handle the $16P^2(q^4)$ and $P^2(q)$ terms, observe that the formulas (4.4) and (4.6) can be combined to give

$$4P(q^4) = \frac{1}{2}\left(12\mathrm{D}Z + 5BZ + 3Z\right) \quad \text{and} \quad P(q) = \frac{1}{2}\left(12\mathrm{D}Z + 5BZ - 3Z\right).$$

Squaring and adding gives

$$\begin{aligned} 16P^2(q^4) + P^2(q) &= \frac{1}{2}(12\mathrm{D}Z + 5BZ)^2 + \frac{9}{2}Z^2 \\ &= 72(\mathrm{D}Z)^2 + 60BZ(\mathrm{D}Z) + \frac{25}{2}B^2Z^2 + \frac{9}{2}Z^2, \end{aligned}$$

and substituting back in (4.8) gives

$$12\mathrm{D}^2 Z + 4\mathrm{D}(BZ) = \frac{1}{12Z}\left(72(\mathrm{D}Z)^2 + 48BZ(\mathrm{D}Z) + 8B^2Z^2 - 8Z^2\right). \tag{4.9}$$

Now divide by 12 and rearrange terms to obtain

$$\begin{aligned}
\mathrm{D}^2 Z &= \frac{1}{2Z}(\mathrm{D}Z)^2 + \frac{5}{12}B(\mathrm{D}Z) - \frac{5}{12}\mathrm{D}(BZ) + \frac{Z}{288}(25B^2 - 25 + 64X) \\
&= \frac{1}{2Z}(\mathrm{D}Z)^2 + 8XZ
\end{aligned}$$

where the last step follows by simple calculation using the definitions of B and D. This completes the proof of (c).

Finally we prove (d). By the result of (c) and application of Theorem 3.1 with $B^2 = 1 - 64X$ and $H = 8X$ it follows that the coefficients $T(n)$ in the expansion $Z = \sum_{n=0}^{\infty} T(n)X^n$ satisfy the two-term recurrence relation

$$(n+1)^3 T(n+1) + (n+1)(-64)\left(n+\tfrac{1}{2}\right) nT(n) = 16\left(n+\tfrac{1}{2}\right) T(n)$$

which simplifies to

$$(n+1)^3 T(n+1) = 8(2n+1)^3 T(n).$$

On solving the recurrence and noting that $T(0) = 1$ because $Z = 1$ and $X = 0$ when $q = 0$, we deduce that

$$T(n) = \binom{2n}{n}^3.$$

This completes the proof of Theorem 4.1 and establishes the modular parameterisation in the formula (4.1) that was used by Ramanujan. □

To give another example in explicit detail, here is the result for level 10.

THEOREM 4.2. *Let w, X and Z be defined by*

$$w = \frac{\eta_2^4 \eta_{10}^4}{\eta_1^4 \eta_5^4}, \quad \frac{1}{X} = \frac{1}{w} + 8 + 16w \quad \text{and} \quad Z = \frac{\eta_1 \eta_2 \eta_5 \eta_{10}}{X^{3/4}}.$$

Then:

(a) *The differentiation formula*

$$q\frac{\mathrm{d}X}{\mathrm{d}q} = Z\,X\,B$$

holds, where $B = \sqrt{(1+4X)(1-16X)}$.

(b) *The differential equation*

$$\mathrm{D}^2 Z - \frac{1}{2Z}(\mathrm{D}Z)^2 = HZ$$

holds, where

$$\mathrm{D} = B\,X\frac{\mathrm{d}}{\mathrm{d}X} \quad \text{and} \quad H = 2X(1+15X),$$

and B is as above.

(c) *The coefficients $T(n)$ in the expansion $Z = \sum_{n=0}^{\infty} T(n)X^n$ satisfy the three-term recurrence relation*

$$\begin{aligned}
&(n+1)^3 T(n+1) \\
&\quad = 2(2n+1)(3n^2+3n+1)T(n) + 4n(16n^2-1)T(n-1)
\end{aligned}$$

and initial condition $T(0) = 1$, the single initial condition being sufficient to start the recurrence. Hence, in a neighbourhood of $q = 0$, the following power series expansion holds:

$$Z = \sum_{n=0}^{\infty} \left\{ \sum_{j=0}^{n} \binom{n}{j}^4 \right\} X^n.$$

PROOF. Parts (a) and (c) follow from Theorems 10.15, 10.17, 10.19 and 10.24 in [**16**]. Part (b) follows from Theorem 10.20 in [**16**] by change of variable. A different proof of (c) has been given in [**10**]. □

It may be remarked that the three-term recurrence relation in Part (c) of Theorem 4.2 follows immediately from the result of Part (b) by application of Theorem 3.1. Moreover, following the comments after (1.2), a computer algebra system can be used to show that the sequence defined by

$$T(n) = \sum_{j=0}^{n} \binom{n}{j}^4$$

is a solution to the recurrence relation.

The examples in Theorems 4.1 and 4.2 have some key aspects in common:

(i) They start with a modular function X and a weight two modular form Z. The weight two modular form is normally defined by

$$Z = \left(\prod_{d|\ell} \eta_d^{e_d} \right) \times \frac{1}{X^m}$$

where d ranges over the divisors of the level ℓ, the exponents e_d are integers with $\sum_{d|\ell} e_d = 4$, and

$$m = \frac{1}{24} \sum_{d|\ell} d e_d.$$

(ii) There is a polynomial $G = G(X)$ with $G(0) = 1$ for which the differentiation formula

$$q \frac{\mathrm{d}X}{\mathrm{d}q} = BXZ \tag{4.10}$$

holds with $B = \sqrt{G}$.

(iii) There is a polynomial $H = H(X)$ with $H(0) = 0$ for which the differential equation

$$\mathrm{D}^2 Z - \frac{1}{2Z} (\mathrm{D}Z)^2 = HZ \tag{4.11}$$

holds, where D is the differential operator

$$\mathrm{D} = BX \frac{\mathrm{d}}{\mathrm{d}X}$$

and B is as for (ii).

(iv) If k is the maximum of the degrees of the polynomials G and H, then the coefficients $T(n)$ in the expansion $Z = \sum_{n=0}^{\infty} T(n) X^n$ satisfy a $(k+1)$-term recurrence relation given by (3.11).

Further examples of functions X and Z satisfying (i)–(iv) are given in Tables 3–4 and 5–6. The results of Theorems 4.1 and 4.2 appear in the tables as the entries corresponding to levels 4 and 10, respectively. Tables 3–4 and 5–6 extend [**16**, Tables 14.1 and 14.2]. Explicit formulas for $T(n)$ in terms of sums of binomial coefficients, where such formulas are known, are listed in Tables 3–4. Otherwise, by Theorem 3.1, the number of terms in the recurrence relation is one more than the maximum of the degrees of B^2 and H, and this is recorded in the last column of the table.

In principle it is possible to extend Tables 3–4 and 5–6 further to other subgroups of $\mathrm{SL}_2(\mathbb{R})$ of genus zero as per the following result which we quote exactly from Zagier's survey [**44**, Prop. 21].

THEOREM 4.3. *Let $f(z)$ be a (holomorphic or meromorphic) modular form of positive weight k on some group Γ and $t(z)$ a modular function with respect to Γ. Express $f(z)$ (locally) as $\varphi(t(z))$. Then the function $\varphi(t)$ satisfies a linear differential equation of order $k+1$ with algebraic coefficients, or with polynomial coefficients if $\Gamma \setminus \mathfrak{H}$ has genus 0 and $t(z)$ generates the field of modular functions on Γ.*

The results in Theorems 4.1 and 4.2, and indeed all of the entries in Tables 3–4 and 5–6, correspond to particular and explicit instances of Zagier's Theorem 4.3 in the case $k = 2$. The entries in Tables 1 and 2 are particular instances of the cases $k = 1$ and $k = 2$, respectively, of Zagier's Theorem 4.3.

The entries for w in Tables 2 and 3–4 are identical for the levels labelled 5, 6(A), 6(B), 6(C), 8 and 9. Closer inspection of Tables 2 and 3–4 shows that for those levels, X is related to w by the change of variable

$$X = \frac{w}{(1+\alpha w)^2 + 4\gamma w^2}, \tag{4.12}$$

where (α, β, γ) is the appropriate triple from the set S in (2.4). Moreover, the corresponding modular form Z in Tables 3–4 is related to the corresponding modular form y in Table 2 by

$$X^{1/2} Z = w^{1/2} y. \tag{4.13}$$

Applying this change of variables to the formulas (3.4) and (3.5) leads to the differentiation formula

$$q\frac{\mathrm{d}X}{\mathrm{d}q} = ZXB$$

and the differential equation

$$\tilde{\mathrm{D}}^2 Z - \frac{1}{2Z}\left(\tilde{\mathrm{D}}Z\right)^2 = \tilde{H}Z, \tag{4.14}$$

where

$$B^2 = 1 - 4\alpha X - 16\gamma X^2, \quad \tilde{H} = 2X(\beta + 3\gamma X) \quad \text{and} \quad \tilde{\mathrm{D}} = BX\frac{\mathrm{d}}{\mathrm{d}X}. \tag{4.15}$$

Application of Theorem 3.1, with B^2 and $\tilde{H}$ as above, shows that the coefficients $T(n)$ in the expansion $Z = \sum_{n=0}^{\infty} T(n) X^n$ satisfy the three term recurrence relation

$$(n+1)^3 T(n+1) = 2(2n+1)(\alpha n^2 + \alpha n + \beta)T(n) + 4\gamma n(4n^2-1)T(n-1). \tag{4.16}$$

TABLE 3. Data for $Z=\sum_{n=0}^{\infty}T(n)X^n$. The notation is:

$$Q=1+240\sum_{j=1}^{\infty}\frac{j^3q^j}{1-q^j},\quad R=1-504\sum_{j=1}^{\infty}\frac{j^5q^j}{1-q^j},$$
$$\eta_N=q^{N/24}\prod_{j=1}^{\infty}(1-q^{jN}),\quad \theta_{a,b,c}=\sum_{j=-\infty}^{\infty}\sum_{k=-\infty}^{\infty}q^{aj^2+bjk+ck^2}.$$

level	w	X	Z	$T(n)$
1	$\frac{1}{432}\left(\frac{Q^{3/2}-R}{Q^{3/2}+R}\right)$	$\frac{w}{(1+432w)^2}$	$\frac{\eta_1^4}{X^{1/6}}$	$\binom{6n}{3n}\binom{3n}{n}\binom{2n}{n}$
2	$\frac{\eta_2^{24}}{\eta_1^{24}}$	$\frac{w}{(1+64w)^2}$	$\frac{\eta_1^2\eta_2^2}{X^{1/4}}$	$\binom{4n}{2n}\binom{2n}{n}^2$
3	$\frac{\eta_3^{12}}{\eta_1^{12}}$	$\frac{w}{(1+27w)^2}$	$\frac{\eta_1^2\eta_3^2}{X^{1/3}}$	$\binom{3n}{n}\binom{2n}{n}^2$
4	$\frac{\eta_4^8}{\eta_1^8}$	$\frac{w}{(1+16w)^2}$	$\frac{\eta_1^2\eta_4^2}{X^{5/12}}$	$\binom{2n}{n}^3$
5	$\frac{\eta_5^6}{\eta_1^6}$	$\frac{w}{1+22w+125w^2}$	$\frac{\eta_1^2\eta_5^2}{X^{1/2}}$	$\binom{2n}{n}\sum_j\binom{n}{j}^2\binom{n+j}{j}$
6 (A)	$\frac{\eta_1^{12}\eta_6^{12}}{\eta_2^{12}\eta_3^{12}}$	$\frac{w}{1-34w+w^2}$	$\frac{\eta_1\eta_2\eta_3\eta_6}{X^{1/2}}$	$\binom{2n}{n}\sum_{j,\ell}(-8)^{n-j}\binom{n}{j}\binom{j}{\ell}^3$
6 (B)	$\frac{\eta_2^6\eta_6^6}{\eta_1^6\eta_3^6}$	$\frac{w}{1+20w+64w^2}$	$\frac{\eta_1\eta_2\eta_3\eta_6}{X^{1/2}}$	$\binom{2n}{n}\sum_j\binom{n}{j}^2\binom{2j}{j}$
6 (C)	$\frac{\eta_3^4\eta_6^4}{\eta_1^4\eta_2^4}$	$\frac{w}{1+14w+81w^2}$	$\frac{\eta_1\eta_2\eta_3\eta_6}{X^{1/2}}$	$\binom{2n}{n}\sum_j\binom{n}{j}^3$
7	$\frac{\eta_7^4}{\eta_1^4}$	$\frac{w}{1+13w+49w^2}$	$\frac{\eta_1^2\eta_7^2}{X^{2/3}}$	$\sum_j\binom{n}{j}^2\binom{2j}{n}\binom{n+j}{j}$
8	$\frac{\eta_1^8\eta_8^8}{\eta_2^8\eta_4^8}$	$\frac{1}{1-24w+16w^2}$	$\frac{\eta_2^2\eta_4^2}{X^{1/2}}$	$\binom{2n}{n}(-1)^n\sum_j\binom{n}{j}\binom{2j}{j}\binom{2n-2j}{n-j}$
9	$\frac{\eta_1^6\eta_9^6}{\eta_3^{12}}$	$\frac{w}{1-18w-27w^2}$	$\frac{\eta_3^4}{X^{1/2}}$	$\binom{2n}{n}\sum_j(-3)^{n-3j}\binom{n}{j}\binom{n-j}{j}\binom{n-2j}{j}$
10	$\frac{\eta_5^2\eta_{10}^2}{\eta_1^2\eta_2^2}$	$\frac{w}{1+6w+25w^2}$	$\frac{\eta_1\eta_2\eta_5\eta_{10}}{X^{3/4}}$	$\sum_j\binom{n}{j}^4$
11		$\left(\frac{\eta_1\eta_{11}}{\theta_{1,1,3}}\right)^2$	$\frac{\eta_1^2\eta_{11}^2}{X}$	4-term recurrence relation
12	$\frac{\eta_1^4\eta_{12}^4}{\eta_3^4\eta_4^4}$	$\frac{w}{(1+w)^2}$	$\frac{\eta_1\eta_3\eta_4\eta_{12}}{X^{5/6}}$	$\sum_j\binom{n}{j}^2\binom{2j}{j}\binom{2n-2j}{n-j}$

But it is easy to check that if the coefficients $t(n)$ satisfy the quadratic three-term recurrence relation (2.3), then the coefficients $\binom{2n}{n}t(n)$ will satisfy the cubic three-term recurrence relation (4.16). It follows that for each of the six parameter sets

TABLE 4. A continuation of Table 3.

level	w	X	Z	$T(n)$
13	$\frac{\eta_{13}^2}{\eta_1^2}$	$\frac{w}{1+5w+13w^2}$	$\frac{\eta_1^2\eta_{13}^2}{X^{7/6}}$	4-term recurrence relation
14 (A)	$\frac{\eta_1^4\eta_{14}^4}{\eta_2^4\eta_7^4}$	$\frac{w}{(1-w)^2}$	$\frac{\eta_1\eta_2\eta_7\eta_{14}}{X}$	$\sum_{j,k}\binom{n+j}{2j+2k}\binom{2j+2k}{j+k}\binom{2k}{k}^2\binom{k}{j}$
14 (B)	$\frac{\eta_1^4\eta_{14}^4}{\eta_2^4\eta_7^4}$	$\frac{w}{(1+w)^2}$	$\frac{\eta_1\eta_2\eta_7\eta_{14}}{X}$	4-term recurrence relation
15 (A)	$\frac{\eta_3^2\eta_{15}^2}{\eta_1^2\eta_5^2}$	$\frac{w}{(1+3w)^2}$	$\frac{\eta_1\eta_3\eta_5\eta_{15}}{X}$	4-term recurrence relation
15 (B)	$\frac{\eta_3^2\eta_{15}^2}{\eta_1^2\eta_5^2}$	$\frac{w}{(1-3w)^2}$	$\frac{\eta_1\eta_3\eta_5\eta_{15}}{X}$	4-term recurrence relation
18	$\frac{\eta_1^2\eta_2^2\eta_9^2\eta_{18}^2}{\eta_3^4\eta_6^4}$	$\frac{w}{(1+3w)^2}$	$\frac{\eta_3^2\eta_6^2}{X^{3/4}}$	$\sum_j(-1)^j\binom{n}{j}\binom{2j}{j}\binom{2n-2j}{n-j}\binom{2n-3j}{n}$
20	$\frac{\eta_1^2\eta_{20}^2}{\eta_4^2\eta_5^2}$	$\frac{w}{(1+w)^2}$	$\frac{\eta_2^2\eta_{10}^2}{X}$	4-term recurrence relation
21	$\frac{\eta_1^2\eta_{21}^2}{\eta_3^2\eta_7^2}$	$\frac{w}{(1-w)^2}$	$\frac{\eta_1\eta_3\eta_7\eta_{21}}{X^{4/3}}$	4-term recurrence relation
22	$\frac{\eta_2^2\eta_{22}^2}{\eta_1^2\eta_{11}^2}$	$\frac{w}{(1+2w)^2}$	$\frac{\eta_1\eta_2\eta_{11}\eta_{22}}{X^{3/2}}$	5-term recurrence relation
23		$\frac{2\eta_1\eta_{23}}{\theta_{1,1,6}+\theta_{2,1,3}}$	$\frac{\eta_1^2\eta_{23}^2}{X^2}$	7-term recurrence relation
24	$\frac{\eta_1^2\eta_3^2\eta_8^2\eta_{24}^2}{\eta_2^2\eta_4^2\eta_6^2\eta_{12}^2}$	$\frac{w}{1+4w^2}$	$\frac{\eta_2\eta_4\eta_6\eta_{12}}{X}$	$\sum_j\binom{n}{2j}\binom{2j}{j}^2\binom{2n-4j}{n-2j}$
33	$\frac{\eta_3\eta_{33}}{\eta_1\eta_{11}}$	$\frac{w}{1+w+3w^2}$	$\frac{\eta_1\eta_3\eta_{11}\eta_{33}}{X^2}$	6-term recurrence relation
35	$\frac{\eta_1\eta_{35}}{\eta_5\eta_7}$	$\frac{w}{1+w-w^2}$	$\frac{\eta_1\eta_5\eta_7\eta_{35}}{X^2}$	6-term recurrence relation

$(\alpha,\beta,\gamma)\in S$ we have

$$T(n)=\binom{2n}{n}t(n)$$

as can be seen in Table 3 for the levels labelled 5, 6(A), 6(B), 6(C), 8 and 9. The discussion above provides proof of the following result of of H. H. Chan, Y. Tanigawa, Y. Yang and W. Zudilin (see [**10**] or [**16**, p. 342]).

TABLE 5. Data for B^2 and H, where X and Z are as for Tables 3–4,

$$B = \frac{q}{ZX}\frac{\mathrm{d}X}{\mathrm{d}q}, \quad D^2Z - \frac{1}{2Z}(DZ)^2 = HZ, \quad \text{and} \quad D = BX\frac{\mathrm{d}}{\mathrm{d}X}.$$

level	B^2	H	References
1	$1-1728X$	$120X$	[16], [34]
2	$1-256X$	$24X$	[16], [34]
3	$1-108X$	$12X$	[16], [34]
4	$1-64X$	$8X$	[16], [34]
5	$1-44X-16X^2$	$6X(1+X)$	[10], [16]
6 (A)	$(1+32X)(1+36X)$	$-12X(1+36X)$	[10], [16]
6 (B)	$(1-4X)(1-36X)$	$6X(1-9X)$	[10], [16]
6 (C)	$(1+4X)(1-32X)$	$4X(1+12X)$	[10], [16]
7	$(1+X)(1-27X)$	$4X(1+3X)$	[14], [16]
8	$(1+16X)(1+32X)$	$-8X(1+24X)$	[10], [16]
9	$1+36X+432X^2$	$-6X(1+27X)$	[10], [16]
10	$(1+4X)(1-16X)$	$2X(1+15X)$	[10], [13], [16]
11	$1-20X+56X^2-44X^3$	$4X(1-8X+11X^2)$	[16], [17]
12	$(1-4X)(1-16X)$	$4X(1-8X)$	[16], [21]

THEOREM 4.4. *Let $\{t(n)\}$ be the sequence that satisfies the initial condition $t(0) = 1$ and the three-term recurrence relation*

$$(n+1)^2t(n+1) = (\alpha n^2 + \alpha n + \beta)t(n) + \gamma n^2 t(n-1)$$

TABLE 6. A continuation of Table 5.

level	B^2	H	References
13	$(1+X)(1-10X-27X^2)$	$\frac{X}{8}(12+175X+231X^2)$	[**19**]
14 (A)	$(1+4X)(1-10X-7X^2)$	$\frac{X}{2}(2+51X+56X^2)$	[**18**]
14 (B)	$(1-4X)(1-18X+49X^2)$	$\frac{X}{2}(10-141X+392X^2)$	[**20**], [**25**]
15 (A)	$(1-12X)(1-2X+5X^2)$	$\frac{3X}{2}(2-11X+40X^2)$	[**20**]
15 (B)	$(1+12X)(1+22X+125X^2)$	$-\frac{3X}{2}(2+25X)(3+40X)$	[**15**]
18	$(1-12X)(1-16X)$	$6X(1-15X)$	[**14**], [**16**]
20	$(1-4X)(1-12X+16X^2)$	$4X(1-10X+18X^2)$	[**29**]
21	$(1+4X)(1-2X-27X^2)$	$-\frac{X}{2}(2-47X-240X^2)$	[**3**]
22	$(1-8X)(1-4X^2+4X^3)$	$2X(1-3X)(1+4X-10X^2)$	[**3**]
23	$(1-X^2+X^3)(1-8X+3X^2-7X^3)$	$2X(1-X-X^2+12X^3-15X^4+14X^5)$	[**17**]
24	$(1+4X)(1-4X)(1-8X)$	$2X(1+5X-64X^2)$	[**18**], [**42**]
33	$(1-2X-11X^2)(1+4X+8X^2+4X^3)$	$-\frac{X}{2}(2-15X-152X^2-404X^3-264X^4)$	[**3**]
35	$(1-2X+5X^2)(1-8X+16X^2-28X^3)$	$\frac{X}{2}(6-61X+296X^2-580X^3+840X^4)$	[**3**]

for $n \geqslant 0$. Then in a neighborhood of $x=0$, the following identity holds:

$$\left(\sum_{n=0}^{\infty} t(n)x^n\right)^2 = \frac{1}{1+\gamma x^2}\sum_{n=0}^{\infty}\binom{2n}{n}t(n)\left(\frac{x(1-\alpha x-\gamma x^2)}{(1+\gamma x^2)^2}\right)^n. \tag{4.17}$$

PROOF. Note that the change of variables (3.3) along with (4.12) and (4.13) can be combined to give the overall change of variable

$$Z=(1+\gamma x^2)z^2 \quad \text{and} \quad X=\frac{x(1-\alpha x-\gamma x^2)}{(1+\gamma x^2)^2}.$$

Then apply Theorem 3.1 for the data in (4.14) and (4.15). □

As already mentioned, the data in Tables 3 and 5 for level 4 corresponds to Ramanujan's example worked out in detail in Theorem 4.1. It is called the classical theory. The data for levels 1, 2 and 3 correspond to what are called Ramanujan's theories of elliptic functions to alternative bases, e.g., see [**6**] or [**16**, Chapter 4].

The recurrence relations for the sequences $T(n)$ in Tables 3–4 for levels 7, 10 and 18 are of the form

$$(n+1)^3T(n+1) = (2n+1)(an^2+an+b)T(n) + n(cn^2+d)T(n-1) \tag{4.18}$$

for $(a,b,c,d) = (6,2,64,-4)$, $(13,4,27,-3)$ and $(14,6,-192,12)$, respectively. The recurrence (4.18) is a generalisation of the recurrences (3.16) and (4.16). Precisely, the recurrences (3.16) and (4.16) correspond to the special cases $d=0$ and $c=-4d$ respectively of (3.16), and that is why the level 7, 10 and 18 sequences do not occur in Zagier's examples in Section 2. The sequences $T(n)$ nevertheless have interesting number theoretic properties. For example, A. Malik and A. Straub [**31**] showed that the sequences $T(n)$ in Tables 3–4 corresponding to levels 7, 10 and 18, as well as all of the sequences $t(n)$ and $s(n)$ in Tables 1 and 2, all satisfy Lucas congruences of the form (1.4) for every prime p.

All of the sequences $\{T(n)\}$ in Tables 3–4 are integer valued, except for the sequence corresponding to level 13 which will be discussed in detail in Section 6. Before doing that, the level 11 theory will be discussed in Section 5. In Sections 7 and 8 a more detailed investigation is performed for the level 14 and level 15 theories. This will result in new Apéry-like sequences that take values in $\mathbb{Z}[\sqrt{2}]$ for level 14, and values in $\mathbb{Z}[i]$ for level 15. The remaining sequences from Tables 3–4 that satisfy four-term recurrence relations are investigated for Lucas congruences in Section 9.

5. The level 11 theory

In this section we analyse the data in Tables 3 and 5 for level 11 in detail and make a conjecture. According to those tables, Theorem 3.1 can be applied to the functions

$$Z = \left(\sum_{j=-\infty}^{\infty}\sum_{k=-\infty}^{\infty} q^{j^2+jk+3k^2}\right)^2 \quad \text{and} \quad X = \frac{\eta_1^2\eta_{11}^2}{Z}$$

with data given by

$$B^2 = 1-20X+56X^2-44X^3 \quad \text{and} \quad H = 4X(1-8X+11X^2)$$

to deduce that the coefficients $T(n)$ in the expansion

$$Z = \sum_{n=0}^{\infty} T(n)X^n$$

satisfy the four-term recurrence relation

$$\begin{aligned}(n+1)^3T(n+1) &= 2(2n+1)(5n^2+5n+2)T(n)\\ &\quad - 8n(7n^2+1)T(n-1)\\ &\quad + 22n(2n-1)(n-1)T(n-2).\end{aligned}$$

The single initial condition $T(0)=1$ is enough to start the recurrence. It follows from the q-expansions for X and Z that $T(n)$ is always an integer. The first few terms $\{T(n)\}_{n=0}^{10}$ are given by

$$1, 4, 28, 268, 3004, 36784, 476476, 6418192, 88986172, 1261473136, 18200713168.$$

Numerical computations suggest:

CONJECTURE 5.1. *The sequence $\{T(n)\}$ satisfies a Lucas congruence for every prime p.*
That is, for every prime p and any positive integer n,

$$T(n) \equiv T(n_0)T(n_1)\cdots T(n_r) \pmod{p}$$

where $n = n_0 + n_1 p + \cdots + n_r p^r$ is the p-adic expansion of n in base p.

The conjecture has been checked for all primes $p < 1000$ and all $n < 50000$. One way to approach the conjecture would be to find a formula for $T(n)$ as a sum of binomial coefficients and then apply the methods of Malik and Straub [**31**]. At present, such a binomial sum is unknown.

6. The level 13 theory

The entries in Tables 4 and 6 for level 13 are slightly different from, and can be deduced from, the results in [**19**]. In that reference the underlying functions, which we shall denote by X^* and Z^*, are defined by

$$X^* = \frac{w}{1+6w+13w^2} \quad \text{and} \quad Z^* = \frac{1}{12}\left(13P(q^{13}) - P(q)\right) \tag{6.1}$$

where $w = \eta_{13}^2/\eta_1^2$. By [**19**, Th. 8.1] we can deduce the alternative formula

$$Z^* = \eta_1^2\eta_{13}^2 \times \frac{(1-X^*)^{2/3}}{(X^*)^{7/6}} \tag{6.2}$$

which resembles the other formulas in the column for Z in Tables 3–4 apart from the factor of $(1-X^*)^{2/3}$. Further, by [**19**, Theorems 6.3 and 7.2] it follows that the differentiation formula (4.10) and the differential equation (4.11) hold with

$$(B^*)^2 = 1 - 12X^* - 16X^{*2} \quad \text{and} \quad H^* = \frac{2X^*(1+5X^* - 3X^{*2} + 3X^{*3})}{(1-X^*)^2}. \tag{6.3}$$

Observe that H^* is not a polynomial but is a rational function instead. The coefficients $T^*(n)$ in the expansion

$$Z^* = \sum_{n=0}^{\infty} T^*(n) X^{*n}$$

are integers and were shown in [**20**] to satisfy a six-term recurrence relation. By contrast, the data corresponding to level 13 in Tables 4 and 6 are for the functions

$$X = \frac{w}{1+5w+13w^2} \quad \text{and} \quad Z = \frac{\eta_1^2\eta_{13}^2}{X^{7/6}} \tag{6.4}$$

with $w = \eta_{13}^2/\eta_1^2$ as before. On comparing (6.4) with (6.1) and (6.2) we readily obtain the change of variable formulas

$$X = \frac{X^*}{1-X^*} \quad \text{and} \quad Z = (1-X^*)^{1/2} Z^*.$$

On making the change of variable and applying (6.1)–(6.3), it follows that X and Z satisfy the differentiation formula (4.10) and the differential equation (4.11) with B^2 and H given by

$$B^2 = (1+X)(1-10X-27X^2) \quad \text{and} \quad H = \frac{X}{8}\left(12 + 175X + 231X^2\right).$$

It then follows by Theorem 3.1 that the coefficients $T(n)$ in the power series expansion

$$Z = \sum_{n=0}^{\infty} T(n) X^n$$

satisfy the four-term recurrence relation

$$\begin{aligned}(n+1)^3 T(n+1) &= \frac{3}{2}(2n+1)(3n^2+3n+1)T(n) \\ &\quad + \frac{n}{4}(148n^2+27)T(n-1) \\ &\quad + \frac{3}{8}(6n-5)(6n-1)(2n-1)T(n-2)\end{aligned}$$

with initial conditions $T(0)=1$, $T(-1)=0$, $T(-2)=0$. However, the first few terms are given by

$$Z = 1 + \frac{3}{2}X + \frac{91}{8}X^2 + \frac{1287}{16}X^3 + \frac{86931}{128}X^4 + \frac{1566669}{256}X^5 + \frac{59494799}{1024}X^6 + \cdots,$$

and the terms $T(n)$ are not integers. We claim that $4^n T(n)$ is always an integer. To show this, first recall from [**19**, Lemma 5.4] that

$$\eta_1^{24} = U^6 \times \frac{w}{(1+5w+13w^2)^4} \quad \text{and} \quad \eta_{13}^{24} = U^6 \times \frac{w^{13}}{(1+5w+13w^2)^4}, \tag{6.5}$$

where

$$U = 1 - \sum_{j=1}^{\infty} \left(\frac{j}{13}\right) \frac{jq^j}{1-q^j},$$

and $\left(\frac{j}{13}\right)$ is the Legendre symbol. On using (6.5) in (6.4) we obtain

$$Z^2 = U^2 \times (1+5w+13w^2) \tag{6.6}$$

and it follows from the definitions of U and w that the coefficients in the q-expansion of Z^2 are integers. Next, since

$$X = \frac{w}{1+5w+13w^2}, \quad \text{with} \quad w = q\prod_{j=1}^{\infty} \frac{(1-q^{13j})^2}{(1-q^j)^2},$$

it follows that the coefficients in the q-expansion of X are integers. Further, it is clear from the above that

$$Z^2 = 1 + O(q) \quad \text{and} \quad X = q + O(q^2).$$

Hence, the coefficients $z(n)$ in the expansion

$$Z^2 = \sum_{n=0}^{\infty} z(n) X^n = 1 + 3X + 25X^2 + 195X^3 + 1729X^4 + 16107X^5 + 156481X^6 + \cdots$$

are integers. Now write

$$Z^2 = 1 + I \quad \text{where } I = \sum_{n=1}^{\infty} z(n) X^n = 3X + 25X^2 + 195X^3 + 1729X^4 + \cdots$$

so that I is a power series with integer coefficients and no constant term. By the binomial theorem we have

$$Z = Z^2 \times Z^{-1} = Z^2 \times (1+I)^{-1/2} = Z^2 \times \sum_{j=0}^{\infty} \binom{2j}{j} \left(-\frac{I}{4}\right)^j.$$

When the expression on the right hand side is expanded in powers of X, the only terms in the sum that contribute to the coefficient of X^n are those $Z^2\binom{2j}{j}(-\frac{I}{4})^j$ with $0 \leqslant j \leqslant n$. Since Z^2 and I can each be expanded in power series in X with integer coefficients, it follows that $4^n T(n)$ is an integer.

Let $S(n) = 4^n T(n)$. Then all of the above can be summarised as follows.

THEOREM 6.1. *Let $\{S(n)\}$ be the sequence defined by the recurrence relation*

$$\begin{aligned}(n+1)^3 S(n+1) &= 6(2n+1)(3n^2+3n+1)S(n) \\ &\quad + 4n(148n^2+27)S(n-1) \\ &\quad + 24(6n-5)(6n-1)(2n-1)S(n-2)\end{aligned}$$

and initial conditions $S(0) = 1$, $S(-1) = 0$, $S(-2) = 0$. Let X and Z be defined by (6.4), i.e.,

$$X = \frac{w}{1+5w+13w^2} \quad \text{and} \quad Z = \frac{\eta_1^2\eta_{13}^2}{X^{7/6}},$$

with $w = \eta_{13}^2/\eta_1^2$. Then $S(n)$ is always an integer, and

$$Z = \sum_{n=0}^{\infty} S(n)\left(\frac{X}{4}\right)^n.$$

The first few terms $\{S(n)\}_{n=0}^{10}$ are given by:

1, 6, 182, 5148, 173862, 6266676, 237979196, 9366227832,
378768328198, 15643121895492, 657035290739412.

I did not find any Lucas property satisfied by the sequence $\{S(n)\}$. It is an open question as to whether this sequence has any interesting number theoretic properties.

7. Irrational sequences arising from the theory of level 14

Apéry-like sequences that arise from the level 14 theory have been studied in [**18**, Sec. 7], [**20**] and [**25**, Ex. 6]. In this section we present the detailed modular background. Four sequences emerge, including two irrational sequences which are new.

Let w_{+2}, w_{+7} and w_{+14} be the modular functions defined by

$$w_{+2} = \frac{\eta_7^4\eta_{14}^4}{\eta_1^4\eta_2^4}, \quad w_{+7} = \frac{\eta_2^3\eta_{14}^3}{\eta_1^3\eta_7^3}, \quad w_{+14} = \frac{\eta_1^4\eta_{14}^4}{\eta_2^4\eta_7^4}. \tag{7.1}$$

The notation w_{+m}, where $m \in \{2, 7, 14\}$, is an abbreviation for w_{14+m} which denotes a fixed Hauptmodul for $\Gamma_0(14) + m$. Let y_{+2}, y_{+7} and y_{+14} be the corresponding weight two modular forms defined by

$$y_{+2} = \frac{\eta_1^5\eta_2^5}{\eta_7^3\eta_{14}^3}, \quad y_{+7} = \frac{\eta_1^4\eta_7^4}{\eta_2^2\eta_{14}^2}, \quad y_{+14} = \frac{\eta_2^5\eta_7^5}{\eta_1^3\eta_{14}^3}.$$

Observe that

$$w_{+2}y_{+2} = w_{+7}y_{+7} = w_{+14}y_{+14} = \eta_1\eta_2\eta_7\eta_{14}. \tag{7.2}$$

The q-expansions

$$w_{+2} = q^3 + O(q^4), \quad w_{+7} = q + O(q^2) \quad \text{and} \quad w_{+14} = q + O(q^2)$$

and

$$y_{+2} = \frac{1}{q^2} + O(1), \quad y_{+7} = 1 + O(q) \quad \text{and} \quad y_{+14} = 1 + O(q),$$

indicate that the pair of functions w_{+2} and y_{+2} may have different properties from the pairs w_{+7}, y_{+7} and w_{+14}, y_{+14}. Indeed, the functions w_{+7} and w_{+14} are related by [**20**, Thm. 3.3]

$$\frac{1}{w_{+7}} + 8w_{+7} = \frac{1}{w_{+14}} + w_{+14} - 7, \tag{7.3}$$

whereas w_{+2} is related to w_{+14} by the more complicated identity [**20**, Thm. 3.4]

$$\frac{1}{w_{+2}} + 2401w_{+2} = \frac{1}{w_{+14}^3} - \frac{16}{w_{+14}^2} + \frac{48}{w_{+14}} + 32 + 48w_{+14} - 16w_{+14}^2 + w_{+14}^3. \tag{7.4}$$

The identity (7.3) could have been written in the form

$$\frac{1}{w_{+7}} + 8w_{+7} + \epsilon = \frac{1}{w_{+14}} + w_{+14} - 7 + \epsilon, \tag{7.5}$$

for any constant ϵ. This suggests considering the functions X_ϵ and Z_ϵ defined by

$$\frac{1}{X_\epsilon} = \frac{1}{w_{+7}} + 8w_{+7} + \epsilon = \frac{1}{w_{+14}} + w_{+14} - 7 + \epsilon \tag{7.6}$$

and

$$Z_\epsilon = \frac{\eta_1\eta_2\eta_7\eta_{14}}{X_\epsilon}, \tag{7.7}$$

so that

$$X_\epsilon = \frac{1}{q} + (\epsilon - 3) + 11q + 20q^2 + 57q^3 + 92q^4 + 207q^5 + O(q^6)$$

and

$$Z_\epsilon = y_{+7} \times (1 + \epsilon w_{+7} + 8w_{+7}^2) = y_{+14} \times (1 + (\epsilon - 7)w_{+14} + w_{+14}^2). \tag{7.8}$$

The next result suggests that certain values of ϵ stand out as special.

THEOREM 7.1. *Let B_ϵ and H_ϵ be defined by*

$$B_\epsilon = \Big(1 - (\epsilon - 9)X_\epsilon\Big)^{1/2}\Big(1 - (\epsilon - 5)X_\epsilon\Big)^{1/2}\Big(1 - 2\epsilon X_\epsilon + (\epsilon^2 - 32)X_\epsilon^2\Big)^{1/2}$$

and

$$H_\epsilon = X_\epsilon\Bigg(\epsilon - 4 - \frac{1}{2}\left(7\epsilon^2 - 50\epsilon + 24\right)X_\epsilon + \left(4\epsilon^3 - 42\epsilon^2 + 26\epsilon + 448\right)X_\epsilon{}^2 - \frac{3}{2}\left(\epsilon - 9\right)\left(\epsilon - 5\right)\left(\epsilon^2 - 32\right)X_\epsilon{}^3\Bigg).$$

Then the differentiation formula (4.10) *and the differential equation* (4.11) *hold for $B = B_\epsilon$, $H = H_\epsilon$, $X = X_\epsilon$ and $Z = Z_\epsilon$.*

PROOF. The results were proved for the case $\epsilon = 9$ in [**20**, Lemma 4.1 and Theorem 4.2]. The proofs in the other cases follow by making the change of variables

$$\frac{1}{X_\epsilon} - \epsilon = \frac{1}{X_{\epsilon=9}} - 9 \quad \text{and} \quad X_\epsilon Z_\epsilon = X_{\epsilon=9}Z_{\epsilon=9}.$$

□

Now consider the expansion

$$Z_\epsilon = \sum_{n=0}^{\infty} T_\epsilon(n) X_\epsilon^n.$$

By Theorem 3.1, the coefficients $T_\epsilon(n)$ satisfy the five-term recurrence relation

$$\begin{aligned}(n+1)^3 T_\epsilon(n+1) &= (2n+1)\big((2\epsilon-7)(n^2+n)+\epsilon-4\big)T_\epsilon(n)\\ &\quad - n\big((6\epsilon^2-42\epsilon+13)n^2+\epsilon^2-8\epsilon+11\big)T_\epsilon(n-1)\\ &\quad + n(2n-1)(n-1)(2\epsilon^3-21\epsilon^2+13\epsilon+224)T_\epsilon(n-2)\\ &\quad - n(n-1)(n-2)(\epsilon-5)(\epsilon-9)(\epsilon^2-32)T_\epsilon(n-3).\end{aligned} \tag{7.9}$$

The single initial condition $T_\epsilon(0) = 1$ is enough to start the recurrence.
If $\epsilon \in \{5, 9, \pm\sqrt{32}\}$ then the degree of H_ϵ drops by one, the function B_ϵ also simplifies, and the recurrence relation (7.9) reduces to a four-term relation of the form

$$\begin{aligned}(n+1)^3 T_\epsilon(n+1) &= (2n+1)(an^2+an+b)T_\epsilon(n)\\ &\quad + n(cn^2+d)T_\epsilon(n-1)\\ &\quad + en(2n-1)(n-1)T_\epsilon(n-2).\end{aligned} \tag{7.10}$$

It may be noted that the values $\epsilon = 5,\ 9,\ \sqrt{32}$ and $-\sqrt{32}$ are precisely the values that make one of the sides of (7.5) a square, and we will write $T_{14A}(n)$, $T_{14B}(n)$, $T_{14C}(n)$ and $T_{14\overline{C}}(n)$, respectively, for $T_\epsilon(n)$ in these cases.

The sequence $\{T_{14B}(n)\}$ was first studied in [**25**] and investigated further in [**20**]. The sequence $\{T_{14A}(n)\}$ seems to have first arisen in [**18**] in the course of proving some conjectures of Z.-W. Sun [**40**]. The sequences $\{T_{14A}(n)\}$ and $\{T_{14B}(n)\}$ are listed in Table 4 along with the corresponding modular forms. The sequences $\{T_{14C}(n)\}$ and $\{T_{14\overline{C}}(n)\}$ are new.

The first few terms of the sequences $\{T_{14A}(n)\}$, $\{T_{14B}(n)\}$ and $\{T_{14C}(n)\}$ are listed in Table 7. If

$$T_{14C}(n) = u + v\sqrt{2}$$

for some integers u and v then

$$T_{14\overline{C}}(n) = u - v\sqrt{2},$$

so that $T_{14\overline{C}}(n)$ is obtained from $T_{14C}(n)$ by conjugation.

The values of the constants a, b, c, d and e corresponding to the recurrence relation (7.10) for the values $\epsilon = 5,\ 9,\ \pm\sqrt{32}$ are given in Table 9 under the heading level 14.

The following formulas, involving binomial sums, were proved in [**18**]:

$$\begin{aligned}T_{14A}(n) &= \sum_{j,k} \binom{n+j}{2j+2k}\binom{2j+2k}{j+k}\binom{2k}{k}^2\binom{k}{j}\\ &= \sum_{j,k} (-1)^{n-j}\binom{n+j}{2j+2k}\binom{2j+2k}{j+k}\binom{2k}{k}\binom{j+2k}{k}\binom{j+k}{k}\\ &= \sum_{j,k} \binom{n+j-k}{2j+2k}\binom{2j+2k}{j+k}\binom{j+k}{k}^2\binom{2j}{j+k}(-3)^{n-j-3k}.\end{aligned}$$

From (7.7) we have

$$Z_\epsilon X_\epsilon = \eta_1\eta_2\eta_7\eta_{14}$$

TABLE 7. The first few terms of the level 14 sequences $\{T_\epsilon(n)\}$.

n	$T_{14A}(n):\ \epsilon=5$	$T_{14B}(n):\epsilon=9$	$T_{14C}(n):\ \epsilon=\sqrt{32}$
0	1	1	1
1	1	5	$-4+4\sqrt{2}$
2	9	33	$56-32\sqrt{2}$
3	49	269	$-520+416\sqrt{2}$
4	385	2545	$6512-4224\sqrt{2}$
5	2961	26565	$-69664+52416\sqrt{2}$
6	24801	295785	$862904-582400\sqrt{2}$
7	212409	3441765	$-9870928+7232544\sqrt{2}$
8	1878129	41336145	$123164432-84724224\sqrt{2}$
9	16924945	508419125	$-1472036416+1063509568\sqrt{2}$
10	155204329	6370849633	$18601926816-12933544448\sqrt{2}$

is independent of ϵ, and on using (7.6) it follows further that

$$\sum_{n=0}^{\infty} T_\epsilon(n)\left(\frac{w}{1+\epsilon w+8w^2}\right)^{n+1} = w-4w^2+16w^3-72w^4+368w^5-2080w^6+\cdots$$

is also independent of ϵ. In particular, the generating functions for the four sequences $\{T_\epsilon(n)\}$ for $\epsilon\in\{5,9,\pm\sqrt{32}\}$ are related as follows.

THEOREM 7.2. *The following identity holds, in a neighborhood of $w=0$:*

$$\begin{aligned}\sum_{n=0}^{\infty} T_{14A}(n)\left(\frac{w}{1+5w+8w^2}\right)^{n+1} &= \sum_{n=0}^{\infty} T_{14B}(n)\left(\frac{w}{1+9w+8w^2}\right)^{n+1}\\ &= \sum_{n=0}^{\infty} T_{14C}(n)\left(\frac{w}{1+\sqrt{32}w+8w^2}\right)^{n+1}\\ &= \sum_{n=0}^{\infty} T_{14\overline{C}}(n)\left(\frac{w}{1-\sqrt{32}w+8w^2}\right)^{n+1}.\end{aligned}$$

We end the section with a conjectural congruence identity that has been tested extensively by computer.

CONJECTURE 7.3. *The sequences $\{T_{14A}(n)\}$ and $\{T_{14B}(n)\}$ satisfy the Lucas congruence (1.4) for every prime p. The sequences $\{T_{14C}(n)\}$ and $\{T_{14\overline{C}}(n)\}$ satisfy the Lucas congruence for a prime p if and only if $p=2$ or p is congruent to 1 or 7 modulo 8.*

8. Complex valued sequences arising from the level 15 theory

Despite having many similarities, the level 15 theory has only been studied in [**20**]. For example, in the last section of [**18**], it was noted with surprise that while the level 14 theory occurred several times in that work, the level 15 theory did not show up at all. In this section we develop the modular background and obtain four sequences, three of which are new.

Using similar notation as for the previous section, let w_{+3}, w_{+5} and w_{+15} be the modular functions defined by

$$w_{+3} = \frac{\eta_5^3\eta_{15}^3}{\eta_1^3\eta_3^3}, \quad w_{+5} = \frac{\eta_3^2\eta_{15}^2}{\eta_1^2\eta_5^2}, \quad w_{+15} = \frac{\eta_1^3\eta_{15}^3}{\eta_3^3\eta_5^3}, \tag{8.1}$$

and let y_{+3}, y_{+5} and y_{+15} be the corresponding weight two modular forms defined by

$$y_{+3} = \frac{\eta_1^4\eta_3^4}{\eta_5^2\eta_{15}^2}, \quad y_{+5} = \frac{\eta_1^3\eta_5^3}{\eta_3\eta_{15}}, \quad y_{+15} = \frac{\eta_3^4\eta_5^4}{\eta_1^2\eta_{15}^2}.$$

Analogous to (7.2) we have

$$w_{+3}y_{+3} = w_{+5}y_{+5} = w_{+15}y_{+15} = \eta_1\eta_3\eta_5\eta_{15}. \tag{8.2}$$

Proceeding as in the previous section, the q-expansions

$$w_{+3} = q^2 + O(q^3), \quad w_{+5} = q + O(q^2) \quad \text{and} \quad w_{+15} = q + O(q^2)$$

and

$$y_{+3} = \frac{1}{q} + O(1), \quad y_{+5} = 1 + O(q) \quad \text{and} \quad y_{+15} = 1 + O(q),$$

suggest that the pair of functions w_{+3} and y_{+3} may have different properties from the pairs w_{+5}, y_{+5} and w_{+15}, y_{+15}. The functions w_{+5} and w_{+15} are related by [**20**, Thm. 5.3]

$$\frac{1}{w_{+5}} + 9w_{+5} + 5 = \frac{1}{w_{+15}} - w_{+15}, \tag{8.3}$$

whereas w_{+3} is related to w_{+5} by the more complicated identity [**20**, Thm. 5.4]

$$\frac{1}{w_{+3}} - 125w_{+3} = \frac{1}{w_{+5}^2} + \frac{1}{w_{+5}} - 9w_{+5} - 81w_{+5}^2. \tag{8.4}$$

The identity (8.3) could have been written in the form

$$\frac{1}{w_{+5}} + 9w_{+5} + 5 + \epsilon = \frac{1}{w_{+15}} - w_{+15} + \epsilon \tag{8.5}$$

for any constant ϵ. This suggests considering the functions X_ϵ and Z_ϵ defined by

$$\frac{1}{X_\epsilon} = \frac{1}{w_{+15}} - w_{+15} + \epsilon \quad \text{and} \quad X_\epsilon Z_\epsilon = \eta_1\eta_3\eta_5\eta_{15} \tag{8.6}$$

so that

$$Z_\epsilon = y_{+15} \times (1 + \epsilon w_{+15} - w_{+15}^2). \tag{8.7}$$

To help determine interesting values of ϵ, we have the following analogue of Theorem 7.1.

THEOREM 8.1. *Let B_ϵ and H_ϵ be defined by*

$$B_\epsilon = \Big(1 - (\epsilon - 1)X_\epsilon\Big)^{1/2}\Big(1 - (\epsilon + 11)X_\epsilon\Big)^{1/2}\Big(1 - 2\epsilon X_\epsilon + (\epsilon^2 + 4)X_\epsilon^2\Big)^{1/2}$$

and

$$H_\epsilon = X_\epsilon\Bigg(\epsilon+2-\frac{1}{2}\left(7\,\epsilon^2+34\,\epsilon-8\right)X_\epsilon+\left(4\,\epsilon^3+30\,\epsilon^2-14\,\epsilon+40\right){X_\epsilon}^2 \\ -\frac{3}{2}\left(\epsilon-1\right)\left(\epsilon+11\right)\left(\epsilon^2+4\right){X_\epsilon}^3\Bigg).$$

Then the differentiation formula (4.10) *and the differential equation* (4.11) *hold for* $B=B_\epsilon$, $H=H_\epsilon$, $X=X_\epsilon$ *and* $Z=Z_\epsilon$.

PROOF. The identities were proved for the case $\epsilon=1$ in [**20**, Lemma 6.1 and Theorem 6.2]. The proofs in the other cases follow by the change of variables

$$\frac{1}{X_\epsilon}-\epsilon=\frac{1}{X_{\epsilon=1}}-1 \quad\text{and}\quad X_\epsilon Z_\epsilon = X_{\epsilon=1}Z_{\epsilon=1}. \qquad \square$$

Now consider the expansion

$$Z_\epsilon=\sum_{n=0}^{\infty}T_\epsilon(n)X_\epsilon^n.$$

By Theorem 3.1, the coefficients $T_\epsilon(n)$ satisfy the five-term recurrence relation

$$\begin{aligned}(n+1)^3T_\epsilon(n+1) &= (2n+1)((2\epsilon+5)(n^2+n)+\epsilon+2)T_\epsilon(n)\\ &\quad -n((6\epsilon^2+30\epsilon-7)n^2+\epsilon^2+4\epsilon-1)T_\epsilon(n-1)\\ &\quad +n(2n-1)(n-1)(2\epsilon^3+15\epsilon^2-7\epsilon+20)T_\epsilon(n-2)\\ &\quad -n(n-1)(n-2)(\epsilon-1)(\epsilon+11)(\epsilon^2+4)T_\epsilon(n-3).\end{aligned}$$

If $\epsilon\in\{1,-11,\pm 2i\}$ then the degree of H_ϵ drops by one, the function B_ϵ also simplifies, and the recurrence relation (8) reduces to a four-term relation of the form (7.10). It may be noted that the values $\epsilon=1$, -11, $2i$ and $-2i$ are precisely the values that make one of the sides of (8.5) a square. The terms in the sequences in these cases will be denoted by $T_{15A}(n)$, $T_{15B}(n)$, $T_{15C}(n)$ and $T_{15\overline{C}}(n)$, respectively. The sequence $T_{15A}(n)$ was studied in [**20**]. The other three sequences are (until now) unpublished; I found them in 2015 and presented them at the NTDU conference in 2015, [**15**].

The values of the contants a, b, c, d and e are given in Table 9 under the heading level 15. The first few terms in the sequences corresponding to $\epsilon\in\{1,-11,2i\}$ are listed in Table 8. The case $\epsilon=-2i$ is not listed in Table 8 as the corresponding data will just be the complex conjugate of the data obtained for $\epsilon=2i$. For each sequence, the single initial condition $T_\epsilon(0)=1$ is enough to start the recurrence. From (8.6) we have

$$Z_\epsilon X_\epsilon=\eta_1\eta_3\eta_5\eta_{15}$$

is independent of ϵ, and on using (7.6) it follows further that

$$\sum_{n=0}^{\infty}T_\epsilon(n)\left(\frac{w}{1+\epsilon w-w^2}\right)^{n+1}=w+2w^2+11w^3+72w^4+545w^5+4450w^6+\cdots$$

is also independent of ϵ. In particular, the generating functions for the four sequences $\{T_\epsilon(n)\}$ for $\epsilon\in\{1,-11,\pm 2i\}$ are related as follows.

TABLE 8. The first few terms of the level 15 sequences $\{T_\epsilon(n)\}$.

n	$T_{15A}(n)$: $\epsilon = 1$	$T_{15B}(n)$: $\epsilon = -11$	$T_{15C}(n)$: $\epsilon = 2i$
0	1	1	1
1	3	−9	$2 + 2i$
2	15	87	$6 + 8i$
3	105	−867	$44 + 52i$
4	855	8775	$290 + 480i$
5	7533	−89559	$1612 + 4372i$
6	69909	918141	$7140 + 39568i$
7	673515	−9432873	$2536 + 361688i$
8	6673095	96984423	$-559166 + 3303552i$
9	67565445	−997061295	$-10693900 + 29823140i$
10	696024945	10245169737	$-151732284 + 264070928i$

THEOREM 8.2. *The following identity holds in a neighborhood of* $w = 0$:

$$\begin{aligned}\sum_{n=0}^{\infty} T_{15A}(n)\left(\frac{w}{1+w-w^2}\right)^{n+1} &= \sum_{n=0}^{\infty} T_{15B}(n)\left(\frac{w}{1-11w-w^2}\right)^{n+1}\\ &= \sum_{n=0}^{\infty} T_{15C}(n)\left(\frac{w}{1+2iw-w^2}\right)^{n+1}\\ &= \sum_{n=0}^{\infty} T_{15\overline{C}}(n)\left(\frac{w}{1-2iw-w^2}\right)^{n+1}.\end{aligned}$$

We end this section with a conjecture that is the analogue of Conjecture 7.3.

CONJECTURE 8.3. *The sequences* $\{T_{15A}(n)\}$ *and* $\{T_{15B}(n)\}$ *satisfy the Lucas congruence for every prime* p*. The sequences* $\{T_{15C}(n)\}$ *and* $\{T_{15\overline{C}}(n)\}$ *satisfy the Lucas congruence for a prime* p *if and only if* $p = 2$ *or* p *is congruent to* 1 *modulo* 4.

9. Levels 20, 21 and 24, and a general conjecture

In this section we will consider the remaining entries in Table 4 for which the coefficients $T(n)$ satisfy a four-term recurrence relation, namely levels 20, 21 and 24.

First, by applying Theorem 3.1 to the data in Tables 4 and 6 for levels 20 and 21, we obtain the four-term recurrence relations

$$\begin{aligned}(n+1)^3T(n+1) &= 4(2n+1)(2n^2+2n+1)T(n)\\ &\quad - 16n(4n^2+1)T(n-1) + 8(2n-1)^3T(n-2)\end{aligned}$$

and

$$\begin{aligned}(n+1)^3T(n+1) &= -(2n+1)(n^2+n+1)T(n)\\ &\quad + n(35n^2+12)T(n-1) + 6(2n-1)(3n-1)(3n-2)T(n-2),\end{aligned}$$

respectively. Neither of these recurrence relations is self-starting in the sense that specifying the single initial condition $T(0) = 1$ is not enough to start the recurrence; in order to handle the cases $n = 0$ and $n = 1$ it is also necessary to stipulate that $T(-1) = 0$ and $T(-2) = 0$. Moreover, no Lucas congruences for these sequences have been found.

On the other hand, applying Theorem 3.1 to the data in Tables 4 and 6 for level 24 gives the recurrence relation

$$\begin{aligned}(n+1)^3T(n+1) &= 2(2n+1)(2n^2+2n+1)T(n) \\ &\quad + 4n(4n^2+1)T(n-1) - 64n(n-1)(2n-1)T(n-2),\end{aligned} \tag{9.1}$$

and the single initial condition $T(0) = 1$ is enough to start the recurrence relation. That is because the coefficient of $T(n-1)$ contains a factor of n, and the coefficient of $T(n-2)$ contains a factor of $n(n-1)$. Furthermore, numerical evidence suggests:

CONJECTURE 9.1. *The sequence defined by the recurrence relation* (9.1) *and initial condition* $T(0) = 1$ *satisfies a Lucas congruence for all primes* p.

Now that we have investigated all of the four-term recurrence relations in Table 4, it appears that the sequences which satisfy Lucas congruences are the self-starting sequences. By Theorem 3.1, the sequence $T(n)$ satisfies a four-term recurrence relation if the corresponding polynomials B^2 and H are of degree 3, i.e.,

$$B^2 = 1 + g_1X + g_2X^2 + g_3X^3 \quad \text{and} \quad H = h_1X + h_2X^2 + h_3X^3$$

and the recurrence relation is given by

$$\begin{aligned}&(n+1)^3T(n+1) \\ &\quad + \left(n+\tfrac{1}{2}\right)\left(g_1n^2 + g_1n - 2h_1\right)T(n) \\ &\quad + n\left(g_2n^2 - g_2 - 2h_2\right)T(n-1) \\ &\quad + \left(n-\tfrac{1}{2}\right)\left(g_3n^2 - g_3n - 2g_3 - 2h_3\right)T(n-2) = 0.\end{aligned}$$

The sequence will be self-starting provided the single initial condition $T(0) = 1$ can be used to start the recurrence, and this will be the case when n divides the coefficient of $T(n-1)$ and when $n(n-1)$ divides the coefficient of $T(n-2)$. The first condition is already satisfied, and the second condition will be satisfied if and only if $g_3 = -h_3$, in which case the recurrence relation becomes

$$\begin{aligned}&(n+1)^3T(n+1) \\ &\quad + \left(n+\tfrac{1}{2}\right)\left(g_1n^2 + g_1n - 2h_1\right)T(n) \\ &\quad + n\left(g_2n^2 - g_2 - 2h_2\right)T(n-1) \\ &\quad + g_3n(n-1)\left(n-\tfrac{1}{2}\right)T(n-2) = 0\end{aligned}$$

or equivalently

$$\begin{aligned}(n+1)^3T(n+1) &= (2n+1)(an^2+an+b)T(n) \\ &\quad + n(cn^2+d)T(n-1) \\ &\quad + en(2n-1)(n-1)T(n-2),\end{aligned} \tag{9.2}$$

where

$$a = -\frac{g_1}{2}, \quad b = h_1, \quad c = -g_2, \quad d = g_2 + 2h_2, \quad e = -\frac{g_3}{2} = \frac{h_3}{2}.$$

TABLE 9. Parameters for the four-term recurrence relation

$$(n+1)^3T(n+1) = (2n+1)(an^2+an+b)T(n) + n(cn^2+d)T(n-1) + en(2n-1)(n-1)T(n-2)$$

and initial condition $T(0) = 1$ that produce sequences which are conjectured to satisfy Lucas congruences.

Level	Name	a	b	c	d	e
11	$T_{11}(n)$	10	4	-56	-8	22
14	$T_{14A}(n)$	3	1	47	4	14
	$T_{14B}(n)$	11	5	-121	-20	98
	$T_{14C}(n)$	$8\sqrt{2}-7$	$4\sqrt{2}-4$	$168\sqrt{2}-205$	$32\sqrt{2}-43$	$28(11\sqrt{2}-16)$
	$T_{14\overline{C}}(n)$	$-8\sqrt{2}-7$	$-4\sqrt{2}-4$	$-168\sqrt{2}-205$	$-32\sqrt{2}-43$	$-28(11\sqrt{2}+16)$
15	$T_{15A}(n)$	7	3	-29	-4	30
	$T_{15B}(n)$	-17	-9	-389	-76	-750
	$T_{15C}(n)$	$5+4i$	$2+2i$	$31-60i$	$5-8i$	$-40-30i$
	$T_{15\overline{C}}(n)$	$5-4i$	$2-2i$	$31+60i$	$5+8i$	$-40+30i$
24	$T_{24}(n)$	4	2	16	4	-64

Table 9 contains the values of the parameters (a, b, c, d, e) for which the sequence defined by the recurrence relation (9.2) and initial condition $T(0) = 1$ is conjectured to satisfy Lucas congruences.

10. Asymptotic expansions

An asymptotic formula for the Apéry numbers was mentioned in (1.5). We seek analogous asymptotic expansions for sequences defined by the recurrence relation (3.11) in Theorem 3.1, viz.

$$(n+1)\sum_{j=0}^{k} g_j(n+1-\tfrac{j}{2})(n+1-j)T(n+1-j) = 2\sum_{j=1}^{k} h_j(n+1-\tfrac{j}{2})T(n+1-j), \tag{10.1}$$

where g_j and h_j are constants and $g_0 = 1$. We start with the ansatz

$$T(n) \sim C n^{\alpha} R^n \left(1 + \frac{b_1}{n} + \frac{b_2}{n^2} + \cdots\right) \qquad \text{as } n \to \infty \tag{10.2}$$

and substitute in (10.1). To compute the expansion, note that for $j=0,1,2,\ldots$ we have

$$\begin{aligned}\frac{T(n+1-j)}{n^\alpha R^n} &= \frac{(n+1-j)^\alpha}{n^\alpha}\frac{R^{n+1-j}}{R^{n+1}}\left(1+\frac{b_1}{n+1-j}+\frac{b_2}{(n+1-j)^2}+\cdots\right)\\ &= \frac{1}{R^j}\left(1-\frac{\alpha(j-1)}{n}+\frac{\alpha(\alpha-1)}{2}\frac{(j-1)^2}{n^2}+\cdots\right)\\ &\quad\times\left(1+\frac{b_1}{n}+\frac{(j-1)b_1+b_2}{n^2}+\cdots\right).\end{aligned}$$

Now divide (10.1) by $n^\alpha R^{n+1}n^2(n+1)$ and substitute the above expansion to obtain

$$\begin{aligned}&\sum_{j=0}^{k}\frac{g_j}{R^j}\left(1-\frac{j-1}{n}\right)\left(1-\frac{j-2}{2n}\right)\left(1-\frac{\alpha(j-1)}{n}+\frac{\alpha(\alpha-1)}{2}\frac{(j-1)^2}{n^2}+\cdots\right)\\ &\qquad\times\left(1+\frac{b_1}{n}+\frac{(j-1)b_1+b_2}{n^2}+\cdots\right)\\ &=2\sum_{j=1}^{k}\frac{h_j}{R^j}\left(\frac{1}{n^2}+\cdots\right),\end{aligned}$$

where we have expanded as far as terms involving n^{-2}.

Equating coefficients of n^0 gives

$$\sum_{j=0}^{k}\frac{g_j}{R^j}=0.$$

Hence, R^{-1} is a root of the polynomial $G(x)=1+\sum_{j=1}^{k}g_jx^j$.

Next, equating coefficients of n^{-1} gives

$$\sum_{j=0}^{k}\frac{g_j}{R^j}\left(-(j-1)-\frac{(j-1)}{2}-\alpha(j-1)\right)+b_1\sum_{j=0}^{k}\frac{g_j}{R^j}=0.$$

On using the fact that R^{-1} is a root of $G(x)$ and simplifying, we obtain

$$\sum_{j=0}^{k}\frac{g_j}{R^j}\left(-\frac{3j}{2}\right)=\alpha\sum_{j=0}^{k}\frac{jg_j}{R^j},\quad\text{i.e.,}\quad -\frac{3}{2}G'(R^{-1})=\alpha G'(R^{-1}).$$

If we assume further that R^{-1} is a simple root of G, then it follows that $\alpha=-3/2$.

Setting $\alpha=-3/2$ and equating coefficients of n^{-2} gives

$$\sum_{j=0}^{k}\frac{g_j}{R^j}\left(\frac{j^2}{8}+b_1j-\frac{b_1}{2}+b_2-\frac{1}{8}\right)=2\sum_{j=1}^{k}\frac{h_j}{R^j}.$$

Once again we use the fact that R^{-1} is a root of $G(x)$ to simplify and obtain

$$b_1\sum_{j=0}^{k}\frac{jg_j}{R^j}+\frac{1}{8}\sum_{j=0}^{k}\frac{j^2g_j}{R^j}=2\sum_{j=1}^{k}\frac{h_j}{R^j},$$

and this can be solved to give

$$b_1 = \frac{16H(R^{-1}) - G^{**}(R^{-1})}{8G^{*}(R^{-1})},$$

where

(10.3)

$$H(x) = \sum_{j=1}^{k} h_j x^j, \quad G^{*}(x) = x\frac{\mathrm{d}}{\mathrm{d}x}G(x) \quad \text{and} \quad G^{**}(x) = x\frac{\mathrm{d}}{\mathrm{d}x}\left(x\frac{\mathrm{d}}{\mathrm{d}x}G(x)\right).$$

If needed, additional coefficients $b_2, b_3, \ldots$ could be computed by expanding further and equating coefficients of n^{-3}, $n^{-4}, \ldots$.

It is well-known that the value of C in (10.2) cannot be determined solely from the recurrence relation. Instead, forward differences can be used to determine the value of C accurately, e.g., see [**43**, p. 954] for a description of the method. Then the PSLQ algorithm can be used to conjecture an exact value of the constant. Alternatively, if additional facts are known about the sequence, for example if a representation as a binomial sum is available, then it may be possible to compute the value of C analytically, e.g., see Hirschhorn [**26**], [**28**]. In principle, Hirschhorn's approach could be applied to the sequences T_{14A} and T_{24}, but binomial sums are not known for the other sequences in Table 9 so the values of C are determined numerically and hence are conjectural.

The calculations and discussions above can be used in conjunction with Theorem 3.1 and summarised as follows.

THEOREM 10.1. *Suppose $Z = Z(X)$ is analytic in a neighbourhood of $X = 0$ with $Z(0) = Z_0 \neq 0$, and satisfies the second order non-linear differential equation*

$$\mathrm{D}^2 Z - \frac{1}{2Z}(\mathrm{D}Z)^2 = HZ,$$

where D *is the differential operator*

$$\mathrm{D} = BX\frac{\mathrm{d}}{\mathrm{d}X},$$

where

$$B = \sqrt{G} \quad \text{and} \quad G = 1 + \sum_{j=1}^{k} g_j X^j,$$

and

$$H = \sum_{j=1}^{k} h_j X^j.$$

Suppose $G(x)$ has a unique root r_0 of smallest modulus and assume that r_0 is simple. Then the coefficients $T(n)$ in the power series expansion

$$Z = \sum_{n=0}^{\infty} T(n) X^n$$

have an asymptotic expansion given by

$$T(n) \sim C\,n^{-3/2}\,R^n\left(1 + \frac{b_1}{n} + O\left(\frac{1}{n^2}\right)\right) \qquad \text{as } n \to \infty, \tag{10.4}$$

where $R = 1/r_0$ and $b_1 = \dfrac{16H(r_0) - G^{**}(r_0)}{8G^{*}(r_0)}$, and G^* and G^{**} are defined by (10.3).

It is useful to do some checks of this formula with known results. By the discussion preceding and following Theorem 3.2 we have corresponding to the Apéry numbers

$$G(x) = 1 + 2\alpha x + (\alpha^2 + 4\gamma)x^2 \quad \text{and} \quad H(x) = (2\beta - \alpha)x - \frac{(\alpha^2 + 4\gamma)x^2}{2}$$

with $(\alpha, \beta, \gamma) = (-17, -6, -72)$. Solving $G(x) = 0$ gives smallest root $r_0 = 17 - 12\sqrt{2}$ and hence

$$R = \frac{1}{r_0} = 17 + 12\sqrt{2} = (1 + \sqrt{2})^4.$$

Furthermore,

$$b_1 = \frac{16H(r_0) - G^{**}(r_0)}{8G^*(r_0)} = \left.\frac{57 - 6x}{8x - 136}\right|_{x=17-12\sqrt{2}} = -\frac{3}{4} + \frac{15\sqrt{2}}{64}.$$

Hence we obtain for the Apéry numbers

$$A(n) \sim \frac{C}{n^{3/2}}(1 + 2\sqrt{2})^4 \left(1 + \frac{15\sqrt{2} - 48}{64n} + O\left(\frac{1}{n^2}\right)\right) \qquad \text{as } n \to \infty. \tag{10.5}$$

The value of C may be determined numerically and we find that $C = \frac{(1+2\sqrt{2})^2}{2^{9/4}\pi^{3/2}}$. These results are in agreement with Hirschhorn [**26**] and [**27**], where coefficients of $1/n^2$ and $1/n^3$ are also determined.

As another example, consider the case of level 7. From Table 5 we have that $G(x) = (1 + x)(1 - 27x)$ and $H(x) = 4x(1+3x)$. The root of G of smallest modulus is $r_0 = 1/27$ and therefore $R = 27$. Moreover,

$$b_1 = \frac{16H(r_0) - G^{**}(r_0)}{8G^*(r_0)} = \left.\frac{-45 - 150x}{104 + 216x}\right|_{x=1/27} = -\frac{65}{144}.$$

Hence we obtain

$$T(n) \sim \frac{C}{n^{3/2}} 27^n \left(1 - \frac{65}{144n} + O\left(\frac{1}{n^2}\right)\right) \qquad \text{as } n \to \infty.$$

Once again, the value of C may be determined numerically and we find that $C = \frac{\sqrt{27}}{4\pi^{3/2}}$, and all of this is in agreement with another result of Hirschhorn [**28**].

Theorem 10.1 can be used to determine the parameters R and b_1 in the asymptotic expansions for the eight sequences in Table 9. The polynomials G and H for the sequences T_{11}, T_{14A}, T_{14B}, T_{15A}, T_{15B}, and T_{24} can be found in Tables 5–6, and the polynomials for the other sequences can be extracted from Theorems 7.1 and 8.1. The corresponding parameter C in each asymptotic expansion is determined numerically using finite differences. The results are summarised in Table 10. The result in the table for $T_{14B}(n)$ agrees with the formula given by Guillera and Zudilin [**25**, Ex. 6].

The point to note is that the dominant term in the asymptotic expansion can be determined, up to the constant C, simply by inspecting the polynomials $G = B^2$ in Table 5–6. And if a finer approximation such as the $1/n$ term is required, then the polynomial H in Tables 5–6 comes into play.

TABLE 10. Parameters in the asymptotic expansion

$$T(n) \sim C\,\frac{R^n}{n^{3/2}}\left(1+\frac{b_1}{n}+O\left(\frac{1}{n^2}\right)\right) \quad \text{as } n\to\infty.$$

Sequence	R	b_1	Conjectured value of C
$T_{11}(n)$	$1-\frac{20}{R}+\frac{56}{R^2}-\frac{44}{R^3}=0$ so $R=16.82750\ldots$	$-\frac{1}{8}\left(\frac{275R^{-2}-184R^{-1}+21}{33R^{-2}-28R^{-1}+5}\right)$ $=-0.3995791\cdots$	$\frac{xR}{\pi^{3/2}}=0.3287368\ldots$, where $2^{10}\cdot 11x^6+2^5\cdot 29x^2-11=0$
$T_{14A}(n),\ \epsilon=5$	$5+4\sqrt{2}$	$-\frac{223}{196}+\frac{1025\sqrt{2}}{3136}$	$\frac{(5+4\sqrt{2})}{4\pi^{3/2}}\left(\frac{9\sqrt{2}-8}{14}\right)^{1/2}$
$T_{14B}(n),\ \epsilon=9$	$(1+2\sqrt{2})^2$	$-\frac{7}{4}+\frac{69\sqrt{2}}{64}$	$\frac{(1+2\sqrt{2})^2}{4\pi^{3/2}}\left(\frac{8-5\sqrt{2}}{2}\right)^{1/2}$
$T_{14C}(n),\ \epsilon=\sqrt{32}$	$8\sqrt{2}$	$\frac{443}{392}-\frac{60\sqrt{2}}{49}$	$\frac{8}{\pi^{3/2}}\left(\frac{8\sqrt{2}-11}{7}\right)^{1/2}$
$T_{14\overline{C}}(n),\ \epsilon=-\sqrt{32}$	$-(1+2\sqrt{2})^2$	$\frac{339}{784}+\frac{201\sqrt{2}}{196}$	$\frac{(1+2\sqrt{2})^2}{\pi^{3/2}}\left(\frac{\sqrt{14}}{7}+\frac{\sqrt{7}}{14}\right)$
$T_{15A}(n),\ \epsilon=1$	12	$-\frac{489}{1000}$	$\frac{6\sqrt{3}}{5\pi^{3/2}}$
$T_{15B}(n),\ \epsilon=-11$	-12	$\frac{3}{8}$	$\frac{12\sqrt{3}}{\pi^{3/2}}$
$T_{15C}(n),\ \epsilon=2i$	$11+2i$	$-\frac{1209}{2000}+\frac{231}{1000}i$	$\frac{(11+2i)^{3/2}}{20\pi^{3/2}}$
$T_{15\overline{C}}(n),\ \epsilon=-2i$	$11-2i$	$-\frac{1209}{2000}-\frac{231}{1000}i$	$\frac{(11-2i)^{3/2}}{20\pi^{3/2}}$
$T_{24}(n)$	8	$-\frac{3}{8}$	$\frac{\sqrt{8}}{\pi^{3/2}}$

11. Further conjectures: supercongruences

Suppose $T(0)=1$ and that $\{T(n)\}$ satisfies a Lucas congruence for a prime p; that is, for all positive integers n,

$$T(n)\equiv T(n_0)T(n_1)\cdots T(n_r) \pmod{p},$$

where $n=n_0+n_1p+\cdots+n_rp^r$ is the p-adic expansion of n in base p. Then since $pn=n_0p+n_1p^2+\cdots+n_rp^{r+1}$ it follows immediately that the congruence

$$T(pn)\equiv T(n) \pmod{p}$$

holds for all positive integers n. For the Apéry numbers it was shown by Gessel [**23**] that a stronger congruence holds, namely for all primes $p\geqslant 5$ and all positive integers n,

$$T(pn)\equiv T(n) \pmod{p^3}, \tag{11.1}$$

where $T(n)=A(n)$ are the Apéry numbers. Such a congruence, involving a power of the prime p, is called a supercongruence. The Apéry numbers are the sequence labelled 6(A) in Table 2 and it is natural to check if other sequences in Table 9 satisfy this type of congruence. Chan et al. [**9**, Theorem 2.3] proved that the congruence (11.1) also holds for the sequence labelled 6(B) in Table 2.[1] They

[1] The absolute values of the numbers $s(n)$ labelled 6(B) in Table 2 are called Domb numbers.

conjectured [**9**, Conjecture 3.1] that the congruence also holds for the Almkvist–Zudilin numbers labelled 6(C) and this has been proved by T. Amdeberhan and R. Tauraso [**2**]. Further results about supercongruences satisfied by the sequences in Tables 1 and 2, as well as the sequences in Tables 3–4 corresponding to levels 7, 10 and 18, are summarised by Straub in [**39**].

Let us briefly discuss supercongruences for level 7. Let $T(n)$ be the level 7 sequence in Tables 3–4 and let $t(n)$ be defined by

$$\left(\sum_{n=0}^{\infty} t(n)X^n\right)^2 = \sum_{n=0}^{\infty} T(n)X^n$$

so that

$$\sum_{j=0}^{n} t(j)t(n-j) = T(n) = \sum_{j} \binom{n}{j}^2 \binom{2j}{n}\binom{n+j}{j}.$$

Chan et al. [**9**, Conjecture 5.4] conjectured that

$$t(pn) \equiv t(n) \pmod{p^2} \tag{11.2}$$

for all positive integers n and all primes p that are quadratic residues modulo 7, and L. O'Brien [**32**, Conjecture 9.3] conjectured that the congruence also holds for $p = 7$. As far as I know, this conjecture is still open. For the other sequence it was conjectured [**14**, Conjecture 5.1] that

$$T(pn) \equiv T(n) \pmod{p^3} \tag{11.3}$$

for all positive integers n and all primes $p \geqslant 3$. The conjecture (11.3) has been proved for primes $p \geqslant 5$ by R. Osburn, B. Sahu and A. Straub [**33**, Theorem 1.2]. Thus, the weight one sequence $\{t(n)\}$ satisfies a mod p^2 congruence for primes p in half of the residue classes modulo 7, whereas the weight two sequence $\{T(n)\}$ satisfies the stronger mod p^3 congruence for all primes $p \geqslant 5$ (and conjecturally for $p = 3$).

The discussion above prompts the question of whether supercongruences exist for any of the self-starting sequences defined by four-term recurrence relations in Table 9. Although mod p^3 congruences seem to be rare (I found some for the prime $p = 2$ and one for the prime $p = 3$), I did find several curious mod p^2 congruences which are summarised in the following conjectures.

11.1. Level 11. Here is the conjecture for level 11.

CONJECTURE 11.1. *Let $\{T_{11}(n)\}$ be the level 11 sequence in Table 9. For all positive integers n, we have*

$$T_{11}(pn) \equiv T_{11}(n) \pmod{p^2} \quad \text{if } p \in \{2,\ 59,\ 5581\}.$$

Moreover, in the case $p = 2$ we also have

$$T_{11}(2n) \equiv T_{11}(n) \pmod{2^6}$$

unless $n = 1$ or $n = 1 + 2^{j-1}$ or $n = 1 + 3 \times 2^j$ for some $j \in \mathbb{Z}^+$.

The primes 59 and 5581 in Conjecture 11.1 are surprising. I have tested the primes $p < 10000$ and did not find any others with this property, apart from $p = 2$. There is not enough information to speculate about whether the mod p^2 congruences hold for any other primes $p > 10000$. The conjectures have been tested for $pn \leqslant 10^6$, in other words for $1 \leqslant n \leqslant 1694$ in the case $p = 59$, and for $1 \leqslant n \leqslant 17$ in the

TABLE 11. Values of
$$c(p) := \#\{n : T_{11}(pn) \equiv T_{11}(n) \pmod{p^2} \text{ and } 1 \leqslant n \leqslant 1000\}$$
for primes $p \leqslant 101$. Note the values of $c(2)$ and $c(59)$.

p	2	3	5	7	11	13	17	19	23	29	31	37	41
$c(p)$	1000	333	200	750	875	274	222	286	129	62	32	27	48

p	43	47	53	59	61	67	71	73	79	83	89	97	101
$c(p)$	87	87	18	1000	49	136	56	27	63	24	11	10	9

case $p = 5581$. While 17 data points in the latter case may not seem like much evidence, the 1694 data points is quite compelling evidence for the prime $p = 59$. Some statistics for other primes are given in Table 11. The table suggests that the mod p^2 congruences might hold for particular values of n for other primes. For example, we have:

CONJECTURE 11.2.
$$T_{11}(3n) \equiv T_{11}(n) + 3n \pmod{3^2}$$
and
$$T_{11}(5n) \equiv T_{11}(n) \pmod{5^2} \qquad \text{if and only if } 5|n.$$

This conjecture implies the values $c(3) = 333$ and $c(5) = 200$ in Table 11. The reader should be warned, however, that the values $c(7) = 750$ and $c(11) = 875$ should not to be taken as any suggestion that 3/4 of the integers satisfy the mod p^2 congruence for the primes $p = 7$, or that the proportion is 7/8 of the integers in the case $p = 11$. It is pure coincidence that these proportions match with simple rational numbers, and those proportions change as the range of n increases.

It is worth comparing Conjecture 11.1 with the corresponding level 11 conjecture of Chan et al.:

CONJECTURE 11.3 (Conjecture 5.5 in [**9**]). *Let $t_{11}(n)$ be defined by*
$$\left(\sum_{n=0}^{\infty} t_{11}(n)x^n\right)^2 = \sum_{n=0}^{\infty} T_{11}(n)x^n,$$
where $T_{11}(n)$ is the level 11 sequence in Table 9. Equivalently,
$$\sum_{j=0}^{n} t_{11}(j)t_{11}(n-j) = T_{11}(n)$$
or
$$(n+1)^2 t_{11}(n+1) = 2(10n^2+5n+1)t_{11}(n) \\ - 8(7n^2-7n+2)t_{11}(n-1) + 22(n-1)(2n-3)t_{11}(n-2)$$
with $t_{11}(0) = 1$ and $t_{11}(-1) = t_{11}(-2) = 0$. Then
$$t_{11}(pn) \equiv t_{11}(n) \pmod{p^2}$$
for all positive integers n and all primes p with $\left(\frac{p}{11}\right) = 1$.

In other words, the weight one sequence $\{t_{11}(n)\}$ is conjectured to satisfy a mod p^2 congruence for for the primes that are quadratic residues modulo 11. This is analogous to the level 7 conjecture given by (11.2). However, the weight two sequence $\{T_{11}(n)\}$ appears to satisfy a mod p^2 congruence for a much smaller and as yet undetermined set of primes. This is much weaker than the corresponding level 7 result in (11.3), both in terms of the strength of the congruence and also in terms of the set of primes p.

11.2. Level 14. Here are the corresponding conjectures for level 14.

CONJECTURE 11.4. *Let $\{T_{14A}(n)\}$, $\{T_{14B}(n)\}$, $\{T_{14C}(n)\}$ and $\{T_{14\overline{C}}(n)\}$ be the level 14 sequences in Table 9. For all positive integers n, we have*

$$T_{14A}(pn) \equiv T_{14A}(n) \pmod{p^2} \quad \text{if } p \in \{2,\ 17,\ 263\}$$

and

$$T_{14B}(pn) \equiv T_{14B}(n) \pmod{p^2} \quad \text{if } p \in \{2,\ 7,\ 17,\ 263\}.$$

Moreover, in the case $p = 2$ we also have

$$T_{14A}(2n) - T_{14A}(n) \equiv \begin{cases} 0 \pmod{2^4} & \text{if } n \equiv 0 \text{ or } 3 \pmod 4, \\ 8 \pmod{2^4} & \text{if } n \equiv 1 \text{ or } 2 \pmod 4; \end{cases}$$

$$T_{14B}(2n) - T_{14B}(n) \equiv \begin{cases} 0 \pmod{2^4} & \text{if } n \equiv 0 \pmod 2, \\ 12 \pmod{2^4} & \text{if } n \equiv 1 \pmod 2; \end{cases}$$

and

$$T_{14C}(2n) \equiv T_{14C}(n) \pmod{2^6} \quad \text{and} \quad T_{14\overline{C}}(2n) \equiv T_{14\overline{C}}(n) \pmod{2^6}$$

unless $n = 1, 2, 3$ or $n = 2^j \times 3 + 1$ for some nonnegative integer j.

The conjecture has been tested for primes $p < 10000$ and for values of n satisfying $pn \leqslant 100000$. The most obvious question is why the prime $p = 7$ appears in the conjecture for T_{14B} but not for T_{14A}. As partial answer, we conjecture that

$$T_{14A}(7n) \equiv T_{14A}(n) \pmod{7^2} \quad \text{if } n \equiv 0,\ 4,\ 5 \text{ or } 6 \pmod 7.$$

11.3. Level 15. Here are the corresponding conjectures for level 15. Congruences for $T_{15\overline{C}}(n)$ can be deduced from the conjectures for $T_{15C}(n)$ complex conjugation and hence are not listed.

CONJECTURE 11.5. *Let $\{T_{15A}(n)\}$, $\{T_{15B}(n)\}$ and $\{T_{15C}(n)\}$ be the level 15 sequences in Table 9. Then congruences modulo p^2 exist for $p \in \{2, 3, 5\}$ as follows:*

- *In the case $p = 2$ we have*

$$T_{15A}(2n) - T_{15A}(n) \equiv \begin{cases} 0 \pmod{2^3} & \text{if } n \equiv 0 \pmod 2, \\ 4 \pmod{2^3} & \text{if } n \equiv 1 \pmod 2; \end{cases}$$

$$T_{15B}(2n) - T_{15B}(n) \equiv \begin{cases} 16 \pmod{2^5} & \text{if } n \equiv 2 \pmod 4, \\ 0 \pmod{2^5} & \text{otherwise}, \end{cases}$$

and

$$T_{15C}(2n) \equiv T_{15C}(n) \pmod{2^3}$$

unless $n = 1$ or $n = 1 + 2^j$ for some nonnegative integer j.

- *In the case $p = 3$ we have*
$$T_{15A}(3n) \equiv T_{15A}(n) \pmod{3^3}$$
unless $n = 1$, $n = 2$, $n = 1 + 3^j$, or $n = 1 + 2 \times 3^j$ for some nonnegative integer j. And,
$$T_{15B}(3n) \equiv T_{15B}(n) \pmod{3^3}$$
unless $n = 1$, $n = 1+3^j$, or $n = 1+2\times 3^j$ for some nonnegative integer j. Furthermore,
$$T_{15C}(3n) - T_{15C}(n) \equiv \begin{cases} i \pmod 3 & \text{if } n \equiv 3,\, 5 \text{ or } 6 \pmod 8, \\ 2i \pmod 3 & \text{if } n \equiv 1,\, 2 \text{ or } 7 \pmod 8, \\ 0 \pmod 3 & \text{otherwise.} \end{cases}$$
- *In the case $p = 5$ we have*
$$T_{15B}(5n) \equiv T_{15B}(n) \pmod{5^2}$$
for all positive integers n. Furthermore, if $n - 1 \neq 5^{j_1} + 5^{j_2} + \cdots + 5^{j_k}$ for distinct nonnegative integers $j_1, \ldots, j_k$, i.e., the base 5 expansion of $n - 1$ does not consist entirely of 0's and 1's, then
$$T_{15A}(5n) \equiv T_{15A}(n) \pmod{5^2}$$
and
$$T_{15C}(5n) \equiv T_{15C}(n) \pmod{5^2}.$$

Once again, this conjecture has been tested for primes $p < 10000$ and for values of n satisfying $pn \leqslant 100000$.

11.4. Level 24. Finally, here are the corresponding conjectures for level 24.

CONJECTURE 11.6. *Let $\{T_{24}(n)\}$ be the level 24 sequence in Table 9. For all positive integers n, we have*
$$T_{24}(pn) \equiv T_{24}(n) \pmod{p^2} \quad \text{if } p \in \{2,\, 89,\, 179,\, 643\}.$$
In the case $p = 2$ we also have
$$T_{24}(2n) \equiv T_{24}(n) \pmod{2^5}$$
unless $n = 1$ or $n = 1 + 2^j$ for some nonnegative integer j.

In the case $p = 3$ we have
$$T_{24}(3n) - T_{24}(n) \equiv \begin{cases} 3 \pmod 9 & \text{if } n \equiv 4 \text{ or } 5 \pmod 6, \\ 6 \pmod 9 & \text{if } n \equiv 1 \text{ or } 2 \pmod 6, \\ 0 \pmod 9 & \text{otherwise.} \end{cases}$$

The conjecture has been tested for primes $p < 10000$ and for values of n satisfying $pn \leqslant 100000$.

Acknowledgements

It is a pleasure to recall and acknowledge helpful discussions with Wadim Zudilin back when this work started in 2015. I am grateful to the anonymous referee for suggestions that have improved the presentation of the work.

References

[1] G. Almkvist, D. van Straten, and W. Zudilin, *Generalizations of Clausen's formula and algebraic transformations of Calabi-Yau differential equations*, Proc. Edinb. Math. Soc. (2) **54** (2011), no. 2, 273–295, DOI 10.1017/S0013091509000959. MR2794653

[2] T. Amdeberhan and R. Tauraso, *Supercongruences for the Almkvist-Zudilin numbers*, Acta Arith. **173** (2016), no. 3, 255–268, DOI 10.4064/aa8215-2-2016. MR3512855

[3] T. Anusha, E. N. Bhuvan, S. Cooper, and K. R. Vasuki, *Elliptic integrals and Ramanujan-type series for* $1/\pi$ *associated with* $\Gamma_0(N)$*, where* N *is a product of two small primes*, J. Math. Anal. Appl. **472** (2019), no. 2, 1551–1570, DOI 10.1016/j.jmaa.2018.12.008. MR3906440

[4] R. Apéry, *Irrationalité de* $\zeta 2$ *et* $\zeta 3$ (French), Astérisque **61** (1979), 11–13. Luminy Conference on Arithmetic. MR3363457

[5] B. C. Berndt, *Ramanujan's notebooks. Part III*, Springer-Verlag, New York, 1991, DOI 10.1007/978-1-4612-0965-2. MR1117903

[6] B. C. Berndt, S. Bhargava, and F. G. Garvan, *Ramanujan's theories of elliptic functions to alternative bases*, Trans. Amer. Math. Soc. **347** (1995), no. 11, 4163–4244, DOI 10.2307/2155035. MR1311903

[7] F. Beukers, *Another congruence for the Apéry numbers*, J. Number Theory **25** (1987), no. 2, 201–210, DOI 10.1016/0022-314X(87)90025-4. MR873877

[8] A. Bostan, S. Boukraa, G. Christol, S. Hassani, and J.-M. Maillard, *Ising* n*-fold integrals as diagonals of rational functions and integrality of series expansions*, J. Phys. A **46** (2013), no. 18, 185202, 44 pp., DOI 10.1088/1751-8113/46/18/185202. MR3055667

[9] H. H. Chan, S. Cooper, and F. Sica, *Congruences satisfied by Apéry-like numbers*, Int. J. Number Theory **6** (2010), no. 1, 89–97, DOI 10.1142/S1793042110002879. MR2641716

[10] H. H. Chan, Y. Tanigawa, Y. Yang, and W. Zudilin, *New analogues of Clausen's identities arising from the theory of modular forms*, Adv. Math. **228** (2011), no. 2, 1294–1314, DOI 10.1016/j.aim.2011.06.011. MR2822224

[11] H. H. Chan and H. Verrill, *The Apéry numbers, the Almkvist-Zudilin numbers and new series for* $1/\pi$, Math. Res. Lett. **16** (2009), no. 3, 405–420, DOI 10.4310/MRL.2009.v16.n3.a3. MR2511622

[12] G. Christol, *Diagonales de fractions rationnelles et equations différentielles* (French), Study group on ultrametric analysis, 10th year: 1982/83, No. 2, Inst. Henri Poincaré, Paris, 1984, pp. Exp. No. 18, 10. MR747499

[13] S. Cooper, *Level 10 analogues of Ramanujan's series for* $1/\pi$, J. Ramanujan Math. Soc. **27** (2012), no. 1, 59–76. MR2933486

[14] S. Cooper, *Sporadic sequences, modular forms and new series for* $1/\pi$, Ramanujan J. **29** (2012), no. 1-3, 163–183, DOI 10.1007/s11139-011-9357-3. MR2994096

[15] S. Cooper, *The Rogers–Ramanujan continued fraction*, conference presentation, Number Theory Down Under (NTDU), The University of Newcastle, Australia, September 18, 2015.

[16] S. Cooper, *Ramanujan's theta functions*, Springer, Cham, 2017, DOI 10.1007/978-3-319-56172-1. MR3675178

[17] S. Cooper, J. Ge, and D. Ye, *Hypergeometric transformation formulas of degrees 3, 7, 11 and 23*, J. Math. Anal. Appl. **421** (2015), no. 2, 1358–1376, DOI 10.1016/j.jmaa.2014.07.061. MR3258323

[18] S. Cooper, J. G. Wan, and W. Zudilin, *Holonomic alchemy and series for* $1/\pi$, Analytic number theory, modular forms and q-hypergeometric series, Springer Proc. Math. Stat., vol. 221, Springer, Cham, 2017, pp. 179–205, DOI 10.1007/978-3-319-68376-8_12. MR3773919

[19] S. Cooper and D. Ye, *The Rogers-Ramanujan continued fraction and its level 13 analogue*, J. Approx. Theory **193** (2015), 99–127, DOI 10.1016/j.jat.2014.01.008. MR3324566

[20] S. Cooper and D. Ye, *Level 14 and 15 analogues of Ramanujan's elliptic functions to alternative bases*, Trans. Amer. Math. Soc. **368** (2016), no. 11, 7883–7910, DOI 10.1090/tran6658. MR3546787

[21] S. Cooper and D. Ye, *The level 12 analogue of Ramanujan's function* k, J. Aust. Math. Soc. **101** (2016), no. 1, 29–53, DOI 10.1017/S1446788715000531. MR3518321

[22] J. Franel, *Solution to a question of Laisant,* L'intermédiaire des mathématiciens, **1** (1894), 45–47.

[23] I. Gessel, *Some congruences for Apéry numbers*, J. Number Theory **14** (1982), no. 3, 362–368, DOI 10.1016/0022-314X(82)90071-3. MR660381

[24] O. Gorodetsky, *New representations for all sporadic Apéry-like sequences, with applications to congruences*, Exp. Math. **32** (2023), no. 4, 641–656, DOI 10.1080/10586458.2021.1982080. MR4669285

[25] J. Guillera and W. Zudilin, *Ramanujan-type formulae for $1/\pi$: the art of translation*, The legacy of Srinivasa Ramanujan, Ramanujan Math. Soc. Lect. Notes Ser., vol. 20, Ramanujan Math. Soc., Mysore, 2013, pp. 181–195. MR3221311

[26] M. D. Hirschhorn, *Estimating the Apéry numbers*, Fibonacci Quart. **50** (2012), no. 2, 129–131. MR2927930

[27] M. D. Hirschhorn, *Estimating the Apéry numbers II*, Fibonacci Quart. **51** (2013), no. 3, 215–217. MR3093673

[28] M. D. Hirschhorn, *A sequence considered by Shaun Cooper*, New Zealand J. Math. **43** (2013), 37–42. MR3095664

[29] T. Huber, D. Schultz, and D. Ye, *Series for $1/\pi$ of level 20*, J. Number Theory **188** (2018), 121–136, DOI 10.1016/j.jnt.2017.12.010. MR3778626

[30] L. Lipshitz and A. J. van der Poorten, *Rational functions, diagonals, automata and arithmetic*, Number theory (Banff, AB, 1988), de Gruyter, Berlin, 1990, pp. 339–358. MR1106672

[31] A. Malik and A. Straub, *Divisibility properties of sporadic Apéry-like numbers*, Res. Number Theory **2** (2016), Paper No. 5, 26 pp., DOI 10.1007/s40993-016-0036-8. MR3501018

[32] L. O'Brien, *Modular Forms and two new integer sequences at level* 7, Master of Science Thesis, Massey University, New Zealand, 2016.

[33] R. Osburn, B. Sahu, and A. Straub, *Supercongruences for sporadic sequences*, Proc. Edinb. Math. Soc. (2) **59** (2016), no. 2, 503–518, DOI 10.1017/S0013091515000255. MR3509244

[34] S. Ramanujan, *Modular equations and approximations to π [Quart. J. Math.* **45** *(1914), 350–372]*, Collected papers of Srinivasa Ramanujan, AMS Chelsea Publ., Providence, RI, 2000, pp. 23–39. MR2280849

[35] S. Ramanujan, *Collected papers of Srinivasa Ramanujan* (G. H. Hardy, P. V. Seshu Aiyar, and B. M. Wilson, eds.), AMS Chelsea Publishing, Providence, RI, 2000. Edited by G. H. Hardy, P. V. Seshu Aiyar and B. M. Wilson. Third printing of the 1927 original. With a new preface and commentary by Bruce C. Berndt. MR2280843

[36] T. Sato (Takeshi Sato), *Apéry numbers and Ramanujan's series for* $1/\pi$, Abstract of a talk presented the the Annual Meeting of the Mathematical Society of Japan, 28–31 March, 2002.

[37] N. J. A. Sloane, *The On-Line Encyclopedia of Integer Sequences,* (OEIS) `http://oeis.org` (webpage accessed: Feb. 24, 2025).

[38] A. Straub, *Multivariate Apéry numbers and supercongruences of rational functions*, Algebra Number Theory **8** (2014), no. 8, 1985–2007, DOI 10.2140/ant.2014.8.1985. MR3285621

[39] A. Straub, *Gessel-Lucas congruences for sporadic sequences*, Monatsh. Math. **203** (2024), no. 4, 883–898, DOI 10.1007/s00605-023-01894-3. MR4718684

[40] Z.-W. Sun, *List of conjectural series for powers of π and other constants*, (2014), 33 pp. `arXiv:1102.5649v47` (webpage accessed: December 23, 2022).

[41] A. van der Poorten, *A proof that Euler missed...Apéry's proof of the irrationality of* $\zeta(3)$, Math. Intelligencer **1** (1978/79), no. 4, 195–203, DOI 10.1007/BF03028234. An informal report. MR547748

[42] D. Ye, Private communication to S. Cooper (11 November, 2013).

[43] D. Zagier, *Vassiliev invariants and a strange identity related to the Dedekind eta-function*, Topology **40** (2001), no. 5, 945–960, DOI 10.1016/S0040-9383(00)00005-7. MR1860536

[44] D. Zagier, *Elliptic modular forms and their applications*, The 1-2-3 of modular forms, Universitext, Springer, Berlin, 2008, pp. 1–103, DOI 10.1007/978-3-540-74119-0_1. MR2409678

[45] D. Zagier, *Integral solutions of Apéry-like recurrence equations*, Groups and symmetries, CRM Proc. Lecture Notes, vol. 47, Amer. Math. Soc., Providence, RI, 2009, pp. 349–366, DOI 10.1090/crmp/047/22. MR2500571

School of Mathematical and Computational Sciences, Massey University, Private Bag 102904, North Shore Mail Centre, Auckland, New Zealand

Email address: `s.cooper@massey.ac.nz`

Contemporary Mathematics
Volume **818**, 2025
https://doi.org/10.1090/conm/818/16374

Counting matrix points on certain varieties over finite fields

Yifeng Huang, Ken Ono, and Hasan Saad

ABSTRACT. Classical hypergeometric functions are well known to play an important role in arithmetic algebraic geometry. These functions offer solutions to ordinary differential equations, and special cases of such solutions are periods of Picard–Fuchs varieties of Calabi–Yau type. Gauss's ${}_2F_1$ includes the celebrated case of elliptic curves through the theory of elliptic functions. In the 80s, Greene defined finite field hypergeometric functions that can be used to enumerate the number of finite field points on such varieties. We extend some of these results to count finite field "matrix points." For example, for every $n \geqslant 1$, we consider the matrix elliptic curves

$$B^2 = A(A - I_n)(A - aI_n),$$

where (A, B) are commuting $n \times n$ matrices over a finite field $\mathbb{F}_q$ and $a \neq 0, 1$ is fixed. Our formulas are assembled from Greene's hypergeometric functions and q-multinomial coefficients. We use these formulas to prove Sato–Tate distributions for the error terms for matrix point counts for these curves and some families of $K3$ surfaces.

1. Introduction and statement of results

Classical hypergeometric functions are well known to give periods of elliptic curves. To be precise, if n is a nonnegative integer, then define $(\gamma)_n$ by

$$(\gamma)_n := \begin{cases} 1 & \text{if } n = 0, \\ \gamma(\gamma+1)(\gamma+2)\cdots(\gamma+n-1) & \text{if } n \geqslant 1. \end{cases}$$

The *classical hypergeometric function* in parameters $\alpha_1, \ldots, \alpha_h, \beta_1, \ldots, \beta_j \in \mathbb{C}$ is defined by

$${}_hF_j^{\mathrm{cl}}\left(\begin{matrix}\alpha_1 & \alpha_2 & \ldots & \alpha_h \\ & \beta_1 & \ldots & \beta_j\end{matrix} \,\middle|\, x\right) := \sum_{n=0}^{\infty} \frac{(\alpha_1)_n(\alpha_2)_n(\alpha_3)_n \cdots (\alpha_h)_n}{(\beta_1)_n(\beta_2)_n \cdots (\beta_j)_n} \cdot \frac{x^n}{n!}.$$

Perhaps the most famous example illustrating the role of these functions in geometry involves the Legendre elliptic curves

$$(1.1) \qquad E_{\mathrm{L}}(a)\colon y^2 = x(x-1)(x-a), \qquad a \in \mathbb{C} \setminus \{0, 1\}.$$

2020 *Mathematics Subject Classification.* Primary 33C70; 14Gxx.

Key words and phrases. Hypergeometric functions, matrix points, elliptic curves, $K3$ surfaces.

Y. H. thanks the AMS–Simons Travel Grant for making this collaboration possible.

K. O. thanks the Thomas Jefferson Fund and the NSF (DMS-2002265 and DMS-2055118) for their support.

The theory of elliptic integrals shows, for $0 < a < 1$, that the function ${}_2F_1^{\mathrm{cl}}(x) := {}_2F_1^{\mathrm{cl}}\left(\begin{matrix}\frac{1}{2} & \frac{1}{2} \\ & 1\end{matrix} \middle| x\right)$ (for example, see page 184 of [**11**]) gives the real period $\Omega_{\mathrm{L}}(a)$ of $E_{\mathrm{L}}(a)$ by the formula

$$\Omega_{\mathrm{L}}(a) = \pi \cdot {}_2F_1^{\mathrm{cl}}(a). \tag{1.2}$$

There is another kind of hypergeometric function, the finite field hypergeometric function, that gives further information about these elliptic curves and higher dimensional varieties. These functions count points over finite fields. To make this precise, we first recall their definition, which is due to Greene [**8**]. If q is a prime power and A and B are two Dirichlet characters on $\mathbb{F}_q$ (extended so that $A(0) = B(0) = 0$), then let $\binom{A}{B}$ be the normalized Jacobi sum

$$\binom{A}{B} := \frac{B(-1)}{q} J(A, \overline{B}) = \frac{B(-1)}{q} \sum_{x \in \mathbb{F}_q} A(x)\overline{B}(1-x).$$

Here $\overline{B}$ is the complex conjugate of B. If $A_0, \ldots, A_n$, and $B_1, \ldots, B_n$ are characters on $\mathbb{F}_q$, then the *finite field hypergeometric function* in these parameters is defined by

$${}_{n+1}F_n^{\mathrm{ff}}\left(\begin{matrix}A_0 & A_1 & \ldots & A_n \\ & B_1 & \ldots & B_n\end{matrix} \middle| x\right)_q := \frac{q}{q-1} \sum_{\chi} \binom{A_0\chi}{\chi}\binom{A_1\chi}{B_1\chi} \cdots \binom{A_n\chi}{B_n\chi} \chi(x).$$

Here $\sum_\chi$ denotes the sum over all characters χ of $\mathbb{F}_q$.

It has been observed by many authors (see [**8**], [**9**], [**12**], [**13**], [**16**], and [**18**], to name a few) that the Gaussian analogue of a classical hypergeometric series with rational parameters is obtained by replacing each $\frac{1}{n}$ with a character χ_n of order n (and $\frac{a}{n}$ with χ_n^a). Let q be a power of an odd prime, ϵ_q be the trivial character on $\mathbb{F}_q$ and let ϕ_q be the character of order 2. Then the finite field analogue of ${}_2F_1^{\mathrm{cl}}(x)$ is

$${}_2F_1^{\mathrm{ff}}(x)_q := {}_2F_1^{\mathrm{ff}}\left(\begin{matrix}\phi_q & \phi_q \\ & \epsilon_q\end{matrix} \middle| x\right)_q.$$

More generally, we let

$${}_{n+1}F_n^{\mathrm{ff}}(x)_q := {}_{n+1}F_n^{\mathrm{ff}}\left(\begin{matrix}\phi_q & \phi_q & \ldots \phi_q \\ & \epsilon_q & \ldots \epsilon_q\end{matrix} \middle| x\right)_q. \tag{1.3}$$

M. Koike proved [**13**] that if p is an odd prime, q is a power of p and $a \in \mathbb{F}_q \setminus \{0, 1\}$, then

$${}_2F_1^{\mathrm{ff}}(a)_q = -\frac{\phi_q(-1)}{q} \cdot a_{\mathrm{L}}(a; q), \tag{1.4}$$

where $q + 1 - a_{\mathrm{L}}(a; q)$ counts the number of $\mathbb{F}_q$-points on $E_{\mathrm{L}}(a)$. This expression is the finite field analogue of Gauss's period formula (1.2).

Motivated by (1.2) and (1.4), it is natural for us to ask whether other finite field hypergeometric function evaluations give point counts for other varieties. This is indeed the case, and perhaps the most beautiful example involves the analogue of the celebrated classical Clausen identity [**5**]

$${}_3F_2^{\mathrm{cl}}\left(\begin{matrix}b+c & 2b & 2c \\ & b+c+\frac{1}{2} & 2b+2c\end{matrix} \middle| x\right) = {}_2F_1^{\mathrm{cl}}\left(\begin{matrix}b & c \\ & b+c+\frac{1}{2}\end{matrix} \middle| x\right)^2. \tag{1.5}$$

Using this identity, D. McCarthy [**14**] proved that if $a > 0$, then

$$ {}_3F_2^{\mathrm{cl}}\left(\begin{matrix}\frac{1}{2} & \frac{1}{2} & \frac{1}{2} \\ & 1 & 1\end{matrix}\,\middle|\, \frac{a}{a+1}\right) = \frac{\sqrt{1+a}}{\pi^2}\cdot \Omega_{\mathrm{CL}}(a)^2, $$

where $\Omega_{\mathrm{CL}}(a)$ is the real period of the Clausen elliptic curve

$$ E_{\mathrm{CL}}(a)\colon y^2 = (x-1)(x^2+a). $$

In the finite field case, the second author proved [Theorem 5 of [**16**]] that if $\mathbb{F}_q$ is a finite field of characteristic $\mathrm{char}(\mathbb{F}_q) \geqslant 3$ and $a \in \mathbb{F}_q \setminus \{0,-1\}$, then

$$ (1.6)\qquad q + q^2\phi_q(a+1)\cdot {}_3F_2^{\mathrm{ff}}\left(\frac{a}{a+1}\right)_q = a_{\mathrm{CL}}(a;q)^2 = q^2\cdot {}_2F_1^{\mathrm{ff}}\left(\frac{1-\sqrt{-a}}{1+\sqrt{-a}}\right)_q^2, $$

where $q+1-a_{\mathrm{CL}}(a;q)$ is the number of $\mathbb{F}_q$ points on $E_{\mathrm{CL}}(a)$, and where the second equality holds whenever $-a$ is a square in $\mathbb{F}_q$, by noting that $E_{\mathrm{CL}}(a)$ is a quadratic twist of $E_{\mathrm{L}}\left(\frac{1-\sqrt{-a}}{1+\sqrt{-a}}\right)$ and using (1.4). This equality is an analogue of a special case of Clausen's identity. Furthermore, this identity can be interpreted in terms of $K3$ surfaces whose function fields are given by

$$ X_a\colon s^2 = xy(x+1)(y+1)(x+ay), $$

where $a \in \mathbb{F}_q \setminus \{0,-1\}$. In this notation, it is known (see Theorem 11.18 of [**15**] and Proposition 4.1 of [**1**]) that

$$ (1.7)\qquad |X_a(\mathbb{F}_q)| = 1 + q^2 + 19q + q^2\cdot {}_3F_2^{\mathrm{ff}}(-a)_q. $$

In this note we show that the hypergeometric identities (1.4) and (1.7), combined with the combinatorial input from partitions and q-multinomial coefficients, count suitable "matrix points" on these curves and surfaces. To make this precise, we first introduce some notation. If n, m are positive integers and K is a field, then let $C_{n,m}(K)$ denote the set of pairwise-commuting m-tuples of $n\times n$ matrices over K. Due to the noncommutativity of matrix multiplication, geometric problems related to studying matrix rational points on curves and higher varieties only make sense when the matrices are commuting, that is, when the matrix points are in $C_{n,m}(K)$. We will be interested in counting tuples in $C_{n,m}(K)$ which satisfy the equations defining some affine varieties. More precisely, we will consider the sets

$$ \{(A,B)\in C_{n,2}(\mathbb{F}_q) : B^2 = A(A-I_n)(A-aI_n)\} $$

and

$$ \{(A,B,C)\in C_{n,3}(\mathbb{F}_q) : C^2 = AB(A+I_n)(B+I_n)(A+aB), C\in \mathrm{GL}_n(\mathbb{F}_q)\} $$

as matrix analogues of (the smooth affine parts of) the Legendre elliptic curves E_{L} and the $K3$ surfaces X_a considered above.

To express our results, we introduce some notation. If λ is a partition of a nonnegative integer k, we write $n(\lambda;i)$ to denote the number of times i is repeated in λ. Furthermore, we write $|\lambda| = k$, and write $l(\lambda) = \sum n(\lambda;i)$ to denote the number of parts of λ. Additionally, we introduce certain polynomials in q. More precisely, if z and q are any complex numbers, and n is any positive integer, then we define the q-Pochhammer symbol

$$ (1.8)\qquad (z;q)_n := (1-z)(1-zq)\dots(1-zq^{n-1}) $$

with $(z;q)_0 = 1$. The series expansion of $(q;q)_5$ will play a special role in our results. For clarity of results, we define b_r for integers $r \geqslant 0$ by

$$\sum_{r=0}^{\infty} b_r q^r = (q;q)_\infty^5 = \left(1+\sum_{m=1}^{\infty}(-1)^m q^{m(3m-1)/2}(1+q^m)\right)^5, \tag{1.9}$$

where the last expression due to Euler's pentagonal number theorem [**2**, Corollary 1.7] allows explicit computation for each b_r.

Finally, for an integer $n \geqslant 0$ and $m_1 + \ldots + m_k = n$ a partition of n, we define the q-multinomial factor

$$\binom{n}{m_1, m_2, \ldots, m_k}_q := \frac{(q;q)_n}{(q;q)_{m_1}(q;q)_{m_2}\cdots(q;q)_{m_k}}.$$

It is known that $\binom{n}{m_1,\ldots,m_k}_q$ is a monic polynomial in q and that $\binom{n}{m_1,\ldots,m_k}_q$ approaches the usual multinomial coefficient as $q \to 1$.

We start by expressing the number of commuting matrices on a Legendre elliptic curve. More precisely, if n is a positive integer, q is a prime power and $a \in \mathbb{F}_q$, we let

$$N_{n,2}(a;q) := |\{(A,B) \in C_{n,2}(\mathbb{F}_q) : B^2 = A(A-I_n)(A-aI_n)\}|. \tag{1.10}$$

In this notation, we have the following theorem that determines these counts, and also explains the connection with the classical ${}_2F_1^{\mathrm{cl}}$-hypergeometric function.

THEOREM 1.1. *If $q = p^r$ is a prime power with $p \geqslant 3$ and $a \in \mathbb{F}_q \setminus \{0,1\}$, then*

$$N_{n,2}(a;q) = P(n,0)_q - \sum_{k=1}^{n} \phi_{q^k}(-1) \cdot P(n,k)_q \cdot {}_2F_1^{\mathrm{ff}}(a)_{q^k},$$

where

$$P(n,k)_q := (-1)^k q^{n(n-k)+\frac{k(k+1)}{2}} \sum_{s=0}^{\lfloor \frac{n-k}{2} \rfloor} q^{2s(s-n+k)} \binom{n}{s, n-k-2s, k+s}_q.$$

Moreover, $P(n,k)_q$ is a polynomial in q with leading term $(-1)^k \cdot q^{n^2-\frac{k(k-1)}{2}}$ and

$$\lim_{q\to 1} P(n,k)_q = (-1)^k \binom{n}{k} \cdot {}_2F_1^{\mathrm{cl}}\left(\begin{matrix}\frac{k-n}{2} & \frac{k+1-n}{2} \\ & k+1\end{matrix} \,\middle|\, 4\right).$$

REMARK. Although the number theoretic results in this paper require that each q is a prime power, we included the limiting behavior $\lim_{q\to 1} P(n,k)_q$ to illustrate that these number theoretic quantities are q-analogues of a classical ${}_2F_1^{\mathrm{cl}}$ evaluation.

As a corollary, we consider the matrix analogue of the Sato–Tate distribution for point counts for elliptic curves over finite fields. In direct analogy, we find that the limiting distribution of the "random part" of matrix point counts on Legendre elliptic curves is semicircular. More precisely, if n is a positive integer and q is a prime power, then we let

$$a_{\mathrm{L},n}(a;q) := N_{n,2}(a;q) - P(n,0)_q. \tag{1.11}$$

In this notation, we have the following result.

COROLLARY 1.2. *If $-2 \leqslant b < c \leqslant 2$ and n and r are fixed positive integers, then we have*

$$\lim_{p\to\infty} \frac{|\{a \in \mathbb{F}_{p^r} : p^{\frac{r}{2}-rn^2} a_{\mathrm{L},n}(a;p^r) \in [b,c]\}|}{p^r} = \frac{1}{2\pi}\int_b^c \sqrt{4-t^2}dt.$$

EXAMPLE 1.3. For the prime $p = 93283$, we compare the histogram of the distribution of $p^{-7/2}a_{L,2}(a;p)$ for $a \in \mathbb{F}_p$ with the limiting distribution. They are shown in Figure 1.

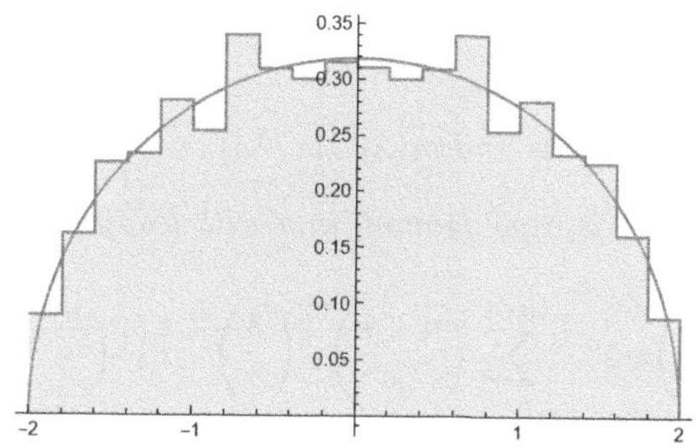

FIGURE 1. The $p^{-7/2}a_{L,2}(a;p)$ histogram for $p = 93283$.

We also consider the matrix version of the $K3$ surfaces described above. If n is a positive integer, q is a prime power, and $a \in \mathbb{F}_q$, then we let

(1.12)

$$N_{n,3}(a;q) := |\{(A,B,C) \in C_{n,3}(\mathbb{F}_q) : C^2 = AB(A+I_n)(B+I_n)(A+aB),\ C \in \mathrm{GL}_n(\mathbb{F}_q)\}|.$$

In this notation, we have the following theorem that gives matrix point counts in terms of the ${}_3F_2^{\mathrm{ff}}$-hypergeometric function, 6-tuples of integer partitions, the coefficients b_r in (1.9), and q-multinomial coefficients.

THEOREM 1.4. *If $q = p^r$ is a prime power with $p \geqslant 3$ and $a \in \mathbb{F}_q \setminus \{0,-1\}$, then we have*

$$N_{n,3}(a;q) = R(n,\phi_q(a+1))_q + Q\left(n,0,\phi_q(a+1)\right)_q + \sum_{k=1}^{n} Q\left(n,k,\phi_q(a+1)\right)_q \cdot {}_3F_2^{\mathrm{ff}}\left(-a\right)_{q^k},$$

where

$$Q(n,k,\gamma)_q := q^{\frac{n(n-1)}{2}} \sum_{r=k}^{n} b_{n-r} \sum_{\substack{\lambda_1,\ldots,\lambda_6 \\ |\lambda_1|+\ldots+|\lambda_6|=r \\ l(\lambda_5)-l(\lambda_6)=k}} q^{2l(\lambda_3)+l(\lambda_4)+2l(\lambda_5)}\gamma^{l(\lambda_4)}(-1)^{n-m(\lambda_1,\ldots,\lambda_6)} \cdot (q;q)_{n-m(\lambda_1,\ldots,\lambda_6)}\, q^{\sum \frac{n(\lambda_i,j)(n(\lambda_i,j)-1)}{2}} \binom{n}{n(\lambda_i,j),\, n-m(\lambda_1,\ldots,\lambda_6)}_q$$

and

$$R(n,\gamma)_q := \\ -q^{\frac{n(n-1)}{2}}\sum_{k=1}^{n} q^k \sum_{r=1}^{n} b_{n-r} \\ \sum_{\substack{\lambda_1,\ldots,\lambda_6 \\ |\lambda_1|+\ldots+|\lambda_6|=rl(\lambda_5)-l(\lambda_6)=k}} \gamma^k q^{2l(\lambda_3)+l(\lambda_4)+2l(\lambda_6)}\gamma^{l(\lambda_4)}(-1)^{n-m(\lambda_1,\ldots,\lambda_6)} \\ \cdot (q;q)_{n-m(\lambda_1,\ldots,\lambda_6)}\, q^{\sum \frac{n(\lambda_i,j)(n(\lambda_i,j)-1)}{2}} \binom{n}{n(\lambda_i,j),\, n-m(\lambda_1,\ldots,\lambda_6)}_q$$

with $\lambda_1,\ldots,\lambda_6$ *being partitions and* $m(\lambda_1,\ldots,\lambda_6)=\sum\limits_{i=1}^{6} l(\lambda_i)$.

Moreover, $Q(n,k,\gamma)_q$ *is a polynomial in* q *with leading term* q^{n^2+n} *and*

(1.13)
$$\lim_{q\to 1} Q(n,k,\gamma)_q = \binom{n}{k}\cdot\gamma^{n-k}\sum_{s=0}^{n-k}\binom{n-k}{s}\cdot\left(\frac{3}{\gamma}\right)^s {}_2F_1^{\mathrm{cl}}\left(\begin{matrix}\frac{k-n+s}{2} & \frac{k+1-n+s}{2} \\ & k+1\end{matrix}\,\middle|\,\frac{4}{\gamma^2}\right)$$

when $\gamma \neq 0$, *and*

$$\lim_{q\to 1} Q(n,k,0)_q = \binom{n}{k}\cdot 3^{n-k}\cdot {}_2F_1^{\mathrm{cl}}\left(\begin{matrix}\frac{k-n}{2} & \frac{k+1-n}{2} \\ & k+1\end{matrix}\,\middle|\,\frac{4}{9}\right).$$

REMARK. Although the number theoretic results in this paper require that each q is a prime power, we included the limiting behavior $\lim_{q\to 1} Q(n,k,\gamma)_q$ to illustrate that these formulas are q-analogues of natural weighted sums of ${}_2F_1^{\mathrm{cl}}$ evaluations (i.e., see (1.13)) that enjoy nice recurrence relations. Namely, if we let

$$F(z;m,k) := z^m\sum_{s=0}^{m}\binom{m}{s}\left(\frac{3}{z}\right)^s {}_2F_1^{\mathrm{cl}}\left(\begin{matrix}\frac{s-m}{2} & \frac{s-m+1}{2} \\ & k+1\end{matrix}\,\middle|\,\frac{4}{z^2}\right)$$

then we have

$$z\frac{d}{dz}F(z;m,k) = mF(z;m,k) - 3mF(z;m-1,k) - \frac{2m(m-1)}{k+1}F(z;m-2,k+1),$$

where $F(z;0,k)=1$ for $k \geqslant 0$. This follows immediately from the fact that

$$\frac{d}{dz}{}_2F_1^{\mathrm{cl}}\left(\begin{matrix}a & b \\ & c\end{matrix}\,\middle|\,z\right) = \frac{ab}{c}\cdot {}_2F_1^{\mathrm{cl}}\left(\begin{matrix}a+1 & b+1 \\ & c+1\end{matrix}\,\middle|\,z\right).$$

Theorem 1.4 allows us to determine the Sato–Tate type limiting distribution of the “random part” of matrix point counts on the $K3$ surfaces X_a. More precisely, if n is a positive integer and q is a prime power, then we let

(1.14) $$A_n(a;q) := N_{n,3}(a;q) - Q(n,0,\phi_q(a+1))_q - R(n,\phi_q(a+1))_q.$$

In this notation, we have the following result.

COROLLARY 1.5. *If* $-3 \leqslant b < c \leqslant 3$ *and* n *and* r *are fixed positive integers, then we have*

$$\lim_{p\to\infty}\frac{\{a\in\mathbb{F}_{p^r} : p^{r-rn^2-rn}A_n(a;p^r)\in[b,c]\}}{p^r} = \frac{1}{4\pi}\int_b^c f(t)\,dt,$$

where

$$f(t) := \begin{cases} \sqrt{\frac{3-|t|}{1+|t|}} & \textit{if } 1 < |t| < 3, \\ \sqrt{\frac{3-t}{1+t}} + \sqrt{\frac{3+t}{1-t}} & \textit{if } |t| < 1, \\ 0 & \textit{otherwise.} \end{cases}$$

EXAMPLE 1.6. For the prime $p = 93283$, we compare the histogram of the distribution of $p^{-5}A_2(a;p)$ for $a \in \mathbb{F}_p$ with the limiting distribution. They are shown in Figure 2.

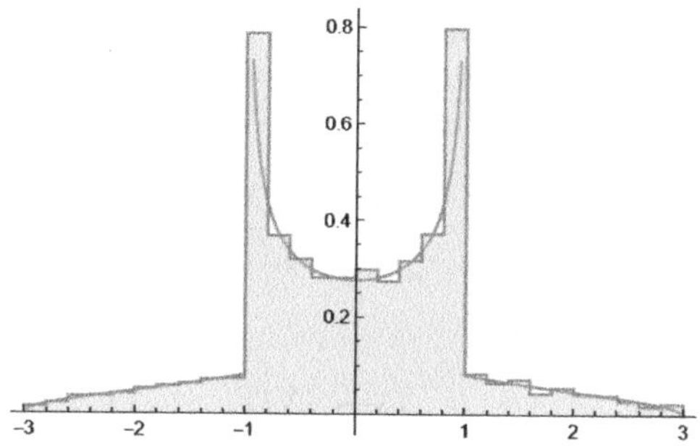

FIGURE 2. The $p^{-5}A_2(a;p)$ histogram for $p = 93283$.

REMARK. The results of this paper can be extended almost *mutatis mutandis* to other hypergeometric families of varieties of dimensions 1 and 2, such as those introduced by Beukers, Cohen, and Mellit in [**3**].

This paper is organized as follows. In Section 2 we recall properties of zeta functions for curves and surfaces in the commuting matrix situation. These results [**10**] are due to the first author. In Section 3 we recall results of the second two authors, which we then combine with these zeta functions to obtain our results.

2. Some zeta functions

Let q be a prime power. Recall that $\mathrm{GL}_n(\mathbb{F}_q)$ is the group of $n \times n$ invertible matrices over the finite field $\mathbb{F}_q$ with q elements. It will be repetitively used in this paper that

$$|\mathrm{GL}_n(\mathbb{F}_q)| = (-1)^n q^{\frac{n(n-1)}{2}} (q;q)_n. \tag{2.1}$$

Now, let $X = \operatorname{Spec} R$ be an affine variety over $\mathbb{F}_q$. Say

$$R := \frac{\mathbb{F}_q[T_1, \ldots, T_m]}{(f_1, \ldots, f_r)}. \tag{2.2}$$

Following the work [**10**] of the first author, we define the set of $n \times n$ *matrix points* on X as the set of commuting tuples of matrices satisfying the defining equations for X:

$$C_n(X) := \left\{ \underline{A} = (A_1, \ldots, A_m) \in \mathrm{Mat}_n(\mathbb{F}_q)^m : [A_i, A_j] = 0,\ f_i(\underline{A}) = 0 \right\}. \tag{2.3}$$

Note that $C_1(X) \cong X(\mathbb{F}_q)$. Though not needed in this paper, it is worth pointing out that the cardinality of $C_n(X)$ is independent of the choice of defining equations for X; in fact, by comparing [**10**, Eq. 4.2] and [**10**, Eq. 4.15], there is an equation-free equivalent characterization for the cardinality of $C_n(X)$:

$$\frac{|C_n(X)|}{|\mathrm{GL}_n(\mathbb{F}_q)|} = \sum_{\dim_{\mathbb{F}_q} H^0(X;M)=n} \frac{1}{|\mathrm{Aut}\, M|}, \tag{2.4}$$

where the sum ranges over all isomorphism classes of zero-dimensional coherent sheaves on X of degree n. This characterization also makes $|C_n(X)|$ well-defined for any variety X over $\mathbb{F}_q$.

The number of matrix points on a smooth curve or a smooth surface is given by infinite product formulas for a zeta function associated to it. For any (affine) variety X over $\mathbb{F}_q$, consider its *Cohen–Lenstra series* (terminology of [**10**])

$$\hat{Z}_X(t) := \sum_{n=0}^{\infty} \frac{|C_n(X)|}{|\mathrm{GL}_n(\mathbb{F}_q)|} t^n, \tag{2.5}$$

and recall the local zeta function

$$Z_X(t) := \exp\left(\sum_{n=1}^{\infty} \frac{|X(\mathbb{F}_{q^n})|}{n} t^n\right). \tag{2.6}$$

PROPOSITION 2.1 ([**10**, Proposition 4.6(a)]). *If X is a smooth curve over $\mathbb{F}_q$, then*

$$\hat{Z}_X(t) = \prod_{j\geqslant 1} Z_X(tq^{-j}). \tag{2.7}$$

PROPOSITION 2.2 ([**10**, Proposition 4.6(b)]). *If X is a smooth surface over $\mathbb{F}_q$, then*

$$\hat{Z}_X(t) = \prod_{i,j\geqslant 1} Z_X(t^i q^{-j}). \tag{2.8}$$

Proposition 2.1 is essentially due to Cohen and Lenstra [**6**], and Proposition 2.2 is essentially due to the Feit–Fine formula [**7**] for counting commuting matrices and ideas of Bryan and Morrison [**4**]. We remark that both formulas heavily exploit the local geometry of X, namely, smoothness of dimension 1 or 2. In fact, in light of the main theorem of [**10**], Proposition 2.1 ceases to hold if X is a multiplicative reduction of an elliptic curve over a number field (but holds if it is a good reduction).

3. Proofs of Theorems 1.1 and 1.4

Here we use the results of the previous section to prove Theorems 1.1 and 1.4 and their corollaries.

3.1. Proof of Theorem 1.1. Fix a prime power $q = p^r$ with $p \geqslant 3$ and $r \geqslant 1$, and fix $a \in \mathbb{F}_q \setminus \{0,1\}$. Then, denoting by X the affine part of $E_{\mathrm{L}}(a)$, Theorem V.2.4 of [**19**] states that

$$Z_X(t) = \frac{(1-\alpha t)(1-\overline{\alpha} t)}{1-qt},$$

where α and $\overline{\alpha}$ are the eigenvalues of Frobenius acting on the Tate module of $E_{\mathrm{L}}(a)$. Note that there is a missing factor of $\frac{1}{1-t}$ in this expression since we are only considering the affine part of $E_{\mathrm{L}}(a)$.

By Proposition 2.1, we then have that

$$\hat{Z}_X(t) = \prod_{j\geqslant 1} \frac{(1-\alpha t q^{-j})(1-\overline{\alpha} t q^{-j})}{1-tq^{1-j}}.$$

It is well-known due to Euler [**2**, Corollary 2.2] that

$$\prod_{j\geqslant 1}(1-ctq^{-j}) = \sum_{m\geqslant 0} \frac{(ct)^m}{(q;q)_m} \tag{3.1}$$

and

$$\prod_{j\geqslant 1}(1-ctq^{-j})^{-1} = \sum_{m\geqslant 0} \frac{(-1)^m q^{m(m-1)/2}\cdot (ct)^m}{(q;q)_m}. \tag{3.2}$$

This implies that

$$\hat{Z}_X(t) = \left(\sum_{r\geqslant 0}\frac{(\alpha t)^r}{(q;q)_r}\right)\cdot\left(\sum_{s\geqslant 0}\frac{(\overline{\alpha} t)^s}{(q;q)_s}\right)\cdot\left(\sum_{u\geqslant 0}\frac{(-1)^u q^{u(u+1)/2}\cdot t^u}{(q;q)_u}\right).$$

By the definition of $\hat{Z}_X(t)$ and by (2.1), we then have

$$N_{n,2}(a;q) = (-1)^n q^{n(n-1)/2}(q;q)_n \cdot \sum_{\substack{r+s+u=n\\ r,s,u\geqslant 0}} \frac{\alpha^r\overline{\alpha}^s(-1)^u q^{\frac{u(u+1)}{2}}}{(q;q)_r(q;q)_s(q;q)_u}.$$

Furthermore, again by Theorem V.2.4 of [**19**], we have that $\alpha\overline{\alpha} = q$ and therefore, we can rewrite this sum as

$$N_{n,2}(a;q) = (-1)^n q^{n(n-1)/2}(q;q)_n \sum_{\substack{r+s+u=n\\ r,s,u\geqslant 0}} \frac{(-1)^u \alpha^{r-s} q^{s+\frac{u(u+1)}{2}}}{(q;q)_r(q;q)_s(q;q)_u}.$$

Dividing this sum according to the value of $r-s$, we then have

$$N_{n,2}(a;q) = (-1)^n q^{n(n-1)/2}(q;q)_n \sum_k \alpha^k \sum_{\substack{s\geqslant 0\\ r=s+k\geqslant 0\\ u=n-2s-k\geqslant 0}} (-1)^u q^{s+u(u+1)/2} \frac{1}{(q;q)_r(q;q)_s(q;q)_u}$$

$$= (-1)^n q^{n(n-1)/2}(q;q)_n \sum_k \alpha^k \sum_{s=\max\{-k,0\}}^{\lfloor \frac{n-k}{2}\rfloor} (-1)^{n-k} \cdot q^{s+\frac{(n-2s-k)(n-2s-k+1)}{2}} \frac{1}{(q;q)_s(q;q)_{s+k}(q;q)_{n-2s-k}}$$

$$= q^{n(n-1)/2} \sum_k \alpha^k(-1)^k \sum_{s=\max\{-k,0\}}^{\lfloor \frac{n-k}{2}\rfloor} q^{\frac{k^2}{2}-kn-\frac{k}{2}+\frac{n^2}{2}+\frac{n}{2}} q^{2ks-2ns+2s^2} \frac{(q;q)_n}{(q;q)_s(q;q)_{s+k}(q;q)_{n-2s-k}}$$

$$= q^{n(n-1)/2} \sum_{s=0}^{\lfloor \frac{n}{2}\rfloor} q^{\frac{n^2}{2}+\frac{n}{2}+2s^2-2ns} \frac{(q;q)_n}{(q;q)_s(q;q)_s(q;q)_{n-2s}}$$
$$+ q^{n(n-1)/2} \sum_{k>0} \alpha^k(-1)^k \sum_{s=0}^{\lfloor \frac{n-k}{2}\rfloor} q^{\frac{k^2}{2}-kn-\frac{k}{2}+\frac{n^2}{2}+\frac{n}{2}} q^{2ks-2ns+2s^2} \frac{(q;q)_n}{(q;q)_s(q;q)_{s+k}(q;q)_{n-2s-k}}$$
$$+ q^{n(n-1)/2} \sum_{k<0} q^k\overline{\alpha}^{-k}(-1)^k \sum_{s=-k}^{\lfloor \frac{n-k}{2}\rfloor} q^{\frac{k^2}{2}-kn-\frac{k}{2}+\frac{n^2}{2}+\frac{n}{2}} q^{2ks-2ns+2s^2} \frac{(q;q)_n}{(q;q)_s(q;q)_{s+k}(q;q)_{n-2s-k}}.$$

Replacing k in the last sum with $-k$ and then s with $s+k$, we then have

$$N_{n,2}(a;q) = P(n,0)_q$$
$$+ \sum_{k>0} \alpha^k(-1)^k \sum_{s=0}^{\lfloor \frac{n-k}{2}\rfloor} q^{\frac{k^2}{2}-kn-\frac{k}{2}+n^2} q^{2ks-2ns+2s^2} \frac{(q;q)_n}{(q;q)_s(q;q)_{s+k}(q;q)_{n-2s-k}}$$
$$+ \sum_{k>0} \overline{\alpha}^k(-1)^k \sum_{s=k}^{\lfloor \frac{n+k}{2}\rfloor} q^{\frac{k^2}{2}+kn-\frac{k}{2}+n^2} q^{-2ks-2ns+2s^2} \frac{(q;q)_n}{(q;q)_s(q;q)_{s-k}(q;q)_{n-2s+k}}$$

$$= P(n,0)_q$$
$$+ \sum_{k>0} \alpha^k(-1)^k \sum_{s=0}^{\lfloor \frac{n-k}{2}\rfloor} q^{\frac{k^2}{2}-kn-\frac{k}{2}+n^2} q^{2ks-2ns+2s^2} \frac{(q;q)_n}{(q;q)_s(q;q)_{s+k}(q;q)_{n-2s-k}}$$
$$+ \sum_{k>0} \overline{\alpha}^k(-1)^k \sum_{s=0}^{\lfloor \frac{n-k}{2}\rfloor} q^{\frac{k^2}{2}+kn-\frac{k}{2}+n^2} q^{-2nk+2ks-2ns+2s^2} \frac{(q;q)_n}{(q;q)_s(q;q)_{s+k}(q;q)_{n-2s-k}}$$

$$= P(n,0)_q$$
$$+ \sum (-1)^k(\alpha^k+\overline{\alpha}^k) q^{\frac{k^2}{2}-kn-\frac{k}{2}+n^2} \sum_{s=0}^{\lfloor \frac{n-k}{2}\rfloor} q^{2ks-2ns+2s^2} \frac{(q;q)_n}{(q;q)_s(q;q)_{s+k}(q;q)_{n-2s-k}}.$$

Since $a_{\mathrm{L}}(a;q^k) = \alpha^k + \overline{\alpha}^k$, (1.4) implies that

$$N_{n,2}(a;q) = P(n,0)_q - \sum_{k=1}^{n} \phi_{q^k}(-1)\cdot P(n,k)_q \cdot {}_2F_1^{\mathrm{ff}}(a)_{q^k},$$

where

$$P(n,k)_q = (-1)^k \cdot q^{n(n-k)+\frac{k(k+1)}{2}} \sum_{s=0}^{\lfloor \frac{n-k}{2} \rfloor} q^{2s(s-n+k)} \cdot \frac{(q;q)_n}{(q;q)_s (q;q)_{s+k} (q;q)_{n-2s-k}}.$$

The leading coefficient of $P(n,k)_q$ is clear from the expression. Since the q-multinomial approaches the usual multinomial as $q \to 1$, we have

$$\begin{aligned}\lim_{q\to 1} P(n,k)_q &= (-1)^k \sum_{s=0}^{\lfloor \frac{n-k}{2} \rfloor} \binom{n}{s, n-k-2s, k+s} \\ &= (-1)^k \binom{n}{k} \sum_{s=0}^{\lfloor \frac{n-k}{2} \rfloor} \frac{k!(n-k)!}{s!(k+s)!(n-k-2s)!}.\end{aligned}$$

It is easy to see by induction that if m and s are integers with $m < 0$ and $2s+m \leqslant 0$, then

$$\left(\frac{m}{2}\right)_s \left(\frac{m}{2}+\frac{1}{2}\right)_s = \frac{(-m)!}{(-m-2s)!4^s}.$$

Furthermore, it is evident by definition that for $k \geqslant 0$, we have $(k+1)_s = \frac{(k+s)!}{k!}$. Applying this above with $m = k-n$, we have

$$\lim_{q\to 1} P(n,k)_q = (-1)^k \binom{n}{k} \sum_{s=0}^{\lfloor \frac{n-k}{2} \rfloor} \frac{\left(\frac{k-n}{2}\right)_s \left(\frac{k-n+1}{2}\right)_s}{(k+1)_s} \cdot \frac{4^s}{s!},$$

which is our statement since the summand vanishes for $s > \lfloor \frac{n-k}{2} \rfloor$.

3.2. Proof of Corollary 1.2. We prove this corollary by implementing the method of moments, as employed in previous work by the second two authors in [**17**]. By Theorem 1.1 and the fact that $p^{\frac{r}{2}} {}_2F_1^{\mathrm{ff}}(a)_{p^r} \in [-2,2]$, we have

$$p^{\frac{r}{2}-rn^2} a_{\mathrm{L},n}(a;p^r) = \phi_{p^r}(-1) \cdot p^{\frac{r}{2}} {}_2F_1^{\mathrm{ff}}(a)_{p^r} + O_{r,n}(p^{-\frac{r}{2}}).$$

Therefore, if m is a nonnegative integer, we have that

$$\begin{aligned}&\frac{1}{p^r} \sum_{a\in\mathbb{F}_{p^r}\setminus\{0,1\}} \left(p^{\frac{r}{2}-rn^2} a_{\mathrm{L},n}(a;p^r)\right)^m \\ &= \frac{1}{p^r} \sum_{a\in\mathbb{F}_{p^r}\setminus\{0,1\}} (\phi_{p^r}(-1) p^{\frac{r}{2}} {}_2F_1^{\mathrm{ff}}(a)_{p^r})^m \\ &\quad + \frac{1}{p^r} \sum_{k=1}^{m} \frac{1}{p^{\frac{rk}{2}}} \cdot \binom{m}{k} \frac{1}{p^r} \sum_{a\in\mathbb{F}_{p^r}\setminus\{0,1\}} (\phi_{p^r}(-1) p^{\frac{r}{2}} {}_2F_1^{\mathrm{ff}}(a)_{p^r})^{m-k} + o_{m,r,n}(1) \\ &= \frac{1}{p^r} \sum_{a\in\mathbb{F}_{p^r}\setminus\{0,1\}} (\phi_{p^r}(-1) p^{\frac{r}{2}} {}_2F_1^{\mathrm{ff}}(a)_{p^r})^m + o_{m,r,n}(1), \quad \text{as } p \to \infty.\end{aligned}$$

By Theorem 1.1 of [**17**], this implies that as $p \to \infty$ we have

$$\frac{1}{p^r} \sum_{a\in\mathbb{F}_{p^r}\setminus\{0,1\}} \left(p^{\frac{r}{2}-rn^2} a_{\mathrm{L},n}(a;p^r)\right)^m = \begin{cases} o_{m,r,n}(1) & \text{if } m \text{ is odd,} \\ \frac{(2l)!}{l!(l+1)!} + o_{m,r,n}(1) & \text{if } m = 2l \text{ is even.} \end{cases}$$

The proof of Corollary 1.2 of [**17**] then implies the limiting distribution.

3.3. Proof of Theorem 1.4. Fix a prime power $q = p^r$ with $p \geqslant 3$ and $r \geqslant 1$, and fix $a \in \mathbb{F}_q \setminus \{0, -1\}$. If A_a denotes the smooth affine surface given by

$$s^2 = xy(x+1)(y+1)(x+ay), \quad s \neq 0,$$

then $X := A_a$ and X_a differ by a connected union of rational curves (see [**1**, §1]). In particular, we have

$$[X_a] = [X] + 24\mathbb{L} - 6$$

in the Grothendieck ring of $\mathbb{F}_q$-varieties, where $\mathbb{L}$ is the class of the affine line (cf. the term $|\psi^{-1}(U_a)| = (24q-6)$ at the end of the proof of [**1**, Proposition 4.1]). Therefore, by Theorem 1.1 of [**1**], the local zeta function of X is given by

$$Z_X(t) = \frac{(1-qt)^5}{(1-t)^7(1-q^2t)(1-\gamma qt)(1-\gamma\alpha^2 t)(1-\gamma\overline{\alpha}^2 t)},$$

where $\gamma = \phi_q(a+1)$ and $\alpha, \overline{\alpha}$ are the Frobenius eigenvalues for the Clausen elliptic curve $E_{\mathrm{CL}}\left(\frac{-a}{a+1}\right)$.

Therefore, by Proposition 2.2, we have that

$$\begin{aligned}
\hat{Z}_X(t) &= \prod_{i,j\geqslant 1} \frac{(1-q^{1-j}t^i)^5}{(1-q^{-j}t^i)^7} \prod_{i,j\geqslant 1} \frac{1}{(1-q^{2-j}t^i)(1-\gamma q^{1-j}t^i)(1-\gamma\alpha^2 q^{-j}t^i)(1-\gamma\overline{\alpha}^2 q^{-j}t^i)} \\
&= \prod_{i\geqslant 1}(1-t^i)^5 \cdot \prod_{i,j\geqslant 1} \frac{1}{(1-q^{-j}t^i)^2} \cdot \prod_{i\geqslant 1} \prod_{b\in\{q^2,\gamma q,\gamma\alpha^2,\gamma\overline{\alpha}^2\}} \prod_{j\geqslant 1} \frac{1}{1-bt^iq^{-j}}.
\end{aligned}$$

By (3.2) and $\alpha\overline{\alpha} = q$, we then have

$$\begin{aligned}
\hat{Z}_X(t) &= (t;t)_\infty^5 \cdot \prod_{i\geqslant 1}\left(\sum_{m\geqslant 0} \frac{(-1)^m q^{\frac{m(m-1)}{2}} t^{mi}}{(q;q)_m}\right)^2 \\
&\quad \cdot \prod_{i\geqslant 1} \prod_{b\in\{q^2,\gamma q,\gamma\alpha^2,\gamma\overline{\alpha}^2\}} \sum_{m\geqslant 0} \frac{(-1)^m q^{\frac{m(m-1)}{2}} b^m t^{im}}{(q;q)_m} \\
&= (t;t)_\infty^5 \prod_{i\geqslant 1} \sum_{m\geqslant 0} t^{im} \\
&\quad \cdot \sum_{m_1+\ldots+m_6=m} \frac{(-1)^m q^{\sum \frac{m_i(m_i-1)}{2}} \gamma^{m_4+m_5+m_6} \cdot q^{2m_3+m_4} \cdot \alpha^{2m_5}\overline{\alpha}^{2m_6}}{(q;q)_{m_1} \cdot \ldots \cdot (q;q)_{m_6}} \\
&= (t;t)_\infty^5 \sum_{r\geqslant 0} t^r \cdot \sum (-1)^{\sum m_{u,v}} \\
&\quad \cdot \frac{q^{\sum \frac{m_{u,v}(m_{u,v}-1)}{2}} q^{\sum 2m_{u,3}+m_{u,4}} \gamma^{\sum m_{u,4}+m_{u,5}+m_{u,6}} \cdot \alpha^{2\sum m_{u,5}} \overline{\alpha}^{2\sum m_{u,6}}}{\prod (q;q)_{m_{u,v}}},
\end{aligned}$$

where the latter sum is over all possible combinations of nonnegative integers $m_{u,v}$ with $v = 1, \ldots, 6$ and positive integers i_u such that $\sum i_u m_{u,v} = r$. To simplify this expression, for each $v = 1, \ldots, 6$, we denote by λ_v the partition given by adding i_u with multiplicity $m_{u,v}$.

Then, in the notation of the theorem, the coefficient of t^n in $\hat{Z}_X(t)$ is given by

$$\sum_{r=0}^{n} b_{n-r} \sum_{\substack{\lambda_1,\ldots,\lambda_6 \\ |\lambda_1|+\ldots+|\lambda_6|=r}} (-1)^{l(\lambda_1)+\ldots+l(\lambda_6)} \cdot \frac{q^{\frac{\sum n(\lambda_i,j)(n(\lambda_i,j)-1)}{2}} q^{2l(\lambda_3)+l(\lambda_4)}}{\prod (q;q)_{n(\lambda_i,j)}} \gamma^{l(\lambda_4)+l(\lambda_5)+l(\lambda_6)} \alpha^{2l(\lambda_5)} \overline{\alpha}^{2l(\lambda_6)}.$$

Dividing those partitions into $l(\lambda_5)-l(\lambda_6)=k$ for $0 \leqslant k \leqslant n$, using the definition[1] of $\hat{Z}_X(t)$, the equation (1.6) and the relations $\alpha\overline{\alpha}=q$ and $\phi_{q^k}(a+1)=\phi_q(a+1)^k$, we have that

$$\begin{aligned} N_{n,3}(a;q) &= S\left(n,0,\phi_q(a+1)\right) + \sum_{k=1}^{n} S(n,k,\phi_q(a+1))_q \left((\alpha^k+\overline{\alpha}^k)^2 - 2\alpha^k\overline{\alpha}^k\right) \\ &= S\left(n,0,\phi_q(a+1)\right) + \sum_{k=1}^{n} S(n,k,\phi_q(a+1))_q \left(a_{\mathrm{CL}}\left(\frac{-a}{a+1};q^k\right)^2 - 2q^k\right) \\ &= S\left(n,0,\phi_q(a+1)\right) \\ &\quad + \sum_{k=1}^{n} S(n,k,\phi_q(a+1))_q \left(q^{2k}\cdot \phi_q(a+1)^k \cdot {}_3F_2^{\mathrm{ff}}\left(-a\right)_{q^k} - q^k\right), \end{aligned}$$

where

$$S(n,k,\gamma)_q := (-1)^n q^{\frac{n(n-1)}{2}} (q;q)_n \cdot \sum_{r=k}^{n} b_{n-r} \sum_{\substack{|\lambda_1|+\ldots+|\lambda_6|=r \\ l(\lambda_5)-l(\lambda_6)=k}} (-1)^{m(\lambda_1,\ldots,\lambda_6)} \gamma^{l(\lambda_4)+k} \cdot \frac{q^{2l(\lambda_3)+l(\lambda_4)+2l(\lambda_6)} q^{\frac{\sum n(\lambda_i;j)(n(\lambda_i;j)-1)}{2}}}{\prod (q;q)_{n(\lambda_i,j)}}.$$

The expression for $N_{n,3}(a;q)$ follows immediately.

To obtain the leading term for $Q(n,k,\gamma)_q$, note that the leading term of $(q;q)_l$ is given by $(-1)^l q^{\frac{l(l+1)}{2}}$. Therefore, the leading term of the summand for fixed r,k and $\lambda_1,\ldots,\lambda_6$ is given by

$$\begin{aligned} & q^{\frac{n(n-1)}{2}} \cdot b_{n-r} \cdot q^{2l(\lambda_3)+l(\lambda_4)+2l(\lambda_5)} \gamma^{l(\lambda_4)} (-1)^{n-m(\lambda_1,\ldots,\lambda_6)} \cdot q^{\sum \frac{n(\lambda_i,j)(n(\lambda_i,j)-1)}{2}} \\ & \quad \cdot (-1)^n q^{\frac{n(n+1)}{2}} \cdot (-1)^{\sum n(\lambda_i,j)} q^{-\sum \frac{n(\lambda_i,j)(n(\lambda_i,j)+1)}{2}} \\ & = b_{n-r}\gamma^{l(\lambda_4)} q^{n^2+2l(\lambda_3)+l(\lambda_4)+2l(\lambda_5)-m(\lambda_1,\ldots,\lambda_6)} \\ & = b_{n-r}\gamma^{l(\lambda_4)} q^{n^2-l(\lambda_1)-l(\lambda_2)+l(\lambda_3)+l(\lambda_5)-l(\lambda_6)}. \end{aligned}$$

The exponent of q is maximized when $l(\lambda_1)=l(\lambda_2)=l(\lambda_6)=0$ and $l(\lambda_3)+l(\lambda_5)$ is maximal. This implies that $l(\lambda_4)=0$, $r=n$ and λ_3 and λ_5 are of the form $(1,\ldots,1)$. However, since $l(\lambda_5)-l(\lambda_6)=n$, we thus have that $\lambda_5=n\cdot 1$, and therefore, the leading term of $Q(n,k,\gamma)_q$ is q^{n^2+n}.

---

[1] The set in (1.12) is indeed $C_n(X)$ as is defined in (2.3): a closed equation for X is $s^2=xy(x+1)(y+1)(x+ay)$, $sz=1$, so the corresponding matrix equation is $(A,B,C,D)\in C_{n,4}(\mathbb{F}_q)$, $C^2=AB(A+I_n)(B+I_n)(A+aB)$, $CD=I_n$, which is equivalent to (1.12) by setting $D=C^{-1}$.

It remains to show the behavior of this polynomial as $q \to 1$. To this end, note that

$$\lim_{q\to 1}(q;q)_{n-m(\lambda_1,\ldots,\lambda_6)} = 0$$

if $m(\lambda_1,\ldots,\lambda_6) < n$. Therefore, the only contributing partitions are those with $r = n$ and $l(\lambda_1)+\ldots+l(\lambda_6) = n$. This implies that the partitions $\lambda_1,\ldots,\lambda_6$ are of the form $\lambda_i = (1,1,\ldots,1)$. Therefore, since $b_0 = 1$, we have

$$\begin{aligned}\lim_{q\to 1} Q(n,k,\gamma)_q &= \sum_{\substack{x_1+\ldots+x_6=n\\ x_5-x_6=k}} \frac{n!}{x_1!\ldots x_6!}\cdot \gamma^{x_4}\\
&= \sum_{x_1+\ldots+x_4+2x_6=n-k} \gamma^{x_4}\frac{n!}{x_1!\ldots x_4!x_6!(x_6+k)!}\\
&= \sum_{s=0}^{n-k}\left(\sum_{x_1+x_2+x_3=s}\binom{n}{x_1,x_2,x_3,n-s}\right)\\
&\qquad \cdot \sum_{x_4+2x_6=n-k-s}\binom{n-s}{x_4,x_6,x_6+k}\gamma^{x_4}\end{aligned}$$

The rest of the computation proceeds exactly as that of Theorem 1.1.

3.4. Proof of Corollary 1.5. We prove this corollary by implementing the method of moments, as employed in previous work by the second two authors in [**17**]. By Theorem 1.4 and the fact that $p^r{}_3F_2^{\mathrm{ff}}(a)_{p^r} \in [-3,3]$ which follows from (1.6) we have

$$p^{r-rn^2-rn}A_n(a;p^r) = p^r{}_3F_2^{\mathrm{ff}}(-a)_{p^r} + O_{r,n}(p^{-r}).$$

Therefore, as in the proof of Corollary 1.5, and using Theorem 1.3 of [**17**], for positive integers m, we have that

$$\begin{aligned}&\frac{1}{p^r}\sum_{a\in\mathbb{F}_{p^r}\setminus\{0,1\}}\left(p^{r-rn^2-rn}A_n(a;p^r)\right)^m\\
&\qquad = \begin{cases} o_{m,r,n}(1) & \text{if } m \text{ is odd,}\\ \sum\limits_{i=0}^{m}(-1)^i\binom{m}{i}\frac{(2i)!}{i!(i+1)!} + o_{m,r,n}(1) & \text{if } m \text{ is even.}\end{cases}\end{aligned}$$

The proof of Corollary 1.4 of [**17**] then implies the limiting distribution.

Acknowledgments

The authors thank the anonymous referees for their remarks and corrections.

References

[1] S. Ahlgren, K. Ono, and D. Penniston, *Zeta functions of an infinite family of K3 surfaces*, Amer. J. Math. **124** (2002), no. 2, 353–368. MR1890996

[2] G. E. Andrews, *The theory of partitions*, Cambridge Mathematical Library, Cambridge University Press, Cambridge, 1998. Reprint of the 1976 original. MR1634067

[3] F. Beukers, H. Cohen, and A. Mellit, *Finite hypergeometric functions*, Pure Appl. Math. Q. **11** (2015), no. 4, 559–589, DOI 10.4310/PAMQ.2015.v11.n4.a2. MR3613122

[4] J. Bryan and A. Morrison, *Motivic classes of commuting varieties via power structures*, J. Algebraic Geom. **24** (2015), no. 1, 183–199, DOI 10.1090/S1056-3911-2014-00657-3. MR3275657

[5] Th. Clausen, *Ueber die Fälle, wenn die Reihe von der Form* $y = 1+\frac{\alpha}{1}\cdot\frac{\beta}{\gamma}x+\frac{\alpha.\alpha+1}{1.2}\cdot\frac{\beta.\beta+1}{\gamma\cdot\gamma+1}x^2+$ etc. *ein Quadrat von der Form* $z = 1+\frac{\alpha'}{1}\cdot\frac{\beta'}{\gamma'}\cdot\frac{\delta'}{\epsilon'}x+\frac{\alpha'.\alpha'+1}{1.2}\cdot\frac{\beta'.\beta'+1}{\gamma'\cdot\gamma'+1}\cdot\frac{\delta'.\delta'+1}{\epsilon'.\epsilon'+1}x^2+$ etc. *hat* (German), J. Reine Angew. Math. **3** (1828), 89–91, DOI 10.1515/crll.1828.3.89. MR1577682

[6] H. Cohen and H. W. Lenstra Jr., *Heuristics on class groups of number fields*, Number theory, Noordwijkerhout 1983 (Noordwijkerhout, 1983), Lecture Notes in Math., vol. 1068, Springer, Berlin, 1984, pp. 33–62, DOI 10.1007/BFb0099440. MR756082

[7] W. Feit and N. J. Fine, *Pairs of commuting matrices over a finite field*, Duke Math. J. **27** (1960), 91–94. MR109810

[8] J. Greene, *Hypergeometric functions over finite fields*, Trans. Amer. Math. Soc. **301** (1987), no. 1, 77–101, DOI 10.2307/2000329. MR879564

[9] J. Greene and D. Stanton, *A character sum evaluation and Gaussian hypergeometric series*, J. Number Theory **23** (1986), no. 1, 136–148, DOI 10.1016/0022-314X(86)90009-0. MR840021

[10] Y. Huang, *Mutually annihilating matrices, and a Cohen-Lenstra series for the nodal singularity*, J. Algebra **619** (2023), 26–50, DOI 10.1016/j.jalgebra.2022.11.021. MR4527427

[11] D. Husemöller, *Elliptic curves*, 2nd ed., Graduate Texts in Mathematics, vol. 111, Springer-Verlag, New York, 2004. MR2024529

[12] M. Ishibashi, H. Sato, and K. Shiratani, *On the Hasse invariants of elliptic curves*, Kyushu J. Math. **48** (1994), no. 2, 307–321, DOI 10.2206/kyushujm.48.307. MR1294533

[13] M. Koike, *Orthogonal matrices obtained from hypergeometric series over finite fields and elliptic curves over finite fields*, Hiroshima Math. J. **25** (1995), no. 1, 43–52. MR1322601

[14] D. McCarthy, ${}_3F_2$ *hypergeometric series and periods of elliptic curves*, Int. J. Number Theory **6** (2010), no. 3, 461–470, DOI 10.1142/S1793042110002946. MR2652891

[15] K. Ono, *The web of modularity: arithmetic of the coefficients of modular forms and q-series*, CBMS Regional Conference Series in Mathematics, vol. 102, Published for the Conference Board of the Mathematical Sciences, Washington, DC, by the American Mathematical Society, Providence, RI, 2004. MR2020489

[16] K. Ono, *Values of Gaussian hypergeometric series*, Trans. Amer. Math. Soc. **350** (1998), no. 3, 1205–1223, DOI 10.1090/S0002-9947-98-01887-X. MR1407498

[17] K. Ono, H. Saad, and N. Saikia, *Distribution of values of Gaussian hypergeometric functions*, Pure Appl. Math. Q. **19** (2023), no. 1, 371–407, DOI 10.4310/pamq.2023.v19.n1.a14. MR4570167

[18] J. Rouse, *Hypergeometric functions and elliptic curves*, Ramanujan J. **12** (2006), no. 2, 197–205, DOI 10.1007/s11139-006-0073-3. MR2286245

[19] J. H. Silverman, *The arithmetic of elliptic curves*, Graduate Texts in Mathematics, vol. 106, Springer-Verlag, New York, 1986, DOI 10.1007/978-1-4757-1920-8. MR817210

Dept. of Mathematics, University of British Columbia, Vancouver, British Columbia, Canada V6T 1Z2

Email address: huangyf@math.ubc.ca

Dept. of Mathematics, University of Virginia, Charlottesville, Virginia 22904

Email address: ken.ono691@virginia.edu

Dept. of Mathematics, University of Virginia, Charlottesville, Virginia 22904

Email address: hs7gy@virginia.edu

Contemporary Mathematics
Volume **818**, 2025
https://doi.org/10.1090/conm/818/16375

Hypergeometry and the AGM over finite fields

Eleanor McSpirit and Ken Ono

ABSTRACT. One of the most celebrated applications of Gauss's ${}_2F_1$ hypergeometric functions is in connection with the rapid convergence of sequences and special values that arise in the theory of arithmetic and geometric means. This theory was the inspiration for a recent paper (see M. J. Griffin, Ken Ono, N. Saikia, and W.-L. Tsai, [AGM and jellyfish swarms of elliptic curves, Amer. Math. Monthly **130** (2023), no. 4, 355-369]) in which a finite field analogue of $\mathrm{AGM}_{\mathbb{R}}$ was defined and then studied using finite field hypergeometric functions. Instead of convergent sequences, one gets directed graphs that combine to form disjoint unions of graphs that individually resemble *jellyfish*. Echoing the connection of hypergeometric functions to periods of elliptic curves, these graphs organize elliptic curves over finite fields. Here we use such "jellyfish swarms" to prove new identities for Gauss's class numbers of positive definite binary quadratic forms. Moreover, we prove that the sizes of jellyfish are in part dictated by the order of the prime above 2 in certain class groups.

1. Introduction and statement of results

Recall that the classical *arithmetic–geometric mean* iteration is defined for positive real numbers a and b by the sequence of pairs

$$\mathrm{AGM}_{\mathbb{R}}(a,b) := \{(a_1,b_1),(a_2,b_2),\dots\},$$

where $a_1 := a$, $b_1 := b$, and successive terms are given by

$$a_n := \frac{a_{n-1}+b_{n-1}}{2} \quad \text{and} \quad b_n := \sqrt{a_{n-1}b_{n-1}}.$$

It is well known that both (a_n) and (b_n) rapidly converge to the same limit [**1**, p. 2]. One of the most famous results on the $\mathrm{AGM}_{\mathbb{R}}$ is due to Gauss, who showed using the theory of elliptic integrals that one can generate extraordinary approximations for π with relatively few iterations by considering the related sequence

$$p_n := \frac{a_n^2}{1-\sum_{i=1}^{n} 2^{i-2}(a_i^2-b_i^2)}.$$

2020 *Mathematics Subject Classification.* Primary 11G20, 14H52, 14K02, 33C90.

E. M. acknowledges the support of a UVa Dean's Doctoral Fellowship.

K. O. thanks the Thomas Jefferson Fund and the NSF (DMS-2002265 and DMS-2055118) for their support.

In [**8**], Griffin, Saikia, Tsai, and the second author defined a finite-field analogue of the $\mathrm{AGM}_{\mathbb{R}}$ sequence over $\mathbb{F}_q$ when $q = p^r \equiv 3 \bmod 4$. In this setting, -1 is not a square mod q, mirroring the fact that -1 is not a square in $\mathbb{R}$. This allows us to choose square roots such that the iterated geometric means are well-defined. That is, there is always a unique choice of $b_n = \sqrt{a_{n-1}b_{n-1}}$ such that $a_n b_n$ is a square mod q when one starts with $a, b \in \mathbb{F}_q^{\times}$, $a \neq \pm b$, and ab a square mod q.

For example, consider $q = 7$ and $(a, b) = (4, 2)$. Then

$$\mathrm{AGM}_{\mathbb{F}_7}(4,2) = \{(4,2), \overline{(3,6),(1,2),(5,3),(4,1),(6,5),(2,4)}, \dots\},$$

where the overlined pairs form a repeating orbit. The pairs $(6,3)$, $(2,1)$, $(3,5)$, $(1,4)$, and $(5,6)$ also enter this orbit after one AGM iteration. In [**8**], the authors first explored the properties of the connected components of the directed graph representing the sequences of $\mathrm{AGM}_{\mathbb{F}_q}(a,b)$ over all admissible pairs (a,b). They showed that all components always consist of one cycle and one "tentacle" of length one connected to each cycle vertex, for which they coined the name *jellyfish*. For example, the following figure shows the unique connected component of the $\mathrm{AGM}_{\mathbb{F}_7}$ graph $\mathcal{J}_{\mathbb{F}_7}$.

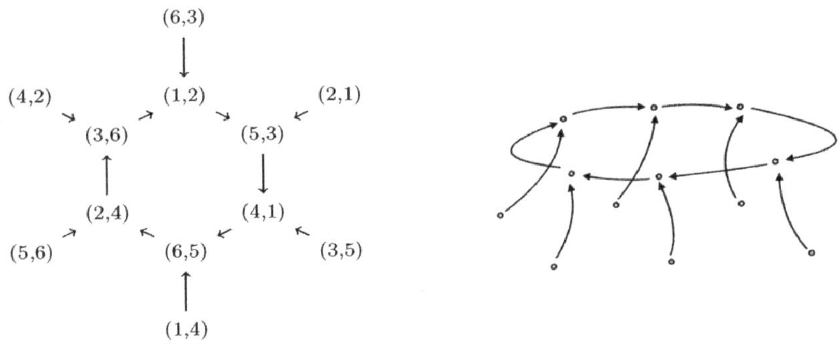

FIGURE 1. Example of the jellyfish comprising $\mathcal{J}_{\mathbb{F}_7}$.

In general, the graph $\mathcal{J}_{\mathbb{F}_q}$ consists of many such jellyfish, which together comprise a *swarm*. In general, jellyfish swarms contain jellyfish of varying sizes and multiplicities, as exemplified by the following figure showing the swarm $\mathcal{J}_{\mathbb{F}_{19}}$.

The theory underlying the classical $\mathrm{AGM}_{\mathbb{R}}$ involves special integrals and their relationship with Gauss's hypergeometric functions. The transformations of elliptic integrals are intrinsically linked to the analysis of limits of arithmetic sequences like the AGM. For $a > b > 0$ let

$$I_{\mathbb{R}}(a,b) := \frac{1}{2a}\int_1^{\infty} \frac{dx}{\sqrt{x(x-1)(x-(1-b^2/a^2))}}. \tag{1.1}$$

A straightforward check shows that $I_{\mathbb{R}}(a,b) = I_{\mathbb{R}}(\frac{a+b}{2}, \sqrt{ab})$, and so the sequence $\mathrm{AGM}_{\mathbb{R}}(a,b) = \{(a_1,b_1),(a_2,b_2),\dots\}$ satisfies

$$I_{\mathbb{R}}(a_1,b_1) = I_{\mathbb{R}}(a_2,b_2) = \cdots = I_{\mathbb{R}}(a_n,b_n) = \dots.$$

This relationship allowed Gauss to derive a formula for limit of the AGM [**1**, Section 1.2].

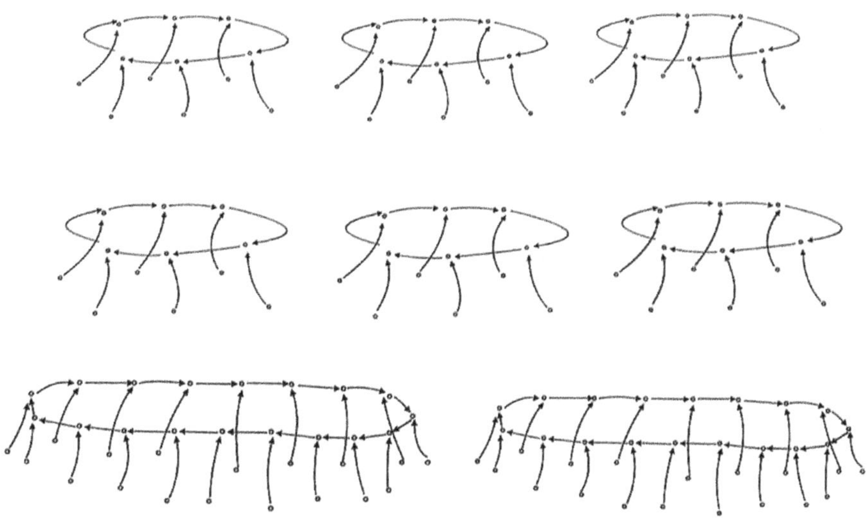

FIGURE 2. Example of the jellyfish swarm $\mathcal{J}_{\mathbb{F}_{19}}$.

It turns out that $I_{\mathbb{R}}(a,b)$ gives an integral representation for Gauss's classical ${}_2F_1$ hypergeometric function:

$$I_{\mathbb{R}}(a,b) = \frac{\pi}{2a} \cdot {}_2F_1{}^{\text{class}}\left(\begin{matrix}\frac{1}{2} & \frac{1}{2} \\ & 1\end{matrix} \,\middle|\, 1 - \frac{b^2}{a^2}\right). \tag{1.2}$$

Further, this elliptic integral and its corresponding hypergeometric function encode data about elliptic curves; for $\lambda \neq 0, 1$ we may define the Legendre elliptic curve

$$E_\lambda \colon y^2 = x(x-1)(x-\lambda).$$

Then the real period of E_λ is computed by the integral

$$\Omega(E_\lambda) = \int_1^\infty \frac{dx}{\sqrt{x(x-1)(x-\lambda)}}.$$

Gauss's ${}_2F_1$ hypergeometric functions offer us a closed formula for the real period of E_λ when $0 < \lambda < 1$:

$$\Omega(E_\lambda) = \pi \cdot {}_2F_1{}^{\text{class}}\left(\begin{matrix}\frac{1}{2} & \frac{1}{2} \\ & 1\end{matrix} \,\middle|\, \lambda\right). \tag{1.3}$$

Just as the classical $\mathrm{AGM}_{\mathbb{R}}$ interacts with hypergeometric functions and elliptic curves, so too does the finite field analogue described above. In analogy with the elliptic integrals in (1.1), one can define

$$I_{\mathbb{F}_q}(a,b) := \frac{1}{2a} \sum_{x \in \mathbb{F}_q} \phi_q(x)\phi_q(x-1)\phi_q(x-(1-b^2/a^2)), \tag{1.4}$$

where $\phi_q(-)$ denotes the quadratic character over $\mathbb{F}_q$. Greene's 1984 PhD thesis [**7**] offers us the appropriate analogue of Gauss's hypergeometric functions with which to draw our connection; for multiplicative characters $\{\alpha_i\}$, $\{\beta_j\}$ over $\mathbb{F}_q^\times$, he defined

$$ {}_nF_{n-1}\left(\begin{matrix}\alpha_1 & \alpha_2 & \dots & \alpha_n \\ & \beta_1 & \dots & \beta_{n-1}\end{matrix}\,\middle|\, x\right)_q := \frac{q}{q-1}\sum_{\chi}\binom{\alpha_1\chi}{\chi}\binom{\alpha_2\chi}{\beta_1\chi}\cdots\binom{\alpha_n\chi}{\beta_{n-1}\chi}\chi(x), $$

where $\binom{\alpha}{\beta}$ is the normalized Jacobi sum $J(\alpha,\beta)$, defined by

$$ \binom{\alpha}{\beta} := \frac{\beta(-1)}{q}J(\alpha,\bar{\beta}) := \frac{\beta(-1)}{q}\sum_{x\in\mathbb{F}_q}\alpha(x)\bar{\beta}(1-x). $$

By comparing their definitions, one finds that for $q = p^r \equiv 3 \bmod 4$,

$$ I_{\mathbb{F}_q}(a,b) = \frac{q}{2a}\cdot {}_2F_1\left(\begin{matrix}\phi_q & \phi_q \\ & \varepsilon_q\end{matrix}\,\middle|\, 1-\frac{b^2}{a^2}\right)_q . \tag{1.5} $$

This expression is the finite field analogue of (1.2). Moreover, a recent result of Evans and Greene (Theorem 2 of [**5**]) offers a quadratic transformation law for the finite field ${}_2F_1$ which implies that the sequence $\mathrm{AGM}_{\mathbb{F}_q}(a,b) = \{(a_1,b_1),(a_2,b_2),\dots\}$ satisfies

$$ I_{\mathbb{F}_q}(a_1,b_1) = I_{\mathbb{F}_q}(a_2,b_2) = \cdots = I_{\mathbb{F}_q}(a_n,b_n) = \dots . $$

We may again consider the relationship of these hypergeometric functions to Legendre elliptic curves, this time analyzing the $\mathbb{F}_q$-points $E_\lambda(\mathbb{F}_q)$. In this case, instead of an elliptic integral telling us the period of E_λ, when $\mathrm{char}(\mathbb{F}_q) \geqslant 5$ we have for $\lambda \in \mathbb{F}_q\backslash\{0,1\}$ that

$$ a_\lambda(q) = -\sum_{x\in\mathbb{F}_q}\phi_q((x)(x-1)(x-\lambda)), $$

where $a_\lambda(q)$ is the *trace of Frobenius* of E_λ (see Theorem 11.10 of [**11**]). Number theoretically, the trace of Frobenius is essentially a formula for the number of points on an elliptic curve over $\mathbb{F}_q$ since

$$ a_\lambda(q) = q+1-\#E_\lambda(\mathbb{F}_q). $$

Cast in terms of Greene's hypergeometric functions, we have in analogy with (1.3)

$$ a_\lambda(q) = -q\cdot {}_2F_1\left(\begin{matrix}\phi_q, & \phi_q \\ & \varepsilon_q\end{matrix}\,\middle|\, \lambda\right)_q . \tag{1.6} $$

The connection between the $\mathrm{AGM}_{\mathbb{F}_q}$ and elliptic curves over $\mathbb{F}_q$ does not stop here, however. The preceding discussion has linked points on the graph of $\mathrm{AGM}_{\mathbb{F}_q}$ to Legendre elliptic curves over $\mathbb{F}_q$ and has shown that two curves represented in the same connected component have the same trace of Frobenius, and hence the same number of $\mathbb{F}_q$-points. The authors of [**8**] are able to prove something stronger. They show that given a sequence $\mathrm{AGM}_{\mathbb{F}_q}(a,b) = \{(a_1,b_1),(a_2,b_2),\dots\}$, if $\lambda_n := b_n^2/a_n^2$, then

$$ E_{\lambda_1}(\mathbb{F}_q) \cong E_{\lambda_2}(\mathbb{F}_q) \cong \cdots \cong E_{\lambda_n}(\mathbb{F}_q) \cong \cdots $$

as abelian groups.

Not only can the vertices of the $\mathrm{AGM}_{\mathbb{F}_q}$ graphs be associated to elliptic curves, but it turns out that each iteration of the arithmetic–geometric mean, and hence each edge of the graph, corresponds to an isogeny of degree 2 between these curves (see Theorem 3 (2) of [**8**]). Following this observation, we can employ tools from the study of elliptic curves in order to understand the structure of $\mathrm{AGM}_{\mathbb{F}_q}$. Our approach will rely on the theory of complex multiplication. In particular, we will study the action of the class group of an imaginary quadratic order $\mathcal{O}$ on the set of isomorphism classes of elliptic curves over $\mathbb{F}_q$ with complex multiplication by $\mathcal{O}$.

Our setup in some ways mirrors the robust theory of "isogeny volcanoes," graphs which organize isomorphism classes of elliptic curves over finite fields and their isogenies. The seminal work on such graphs is due to Kohel,[1] who studied endomorphism rings of elliptic curves over finite fields by understanding the structure of ℓ-isogeny graphs $\mathcal{G}_\ell(\mathbb{F}_q)$, which are visually quite similar to our jellyfish [**9**].

However, there are two main differences between our isogeny graphs and the graphs $\mathcal{G}_2(\mathbb{F}_q)$ studied by Kohel and others. First, only particular 2-isogenies between the represented elliptic curves will appear in our setting. Second, different vertices can correspond to elliptic curves in the same $\mathbb{F}_q$-isomorphism class. What's more, some jellyfish may in fact be identical after identifying nodes with their corresponding elliptic curves. These multiplicities will require attention for all of our results.

In this note, we first show that $\mathrm{AGM}_{\mathbb{F}_q}$ provides new information in the context of Gauss's theory of class numbers of imaginary quadratic fields and class numbers of positive definite binary quadratic forms. To make this connection, we count the $\mathbb{F}_q$-isomorphism classes of elliptic curves that are represented in these graphs. To this end, the authors of [**8**] made use of a correspondence between isomorphism classes of elliptic curves with prescribed torsion and certain class numbers to count the number of distinct j-invariants which appear.

Recall that the Hurwitz class number $H(D)$, introduced by Adolf Hurwitz, is a modification of the class number of binary quadratic forms of discriminant $D \leqslant 0$. If f is a quadratic form, then a matrix

$$\begin{bmatrix} \alpha & \beta \\ \gamma & \delta \end{bmatrix} \in \mathrm{SL}_2(\mathbb{Z})$$

is an automorphism of f if $f(\alpha x + \beta y, \gamma x + \delta y) = f(x, y)$. Then $H(D)$ weights forms of discriminant D by $2/g$, where g is the order of their automorphism group. We additionally declare $H(0) = -1/12$.

Following the approach of [**8**], we obtain new class number formulas that are relatives of classical results like the Hurwitz–Kronecker class number formula

$$\sum_{|t| \leqslant 2\sqrt{N}} H(4N - t^2) = \sum_{d|N} \max\{d, N/d\},$$

which expresses sums of Hurwitz class numbers in terms of divisor sums. The above specializes to

$$\sum_{|t| \leqslant 2\sqrt{p}} H(4p - t^2) = 2p$$

when p is prime. Generalizations of the above were proven by Eichler and Zagier, the latter achieved through the construction of a weight-3/2 non-holomorphic Eisenstein series whose coefficients are Hurwitz class numbers [**4**], [**19**]. More recently, Mertens [**10**] analyzed the holomorphic projection of the Rankin–Cohen bracket of the Harmonic Maass form $\mathcal{H}(\tau)$ with certain theta functions in order to obtain weighted class number formulas and their asymptotics as $q \to \infty$. In [**12**] Saad,

[1]Kohel did not use this language to describe the graphs he studied. The terminology came later, in a paper by Fouquet and Morain, which applied the work of Kohel to the Schoof–Atkin–Elkies point-counting algorithm [**6**].

Saikia, and the second author extended Mertens's approach in their work on the Sato–Tate distribution for a certain family of $K3$ surfaces.

Additionally, many more such identities were recently proven using the trace formula and the combinatorics of j-invariants of elliptic curves over finite fields [**2**].

Our addition to this area is the following:

THEOREM 1.1. *Let $q = p^r \equiv 3 \bmod 4$ where $p > 3$. The following sums are taken over t such that $|t| \leqslant 2\sqrt{q}$.*

(1) *If $q \equiv 3 \bmod 8$, then we have*

$$q = 3 + 4 \sum_{\substack{(t,q)=1 \\ t\equiv q+1(8)}} H\left(\frac{4q - t^2}{4}\right).$$

(2) *If $q \equiv 7 \bmod 8$, then we have*

$$q = 3 + 4 \cdot \left[h(-p) + \sum_{\substack{(t,q)=1 \\ t\equiv q+1(8)}} H\left(\frac{4q - t^2}{4}\right)\right],$$

where $h(D)$ denotes the class number of discriminant D.

REMARK 1.2. The expressions in the two cases of Theorem 1.1 have the same form for prime q.

There are many questions one can immediately ask about $\mathrm{AGM}_{\mathbb{F}_q}$ once the "jellyfish swarm" structure is known. For example, how many jellyfish $\mathcal{J}_i$ appear in a swarm? How large do we expect jellyfish to be?

Interestingly, these questions are not easily answered. To start, the number of jellyfish varies greatly across prime powers q. For example, we give the following table showing some values of $d(q)$, the number of jellyfish in $\mathcal{J}_{\mathbb{F}_q}$:

TABLE 1. Selected values of $d(q)$ for prime powers q.

q	7	11	19	23	27	31	43	47	$\cdots$	161047	161051	161059	161071
$d(q)$	1	3	8	5	39	10	7	4	$\cdots$	6499	25558635	4902	33744

These questions are complicated by the fact that the sizes of jellyfish within a single swarm can vary widely. For example, for $q = 161051 = 11^5$, the swarm $\mathcal{J}_{\mathbb{F}_q}$, one has tiny jellyfish of size 10 alongside those of massive size 7500. We offer the following table showing the size disparities between the smallest and largest jellyfish, which we denote by $\min(q)$ and $\max(q)$ respectively, for small values of q:

TABLE 2. Selected values of $\min(q)$, $\max(q)$ for prime powers q.

q	7	11	19	23	27	31	43	47	59	67	71	79	83
$\min(q)$	12	10	12	22	6	30	42	230	174	18	28	52	410
$\max(q)$	12	20	36	132	12	60	168	276	348	396	280	390	820

Despite this complicated behavior, the authors of [**8**] use the fact that the trace of Frobenius is constant on a jellyfish to give a lower bound for $d(q)$ as a function of q, namely if $\varepsilon > 0$ then for sufficiently large q we have

$$d(q) \geqslant \left(\frac{1}{2} - \varepsilon\right)\sqrt{q}$$

by counting the number of traces that must be obtained by curves in $\mathcal{J}_{\mathbb{F}_q}$.

In addition to size considerations for $d(q)$, one can also ask about the sizes of individual jellyfish. Here we show that these sizes are related to the algebraic properties of the endomorphism rings $\mathrm{End}(E)$ of elliptic curves and their class groups (see Section 2 for background and definitions). Using Theorem 1.1 and the theory of complex multiplication, we are able to show the following:

THEOREM 1.3. *Let $q = p^r \equiv 3 \bmod 4$ where $p > 3$. Suppose (a, b) satisfies the conditions to appear in the $\mathrm{AGM}_{\mathbb{F}_q}$ graph on the jellyfish $\mathcal{J}$, and let $\lambda := b^2/a^2$. Let $\#\mathcal{J}$ denote the number of vertices in $\mathcal{J}$. If $\mathcal{O} := End(E_\lambda)$ and $h_2(\mathcal{O})$ denotes the order of $[\mathfrak{p}_2]$ in the class group $cl(\mathcal{O})$, where $\mathfrak{p}_2$ is a prime above (2) in $\mathcal{O}$, then we have*

$$2 \cdot h_2(\mathcal{O}) \;\Big|\; \#\mathcal{J}.$$

Additionally, if $m(\mathcal{J})$ denotes the multiplicity with which a jellyfish appears, then

$$m(\mathcal{J}) \cdot \#\mathcal{J} = 2(q-1) \cdot h_2(\mathcal{O}).$$

EXAMPLE 1.4. We illustrate the theorem with $q = 271$.

$\mathcal{J}$	t	$h(\mathcal{O})$	$h_2(\mathcal{O})$	$\#\mathcal{J}$	$m(\mathcal{J})$
$\mathcal{J}_1$	-32	2	2	540	2
$\mathcal{J}_2$	-24	5	5	900	3
$\mathcal{J}_3$	-16	6	6	1620	2
$\mathcal{J}_4$	-16	3	3	810	2
$\mathcal{J}_5$	-8	12	6	1620	2
$\mathcal{J}_6$	-8	12	6	1620	2
$\mathcal{J}_7$	0	11	11	2970	2
$\mathcal{J}_8$	8	12	6	1620	2
$\mathcal{J}_9$	8	12	6	1620	2
$\mathcal{J}_{10}$	16	6	6	1620	2
$\mathcal{J}_{11}$	16	3	3	810	2
$\mathcal{J}_{12}$	24	5	5	2700	1
$\mathcal{J}_{13}$	32	2	2	108	10

REMARK 1.5. While $2 \cdot h_2(\mathcal{O}) \mid \#\mathcal{J}$, these two numbers are in general unequal, as illustrated in the previous example. The quantity $\#\mathcal{J}/2 \cdot h_2(\mathcal{O})$ is a divisor of $q-1$ whose appearance we will explain in Section 3. For now, it suffices to remark that it is not determined by the traces or endomorphism rings of the elliptic curves comprising $\mathcal{J}$.

This paper is organized as follows. Section 2 will recall necessary background from the theory of complex multiplication that will allow us to study vertices and edges in $\mathcal{J}_{\mathbb{F}_q}$, using class groups. In Section 3, we give background from [**8**] on the taxonomy of jellyfish. In particular, we characterize the number of vertices in a swarm, as well as the multiplicity of each isomorphism class of elliptic curves appearing on a jellyfish. In Section 4, we prove Theorems 1.1 and 1.3.

2. Complex Multiplication

We first recall the classical theory of complex multiplication of elliptic curves over $\mathbb{C}$ (for example, see Chapter 2 of [**15**]): Let $\mathcal{O}$ be an order in an imaginary quadratic field. If $\mathfrak{a}$ is an invertible $\mathcal{O}$-ideal, then the torus $\mathbb{C}/\mathfrak{a}$ corresponds to an elliptic curve $E(\mathbb{C})$ with complex multiplication by $\mathcal{O}$. Equivalent ideals correspond to isomorphic elliptic curves, and we have a bijection between the ideal class group $cl(\mathcal{O})$ and the set

$$\mathrm{Ell}_{\mathcal{O}}(\mathbb{C}) := \{j(E/\mathbb{C}) \mid \mathrm{End}(E) \cong \mathcal{O}\}$$

of j-invariants of elliptic curves over $\mathbb{C}$ with complex multiplication by $\mathcal{O}$.

Further, another invertible $\mathcal{O}$-ideal $\mathfrak{b}$ uniquely determines a separable isogeny of degree $N(\mathfrak{b})$ with kernel

$$E[\mathfrak{b}] := \{P \in E \mid \alpha \cdot P = O \text{ for all } \alpha \in \mathfrak{b}\}$$

such that the target curve also has multiplication by $\mathcal{O}$. One can check that principal ideals act trivially, and that this defines a faithful $cl(\mathcal{O})$-action on $\mathrm{Ell}_{\mathcal{O}}(\mathbb{C})$.

While this correspondence is pleasing, we are concerned with the case where $\mathbb{C}$ is replaced by the finite field $\mathbb{F}_q$. It turns out that the story in this setting is largely the same.

Let E be an ordinary elliptic curve over $\mathbb{F}_q$, and let π_E denote the Frobenius endomorphism of E. One may compute the trace of Frobenius to be $t = q + 1 - \#E(\mathbb{F}_q)$. Using the characteristic equation for π_E, one derives the *norm equation*

$$t^2 - 4q = v^2 D_K,$$

where $v^2 D_K$ is the discriminant of the imaginary quadratic order $\mathbb{Z}[\pi_E]$ and D_K is the discriminant of its field of fractions K. Then if $\mathcal{O} := \mathrm{End}(E/\mathbb{F}_q)$, we have

$$\mathbb{Z}[\pi_E] \subseteq \mathcal{O} \subseteq \mathcal{O}_K,$$

and $\mathcal{O}$ has discriminant $u^2 D_K$, where $u = [\mathcal{O}_K : \mathcal{O}]$ divides $v = [\mathcal{O}_K : \mathbb{Z}[\pi_E]]$.

Now, consider the set

$$\mathrm{Ell}_t(\mathbb{F}_q) := \{j(E/\mathbb{F}_q) \mid \mathrm{tr}(\pi_E) = t\}$$

of $\overline{\mathbb{F}}_q$-isomorphism classes of elliptic curves over $\mathbb{F}_q$ with trace of Frobenius t. Tate's Isogeny Theorem [**17**] implies that $\mathrm{Ell}_t(\mathbb{F}_q)$ determines an isogeny class. Further, as K is determined by t and q, this set can be written as the disjoint union

$$\mathrm{Ell}_t(\mathbb{F}_q) = \bigsqcup_{\mathbb{Z}[\pi_E] \subseteq \mathcal{O} \subseteq \mathcal{O}_K} \mathrm{Ell}_{\mathcal{O}}(\mathbb{F}_q),$$

where $\mathrm{Ell}_{\mathcal{O}}(\mathbb{F}_q)$ is defined in the same way as $\mathrm{Ell}_{\mathcal{O}}(\mathbb{C})$. As a consequence of the Deuring Lifting Theorem, the norm equation implies that over $\mathbb{F}_q[x]$, the Hilbert class polynomial $H_{u^2 D_K}(x)$ of degree $h(\mathcal{O})$ splits completely and the roots are precisely the set $\mathrm{Ell}_{\mathcal{O}}(\mathbb{F}_q)$. Then so long as $\mathrm{Ell}_{\mathcal{O}}(\mathbb{F}_q)$ is nonempty, the set has cardinality $h(\mathcal{O})$.

Recall the definition of the *Hurwitz class number* for an imaginary quadratic order $\mathcal{O}$:

$$H(\mathcal{O}) := \sum_{\mathcal{O} \subseteq \mathcal{O}' \subseteq \mathcal{O}_K} h(\mathcal{O}'). \tag{2.1}$$

If D is the discriminant of $\mathcal{O}$, we may define $H(D) := H(\mathcal{O})$. This agrees with the definition given earlier. From the results above we immediately have that the cardinality of $\mathrm{Ell}_t(\mathbb{F}_q)$ is equal to $H(t^2 - 4q)$.

As in the characteristic 0 case, we again have a faithful action of $cl(\mathcal{O})$ on $\mathrm{Ell}_{\mathcal{O}}(\mathbb{F}_q)$. What's more, if $\varphi\colon E \to E'$ is an isogeny of degree ℓ such that $\mathcal{O} := \mathrm{End}(E) = \mathrm{End}(E')$, then φ results from the action of an invertible $\mathcal{O}$-ideal $\mathfrak{l}$ of norm ℓ. This action is what will ultimately allow us to compare edges in $\mathcal{J}_{\mathbb{F}_q}$ with elements of class groups.

3. Taxonomy of jellyfish

Here we give several necessary results on the structure of jellyfish swarms. We first recall results from [**8**] about the vertices in a jellyfish swarm.

THEOREM 3.1 (Theorems 1 (2) and 3 (1) of [**8**]). *The following are true.*

(1) *The jellyfish swarm $\mathcal{J}_{\mathbb{F}_q}$ has $(q-3)(q-1)/2$ vertices.*
(2) *Each E_λ for $\lambda \in \mathbb{F}_q^{\times 2}\backslash\{0,1\}$ appears exactly $q-1$ times.*

The proof of (1) counts the number of admissible pairs (a,b), and the proof of (2) relies on the fact that if (a,b) corresponds to an elliptic curve E_λ, then so does every pair (ka, kb) for $k \in \mathbb{F}_q^\times$.

The above imply that the number of distinct λ that occur in $\mathcal{J}_{\mathbb{F}_q}$ is $(q-3)/2$. However, we have alternative ways of counting the λ that appear using the tools developed in Section 3 of [**8**]. In particular, recall that the j-invariants parameterize $\overline{\mathbb{F}}_q$-isomorphism classes of elliptic curves over $\mathbb{F}_q$, and that the trace and j-invariant uniquely determine an isomorphism class over $\mathbb{F}_q$. Further, if two elliptic curves have the same j-invariant, then they are either isomorphic over $\mathbb{F}_q$ or they are nontrivial quadratic twists of one another, in which case their traces differ by a sign. If we are able to count the possible traces and the possible j-invariants corresponding to each trace, as well as the number of distinct λ corresponding to each j-invariant, we will be able to get an alternative formula for the number of λ which occur.

We first characterize the admissible traces of the elliptic curves E_λ. By Lemma 2 of [**8**], one has that as an abelian group, $\mathbb{Z}/2\mathbb{Z} \times \mathbb{Z}/4\mathbb{Z} \subseteq E(\mathbb{F}_q)$. Then $8 \mid \#E(\mathbb{F}_q)$, and in particular $t \equiv q+1 \bmod 8$. We further know that every trace is represented so long as the Hasse bound $|t| \leqslant 2\sqrt{q}$ is satisfied (see [**3**] or [**13**]).

Now we fix a trace t. Since $\mathrm{Ell}_t(\mathbb{F}_q)$ defines an isogeny class, every elliptic curve on a jellyfish must have the same trace. Then consider the set of jellyfish with trace t and define $M_{\mathbb{F}_q}(t)$ to be the number of distinct j-invariants across the union of these jellyfish. The authors of [**8**] prove the following:

THEOREM 3.2 (Theorem 6 of [**8**]). *Suppose $q \equiv 3 \bmod 8$ and $p > 3$. If $|t| \leqslant 2\sqrt{q}$ such that $(t,p) = 1$ and $t \equiv q+1 \bmod 8$, then we have*

$$H\left(\frac{t^2-4q}{4}\right) = M_{\mathbb{F}_q}(t).$$

REMARK 3.3. The careful reader may note that our characterization of Theorem 6 of [**8**] includes the extra assumption that $(t,p) = 1$. This is not strictly necessary in the case where $q \equiv 3 \bmod 8$, as all appearing traces will have this property, but will be important for the case where $q \equiv 7 \bmod 8$ and so we include it here to avoid possible confusion.

The above relies primarily on the correspondence between the number of distinct j-invariants of elliptic curves with trace t and the Hurwitz class number $H(t^2-4q)$ discussed in Section 2, with the extra observation that the removal of a factor of 2 from the conductor u is equivalent to the requirement that $E[2] \subset E(\mathbb{F}_q)$ [**14**], which must be satisfied as all of our elliptic curves admit a Legendre normal form. Since $q \equiv 3 \bmod 4$, one can also show that every elliptic curve over $\mathbb{F}_q$ is of the form E_λ for $\lambda \in \mathbb{F}_q^{\times 2} \backslash \{0,1\}$ by 2-descent.

By the same arguments as that which prove Theorem 3.2, we get the following for the ordinary traces when $q \equiv 7 \bmod 8$:

THEOREM 3.4. *Suppose $q \equiv 7 \bmod 8$ and $p > 3$. If $|t| \leqslant 2\sqrt{q}$ such that $(t,p)=1$ and $t \equiv q+1 \bmod 8$, then we have*

$$H\left(\frac{t^2-4q}{4}\right) = M_{\mathbb{F}_q}(t).$$

We now turn to the case where E is supersingular (i.e., where $t=0$, as we shall see). While we need to amend the correspondence to avoid orders whose conductors are not coprime to p, we end up with a similar result:

LEMMA 3.5. *Suppose $q \equiv 7 \bmod 8$ and $p > 3$. Then we have $h(-q) = M_{\mathbb{F}_q}(0)$.*

PROOF. When $t=0$, the endomorphism ring of E can be identified with an imaginary quadratic order in $\mathbb{Q}(\pi_E)$ with conductor prime to p (Theorem 4.1, [**18**]), containing $\mathcal{O}(-q)$ (Proposition 3.7, [**14**]). Since q is a power of p, there is only one such order, $\mathcal{O}(-p)$ itself. □

Now, to justify our focus on these particular traces, we offer the following classification of the traces of elliptic curves over $\mathbb{F}_q$:

THEOREM 3.6 (Theorem 4.1 of [**18**]). *Let $q = p^k$ be a power of a prime p. Let $t \in \mathbb{Z}$ and let $N = q+1-t$. The integer N is the cardinality of $E(\mathbb{F}_q)$ for some elliptic curve $E/\mathbb{F}_q$ if and only if one of the following conditions is satisfied:*

(1) $|t| \leqslant 2\sqrt{q}$ *and* $(t,p)=1$;
(2) k *is odd and* $t=0$;
(3) k *is odd,* $t = \pm\sqrt{pq}$, *and* $p = 2$ *or* 3;
(4) k *is even,* $t=0$, $p \not\equiv 1 \bmod 4$;
(5) k *is even,* $t = \pm\sqrt{q}$, $p \not\equiv 1 \bmod 3$;
(6) k *is even,* $t = \pm 2\sqrt{q}$.

Since $q \equiv 3 \bmod 4$ and $p > 3$, the only options are (1) and (2), meaning the above completely classify the cases where $M_{\mathbb{F}_q}(t)$ can be nonzero. Now that we have determined the number of j-invariants appearing in $\mathcal{J}_{\mathbb{F}_q}$ using class numbers, we return to the question of how many λ correspond to a particular j. We offer the following answer:

LEMMA 3.7. *Let $L_{\mathbb{F}_q}(t,j)$ denote the number of distinct λ such that E_λ has trace t and j-invariant j. Then for all pairs (t,j) such that $L_{\mathbb{F}_q}(t,j) \neq 0$, we have $L_{\mathbb{F}_q}(t,j) = 2$.*

PROOF. First fix t and j and recall that there are exactly six λ in $\overline{\mathbb{F}}_q \backslash \{0,1\}$ corresponding to each j-invariant not equal to 0 or 1728 (see [**16**], Section III.1).

What's more, given one such λ we can find expressions for the other five by considering the orbit of λ under the group generated by the transformations $\lambda \mapsto 1/\lambda$ and $\lambda \mapsto 1-\lambda$ on $\mathbb{P}^1$. We can write this set as

$$[\lambda] := \{\lambda,\ 1/\lambda,\ 1-\lambda,\ 1/(1-\lambda),\ \lambda/(\lambda-1),\ (\lambda-1)/\lambda\}.$$

Now suppose that $\lambda \in \mathbb{F}_q^{\times 2}\backslash\{1\}$ and that E_λ has trace t and j-invariant j. We first note that all other elements of $[\lambda]$ are also in $\mathbb{F}_q^\times\backslash\{1\}$. We must also check which are squares. If λ is a square, then $1/\lambda$ is a square as well. Moreover, since $E_\lambda^{(\lambda)} \cong E_{1/\lambda}$ and λ is a square, this twist is trivial and the trace and j-invariant are unchanged. Note that only one of $1-\lambda$ and $\lambda-1$ can be a square since -1 is not a square in $\mathbb{F}_q$. Then only one of $\{1-\lambda,\ 1/(1-\lambda)\}$ and $\{\lambda/(\lambda-1),\ (\lambda-1)/\lambda\}$ is a set of squares. One also has that $E_\lambda^{(-1)} \cong E_{1-\lambda}$ and $E_{1/\lambda}^{(-1)} \cong E_{(\lambda-1)/\lambda}$. Since -1 is not a square, both of these are nontrivial twists, and so do not have the same trace as E_λ. Then the only contributions to $L_{\mathbb{F}_q}(t,j)$ are E_λ and $E_{1/\lambda}$.

Now we deal with the two exceptional cases. First, suppose E_λ has trace t and satisfies $j(E_\lambda)=0$. Then λ satisfies $\lambda^2-\lambda+1=0$ by the equation

$$j(E_\lambda) = 2^8\frac{(\lambda^2-\lambda+1)^3}{\lambda^2(\lambda-1)^2}.$$

Moreover, the other solution to $x^2-x+1=0$ is $1/\lambda$, so there are exactly two Legendre elliptic curves with trace t and j-invariant 0.

Finally, when $j(E_\lambda)=1728$, we have that $\lambda \in \{-1,2,1/2\}$. We know that -1 is not a sqaure in our setting, so either both 2 and $1/2$ are squares or neither are. Thus, $L_{\mathbb{F}_q}(t,j) = 0$ or 2. □

4. Proofs of Theorems 1.1 and 1.3

PROOF OF THEOREM 1.1. We will demonstrate the argument for the case when $q \equiv 3 \bmod 8$. The case when $q \equiv 7 \bmod 8$ is the same except for the care needed to deal with the supersingular elliptic curves, for which one applies Lemma 3.5. For $q \equiv 3$ mod 8, the theorem is equivalent to proving that

$$\frac{(q-1)(q-3)}{2} = \sum_{\substack{|t|\leqslant 2\sqrt{q}\\ t\equiv q+1(8)}} 2(q-1)\cdot H\left(\frac{4q-t^2}{4}\right).$$

The left hand side is the number of vertices in $\mathcal{J}_{\mathbb{F}_q}$, so it suffices to show the right hand side is also a count of the vertices in this graph. The sum runs over all the admissible traces of elliptic curves in $\mathcal{J}_{\mathbb{F}_q}$, so it suffices to show that each summand is the count of the number of vertices of trace t. Fixing t, each j invariant defines a curve up to $\mathbb{F}_q$-isomorphism. The number of such j is counted by $H\left(\frac{t^2-4q}{4}\right)$ by Theorem 3.2. Corresponding to each isomorphism class, there are exactly 2 such λ by Lemma 3.7. Finally, by Theorem 3.1 (2), each λ appears in the graph with multiplicity $q-1$. □

PROOF OF THEOREM 1.3. By Theorem 3 (2) of **[8]**, every edge corresponds to the unique isogeny with kernel generated by $\langle(0,0)\rangle$. In particular, this isogeny has degree 2.

By Corollary 4 (1) of **[8]**, the groups $E(\mathbb{F}_q)$ are isomorphic for all E on a jellyfish, and so their endomorphism rings are all the same; call it $\mathcal{O}$. This implies that this isogeny is the image of the class of an ideal $\mathfrak{p}_2$ of norm 2 in $\mathcal{O}$ with $E[\mathfrak{p}_2] = \langle(0,0)\rangle$. Since E_λ and $E_{1/\lambda}$ are isomorphic over $\mathbb{F}_q$, the edge emanating from vertices corresponding to both of these curves must have the same target curve. In particular, every vertex at the end of a tentacle has a corresponding vertex in the cycle with the same j-invariant. Thus, it suffices to consider the action of $[\mathfrak{p}_2]$ on the cycles of the jellyfish with endomorphism ring $\mathcal{O}$. Since this action is faithful and all j-invariants in $\mathrm{Ell}_{\mathcal{O}}(\mathbb{F}_q)$ are represented on some cycle, every cycle must have length divisible by the order of $[\mathfrak{p}_2]$ in $cl(\mathcal{O})$. Since the size of a jellyfish is twice the length of its cycle due to its tentacles of length one, our count must be multiplied by 2.

To see that $m(\mathcal{J})\cdot\#\mathcal{J} = 2(q-1)\cdot h_2(\mathcal{O})$, note that the action of $[\mathfrak{p}_2]$ partitions the set of vertices into $[cl(\mathcal{O}) : \langle[\mathfrak{p}_2]\rangle]$ subsets of size $2(q-1)\cdot h_2(\mathcal{O})$ by Theorem 1.1. Now if n is the order of $[\mathfrak{p}_2]$ in $cl(\mathcal{O})$, then $[\mathfrak{p}_2^n]\cdot E_\lambda = E_\lambda$, and so if $\lambda = b^2/a^2$, the nth pair in the sequence of $\mathrm{AGM}_{\mathbb{F}_q}(a,b)$ is (ka,kb) for some k in $\mathbb{F}_q^\times$. Then if $\mathrm{ord}(k)$ denotes the multiplicative order of k in $\mathbb{F}_q$, each jellyfish must have size $2\cdot h_2(\mathcal{O})\cdot \mathrm{ord}(k)$ and multiplicity $(q-1)/\mathrm{ord}(k)$ by partitioning by the orbits of the action of $\langle k\rangle$. □

References

[1] J. M. Borwein and P. B. Borwein, *Pi and the AGM*, Canadian Mathematical Society Series of Monographs and Advanced Texts, John Wiley & Sons, Inc., New York, 1987. A study in analytic number theory and computational complexity; A Wiley-Interscience Publication. MR877728

[2] B. Brown, N. J. Calkin, T. B. Flowers, K. James, E. Smith, and A. Stout, *Elliptic curves, modular forms, and sums of Hurwitz class numbers*, J. Number Theory **128** (2008), no. 6, 1847–1863, DOI 10.1016/j.jnt.2007.10.008. MR2419197

[3] M. Deuring, *Die Typen der Multiplikatorenringe elliptischer Funktionenkörper* (German), Abh. Math. Sem. Hansischen Univ. **14** (1941), 197–272, DOI 10.1007/BF02940746. MR5125

[4] M. Eichler, *On the class of imaginary quadratic fields and the sums of divisors of natural numbers*, J. Indian Math. Soc. (N.S.) **19** (1955), 153–180 (1956). MR80769

[5] R. Evans and J. Greene, *A quadratic hypergeometric ${}_2F_1$ transformation over finite fields*, Proc. Amer. Math. Soc. **145** (2017), no. 3, 1071–1076, DOI 10.1090/proc/13303. MR3589307

[6] M. Fouquet and F. Morain, *Isogeny volcanoes and the SEA algorithm*, Algorithmic number theory (Sydney, 2002), Lecture Notes in Comput. Sci., vol. 2369, Springer, Berlin, 2002, pp. 276–291, DOI 10.1007/3-540-45455-1_23. MR2041091

[7] J. Greene, *Hypergeometric functions over finite fields*, Trans. Amer. Math. Soc. **301** (1987), no. 1, 77–101, DOI 10.2307/2000329. MR879564

[8] M. J. Griffin, K. Ono, N. Saikia, and W.-L. Tsai, *AGM and jellyfish swarms of elliptic curves*, Amer. Math. Monthly **130** (2023), no. 4, 355–369, DOI 10.1080/00029890.2022.2160157. MR4567422

[9] D. R. Kohel, *Endomorphism rings of elliptic curves over finite fields*, ProQuest LLC, Ann Arbor, MI, 1996. Thesis (Ph.D.)–University of California, Berkeley. MR2695524

[10] M. H. Mertens, *Mock modular forms and class number relations*, Res. Math. Sci. **1** (2014), Art. 6, 16 pp., DOI 10.1186/2197-9847-1-6. MR3377995

[11] K. Ono, *The web of modularity: arithmetic of the coefficients of modular forms and q-series*, CBMS Regional Conference Series in Mathematics, vol. 102, Published for the Conference Board of the Mathematical Sciences, Washington, DC; by the American Mathematical Society, Providence, RI, 2004. MR2020489

[12] K. Ono, H. Saad, and N. Saikia, *Distribution of values of Gaussian hypergeometric functions*, Pure Appl. Math. Q. **19** (2023), no. 1, 371–407, DOI 10.4310/pamq.2023.v19.n1.a14. MR4570167

[13] H.-G. Rück, *A note on elliptic curves over finite fields*, Math. Comp. **49** (1987), no. 179, 301–304, DOI 10.2307/2008268. MR890272

[14] R. Schoof, *Nonsingular plane cubic curves over finite fields*, J. Combin. Theory Ser. A **46** (1987), no. 2, 183–211, DOI 10.1016/0097-3165(87)90003-3. MR914657

[15] J. H. Silverman, *Advanced topics in the arithmetic of elliptic curves*, Graduate Texts in Mathematics, vol. 151, Springer-Verlag, New York, 1994, DOI 10.1007/978-1-4612-0851-8. MR1312368

[16] J. H. Silverman, *The arithmetic of elliptic curves*, 2nd ed., Graduate Texts in Mathematics, vol. 106, Springer, Dordrecht, 2009, DOI 10.1007/978-0-387-09494-6. MR2514094

[17] J. Tate, *Endomorphisms of abelian varieties over finite fields*, Invent. Math. **2** (1966), 134–144, DOI 10.1007/BF01404549. MR206004

[18] W. C. Waterhouse, *Abelian varieties over finite fields*, Ann. Sci. École Norm. Sup. (4) **2** (1969), 521–560. MR265369

[19] D. Zagier, *Nombres de classes et formes modulaires de poids* 3/2 (French, with English summary), C. R. Acad. Sci. Paris Sér. A-B **281** (1975), no. 21, Ai, A883–A886. MR429750

DEPT. OF MATHEMATICS, UNIVERSITY OF VIRGINIA, CHARLOTTESVILLE, VIRGINIA 22903
Email address: egm3zq@virginia.edu

DEPT. OF MATHEMATICS, UNIVERSITY OF VIRGINIA, CHARLOTTESVILLE, VIRGINIA 22903
Email address: ko5wk@virginia.edu

Contemporary Mathematics
Volume **818**, 2025
https://doi.org/10.1090/conm/818/16376

The Heun eigenvalue problem and some of its special cases

Hans Volkmer

ABSTRACT. The standard eigenvalue problem for the Heun differential equation is studied. In particular, its numerical treatment is addressed. Several eigenvalue problems for the Lamé equation are shown to be special cases of the Heun eigenvalue problem. These Lamé eigenvalue problems arise in the solution of the Laplace equation in three dimensional Euclidean space by the method of separation of variables in cyclide coordinate systems. These are the flat-ring, bicyclide and flat-disk coordinates.

1. Introduction

The Heun differential equation [**6**, Chapter 31], [**17**] is

$$\frac{d^2y}{dz^2} + \left(\frac{\gamma}{z} + \frac{\delta}{z-1} + \frac{\epsilon}{z-a}\right)\frac{dy}{dz} + \frac{\alpha\beta z + \lambda}{z(z-1)(z-a)}y = 0, \tag{1.1}$$

where

$$\gamma + \delta + \epsilon = \alpha + \beta + 1.$$

The Heun equation is a Fuchsian equation with four regular singularities at $z = 0, 1, a, \infty$. The exponents at these singularities are respectively $\{0, 1-\gamma\}$, $\{0, 1-\delta\}$, $\{0, 1-\epsilon\}$, $\{\alpha, \beta\}$.

There is a standard eigenvalue problem for the Heun equation [**6**, §31.4], [**17**, Chapter A, §3.5]. The accessory parameter λ is called an eigenvalue if there exists a nontrivial solution $y(z)$ (Heun eigenfunction) of (1.1) which is analytic in a neighborhood of the interval $[0, 1]$, that is, it belongs to the exponent 0 at both singularities 0 and 1. Under the assumptions

$$\gamma > 0, \quad \delta > 0, \quad \alpha + \beta \in \mathbb{R}, \quad \alpha\beta \in \mathbb{R}, \quad a > 1, \tag{1.2}$$

the following facts are known [**4**].

(a) The eigenvalues are real, and they form an increasing sequence $\{\lambda_n\}_{n=0}^\infty$ converging to infinity.

(b) The eigenfunction y_n belonging to the eigenvalue λ_n has exactly n zeros in $(0, 1)$.

2020 *Mathematics Subject Classification.* Primary 34B30, 34L15.

Key words and phrases. Heun's differential equation, Lamé's differential equation, flat-ring coordinates, bicyclide coordinates, flat-disk coordinates.

(c) When properly normalized the eigenfunctions $\{y_n\}$ form an orthonormal basis in the Hilbert space $L^2_w(0,1)$ with weight function

$$w(x) = x^{\gamma-1}(1-x)^{\delta-1}(a-x)^{\epsilon-1}.$$

Since there exist no explicit formulas for eigenvalues or eigenfunctions, it is of interest to provide efficient methods for their numerical computation. This problem will be addressed in Section 3 of this paper.

In Sections 4, 5, 6 we consider the Lamé equation

$$\frac{d^2u}{ds^2} + (h - \nu(\nu+1)k^2 \operatorname{sn}^2(s,k))u = 0. \tag{1.3}$$

In this equation $\operatorname{sn}(s,k)$ denotes the Jacobian elliptic function [**6**, Chapter 22] for the modulus k, h is the eigenvalue parameter, and ν is an additional parameter. In this paper, we always assume that $0 < k < 1$ and $\nu \geqslant -\frac{1}{2}$. The Lamé equation appears when we solve the Laplace equation

$$\frac{\partial^2 v}{\partial x_1^2} + \frac{\partial^2 v}{\partial x_2^2} + \frac{\partial^2 v}{\partial x_3^2} = 0$$

by the method of separation of variables in flat-ring, bicyclide and flat-disk coordinates [**9**, Chapter XV], [**16**]. When we look for solutions (harmonic functions) appropriate for applications we encounter several eigenvalue problems for the Lamé equation. We treat the periodic eigenvalue problem in Section 4 (flat-ring coordinates), the Lamé–Wangerin eigenvalue problem (flat-ring and bicyclide coordinates) in Section 5, and a second kind of Lamé–Wangerin eigenvalue problem (flat-disk coordinates) in Section 6.

Let us add some remarks on the history of these eigenvalue problems. The separation of the Laplace equation in cyclide coordinate systems has been carried out by Wangerin [**25**] for the first time in 1875. Wangerin also obtained the corresponding eigenvalue problems for the Lamé equation. However, the Lamé equation appears in a different form and its treatment is faulty. This is no surprise because spectral theory was not developed well enough at that time to handle these eigenvalue problems. An important contribution was made by Erdélyi, Magnus, and Oberhettinger in Chapter XV of their book [**9**] published in 1955. This chapter treats all eigenvalue problems that we consider in this paper. Several results that we mention in this paper are already contained in [**9**]. However, we believe that many improvements of the theory are presented in this paper. We just mention a few of them.

(i) Although the eigenvalue problem for the Heun equation is mentioned in [**9**, §15.3], it is not explicitly stated that the eigenvalue problems for the Lamé equation are special cases of it.

(ii) In [**9**] the numerical treatment of the eigenvalue problems is based on infinite continued fractions. We prefer the treatment via calculation of eigenvalues and eigenvectors of finite matrices. It is easily implemented using standard software like Maple or Mathematica. Some numerical examples are given.

(iii) Based on [**4**] we embed the eigenvalue problems in general spectral theory. This point is missing in [**9**].

(iv) In [**9**] the eigenvalue problem for the Lamé equation arising in flat-disk coordinates is handled by allowing complex values of the modulus k of

Jacobian elliptic functions. This leads to a somewhat awkward analysis. We present a purely real treatment of this problem in Section 6.

In this paper we consider eigenvalue problems for the Heun equation only in the real case, that is, if the conditions (1.2) are satisfied. It appears that these eigenvalue problems have not yet been treated if we allow complex values for $a, \alpha, \beta, \gamma, \delta$. Such eigenvalue problems could be tackled using methods from perturbation theory following the treatment of eigenvalue problems involving complex parameters for the Mathieu and spheroidal differential equations in [**15**].

It should also be pointed out that there are many papers discussing various aspects of the theory of the Heun equation, for example, see [**5**], [**12**], [**14**] and the references therein.

2. The Heun eigenvalue problem

In this section we study the Heun eigenvalue problem under the assumptions (1.2). This eigenvalue problem is treated in [**4**], [**7**], [**9**, §15.3] [**17**, §3.5], [**19**] and [**21**].

2.1. A singular Sturm–Liouville problem. Consider the Heun differential expression [**4**, §5]

$$L[f] := -(pf')' + qf, \quad f, f' \in \mathrm{AC}_{\mathrm{loc}}(0,1), \tag{2.1}$$

where, for $x \in (0,1)$,

$$\begin{aligned} p(x) &:= x^\gamma (1-x)^\delta (a-x)^\epsilon, \\ w(x) &:= x^{\gamma-1}(1-x)^{\delta-1}(a-x)^{\epsilon-1}, \\ q(x) &:= -\alpha\beta x w(x). \end{aligned}$$

A function $(0,1) \to \mathbb{C}$ belongs to $\mathrm{AC}_{\mathrm{loc}}(0,1)$ if it is absolutely continuous on every compact subinterval of $(0,1)$. Then

$$L[f] = \lambda w f$$

is the Heun equation. Let L^2_w be the Hilbert space of measurable functions $f\colon (0,1) \to \mathbb{C}$ with $\int_0^1 w(x)\,|f(x)|^2\,dx < \infty$. The inner product is

$$\langle f, g\rangle_w := \int_0^1 w(x) f(x)\overline{g(x)}\,dx.$$

The maximal operator [**4**, §5] generated by (2.1) is defined on

$$\Delta := \{f \in L^2_w : f, f' \in \mathrm{AC}_{\mathrm{loc}}(0,1), w^{-1}Lf \in L^2_w\}. \tag{2.2}$$

Since w is integrable on $(0,1)$, the constant function 1 belongs to Δ. It follows [**4**, §7] that the limits

$$\lim_{x\to 0+} p(x)f'(x), \quad \lim_{x\to 1-} p(x)f'(x) \tag{2.3}$$

exist for all $f \in \Delta$. We introduce the domain

$$D := \{f \in \Delta : \lim_{x\to 0+} p(x)f'(x) = \lim_{x\to 1-} p(x)f'(x) = 0\}. \tag{2.4}$$

If $\gamma \geqslant 2$ the condition $\lim_{x\to 0+} p(x)f'(x) = 0$ is automatically true [**4**, §6]. Similarly, if $\delta \geqslant 2$ then the condition $\lim_{x\to 1-} p(x)f'(x) = 0$ is automatically satisfied.

The following theorem is proved in [**4**, §11].

THEOREM 2.1. *The operator T defined by $Tf := w^{-1}Lf$ for $f \in D$ is self-adjoint, bounded below with compact resolvent, on L^2_w. Its eigenvalues are simple.*

Therefore, the spectrum of T consists entirely of real eigenvalues, and the eigenvalues form a sequence converging to infinity. It is easy to show that λ is an eigenvalue of T if and only if λ is an eigenvalue of the Heun eigenvalue problem. We write the sequence of eigenvalues of T in increasing order as

$$\lambda_0 < \lambda_1 < \lambda_2 < \dots .$$

In order to indicate the dependence of λ_n on the parameters we write

$$\lambda_n = \lambda_n(\alpha, \beta, \gamma, \delta, a). \tag{2.5}$$

The Heun operator T will be compared with a simpler differential operator T_0 of hypergeometric type. We set

$$P(x) := x^\gamma (1-x)^\delta, \quad W(x) := x^{\gamma-1}(1-x)^{\delta-1} \quad \text{for } x \in (0,1),$$

and consider the differential expression

$$L_0[f] = -(Pf')'.$$

Note that $L^2_w = L^2_W$ as vector spaces. Moreover, the inner products induced by the weights w and W generate equivalent norms. Set

$$\Delta_0 := \{f \in L^2_W : f, f' \in \mathrm{AC}_{\mathrm{loc}}(0,1), W^{-1}L_0 f \in L^2_W\}, \tag{2.6}$$

and

$$D_0 := \{f \in \Delta_0 : \lim_{x\to 0+} P(x)f'(x) = \lim_{x \to 1-} P(x)f'(x) = 0\}. \tag{2.7}$$

The operator T_0 defined by $T_0 f := W^{-1}L_0 f$ for $f \in D_0$ is self-adjoint on L^2_W, bounded below with compact resolvent. The eigenvalues and eigenfunctions of T_0 are explicitly known [**20**, Section 4.21]. The eigenvalues are

$$n(n+\gamma+\delta-1), \quad n \in \mathbb{N}_0, \tag{2.8}$$

and corresponding eigenfunctions are the hypergeometric polynomials

$$F(-n, n+\gamma+\delta-1; \gamma; x). \tag{2.9}$$

These polynomials can also be written in terms of Jacobi polynomials [**20**, (4.21.2)]

$$p_n(x) := P_n^{(\gamma-1,\delta-1)}(1-2x) = \frac{(\gamma)_n}{n!} F(-n, n+\gamma+\delta-1; \gamma; x). \tag{2.10}$$

The system of functions $\{p_n\}_{n\in\mathbb{N}_0}$ forms an orthogonal basis for L^2_W. An arbitrary function $f \in L^2_W$ can be expanded in the orthogonal series

$$f = \sum_{r=0}^\infty c_r p_r. \tag{2.11}$$

From [**20**, (4.3.3)] we obtain

$$\langle p_r, p_r \rangle_W = g_r := \frac{1}{2r+\gamma+\delta-1} \frac{\Gamma(r+\gamma)\Gamma(r+\delta)}{\Gamma(r+1)\Gamma(r+\gamma+\delta-1)}. \tag{2.12}$$

To avoid division by zero we set $g_0 := \frac{\Gamma(\gamma)\Gamma(\delta)}{\Gamma(\gamma+\delta)}$. It follows that

$$\lim_{r\to\infty} r \int_0^1 W(x) p_r(x)^2 \, dx = \frac{1}{2}. \tag{2.13}$$

Hence the coefficients $\{c_r\}$ satisfy

$$\sum_{r=0}^{\infty} \frac{1}{r+1} |c_r|^2 < \infty. \tag{2.14}$$

Conversely, if (2.14) holds then f defined by (2.11) is in L^2_W.

The expressions L and L_0 are connected by

$$w^{-1}Lf = (a-x)W^{-1}L_0 f + \epsilon W^{-1}Pf' - \alpha\beta x f. \tag{2.15}$$

LEMMA 2.2. *We have $\Delta = \Delta_0$ and $D = D_0$. Moreover, D_0 consists of all functions $f \in L_W$ with orthogonal expansion* (2.11) *such that $\sum_{r=0}^{\infty} r^3 |c_r|^2 < \infty$.*

PROOF. We show $\Delta \subset \Delta_0$, so let $f \in \Delta$. Using (2.15) and $xf \in L^2_w$, it is enough to show that $W^{-1}Pf' \in L^2_W$. If $\gamma < 2$ then the existence of the limit $\lim_{x\to 0+} p(x)f'(x)$ and the fact that W^{-1} is integrable on $(0, \frac{1}{2})$ implies

$$\int_0^{\frac{1}{2}} W^{-1}(x)\{P(x)f'(x)\}^2\, dx < \infty. \tag{2.16}$$

If $\gamma \geqslant 2$ then $\lim_{x\to 0+} p(x)f'(x) = 0$ so

$$p(x)f'(x) = \int_0^x \frac{d}{d\xi}\left(p(\xi)f'(\xi)\right) d\xi \quad \text{for } x \in (0,1).$$

Since the function $w^{-1}(pf')'$ is in L^2_w, using the Cauchy–Schwarz inequality we obtain

$$\{p(x)f'(x)\}^2 \leqslant C_1 \int_0^x w(\xi)\, d\xi \leqslant C_2 x^{\gamma} \quad \text{for } x \in (0, \tfrac{1}{2}),$$

where C_1, C_2 are constants independent of x. This implies (2.16) also in the case $\gamma \geqslant 2$. By a similar argument, we obtain

$$\int_{\frac{1}{2}}^1 W^{-1}(x)\{P(x)f'(x)\}^2\, dx < \infty, \tag{2.17}$$

so the proof of $\Delta \subset \Delta_0$ is complete. The proof of the inclusion $\Delta_0 \subset \Delta$ is very similar.

The equality $D = D_0$ is now clear and the remaining statement of the lemma follows from (2.13) and the fact that $n(n+\gamma+\delta-1)$, $n \in \mathbb{N}_0$, is the sequence of eigenvalues of T_0. □

2.2. Expansion of Heun eigenfunctions in Jacobi polynomials. Let G be the linear space of all complex sequences $\{c_r\}_{r=0}^{\infty}$ for which (2.14) holds. Then $U\colon G \to L^2_w$ defined by

$$Uc = \sum_{r=0}^{\infty} c_r p_r, \quad c = \{c_r\}$$

is a bijective linear map. We define an inner product on G by

$$\langle c, d\rangle = \langle Uc, Ud\rangle_w.$$

Let $M : G \supset D(M) \to G$ be the operator $M := U^{-1}TU$ defined on

$$D(M) = \{c \in G : Uc \in D\} = \{\{c_r\} : \sum_{r=0}^{\infty} r^3 |c_r|^2 < \infty\}. \tag{2.18}$$

The second equality follows Lemma 2.2.

We now represent M by an infinite tridiagonal matrix. Since $p_0(x) = 1$, $p_1(x) = \gamma - (\gamma+\delta)x$, we have

$$Tp_0 = \sigma_0 p_0 + \rho_1 p_1, \tag{2.19}$$

where

$$\sigma_0 := -\frac{\alpha\beta\gamma}{\gamma+\delta}, \qquad \rho_1 := \frac{\alpha\beta}{\gamma+\delta}. \tag{2.20}$$

Using (2.15) and following [**17**, Chapter A, (4.2.17)], we find that

$$Tp_r = \tau_r p_{r-1} + \sigma_r p_r + \rho_{r+1} p_{r+1}, \quad r \in \mathbb{N}_0, \tag{2.21}$$

where

$$\rho_r := \frac{r(r-1+\omega)(r-1+\alpha)(r-1+\beta)}{(2r-2+\omega)(2r-1+\omega)},$$

$$\sigma_r := ar(r+\omega) - \frac{\epsilon r(r+\omega)(\gamma-\delta) + (r(r+\omega)+\alpha\beta)(2r(r+\omega)+\gamma(\omega-1))}{(2r-1+\omega)(2r+1+\omega)},$$

$$\tau_r := \frac{(r-1+\gamma)(r-1+\delta)(r+\omega-\alpha)(r+\omega-\beta)}{(2r+\omega)(2r-1+\omega)},$$

$$\omega := \gamma+\delta-1.$$

Equations (2.19) and (2.21) lead to the following matrix representation of M:

$$M = \begin{bmatrix} \sigma_0 & \tau_1 & 0 & 0 & 0 & 0 & \dots \\ \rho_1 & \sigma_1 & \tau_2 & 0 & 0 & 0 & \dots \\ 0 & \rho_2 & \sigma_2 & \tau_3 & 0 & 0 & \dots \\ 0 & 0 & \rho_3 & \sigma_3 & \tau_4 & 0 & \dots \\ \vdots & \vdots & \vdots & \vdots & \vdots & \vdots & \end{bmatrix} \tag{2.22}$$

For $c \in D(M)$, Mc can be computed by the usual multiplication of matrix times column vector.

We summarize our results.

THEOREM 2.3. *The matrix operator M is similar to the Heun operator T, so M and T have the same eigenvalues. If $\{c_r\}_{r=0}^\infty$ is an eigenvector of M corresponding to an eigenvalue λ, then the function $H(z)$ defined by*

$$H(z) = \sum_{r=0}^\infty c_r p_r(z) \tag{2.23}$$

is a Heun eigenfunction belonging to the eigenvalue λ. The infinite series in (2.23) converges locally uniformly within the ellipse with foci at $z=0$ and $z=1$ passing through $z=a$.

PROOF. The local uniform convergence of the series follows from [**20**, Thm. 9.1.1]. □

If the expansion (2.23) terminates, that is, it contains only a finite number of terms, then there is $r \in \mathbb{N}$ such that $c_{r-1} \neq 0$ and $c_j = 0$ for $j \geqslant r$. It follows that $\rho_r = 0$, and λ is an eigenvalue of the finite submatrix of M consisting of the first r rows and columns of M. Note that under assumption (1.2) $r-1+\omega$ vanishes only for $r=1$ and $\omega=0$ but then also $2r-2+\omega=0$. If $\omega=0$ the formula for ρ_1 contains the expression $0/0$ that has to be deleted. Therefore, the condition $\rho_r = 0$

is equivalent to $\alpha = 1 - r$ or $\beta = 1 - r$. In this case the eigenfunction $H(z)$ is a polynomial known as Heun polynomial.

We mention that there are other eigenvalue problems for the Heun equation. For example, we can consider eigenfunctions on the interval $(0, 1)$ that are Frobenius solutions at both endpoints but now belonging to the exponents $1-\gamma$ and $1-\delta$ at $z = 0$ and $z = 1$, respectively. These eigenvalue problems can all be easily transformed to the standard eigenvalue problem we discussed in this section [**17**, Chapter A, §3.5].

We consider two transformations of the Heun eigenvalue problem into itself.

First transformation. Setting $y(z) = (a-z)^{1-\epsilon}\tilde{y}(z)$ we obtain another Heun equation for $\tilde{y}$ with parameters

$$\tilde{\alpha} = \gamma + \delta - \alpha, \quad \tilde{\beta} = \gamma + \delta - \beta, \tag{2.24}$$

$$\tilde{\gamma} = \gamma, \quad \tilde{\delta} = \delta, \quad \tilde{\lambda} = \lambda - \gamma(\gamma + \delta - \alpha - \beta), \quad \tilde{a} = a. \tag{2.25}$$

This transformation is often called an F-homotopy [**17**, p. 16]. The Heun eigenvalue problem remains unchanged, so we obtain

$$\lambda_n(\gamma + \delta - \alpha, \gamma + \delta - \beta, \gamma, \delta, a) = \lambda_n(\alpha, \beta, \gamma, \delta, a) - \gamma(\gamma + \delta - \alpha - \beta). \tag{2.26}$$

Note that

$$\tilde{\sigma}_r = \sigma_r - \gamma(\gamma + \delta - \alpha - \beta), \qquad \tilde{\rho}_r\tilde{\tau}_r = \rho_r\tau_r. \tag{2.27}$$

Let $\tilde{M}$ be the matrix operator with entries $\tilde{\rho}_r, \tilde{\sigma}_r, \tilde{\tau}_r$. The eigenvectors of M and $\tilde{M}$ are related as follows.

THEOREM 2.4. *Suppose that $\tau_r\tilde{\tau}_r \neq 0$ for all $r \in \mathbb{N}$. If $\{c_r\}_{r=0}^{\infty}$ is an eigenvector of M belonging to the eigenvalue λ_n, then the sequence $\{\tilde{c}_r\}_{r=0}^{\infty}$ defined by*

$$\tilde{c}_r = c_r \prod_{j=1}^{r} \frac{\tau_j}{\tilde{\tau}_j} = c_r \frac{\Gamma(\alpha)\Gamma(\beta)\Gamma(r + \gamma + \delta - \alpha)\Gamma(r + \gamma + \delta - \beta)}{\Gamma(r + \alpha)\Gamma(r + \beta)\Gamma(\gamma + \delta - \alpha)\Gamma(\gamma + \delta - \beta)}$$

is an eigenvector of $\tilde{M}$ belonging to the eigenvalue $\tilde{\lambda}_n = \lambda_n - \gamma(\gamma + \delta - \alpha - \beta)$.

PROOF. Let $c = \{c_r\}$, $\tilde{c} = \{\tilde{c}_r\}$ be eigenvectors of M and $\tilde{M}$ belonging to eigenvalues λ_n and $\tilde{\lambda}_n$, respectively. Since we assume that $\tau_r\tilde{\tau}_r \neq 0$ for all $r \in \mathbb{N}$, we must have $c_0 \neq 0$ and $\tilde{c}_0 \neq 0$ so we may assume that $c_0 = \tilde{c}_0 = 1$. Then the equation $c_r \prod_{j=1}^{r} \tau_j = \tilde{c}_r \prod_{j=1}^{r} \tilde{\tau}_j$ can be proved by induction using (2.27) and the recursion formulas for the sequences c and $\tilde{c}$. □

The following result [**7**, §9] shows that eigenvectors of the matrix operator $\tilde{M}$ are useful in the computation of the normalization integral $\int_0^1 w(x)H(x)^2\,dx$ for a Heun eigenfunction $H(x)$.

THEOREM 2.5. *Let $\tilde{M}$ denote the matrix operator* (2.22) *for the transformed parameter values $\tilde{\alpha}, \tilde{\beta}$ as given in* (2.24). *Let $\{c_r\}_{r=0}^{\infty}$ be an eigenvector of M belonging to the eigenvalue λ_n, and let $\{\tilde{c}_r\}_{r=0}^{\infty}$ be an eigenvector of $\tilde{M}$ belonging* *to the eigenvalue $\tilde{\lambda}_n$. Set*

$$S_0 = \sum_{r=0}^{\infty} c_r \frac{(\gamma)_r}{r!}, \quad S_1 = \sum_{r=0}^{\infty} c_r (-1)^r \frac{(\delta)_r}{r!}, \tag{2.28}$$

$$\tilde{S}_0 = a^{1-\epsilon} \sum_{r=0}^{\infty} \tilde{c}_r \frac{(\gamma)_r}{r!}, \quad \tilde{S}_1 = (a-1)^{1-\epsilon} \sum_{r=0}^{\infty} \tilde{c}_r (-1)^r \frac{(\delta)_r}{r!}, \tag{2.29}$$

$$S_2 = \sum_{r=0}^{\infty} c_r \tilde{c}_r g_r. \tag{2.30}$$

Then the Heun eigenfunction $H(x)$ defined by (2.23) satisfies

$$\int_0^1 w(x) H(x)^2\, dx = \frac{S_0 S_2}{\tilde{S}_0} = \frac{S_1 S_2}{\tilde{S}_1}.$$

PROOF. Define the function

$$\tilde{H}(x) = (a-x)^{1-\epsilon} \sum_{r=0}^{\infty} \tilde{c}_r p_r(x). \tag{2.31}$$

This is also a Heun eigenfunction corresponding to the eigenvalue λ_n. Therefore, $\tilde{H}(x)$ must be a constant multiple of $H(x)$. If we compare $H(x)$ with $\tilde{H}(x)$ at $x=0$ or at $x=1$ and note that $p_r(0) = \frac{(\gamma)_r}{r!}$, $p_r(1) = (-1)^r \frac{(\delta)_r}{r!}$ then we obtain

$$S_0 \tilde{H}(x) = \tilde{S}_0 H(x), \qquad S_1 \tilde{H}(x) = \tilde{S}_1 H(x) \tag{2.32}$$

so $S_0/\tilde{S}_0 = S_1/\tilde{S}_1$. Therefore,

$$\int_0^1 w(x) H(x)^2\, dx = \frac{S_0}{\tilde{S}_0} \int_0^1 x^{\gamma-1}(1-x)^{\delta-1}(a-x)^{\epsilon-1} \tilde{H}(x) H(x)\, dx.$$

If we now substitute (2.23), (2.31) and use the orthogonality of the polynomials p_r with respect to the weight function $W(x)$ together with (2.12) we obtain the desired statement. □

Second transformation. Setting

$$y(z) = (a-\zeta)^{\alpha} \hat{y}(\zeta), \qquad \zeta = a\frac{z-1}{z-a}, \tag{2.33}$$

we obtain another Heun equation for $\hat{y}(\zeta)$ with parameters

$$\hat{\alpha} = \alpha, \quad \hat{\beta} = \gamma + \delta - \beta,$$
$$\hat{\gamma} = \delta, \quad \hat{\delta} = \gamma, \quad \hat{\lambda} = \lambda - \alpha(\delta - \beta), \quad \hat{a} = a.$$

If α, β are real, then the Heun eigenvalue problem remains unchanged, so

$$\lambda_n(\alpha, \gamma+\delta-\beta, \delta, \gamma, a) = \lambda_n(\alpha, \beta, \gamma, \delta, a) - \alpha(\delta - \beta). \tag{2.34}$$

This is confirmed by looking at the corresponding matrix $\hat{M}$ with entries $\hat{\rho}_r, \hat{\sigma}_r, \hat{\tau}_r$. Indeed, we have

$$\hat{\sigma}_r = \sigma_r - \alpha(\delta - \beta), \qquad \hat{\rho}_r \hat{\tau}_r = \rho_r \tau_r. \tag{2.35}$$

A Heun eigenfunction $H(x)$ can be normalized in at least three different ways. Note that $H(0) \neq 0$.

First normalization. We require

$$\int_0^1 w(x)H(x)^2\,dx = 1, \quad H(0) > 0. \tag{2.36}$$

Under this normalization the system of Heun eigenfunctions of the operator T forms an orthonormal basis in $L^2_w(0,1)$. Theorem 2.5 is useful to carry out this normalization,

Second normalization. We require

$$\int_0^1 W(x)H(x)^2\,dx = 1, \quad H(0) > 0. \tag{2.37}$$

This normalization integral has the advantage that it can be expressed directly in terms of the coefficients c_r in (2.23). The orthogonality of Jacobi polynomials and (2.12) imply

$$\int_0^1 W(x)H(x)^2\,dx = \sum_{r=0}^{\infty} g_r\,|c_r|^2\,.$$

Third normalization. We require $H(0) = 1$.

In our numerical work and in Theorem 3.4 we will adopt the second normalization (2.37). We denote the normalized Heun eigenfunction corresponding to the eigenvalue λ_n by $H_n(x)$, or, more precisely by $H_n(\alpha,\beta,\gamma,\delta,a;x)$.

2.3. Expansions of type I. Erdélyi [**7**] gives two types of expansions of solutions to the Heun equation in series of hypergeometric functions. He calls (2.23) an expansion of type II. In this subsection we consider type I expansions which are also useful in the treatment of the Heun eigenvalue problem. Let $H(\alpha,\beta,\gamma,\delta,a,\lambda;z)$ denote the solution of the Heun equation (1.1) which is analytic at $z=0$ with $H(0) = 1$ [**17**, §A.3.3]. This solution exists provided $-\gamma \notin \mathbb{N}_0$. Although the following theorem holds under more general conditions we continue working under assumptions (1.2).

THEOREM 2.6. *Let the sequence* $\{d_r\}_{r=0}^{\infty}$ *be determined by* $d_0 = 1$ *and the recursion*

$$(\breve{\sigma}_0 - \lambda)d_0 + \breve{\tau}_1 d_1 = 0, \tag{2.38}$$

$$\breve{\rho}_r d_{r-1} + (\breve{\sigma}_r - \lambda)d_r + \breve{\tau}_{r+1}d_{r+1} = 0 \quad \textit{for } r \geqslant 1, \tag{2.39}$$

where

$$\breve{\rho}_r = -\frac{(r-1+\omega)(r-1+\delta)(r+\omega-\beta)(r+\omega-\alpha)(r-1+\alpha)(r-1+\beta)}{(2r+\omega)(2r-1+\omega)^2(2r-2+\omega)},$$

$$\breve{\sigma}_r = \sigma_r, \quad \breve{\tau}_r = -r(r-1+\gamma),$$

with σ_r *and* ω *defined after* (2.21). *Then we have the expansion*

$$H(\alpha,\beta,\gamma,\delta,a,\lambda;z) = \sum_{r=0}^{\infty} d_r \left(\frac{z}{a}\right)^r F\left(\alpha+r,\beta+r;\gamma+\delta+2r;\frac{z}{a}\right) \tag{2.40}$$

valid in the open set

$$D_1 = \left\{z : \left|\frac{a}{z}\right| + \left|\frac{a}{z}-1\right| < 2a-1\right\} \cup \{0\}.$$

The infinite series in (2.40) *converges locally uniformly in* D_1.

PROOF. We start with Erdélyi's expansion [**7**, (3.2)] with $P_m = P_m^5$. It represents a solution $y(z)$ to the Heun equation belonging to the exponent α at $z = \infty$ in a series of hypergeometric functions. This expansion is also given in [**6**, (31.11.12)]. We mention that a short proof of the expansion can be based on [**18**, (2.15)]. The expansion converges locally uniformly outside the ellipse with foci at $z = 0$ and $z = 1$ passing through $z = a$. If we set $z = \frac{a}{\zeta}$ and $y = \zeta^\alpha \eta$ then $\eta(\zeta)$ satisfies a Heun equation with parameters

$$\hat{\alpha} = \alpha, \quad \hat{\beta} = \alpha + 1 - \gamma, \quad \hat{\gamma} = \alpha + 1 - \beta, \quad \hat{\delta} = \epsilon, \quad \hat{\epsilon} = \delta,$$
$$\hat{a} = a, \quad \hat{\lambda} = \lambda + \alpha((\beta - \epsilon)a + \beta - \delta).$$

Up to a constant factor $\eta(\zeta)$ agrees with the function $H(\hat{\alpha}, \hat{\beta}, \hat{\gamma}, \hat{\delta}, \hat{a}, \hat{\lambda}; \zeta)$. This gives (2.40) after changing ζ back to z and adjusting the parameters. The formulas for $\breve{\rho}_r, \breve{\sigma}_r, \breve{\tau}_r$ follow from [**7**, (4.5)]. The proof of $\breve{\sigma}_r = \sigma_r$ requires a lengthy calculation that is best carried out with a computer. □

The region D_1 appearing in Theorem 2.6 is bounded by an elliptic limaçon as depicted in Figure 1 for $a = 2$. This figure also shows the circle centered at the origin with unit radius. The power series expansion of $H(\alpha, \beta, \gamma, \delta, a, \lambda; z)$ at $z = 0$ usually converges only within this circle while expansion (2.40) converges in the larger region D_1.

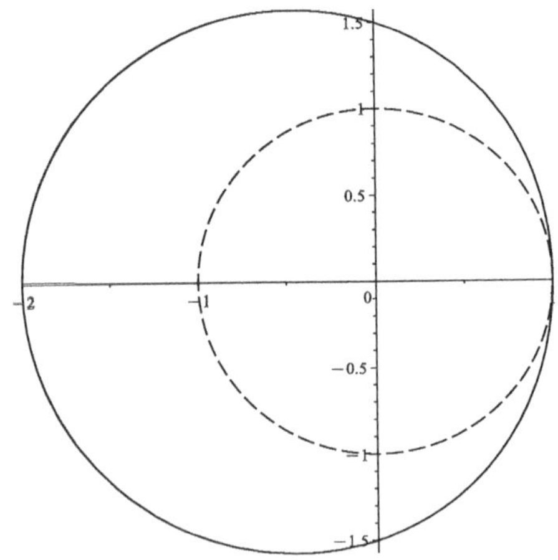

FIGURE 1. Convergence region D_1 of expansion (2.40) for $a = 2$ compared with unit disk.

We mention that $\breve{\rho}_r \breve{\tau}_r = \rho_r \tau_r$ for all $r \in \mathbb{N}$ so that the recursion (2.38), (2.39) can be related to the recursion obtained from the matrix eigenvector equation $Mc = \lambda c$ similar to the statement of Theorem 2.4. If the expansion (2.40) terminates then $\breve{\rho}_r = 0$ for some $r \in \mathbb{N}$ so either $-\alpha$, $-\beta$, $-\tilde{\alpha} = \alpha - \gamma - \delta$ or $-\tilde{\beta} = \beta - \gamma - \delta$ are nonnegative integers.

Erdélyi [**7**, §7] remarks that the type I expansion (2.40) converges in the entire complex plane with a branch cut along the interval $[a, \infty)$ if λ is one of the eigenvalues of the Heun eigenvalue problem we discussed in this section. Note that the hypergeometric function F has a branch cut along the interval $[1, \infty)$. This means that by using (2.40) we obtain a global representation of Heun eigenfunctions. We should recall that the type II expansion (2.23) usually converges only within the ellipse with foci at $z = 0$ and $z = 1$ passing through $z = a$. It should be mentioned that in the representation of internal and external harmonics in flat-ring coordinates [**2**], [**3**] Heun eigenfunctions are evaluated not only on the interval $(0, 1)$ but also outside of this interval.

By numerical investigation the author also noticed that type I expansions lead to more accurate approximations to Heun eigenfunction than type II expansions. However, one should note that type II expansions are in terms of polynomials whereas type I expansions are in terms of transcendental functions which usually cannot be reduced to elementary functions.

3. Numerical treatment of the Heun eigenvalue problem

Let M_N be the $N+1$ by $N+1$ matrix consisting of the entries in the first $N+1$ rows and columns of M. Let $\lambda_{n,N}$ denote the eigenvalues of M_N. Since these eigenvalues might be complex, they are ordered by the size of their real parts:

$$\Re\lambda_{0,N} \leqslant \Re\lambda_{1,N} \leqslant \dots \leqslant \Re\lambda_{N,N}.$$

THEOREM 3.1. *Let $n \in \mathbb{N}_0$. For sufficiently large N, $\lambda_{n,N}$ is real, and $\lambda_{n,N} \to \lambda_n$ as $N \to \infty$.*

PROOF. This follows from Theorem 2 in [**22**]. In that theorem it is assumed that $\tau_r\rho_r \neq 0$ for all $r \in \mathbb{N}$ which is not always true for the matrix (2.22). However, the proof of Theorem 2 in [**22**] just uses that the eigenvalues of the operator M have multiplicity one, and this assumption is satisfied. □

Under additional assumptions on the parameters $\alpha, \beta, \gamma, \delta$ we can say more.

THEOREM 3.2. *Suppose that*

$$(m+\alpha)(m+\beta)(m+\gamma+\delta-\alpha)(m+\gamma+\delta-\beta) > 0 \quad \text{for all } m \in \mathbb{N}_0. \tag{3.1}$$

Then $\lambda_{n,N}$ is real for $n = 0, 1, \dots, N$. Moreover,

$$\lambda_{n,n} > \lambda_{n,n+1} > \lambda_{n,n+2} > \dots. \tag{3.2}$$

PROOF. Assumption (3.1) implies that $\rho_r\tau_r > 0$ for all $r \in \mathbb{N}$. Therefore, every section matrix M_N is similar to a symmetric matrix by means of a diagonal similarity transformation. Thus all eigenvalues of M_N are real. Moreover, (3.2) follows from the maximum-minimum principle for eigenvalues of symmetric matrices, □

Condition (3.1) may or may not be satisfied under assumption (1.2). For example, the condition is satisfied in the special cases of the Heun eigenvalue problem treated in Sections 5 and 6 but not in Section 4. A sufficient condition for (3.1) is that $0 < \alpha < \gamma + \delta$ and $0 < \beta < \gamma + \delta$. If (3.1) holds then every number $\lambda_{n,N}$ is an upper bound for λ_n. By the method of [**22**, §4] we could also derive lower bounds for the eigenvalues λ_n.

We also note that the identities (2.26), (2.34) remain valid if λ_n is replaced by $\lambda_{n,N}$. This follows from (2.27), (2.35).

Let

$$(c_{0,n,N}, c_{1,n,N}, \dots, c_{N,n,N})^T$$

be an eigenvector of M_N corresponding to the eigenvalue $\lambda_{n,N}$. The numbers $c_r = c_{r,n,N}$ satisfy the equations

$$(\sigma_0 - \lambda_{n,N})c_0 + \tau_1 c_1 = 0, \tag{3.3}$$

$$\rho_r c_{r-1} + (\sigma_r - \lambda_{n,N})c_r + \tau_{r+1}c_{r+1} = 0 \quad \text{for } 1 \leqslant r \leqslant N-1, \tag{3.4}$$

$$\rho_N c_{N-1} + (\sigma_N - \lambda_{n,N})c_N = 0. \tag{3.5}$$

If $n \leqslant N$ then the function

$$H_{n,N}(x) = \sum_{r=0}^{N} c_{r,n,N} P_r^{(\gamma-1,\delta-1)}(1-2x) \tag{3.6}$$

when properly normalized is an approximation to the Heun eigenfunction belonging to the eigenvalue λ_n. This statement is made precise by Theorem 3.4 whose proof is prepared by the following lemma.

LEMMA 3.3. *Let $n \in \mathbb{N}_0$, and let $R = 2a - 1 + 2\sqrt{a(a-1)}$ be twice the sum of minor and major semi-axes of the ellipse with foci at $z = 0$ and $z = 1$ passing through a. For every $R_1 \in (1, R)$ there exists $K \geqslant n$ (independent of r, N) such that*

$$|c_{r+1,n,N}| \leqslant R_1^{-1} |c_{r,n,N}| \quad \textit{whenever } K \leqslant r < N. \tag{3.7}$$

PROOF. Since $1 < R_1 < R$ we have

$$\frac{1}{2}(R_1 + R_1^{-1}) < \frac{1}{2}(R + R^{-1}) = 2a - 1.$$

Choose $1 < A < R_1^2$ such that

$$B := \frac{1}{2}(R_1 + AR_1^{-1}) < 2a - 1.$$

Then

$$B + \sqrt{B^2 - A} = R_1. \tag{3.8}$$

We see that

$$\lim_{r\to\infty} \frac{\sigma_r}{\rho_r} = 4a - 2 > 2B, \qquad \lim_{r\to\infty} \frac{\tau_{r+1}}{\rho_r} = 1 < A.$$

Using also Lemma 2.2, it follows that there is $K \geqslant n$ (independent of N) so large that $\rho_r \neq 0$ for $r \geqslant K$, $\lambda_{n,N} \in \mathbb{R}$ for $N \geqslant K$, and

$$\frac{\sigma_r - \lambda_{n,N}}{\rho_r} \geqslant 2B \quad \text{and} \quad \frac{\tau_{r+1}}{\rho_r} \leqslant A \quad \text{for } N \geqslant K, r \geqslant K.$$

For $N \geqslant K$, consider the real finite sequence $c_r = c_{r,n,N}$, $r = 0, 1, \ldots, N$, satisfying (3.3),(3.4) and (3.5). If $c_N = 0$ then $\rho_r \neq 0$ for $r \geqslant K$ implies that $c_r = 0$ for $r = K, \ldots, N$, so (3.7) holds. Now suppose $c_N \neq 0$. We prove the following statement by induction on $j = 0, 1, \ldots, N-K-1$:

$$c_{N-j} \neq 0 \quad \text{and} \quad d_j := -\frac{c_{N-j-1}}{c_{N-j}} \geqslant R_1. \tag{3.9}$$

This is true for $j = 0$ because (3.5) yields

$$d_0 = \frac{\sigma_N - \lambda_{n,N}}{\rho_N} \geqslant 2B \geqslant R_1.$$

Suppose (3.9) is true for $j-1$ in place of j for some $j=1,\dots,N-K-1$. Then $c_{N-j}\neq 0$ by the induction hypothesis and, by (3.4)

$$d_j = \frac{\sigma_{N-j}-\lambda_{n,N}}{\rho_{N-j}} - \frac{\tau_{N-j+1}}{\rho_{N-j}} d_{j-1}^{-1} \geqslant 2B - A d_{j-1}^{-1}, \tag{3.10}$$

or,

$$d_j \geqslant f(d_{j-1}), \quad \text{where } f(x) = 2B - \frac{A}{x}.$$

Now (3.8) gives that $f(x)\geqslant R_1$ for all $x\geqslant R_1$. By the induction hypothesis, $d_{j-1}\geqslant R_1$, so $d_j\geqslant R_1$, and (3.9) holds. This completes the induction proof and also the proof of the lemma. □

Similar to (2.37), let the function $H_{n,N}(x)$ from (3.6) be normalized such that

$$\int_0^1 W(x)\{H_{n,N}(x)\}^2\,dx = 1, \tag{3.11}$$

or equivalently,

$$\sum_{r=0}^{N} g_r \left|c_{r,n,N}\right|^2 = 1. \tag{3.12}$$

We complete the normalization by requiring that $H_{n,N}(x)>0$ for small positive x (note that $H_{n,N}(0)$ may equal 0 by accident). We assume here that N is large enough so that $\lambda_{n,N}$ is real.

THEOREM 3.4. *Let*

$$E := \{z\in\mathbb{C} : |z|+|z-1| < 2a-1\} \tag{3.13}$$

be the region interior to the ellipse with foci at $z=0$ and $z=1$ which passes through $z=a$. Then, for every $n\in\mathbb{N}_0$, the sequence $H_{n,N}(z)$ converges to the Heun eigenfunction $H_n(z)$ as $N\to\infty$ uniformly on every compact subset of E.

PROOF. Let $n\in\mathbb{N}_0$ be fixed. It is enough to show uniform convergence $H_{n,N}(z)\to H_n(z)$ as $N\to\infty$ on the compact set

$$E_0 := \{z\in\mathbb{C} : |z|+|z-1| \leqslant 2a_0-1\}, \quad \text{where } 1<a_0<a.$$

Let

$$R_0 := 2a_0 - 1 + 2\sqrt{a_0(a_0-1)} < R := 2a-1+2\sqrt{a(a-1)}.$$

Select $R_1\in(R_0,R)$ and choose K according to Lemma 3.3 for this value of R_1. In addition, we may assume that $\tau_r\neq 0$ for $r\geqslant K$. The normalization condition (3.12) implies

$$|c_{r,n,N}| \leqslant C := \max_{0\leqslant j\leqslant K} g_j^{-\frac{1}{2}} \quad \text{for } r=0,\dots,K,\ N\geqslant K.$$

Therefore, Lemma 3.3 yields

$$|c_{K+\ell,n,N}| \leqslant C R_1^{-\ell} \quad \text{for } \ell=0,\dots,N-K,\ N\geqslant K. \tag{3.14}$$

There is an increasing sequence $K\leqslant N_1<N_2<N_3<\dots$ such that c_{r,n,N_p} converges to some e_r as $p\to\infty$ for every $r=0,1,\dots,K$. Since $\tau_r\neq 0$ for $r\geqslant K$, the recursion (3.4) shows that

$$c_{r,n,N_p} \text{ converges to some } e_r \text{ as } p\to\infty \text{ for every } r\in\mathbb{N}_0. \tag{3.15}$$

By taking the limit $p \to \infty$ in (3.3), (3.4), using $\lambda_{n,N_p} \to \lambda_n$ as $p \to \infty$, and (3.14), we find that the sequence $\{e_r\}_{r\in\mathbb{N}_0}$ is an eigenvector (if nonzero) of the matrix operator M belonging to the eigenvalue λ_n. Since all eigenvalues of M are simple, there is a constant ω such that $e_r = \omega c_{r,n}$ for all $r \in \mathbb{N}_0$. But the sequence $\{e_r\}$ satisfies the same normalization conditions as $\{c_{r,n}\}$ so $\omega = 1$.

It is known from [**20**, (8.23.1)] that there is a constant D (independent of r and z) such that

$$\left|P_r^{(\gamma-1,\delta-1)}(1-2z)\right| \leqslant DR_0^r \quad \text{for all } r \in \mathbb{N}_0,\ z \in E_0. \tag{3.16}$$

It follows from (3.14), (3.15) with $e_r = c_{r,n}$, (3.16) and $R_0 < R_1$ that

$$H_{n,N_p}(z) \to H_n(z) \quad \text{as } p \to \infty \text{ uniformly for } z \in E_0.$$

So we proved the desired statement except that we needed to select a subsequence of $\{H_{n,N}(z)\}_{N\in\mathbb{N}_0}$ to achieve convergence. However, we can select a subsequence of any given subsequence and still have the same limit H_n. This is enough to prove the theorem. □

According to Theorem 3.4 our numerical approximation of Heun eigenfunctions is valid on the elliptical region (3.13). In the second transformation (2.33) we changed the independent variable from z to ζ. Therefore, our approximations of Heun eigenfunctions are valid inside the ellipse $|\zeta|+|\zeta-1| < 2a-1$. If we transform back to the z-plane we obtain approximations to Heun eigenfunctions valid to the left of the cubic curve

$$(\Im z)^2 = \frac{(a-\Re z)^2(a_2-\Re z)}{\Re z - a_1},$$

where

$$a_1 := \tfrac{1}{2}a + \tfrac{1}{4} < \Re z < a_2 := \frac{a^2}{2a-1}.$$

This cubic curve is shown in Figure 2 for $a = 2$.

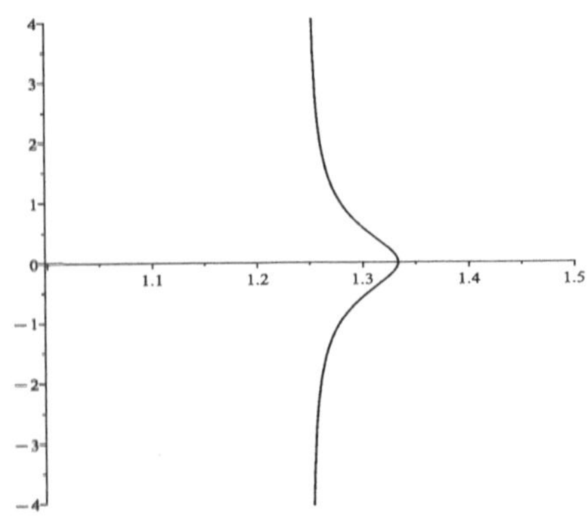

FIGURE 2. Convergence region of numerical computation of Heun eigenfunctions by second transformation is to the left of cubic curve if $a = 2$.

EXAMPLE 3.5. *We take*

$$\alpha = -\tfrac{5}{2},\ \beta = -\tfrac{3}{2},\ \gamma = \tfrac{3}{2},\ \delta = \tfrac{1}{2},\ a = 2,\ n = 2.$$

Then the approximations $\lambda_{2,N}$ *are*

$$\begin{array}{ll} \lambda_{2,2} = 8.5725306516 & \lambda_{2,3} = 8.6343238166 \\ \lambda_{2,4} = 8.6351187567 & \lambda_{2,5} = 8.6351353272 \\ \lambda_{2,6} = 8.6351357254 & \lambda_{2,7} = 8.6351357357 \\ \lambda_{2,8} = 8.6351357359. & \end{array}$$

If we take $N = 8$, *we find the (normalized) eigenvector* $c_r = c_{r,2,8}$ *of* M_N:

$$\begin{array}{llllll} c_0 = & 0.1921813764 & c_1 = & 1.1174716915 & c_2 = & 1.4298040435 \\ c_3 = & 0.0124296777 & c_4 = & -0.0001311313 & c_5 = & 0.0000039142 \\ c_6 = & -0.0000001782 & c_7 = & 0.0000000103 & c_8 = & -0.0000000006. \end{array}$$

Our approximation $H_{2,8}(x)$ *to the Heun eigenfunction* $H_2(x)$ *is*

$$H_{2,8}(x) = \sum_{r=0}^{8} c_{r,2,8} P_r^{(\frac{1}{2},-\frac{1}{2})}(1-2x)$$

whose graph is shown in Figure 3.

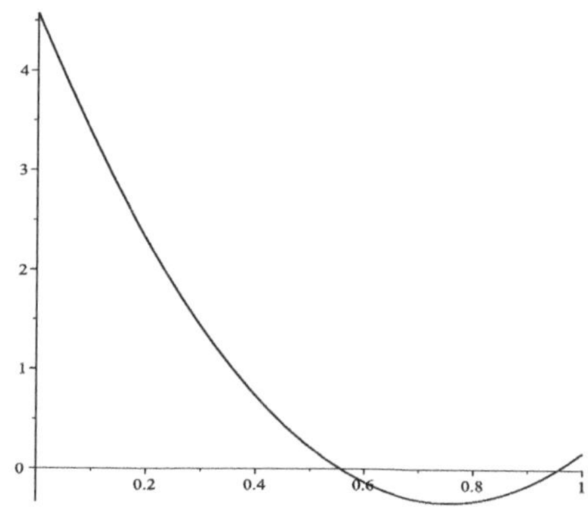

FIGURE 3. Heun eigenfunction $H_2(x)$.

4. The periodic eigenvalue problem for Lamé's equation

This eigenvalue problem appears when we determine internal and external harmonics in flat-ring coordinates [**2**]. It was investigated by Ince [**10**], [**11**] and also appears in [**9**, §15.5.1].

We consider the Lamé equation (1.3). The function $\operatorname{sn}^2(s,k)$ is even and has period $2K$, where

$$K = K(k) = \int_0^{\frac{\pi}{2}} \frac{d\phi}{\sqrt{1-k^2\sin^2\phi}}$$

denotes the complete elliptic integral of the first kind. So (1.3) is an even Hill equation and we have a corresponding periodic eigenvalue problem [**13**]. The parameter h is called an eigenvalue if (1.3) admits a nontrivial solution $u(s)$, $s \in \mathbb{R}$, with period $4K$. The problem splits into four regular Sturm–Liouville problems with separated boundary conditions [**24**, §27] according to Table 1. The eigenvalues are arranged in increasing order and $m \in \mathbb{N}_0$ in all cases. The corresponding eigenfunctions are called periodic Lamé functions.

TABLE 1. Four eigenvalue problems for periodic Lamé functions.

	boundary conditions	eigenvalues	period
1.I	$u(0) = u(K) = 0$	$b_\nu^{2m+2}(k^2)$	$u(s+2K) = u(s)$
1.II	$u'(0) = u(K) = 0$	$b_\nu^{2m+1}(k^2)$	$u(s+2K) = -u(s)$
1.III	$u(0) = u'(K) = 0$	$a_\nu^{2m+1}(k^2)$	$u(s+2K) = -u(s)$
1.IV	$u'(0) = u'(K) = 0$	$a_\nu^{2m}(k^2)$	$u(s+2K) = u(s)$

We show that the eigenvalue problems 1.I, 1.II, 1.III, 1.IV are special cases of the Heun eigenvalue problem.

THEOREM 4.1. *If we put* $x = \operatorname{sn}^2(s,k)$, $0 < s < K$, $0 < x < 1$, *in the Lamé equation* (1.3) *and substitute*

$$\begin{aligned} 1.\mathrm{I}: &\quad u(s) = x^{\frac12}(1-x)^{\frac12} y(x),\\ 1.\mathrm{II}: &\quad u(s) = (1-x)^{\frac12} y(x),\\ 1.\mathrm{III}: &\quad u(s) = x^{\frac12} y(x),\\ 1.\mathrm{IV}: &\quad u(s) = y(x) \end{aligned}$$

then we obtain the Heun equation (1.1) *with parameters*

	α	β	γ	δ	ϵ	λ
1.I	$-\frac12\nu+1$	$\frac12\nu+\frac32$	$\frac32$	$\frac32$	$\frac12$	$\frac14 k^{-2}(h-4-k^2)$
1.II	$-\frac12\nu+\frac12$	$\frac12\nu+1$	$\frac12$	$\frac32$	$\frac12$	$\frac14 k^{-2}(h-1)$
1.III	$-\frac12\nu+\frac12$	$\frac12\nu+1$	$\frac32$	$\frac12$	$\frac12$	$\frac14 k^{-2}(h-k^2-1)$
1.IV	$-\frac12\nu$	$\frac12\nu+\frac12$	$\frac12$	$\frac12$	$\frac12$	$\frac14 k^{-2}h$

and $a = k^{-2}$ *in all four cases. The eigenvalue problems 1.I, 1.II, 1.III, 1.IV correspond to Heun eigenvalue problems so that we obtain*

$$\begin{aligned} 1.\mathrm{I}: &\quad b_\nu^{2m+2}(k^2) = 4k^2\lambda_m(-\tfrac12\nu+1, \tfrac12\nu+\tfrac32, \tfrac32, \tfrac32, k^{-2}) + 4 + k^2,\\ 1.\mathrm{II}: &\quad b_\nu^{2m+1}(k^2) = 4k^2\lambda_m(-\tfrac12\nu+\tfrac12, \tfrac12\nu+1, \tfrac12, \tfrac32, k^{-2}) + 1,\\ 1.\mathrm{III}: &\quad a_\nu^{2m+1}(k^2) = 4k^2\lambda_m(-\tfrac12\nu+\tfrac12, \tfrac12\nu+1, \tfrac32, \tfrac12, k^{-2}) + 1 + k^2,\\ 1.\mathrm{IV}: &\quad a_\nu^{2m}(k^2) = 4k^2\lambda_m(-\tfrac12\nu, \tfrac12\nu+\tfrac12, \tfrac12, \tfrac12, k^{-2}). \end{aligned}$$

PROOF. The first part of the statement follows by direct computation. We now show the stated connection of eigenvalue problem 1.I with the Heun eigenvalue problem. Since $\gamma = \frac32$ the exponents of the regular singularity $z = 0$ of the Heun equation (1.1) are $\{0, -\frac12\}$. Therefore, (1.1) has a solution $y_1(x)$ which is analytic at $x = 0$ with $y_1(0) \neq 0$. The corresponding solution of (1.3) is

$$u_1(s) = \operatorname{cn} s \operatorname{sn} s\, y_1(\operatorname{sn}^2(s,k)) \quad \text{for } 0 < s < K,$$

so $\lim_{s\to 0+} u_1(s) = 0$. There is a second solution of (1.1) of the form $y_2(x) = x^{-\frac{1}{2}} f(x)$ where $f(x)$ is analytic at $x = 0$ with $f(0) \neq 0$. The corresponding solution of (1.3) is

$$u_2(s) = \operatorname{cn} s\, f(\operatorname{sn}^2(s,k))$$

so $\lim_{s\to 0+} u_2(s) \neq 0$. This proves the claimed connection of the eigenvalue problems. The proof for the eigenvalue problems 1.II, 1.III, 1.IV is similar. □

In the following subsections we demonstrate that our approach leads to results which are equivalent to those given in [**6**, Chapter 29] and [**9**, §15.5.1].

4.1. Problem 1.I. In [**9**, page 65, (16)] the eigenfunctions of problem 1.I are expanded in Fourier series:

$$u(s) = \sum_{r=0}^{\infty} B_{2r+2} \sin(2r+2)t, \tag{4.1}$$

where

$$t = \tfrac{1}{2}\pi - \operatorname{am}(s,k).$$

One obtains the recursion formula [**9**, page 66, (22)]

$$\begin{aligned} &(S_0 - H)B_2 + T_1 B_4 = 0, \\ &R_r B_{2r} + (S_r - H)B_{2r+2} + T_{r+1} B_{2r+4} = 0, \quad r \in \mathbb{N}, \end{aligned}$$

where

$$\begin{aligned} R_r &= \tfrac{1}{2}(\nu - 2r)(\nu + 2r + 1)k^2, \\ S_r &= (2r+2)^2(2 - k^2), \\ T_r &= \tfrac{1}{2}(\nu - 2r - 1)(\nu + 2r + 2)k^2, \\ H &= 2h - \nu(\nu+1)k^2. \end{aligned}$$

We used the expansion

$$u(s) = x^{\frac{1}{2}}(1-x)^{\frac{1}{2}} \sum_{r=0}^{\infty} c_r P_r^{(\frac{1}{2},\frac{1}{2})}(1-2x), \quad x = \operatorname{sn}^2(s,k) \tag{4.2}$$

and the recursion formula

$$(\sigma_0 + q)c_0 + \tau_1 c_1 = 0, \tag{4.3}$$

$$\rho_r c_{r-1} + (\sigma_r + q)c_r + \tau_{r+1} c_{r+1} = 0, \quad r \in \mathbb{N}. \tag{4.4}$$

We note that [**20**, page 60]

$$P_n^{(\frac{1}{2},\frac{1}{2})}(-\cos\theta) = (-1)^n P_n^{(\frac{1}{2},\frac{1}{2})}(\cos\theta) = (-1)^n \frac{(\frac{3}{2})_n}{(n+1)!} \frac{\sin(n+1)\theta}{\sin\theta}.$$

Therefore,

$$\begin{aligned} x^{\frac{1}{2}}(1-x)^{\frac{1}{2}} P_n^{(\frac{1}{2},\frac{1}{2})}(1-2x) &= \cos t \sin t (-1)^n \frac{(\frac{3}{2})_n}{(n+1)!} \frac{\sin(2n+2)t}{\sin 2t} \\ &= (-1)^n \frac{(\frac{3}{2})_n}{2(n+1)!} \sin(2n+2)t. \end{aligned}$$

Comparing (4.1) with (4.2) gives

$$B_{2r+2} = (-1)^r \frac{(\frac{3}{2})_r}{2(r+1)!} c_r, \quad r \in \mathbb{N}_0.$$

If we now compare the recursion formulas we obtain

$$
\begin{aligned}
8k^2\rho_r &= -R_r \frac{r+1}{r+\frac{1}{2}}, \quad r \in \mathbb{N}, \\
8k^2\sigma_r + 8 + 2k^2 &= S_r + \nu(\nu+1)k^2, \quad r \in \mathbb{N}_0, \\
8k^2\tau_r &= -T_r \frac{r+\frac{1}{2}}{r+1}, \quad r \in \mathbb{N}.
\end{aligned}
$$

The correctness of these identities can be verified by direct calculation.

EXAMPLE 4.2. *We take $k = \frac{1}{2}$, $\nu = \frac{1}{2}$ and approximate $b_\nu^4(k^2)$. We have*

$$b_\nu^4(k^2) = 4 + k^2 + 4k^2\lambda_1(\tfrac{3}{4}, \tfrac{7}{4}, \tfrac{3}{2}, \tfrac{3}{2}, 4).$$

Note that (3.1) is satisfied in this example. The approximations $\bar{\lambda}_{1,N} = 4 + k^2 + 4k^2\lambda_{1,N}$ to b_ν^4 are

$$
\begin{aligned}
\bar{\lambda}_{1,1} &= 14.1156768669 & \bar{\lambda}_{1,2} &= 13.9909899332 \\
\bar{\lambda}_{1,3} &= 13.9894849636 & \bar{\lambda}_{1,4} &= 13.9894726568 \\
\bar{\lambda}_{1,5} &= 13.9894725720 & \bar{\lambda}_{1,6} &= 13.9894725714.
\end{aligned}
$$

If we take $N = 6$, we find the eigenvector $c_r = c_{r,1,6}$:

$$
\begin{aligned}
c_0 &= 0.2285891125 & c_1 &= 4.8367026122 & c_2 &= -0.3740941978 \\
c_3 &= 0.0263918519 & c_4 &= -0.0017863473 & c_5 &= 0.0001194384 \\
c_6 &= -0.0000079190.
\end{aligned}
$$

Our approximation to the eigenfunction of problem 1.I belonging to the eigenvalue $b_{\frac{1}{2}}^4$ is

$$\operatorname{cn}(s,k)\operatorname{sn}(s,k)\sum_{r=0}^{6} c_{r,1,6} P_r^{(\frac{1}{2},\frac{1}{2})}(1 - 2\operatorname{sn}^2(s,k))$$

whose graph is shown in Figure 4.

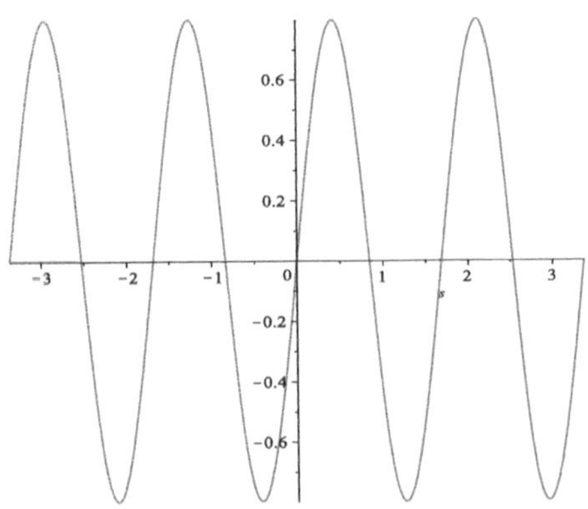

FIGURE 4. Eigenfunction for problem 1.I on $[-2K, 2K]$.

4.2. Problem 1.II. In [**9**, page 65, (17)] the eigenfunctions of problem 1.II are expanded in Fourier series:

$$u(s) = \sum_{r=0}^{\infty} B_{2r+1} \sin(2r+1)t. \tag{4.5}$$

One obtains the recursion formula [**9**, page 66, (24)]

$$\begin{aligned}&(S_0 - H)B_1 + T_1 B_3 = 0,\\ &R_r B_{2r-1} + (S_r - H)B_{2r+1} + T_{r+1}B_{2r+3} = 0, \quad r \in \mathbb{N},\end{aligned}$$

where

$$\begin{aligned}S_0 &= 2 - k^2 - \tfrac{1}{2}\nu(\nu+1)k^2,\\ R_r &= \tfrac{1}{2}(\nu - 2r + 1)(\nu + 2r)k^2, \quad r \in \mathbb{N},\\ S_r &= (2r+1)^2(2-k^2), \quad r \in \mathbb{N},\\ T_r &= \tfrac{1}{2}(\nu - 2r)(\nu + 2r + 1)k^2, \quad r \in \mathbb{N},\\ H &= 2h - \nu(\nu+1)k^2.\end{aligned}$$

We used the expansion

$$u(s) = (1-x)^{\frac{1}{2}} \sum_{r=0}^{\infty} c_r P_r^{(-\frac{1}{2},\frac{1}{2})}(1-2x) \tag{4.6}$$

with recursion formulas (4.3), (4.4). We note that [**20**, page 60]

$$P_n^{(-\frac{1}{2},\frac{1}{2})}(-\cos\theta) = (-1)^n P_n^{(\frac{1}{2},-\frac{1}{2})}(\cos\theta) = (-1)^n \frac{(\frac{1}{2})_n}{n!}\frac{\sin(n+\frac{1}{2})\theta}{\sin\frac{1}{2}\theta}.$$

Therefore,

$$(1-x)^{\frac{1}{2}} P_n^{(-\frac{1}{2},\frac{1}{2})}(1-2x) = (-1)^n \frac{(\frac{1}{2})_n}{n!}\sin(2n+1)t.$$

Comparing (4.5) with (4.6) gives

$$B_{2r+1} = (-1)^r \frac{(\frac{1}{2})_r}{r!} c_r, \quad r \in \mathbb{N}_0.$$

If we now compare the recursion formulas we obtain

$$\begin{aligned}8k^2\rho_r &= -R_r \frac{r}{r-\frac{1}{2}}, \quad r \in \mathbb{N},\\ 8k^2\sigma_r + 2 &= S_r + \nu(\nu+1)k^2, \quad r \in \mathbb{N}_0,\\ 8k^2\tau_r &= -T_r \frac{r-\frac{1}{2}}{r}, \quad r \in \mathbb{N}.\end{aligned}$$

The correctness of these identities can be verified by direct calculation.

4.3. Problem 1.III. In [**9**, page 65, (15)] the eigenfunctions of problem 1.III are expanded in Fourier series:

$$u(s) = \sum_{r=0}^{\infty} A_{2r+1} \cos(2r+1)t. \tag{4.7}$$

One obtains the recursion formula [**9**, page 66, (20)]

$$\begin{aligned}&(S_0-H)A_1+T_1A_3=0,\\&R_rA_{2r-1}+(S_r-H)A_{2r+1}+T_{r+1}A_{2r+3}=0,\quad r\in\mathbb{N},\end{aligned}$$

where

$$\begin{aligned}S_0&=2-k^2+\tfrac12\nu(\nu+1)k^2,\\R_r&=\tfrac12(\nu-2r+1)(\nu+2r)k^2,\quad r\in\mathbb{N},\\S_r&=(2r+1)^2(2-k^2),\quad r\in\mathbb{N},\\T_r&=\tfrac12(\nu-2r)(\nu+2r+1)k^2,\quad r\in\mathbb{N},\\H&=2h-\nu(\nu+1)k^2.\end{aligned}$$

We used the expansion

$$u(s)=\sum_{r=0}^\infty c_rP_r^{(\frac12,-\frac12)}(1-2x) \tag{4.8}$$

with recursion formulas (4.3), (4.4). We note that [**20**, page 60]

$$P_n^{(\frac12,-\frac12)}(-\cos\theta)=(-1)^nP_n^{(-\frac12,\frac12)}(\cos\theta)=(-1)^n\frac{(\frac12)_n}{n!}\frac{\cos(n+\frac12)\theta}{\cos\frac12\theta}.$$

Therefore,

$$x^{\frac12}P_n^{(-\frac12,\frac12)}(1-2x)=(-1)^n\frac{(\frac12)_n}{n!}\cos(2n+1)t.$$

Comparing (4.7) with (4.8) gives

$$A_{2r+1}=(-1)^r\frac{(\frac12)_r}{r!}c_r,\quad r\in\mathbb{N}_0.$$

If we now compare the recursion formulas we obtain

$$\begin{aligned}8k^2\rho_r&=-R_r\frac{r}{r-\frac12},\quad r\in\mathbb{N},\\8k^2\sigma_r+2+2k^2&=S_r+\nu(\nu+1)k^2,\quad r\in\mathbb{N}_0,\\8k^2\tau_r&=-T_r\frac{r-\frac12}{r},\quad r\in\mathbb{N}.\end{aligned}$$

The correctness of these identities can be verified by direct calculation.

4.4. Problem 1.IV. In [**9**, page 65, (14)] the eigenfunctions of problem 1.IV are expanded in Fourier series:

$$u(s)=\tfrac12A_0+\sum_{r=1}^\infty A_{2r}\cos 2rt. \tag{4.9}$$

One obtains the recursion formula [**9**, page 66, (18)]

$$\begin{aligned}&(S_0-H)A_0+T_1A_2=0,\\&R_{r-1}A_{2r-2}+(S_r-H)A_{2r}+T_{r+1}A_{2r+2}=0,\quad r\in\mathbb{N},\end{aligned}$$

where

$$\begin{aligned} R_r &= \tfrac{1}{2}(\nu - 2r + 2)(\nu + 2r - 1)k^2, \quad r \in \mathbb{N}, \\ S_r &= 4r^2(2 - k^2), \quad r \in \mathbb{N}_0, \\ T_1 &= (\nu - 1)(\nu + 2)k^2, \\ T_r &= \tfrac{1}{2}(\nu - 2r + 1)(\nu + 2r)k^2, \quad r \geqslant 2, \\ H &= 2h - \nu(\nu + 1)k^2. \end{aligned}$$

We used the expansion

$$u(s) = \sum_{r=0}^{\infty} c_r P_r^{(-\frac{1}{2},-\frac{1}{2})}(1 - 2x) \tag{4.10}$$

with recursion formulas (4.3), (4.4). We note that [**20**, page 60]

$$P_n^{(-\frac{1}{2},-\frac{1}{2})}(-\cos\theta) = (-1)^n P_n^{(-\frac{1}{2},-\frac{1}{2})}(\cos\theta) = (-1)^n \frac{(\frac{1}{2})_n}{n!} \cos n\theta.$$

Therefore,

$$P_n^{(-\frac{1}{2},\frac{1}{2})}(1 - 2x) = (-1)^n \frac{(\frac{1}{2})_n}{n!} \cos 2nt.$$

Comparing (4.9) with (4.10) gives $\frac{1}{2}A_0 = c_0$ and

$$A_{2r} = (-1)^r \frac{(\frac{1}{2})_r}{r!} c_r, \quad r \in \mathbb{N}.$$

If we now compare the recursion formulas we obtain

$$\begin{aligned} 2k^2\rho_1 &= -R_1, \\ 8k^2\rho_r &= -R_r \frac{r}{r - \frac{1}{2}}, \quad r \geqslant 2, \\ 8k^2\sigma_r &= S_r + \nu(\nu + 1)k^2, \quad r \in \mathbb{N}_0, \\ 8k^2\tau_1 &= -\tfrac{1}{4}T_1, \\ 8k^2\tau_r &= -T_r \frac{r - \frac{1}{2}}{r}, \quad r \geqslant 2. \end{aligned}$$

The correctness of these identities can be verified by direct calculation.

5. The Lamé–Wangerin eigenvalue problem

This eigenvalue problem appears when we determine internal and external harmonics in flat-ring coordinates associated with peanut-shaped coordinate surfaces [**3**]. It also appears in the determination of internal and external harmonics in bicyclide coordinates [**1**]. It has been treated in [**9**, §15.6] and [**23**].

The Lamé equation (1.3) has regular singularities at $s = iK'$ and $s = 2K + iK'$ with exponents $\{-\nu, \nu + 1\}$ at both points. The parameter h in Lamé's equation is called a Lamé–Wangerin eigenvalue if there exists a nontrivial solution along the segment $s = K + iK' + t$, $-K < t < K$, which belongs to the exponent $\nu + 1$ at both singularities $s = iK'$ and $s = 2K + iK'$. In order to simplify notation we will transform the Lamé equation setting $s = K + iK' + t$:

$$\frac{d^2u}{dt^2} + \left(h - \nu(\nu + 1)\frac{\mathrm{dn}^2(t,k)}{\mathrm{cn}^2(t,k)}\right)u = 0, \qquad -K < t < K. \tag{5.1}$$

We notice that (5.1) has regular singularities at $t = \pm K$ with exponents $\{-\nu, \nu+1\}$. Since we assume that $\nu \geqslant -\frac{1}{2}$, the Frobenius solution that belongs to the exponent $\nu+1$ at $t = \pm K$ is the recessive solution at $\pm K$. In the borderline case $\nu = -\frac{1}{2}$ both exponents coincide but it is well known that there is only one solution (apart from a constant factor) which does not involve the logarithm, and this is the recessive solution. Therefore, the parameter h is an eigenvalue if (5.1) has a nontrivial solution $u(t)$, $-K < t < K$, that is recessive at both end points $t = \pm K$. It is easy to see that eigenfunctions have to be even or odd functions. Therefore, the eigenvalue problem splits into two problems according to Table 2.

TABLE 2. Two Lamé–Wangerin eigenvalue problems.

	boundary conditions	eigenvalues
2.I	$u(0) = 0$, u recessive at $t = K$	$c_\nu^{2m+1}(k^2)$
2.II	$u'(0) = 0$, u recessive at $t = K$	$c_\nu^{2m}(k^2)$

The eigenvalues form an increasing sequence

$$c_\nu^0 < c_\nu^1 < c_\nu^2 < \dots .$$

The two eigenvalue problems are singular Sturm–Liouville problems on the interval $[0, K)$ with a regular end point $t = 0$ and a singular end point $t = K$. The corresponding eigenfunctions are called Lamé–Wangerin functions by Erdélyi [**9**].

The eigenvalue problems 2.I and 2.II are special cases of the Heun eigenvalue problem.

THEOREM 5.1. *If we put* $x = \mathrm{sn}^2(t, k)$, $0 < t < K$, $0 < x < 1$, *in Lamé's equation* (1.3) *and substitute*

$$\begin{aligned} 2.\mathrm{I}: &\quad u(t) = x^{\frac{1}{2}}(1-x)^{\frac{1}{2}\nu+\frac{1}{2}} y(x), \\ 2.\mathrm{II}: &\quad u(t) = (1-x)^{\frac{1}{2}\nu+\frac{1}{2}} y(x) \end{aligned}$$

then we obtain the Heun equation (1.1) *for* $y(x)$ *with parameters*

	α	β	γ	δ	ϵ	λ
2.I	$\frac{1}{2}\nu + 1$	$\frac{1}{2}\nu + \frac{3}{2}$	$\frac{3}{2}$	$\nu + \frac{3}{2}$	$\frac{1}{2}$	$\frac{1}{4}k^{-2}(h - k^2 - (\nu+2)^2)$
2.II	$\frac{1}{2}\nu + \frac{1}{2}$	$\frac{1}{2}\nu + 1$	$\frac{1}{2}$	$\nu + \frac{3}{2}$	$\frac{1}{2}$	$\frac{1}{4}k^{-2}(h - (\nu+1)^2)$

and $a = k^{-2}$ *in both cases. The eigenvalue problems 2.I and 2.II correspond to Heun eigenvalue problems so that we obtain*

$$\begin{aligned} 2.\mathrm{I}: &\quad c_\nu^{2m+1}(k^2) = 4k^2 \lambda_m(\tfrac{1}{2}\nu + 1, \tfrac{1}{2}\nu + \tfrac{3}{2}, \tfrac{3}{2}, \nu + \tfrac{3}{2}, k^{-2}) + k^2 + (\nu+2)^2, \\ 2.\mathrm{II}: &\quad c_\nu^{2m}(k^2) = 4k^2 \lambda_m(\tfrac{1}{2}\nu + \tfrac{1}{2}, \tfrac{1}{2}\nu + 1, \tfrac{1}{2}, \nu + \tfrac{3}{2}, k^{-2}) + (\nu+1)^2. \end{aligned}$$

Several of our general results on the Heun eigenvalue problem simplify when applied to the special cases 2.I and 2.II. First of all, since $0 < \alpha < \gamma + \delta$ and $0 < \beta < \gamma + \delta$, condition (3.1) holds, so Theorem 3.2 is applicable. In the series expansion of eigenfunctions of problems 2.I and 2.II we use Jacobi polynomials $P_r^{(\frac{1}{2}, \nu+\frac{1}{2})}$ and $P_r^{(-\frac{1}{2}, \nu+\frac{1}{2})}$, respectively. These polynomials can be expressed in terms of ultra-spherical polynomials [**20**, Thm 4.1]. Finally, we note that the formulations of Theorem 2.4 and Theorem 2.5 simplify because

$$\prod_{j=1}^{r}\frac{\tau_j}{\tilde{\tau}_j}=\begin{cases}\frac{2r+\nu+2}{\nu+2} & \text{in case 2.I,}\\ \frac{2r+\nu+1}{\nu+1} & \text{in case 2.II.}\end{cases}$$

5.1. Problem 2.I. In this case we have

$$\begin{aligned}\rho_r &= \tfrac{1}{4}r(r+\nu+1),\\ \sigma_r &= -\tfrac{1}{8}(2r+\nu+2)^2+\tfrac{1}{8}(\nu+2)(\nu-1)+k^{-2}r(r+\nu+2),\\ \tau_r &= \tfrac{1}{16}(2r+1)(2r+2\nu+1).\end{aligned}$$

Using these sequences the matrix operator M has eigenvalues λ_m. In order to get the eigenvalues c_ν^{2m+1} we consider another matrix $\bar{M}$. We obtain this matrix from M by multiplying each entry by $4k^2$ and then add $k^2+(\nu+2)^2$ to the diagonal sequence, so

$$\begin{aligned}\bar{\rho}_r &:= r(r+\nu+1)k^2,\\ \bar{\sigma}_r &:= \tfrac{1}{2}(2r+\nu+2)^2(2-k^2)+\tfrac{1}{2}\nu(\nu+1)k^2,\\ \bar{\tau}_r &:= \tfrac{1}{4}(2r+1)(2r+2\nu+1)k^2.\end{aligned}$$

It is interesting to compare this result with [**9**, page 79, (16)], where in place of $\bar{\rho}$, $\bar{\sigma}$, $\bar{\tau}$, the following sequences are used:

$$\begin{aligned}\dot{\rho}_r &:= -\tfrac{1}{2}(2r+1)(r+\nu+1)k^2,\\ \dot{\sigma}_r &:= \tfrac{1}{2}(2r+\nu+2)^2(2-k^2)+\tfrac{1}{2}\nu(\nu+1)k^2,\\ \dot{\tau}_r &:= -\tfrac{1}{2}r(2r+2\nu+1)k^2.\end{aligned}$$

We see that

$$\bar{\sigma}_r=\dot{\sigma}_r,\qquad \bar{\rho}_r\bar{\tau}_r=\dot{\rho}_r\dot{\tau}_r.$$

Therefore, the section matrices $\bar{M}_N$ and $\dot{M}_N$ are similar, and so they have the same eigenvalues. However, the approximation of Lamé–Wangerin eigenfunctions given in [**9**] is different from ours. We look at this difference in more detail in the next subsection.

EXAMPLE 5.2. *We take* $k=\frac{1}{2}$, $\nu=\frac{1}{2}$ *and calculate* $c_\nu^3(k^2)$. *We have*

$$c_\nu^3(k^2)=k^2+(\nu+2)^2+4k^2\lambda_1(\tfrac{5}{4},\tfrac{7}{4},\tfrac{3}{2},2,4).$$

The approximations $\bar{\lambda}_{1,N}=k^2+(\nu+2)^2+4k^2\lambda_{1,N}$ *to* c_ν^3 *are*

$$\begin{aligned}\bar{\lambda}_{1,1}&=17.8506465178 & \bar{\lambda}_{1,2}&=17.6818720863\\ \bar{\lambda}_{1,3}&=17.6796076716 & \bar{\lambda}_{1,4}&=17.6795877788\\ \bar{\lambda}_{1,5}&=17.6795876347 & \bar{\lambda}_{1,6}&=17.6795876337.\end{aligned}$$

If we take $N=6$, *we find the eigenvector* $c_r=c_{r,1,6}$:

$$\begin{aligned}c_0&=0.3461428047 & c_1&=5.5923235989 & c_2&=-0.5118017147\\ c_3&=0.0381832823 & c_4&=-0.0026815267 & c_5&=0.0001840715\\ c_6&=-0.0000124460.\end{aligned}$$

Our approximation to the eigenfunction of problem 2.I belonging to the eigenvalue $c^3_{\frac{1}{2}}$ *is*

$$\operatorname{sn}(t,k)\operatorname{cn}^{\nu+1}(t,k)\sum_{r=0}^{6} c_{r,1,6} P_r^{(\frac{1}{2},1)}(1-2\operatorname{sn}^2(t,k))$$

whose graph is shown in Figure 5.

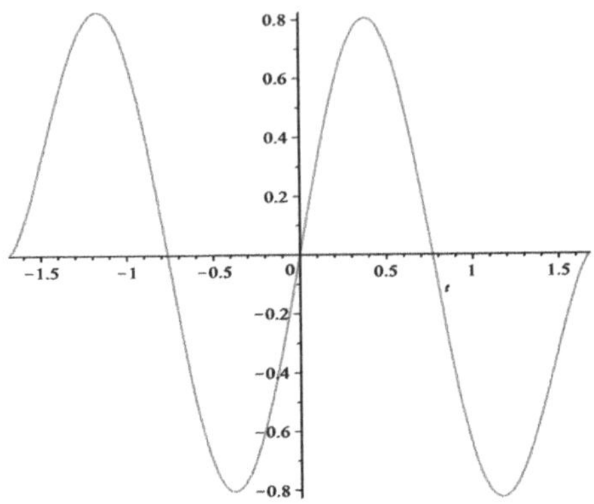

FIGURE 5. Eigenfunction for problem 2.I on $(-K,K)$.

5.2. Problem 2.II. In this case we have

$$\begin{aligned}
\rho_r &= \tfrac{1}{4}r(r+\nu),\\
\sigma_r &= -\tfrac{1}{8}(2r+\nu+1)^2 + \tfrac{1}{8}\nu(\nu+1) + k^{-2}r(r+\nu+1),\\
\tau_r &= \tfrac{1}{16}(2r-1)(2r+2\nu+1).
\end{aligned}$$

Using these sequences the matrix operator M has eigenvalues λ_m. In order to get the eigenvalues c_ν^{2m} we consider the matrix $\bar{M}$. We obtain this matrix from M by multiplying each entry by $4k^2$ and then add $(\nu+1)^2$ to the diagonal sequence, so

$$\begin{aligned}
\bar{\rho}_r &:= r(r+\nu)k^2,\\
\bar{\sigma}_r &:= \tfrac{1}{2}(2r+\nu+1)^2(2-k^2) + \tfrac{1}{2}\nu(\nu+1)k^2,\\
\bar{\tau}_r &:= \tfrac{1}{4}(2r-1)(2r+2\nu+1)k^2.
\end{aligned}$$

In [**9**, page 78, (15)] the following sequences are used

$$\begin{aligned}
\dot{\rho}_r &:= -\tfrac{1}{2}(2r-1)(r+\nu)k^2,\\
\dot{\sigma}_r &:= \tfrac{1}{2}(2r+\nu+1)^2(2-k^2) + \tfrac{1}{2}\nu(\nu+1)k^2,\\
\dot{\tau}_r &:= -\tfrac{1}{2}r(2r+2\nu+1)k^2.
\end{aligned}$$

We see that

$$\bar{\sigma}_r = \dot{\sigma}_r, \qquad \bar{\rho}_r\bar{\tau}_r = \dot{\rho}_r\dot{\tau}_r$$

so the section matrices $\bar{M}_N$ and $\dot{M}_N$ are similar.

Let $\lambda_{n,N}$ be an eigenvalue of the section matrix $\bar{M}_N$, and let $(c_0, c_1, \ldots, c_N)$ denote a corresponding eigenvector. Then our approximation of the corresponding

Lamé–Wangerin function is

$$u(t) \approx \mathrm{cn}^{\nu+1}(t,k) \sum_{r=0}^{N} c_r P_r^{(-\frac{1}{2},\nu+\frac{1}{2})}(1 - 2\,\mathrm{sn}^2(t,k)).$$

Now $\lambda_{n,N}$ is also an eigenvalue of the section matrix $\dot{M}_N$. Let $(d_0, d_1, \ldots, d_N)$ be a corresponding eigenvector. According to [**9**, page 78,(13)] the corresponding Lamé–Wangerin function is approximated by

$$u(t) \approx \sum_{r=0}^{N} d_r \exp(-i\zeta(2r + \nu + 1)),$$

where $\zeta = \frac{\pi}{2} - \mathrm{am}(s,k)$ and $s = t + K + iK'$. Then

$$e^{-i\zeta} = \frac{k\,\mathrm{cn}(t,k)}{\mathrm{dn}(t,k) + k'}.$$

After applying the second transformation from Section 2 these approximations agree with those given in Theorem 2.6 noting that

$$F(\tfrac{1}{2}\nu + \tfrac{1}{2} + r, \tfrac{1}{2}\nu + 1 + r; \nu + 2 + 2r; x) = \left(\frac{2}{1 + \sqrt{1 - x}}\right)^{2r+\nu+1}.$$

6. Another Lamé–Wangerin problem

This eigenvalue problem appears when we determine internal and external harmonics in flat-disk coordinates. Consider the differential equation

$$\frac{d^2u}{dt^2} + \left(h - \nu(\nu+1)\frac{\mathrm{sn}^2(t,k)\,\mathrm{dn}^2(t,k)}{\mathrm{cn}^2(t,k)}\right) u = 0 \tag{6.1}$$

with $0 < k < 1$ and $\nu \geqslant -\frac{1}{2}$.

Equation (6.1) can be transformed to Lamé's equation (1.3) by setting

$$s = (k - ik')t.$$

Using the complex modulus

$$\ell = \frac{k + ik'}{k - ik'},$$

we have

$$\mathrm{sn}(s,\ell) = (k - ik')\frac{\mathrm{sn}(t,k)\,\mathrm{dn}(t,k)}{\mathrm{cn}(t,k)}. \tag{6.2}$$

Equation (6.2) can be derived from [**8**, §13.22]. Applying (6.2), we obtain

$$\frac{d^2u}{ds^2} + \left((k + ik')^2 h - \nu(\nu+1)\ell^2\,\mathrm{sn}^2(s,\ell)\right) u = 0.$$

However, we prefer to work with equation (6.1).

TABLE 3. Another two Lamé–Wangerin eigenvalue problems.

	boundary conditions	eigenvalues
3.I	$u(0) = 0$, u recessive at $t = K$	$d_\nu^{2m+1}(k^2)$
3.II	$u'(0) = 0$, u recessive at $t = K$	$d_\nu^{2m}(k^2)$

Notice that equation (6.1) is the same as (5.1) except for the extra factor $\operatorname{sn}^2(t,k)$. In fact, the treatment of this equation is parallel to that of equation (5.1) with only minor changes. Equation (6.1) has regular singularities at $t=\pm K$ with exponents $\{-\nu,\nu+1\}$. Since we assume that $\nu\geqslant -\frac12$, the Frobenius solution that belongs to the exponent $\nu+1$ at $t=\pm K$ is the recessive solution at $\pm K$. The parameter h is called an eigenvalue if (6.1) has a nontrivial solution $u(t)$, $-K<t<K$, that is recessive at both end points $t=\pm K$. Eigenfunctions have to be even or odd functions. Therefore, the eigenvalue problem splits into two problems according to Table 3. The corresponding eigenfunctions may be called Lamé–Wangerin functions of the second kind.

The eigenvalue problems 3.I and 3.II are special cases of the Heun eigenvalue problem.

THEOREM 6.1. *If we put $x=\operatorname{sn}^2(t,k)$, $0<t<K$, $0<x<1$ in* (6.1) *and substitute*

$$\begin{aligned} 3.\mathrm{I}: &\quad u(t)=x^{\frac12}(1-x)^{\frac12\nu+\frac12}y(x),\\ 3.\mathrm{II}: &\quad u(t)=(1-x)^{\frac12\nu+\frac12}y(x) \end{aligned}$$

then we obtain the Heun equation (1.1) *for $y(x)$ with parameters*

	α	β	γ	δ	ϵ	λ
3.I	1	$\nu+\frac32$	$\frac32$	$\nu+\frac32$	$\frac12$	$\frac14k^{-2}(h-k^2-3\nu-4)$
3.II	$\frac12$	$\nu+1$	$\frac12$	$\nu+\frac32$	$\frac12$	$\frac14k^{-2}(h-\nu-1)$

and $a=k^{-2}$ in both cases. The eigenvalue problems 3.I and 3.II correspond to Heun eigenvalue problems so that we obtain

$$\begin{aligned} 3.\mathrm{I}: &\quad d_\nu^{2m+1}(k^2)=4k^2\lambda_m(1,\nu+\tfrac32,\tfrac32,\nu+\tfrac32,k^{-2})+k^2+3\nu+4,\\ 3.\mathrm{II}: &\quad d_\nu^{2m}(k^2)=4k^2\lambda_m(\tfrac12,\nu+1,\tfrac12,\nu+\tfrac32,k^{-2})+\nu+1. \end{aligned}$$

Again, condition (3.1) holds.

Acknowledgment

The author thanks two referees whose remarks led to several improvements of the paper.

References

[1] B. Alexander, H. S. Cohl, and H. Volkmer, *Internal and external harmonics in bi-cyclide coordinates*, J. Phys. A **56** (2023), no. 32, Paper No. 325203, 26, DOI 10.1088/1751-8121/ace301. MR4617345

[2] L. Bi, H. S. Cohl, and H. Volkmer, *Expansion for a fundamental solution of Laplace's equation in flat-ring cyclide coordinates*, SIGMA Symmetry Integrability Geom. Methods Appl. **18** (2022), Paper No. 041, 31, DOI 10.3842/SIGMA.2022.041. MR4432921

[3] L. Bi, H. S. Cohl, and H. Volkmer, *Peanut harmonic expansion for a fundamental solution of Laplace's equation in flat-ring coordinates*, Anal. Math. **48** (2022), no. 4, 961–989, DOI 10.1007/s10476-022-0175-1. MR4515640

[4] P. B. Bailey, W. N. Everitt, D. B. Hinton, and A. Zettl, *Some spectral properties of the Heun differential equation*, Operator methods in ordinary and partial differential equations (Stockholm, 2000), Oper. Theory Adv. Appl., vol. 132, Birkhäuser, Basel, 2002, pp. 87–110. MR1924973

[5] Y.-M. Chiang, A. Ching, and C.-Y. Tsang, *Symmetries of the Darboux equation*, Kumamoto J. Math. **31** (2018), 15–48. MR3771678

[6] F. W. J. Olver, D. W. Lozier, R. F. Boisvert, and C. W. Clark (eds.), *NIST handbook of mathematical functions*, U.S. Department of Commerce, National Institute of Standards and Technology, Washington, DC; Cambridge University Press, Cambridge, 2010. MR2723248

[7] A. Erdélyi, *Certain expansions of solutions of the Heun equation*, Quart. J. Math. Oxford Ser. **15** (1944), 62–69, DOI 10.1093/qmath/os-15.1.62. MR11506

[8] A. Erdélyi. *Higher transcendental functions, Vol. II.* McGraw-Hill, New York, 1953.

[9] A. Erdélyi. *Higher transcendental functions, Vol. III.* McGraw-Hill, New York, 1955.

[10] E. L. Ince, *The periodic Lamé functions*, Proc. Roy. Soc. Edinburgh **60** (1940), 47–63. MR2399

[11] E. L. Ince, *Further investigations into the periodic Lamé functions*, Proc. Roy. Soc. Edinburgh **60** (1940), 83–99. MR2400

[12] T. A. Ishkhanyan, T. A. Shahverdyan, and A. M. Ishkhanyan, *Expansions of the solutions of the general Heun equation governed by two-term recurrence relations for coefficients*, Adv. High Energy Phys., posted on 2018, Art. ID 4263678, 9 pp., DOI 10.1155/2018/4263678. MR3827824

[13] W. Magnus and S. Winkler, *Hill's equation*, Interscience Tracts in Pure and Applied Mathematics, No. 20, Interscience Publishers John Wiley & Sons, New York-London-Sydney, 1966. MR197830

[14] R. S. Maier, *On reducing the Heun equation to the hypergeometric equation*, J. Differential Equations **213** (2005), no. 1, 171–203, DOI 10.1016/j.jde.2004.07.020. MR2139342

[15] J. Meixner and F. W. Schäfke, *Mathieusche Funktionen und Sphäroidfunktionen mit Anwendungen auf physikalische und technische Probleme* (German), Die Grundlehren der mathematischen Wissenschaften in Einzeldarstellungen mit besonderer Berücksichtigung der Anwendungsgebiete, Band LXXI, Springer-Verlag, Berlin-Göttingen-Heidelberg, 1954. MR66500

[16] P. Moon and D. E. Spencer, *Field theory handbook. Including coordinate systems, differential equations and their solution*, Springer-Verlag, Berlin-Göttingen-Heidelberg, 1961. MR136092

[17] A. Ronveaux (ed.), *Heun's differential equations*, Oxford Science Publications, The Clarendon Press, Oxford University Press, New York, 1995. With contributions by F. M. Arscott, S. Yu. Slavyanov, D. Schmidt, G. Wolf, P. Maroni and A. Duval. MR1392976

[18] D. Schmidt, *Die Lösung der linearen Differentialgleichung 2. Ordnung um zwei einfache Singularitäten durch Reihen nach hypergeometrischen Funktionen* (German), J. Reine Angew. Math. **309** (1979), 127–148, DOI 10.1515/crll.1979.309.127. MR542043

[19] N. Svartholm, *Die Lösung der Fuchsschen Differentialgleichung zweiter Ordnung durch hypergeometrische Polynome* (German), Math. Ann. **116** (1939), no. 1, 413–421, DOI 10.1007/BF01597365. MR1513235

[20] G. Szegö, *Orthogonal polynomials*, American Mathematical Society Colloquium Publications, Vol. 23, American Mathematical Society, Providence, RI, 1959. Revised ed. MR106295

[21] H. Volkmer, *The coexistence problem for Heun's differential equation*, Math. Nachr. **279** (2006), no. 16, 1823–1834, DOI 10.1002/mana.200410458. MR2274837

[22] H. Volkmer, *Approximation of eigenvalues of some differential equations by zeros of orthogonal polynomials*, J. Comput. Appl. Math. **213** (2008), no. 2, 488–500, DOI 10.1016/j.cam.2007.01.019. MR2392572

[23] H. Volkmer, *Eigenvalue problems for Lamé's differential equation*, SIGMA Symmetry Integrability Geom. Methods Appl. **14** (2018), Paper No. 131, 21 pp., DOI 10.3842/SIGMA.2018.131. MR3887203

[24] W. Walter. *Introduction to ordinary differential equations.* Springer-Verlag, New York, Berlin, Heidelberg, 1998.

[25] A. Wangerin. Reduction der Potentialgleichung für gewisse Rotationskörper auf eine gewöhnliche Differentialgleichung. Hirzel, Leipzig, 1875.

Dept. of Mathematical Sciences, University of Wisconsin - Milwaukee
Email address: volkmer@uwm.edu

Contemporary Mathematics
Volume **818**, 2025
https://doi.org/10.1090/conm/818/16377

Hyper-Mahler measures via Goncharov–Deligne cyclotomy

Yajun Zhou

To Prof. Weinan E, on the occasion of his 60th birthday

ABSTRACT. The hyper-Mahler measures $m_k(1+x_1+x_2)$, $k \in \mathbb{Z}_{>1}$ and $m_k(1+x_1+x_2+x_3)$, $k \in \mathbb{Z}_{>1}$, are evaluated in closed form via Goncharov–Deligne periods, namely $\mathbb{Q}$-linear combinations of multiple polylogarithms at cyclotomic points (complex-valued coordinates that are roots of unity). Some infinite series related to these hyper-Mahler measures are also explicitly represented as Goncharov–Deligne periods of levels 1, 2, 3, 4, 6, 8, 10 and 12.

1. Introduction

For an n-variate Laurent polynomial $P \in \mathbb{C}[x_1^{\pm 1}, \dots, x_n^{\pm 1}]$ that does not vanish identically, its k-Mahler measure [**58**, Definition 1] is given by the integral

$$m_k(P) := \int_0^1 \mathrm{d}t_1 \cdots \int_0^1 \mathrm{d}t_n \log^k \left| P\left(e^{2\pi i t_1}, \dots, e^{2\pi i t_n}\right) \right|, \tag{1.1}$$

where k is a positive integer. When $k = 1$, this becomes the logarithmic Mahler measure $m(P) = m_1(P)$ of P. When the integer k is greater than 1, the quantity $m_k(P)$ is a hyper-Mahler measure.

Smyth [**19**, Appendix 1] has evaluated a logarithmic Mahler measure

$$m_1(1+x_1+x_2+x_3) = \frac{7\zeta_3}{2\pi^2} \tag{1.2}$$

in terms of Apéry's constant $\zeta_3 := \sum_{n=1}^{\infty} \frac{1}{n^3}$. Borwein *et al.* (see [**13**, (4.17)]) and [**15**, (6.10)] have computed a hyper-Mahler measure

$$m_2(1+x_1+x_2+x_3) = \frac{24\,\mathrm{Li}_4\left(\frac{1}{2}\right) - \frac{\pi^4}{5} + 21\,\zeta_3 \log 2 - \pi^2 \log^2 2 + \log^4 2}{\pi^2} \tag{1.3}$$

involving a polylogarithmic constant $\mathrm{Li}_4\left(\frac{1}{2}\right) := \sum_{n=1}^{\infty} \frac{1}{2^n n^4}$. If we introduce an alternating double sum $\zeta_{-3,1} := \sum_{m>n>0} \frac{(-1)^m}{m^3 n} \equiv \sum_{m=1}^{\infty} \sum_{n=1}^{m-1} \frac{(-1)^m}{m^3 n}$, then we can condense (1.3) into

$$m_2(1+x_1+x_2+x_3) = \frac{12\,\zeta_{-3,1}}{\pi^2} + \frac{\pi^2}{20}, \tag{1.3$'$}$$

2020 *Mathematics Subject Classification.* Primary 11G55, 11M32, 11R06.
Key words and phrases. Hyper-Mahler measures, multiple polylogarithms.
This research was supported in part by the Applied Mathematics Program within the Department of Energy (DOE) Office of Advanced Scientific Computing Research (ASCR) as part of the Collaboratory on Mathematics for Mesoscopic Modeling of Materials (CM4).

TABLE 1. Selected closed forms for the k-Mahler measures $m_k(1+x_1+x_2)$, where $\omega := e^{2\pi i/3}$.

k	$m_k(1+x_1+x_2)$
1	$\frac{3}{2\pi}\operatorname{Im}\operatorname{Li}_2(\omega)$
2	$-\frac{2\cdot 3}{\pi}\operatorname{Im}\operatorname{Li}_{2,1}(\omega,1)+\frac{\pi^2}{2^2\cdot 3^3}$
3	$-\frac{3^2\cdot 17}{2^3\pi}\operatorname{Im}\operatorname{Li}_4(\omega)+\frac{2^2\cdot 3^2}{\pi}\operatorname{Im}\operatorname{Li}_{2,1,1}(\omega,1,1)+\frac{23\zeta_3}{2\cdot 3}+\frac{\pi}{2^3}\operatorname{Im}\operatorname{Li}_2(\omega)$
4	$-\frac{2^2\cdot 3^2\cdot 5\cdot 11}{13\pi}\operatorname{Im}\operatorname{Li}_{4,1}(\omega,1)-\frac{3^3\cdot 7^2}{13\pi}\operatorname{Im}\operatorname{Li}_{3,2}(\omega,1)$ $-\frac{2^5\cdot 3^2}{\pi}\operatorname{Im}\operatorname{Li}_{2,1,1,1}(\omega,1,1,1)+2^3\cdot 3\operatorname{Re}\operatorname{Li}_{3,1}(\omega,1)-\pi\operatorname{Im}\operatorname{Li}_{2,1}(\omega,1)$ $+\frac{3\zeta_3}{\pi}\operatorname{Im}\operatorname{Li}_2(\omega)+\frac{3^2}{2^2}[\operatorname{Im}\operatorname{Li}_2(\omega)]^2-\frac{5923\pi^4}{2^4\cdot 3^4\cdot 5\cdot 13}$

in light of a proven relation between $\zeta_{-3,1}$ and $\operatorname{Li}_4\left(\frac{1}{2}\right)$ [**17**, (5.5)].

In this work, we study the hyper-Mahler measures $m_k(1+x_1+x_2)$ and $m_k(1+x_1+x_2+x_3)$ in detail, for all $k\in\mathbb{Z}_{>1}$. For every positive integer k, we express the quantity $\pi i\, m_k(1+x_1+x_2)$ as a $\mathbb{Q}$-linear combination of special values for Goncharov's [**46**] multiple polylogarithm[1]

$$\operatorname{Li}_{a_1,\dots,a_n}(z_1,\dots,z_n) := \sum_{\ell_1>\cdots>\ell_n>0}\prod_{j=1}^{n}\frac{z_j^{\ell_j}}{\ell_j^{a_j}}, \tag{1.4}$$

where $a_1,\dots,a_n$ are positive integers and $z_1,\dots,z_n$ are third roots of unity, with $(a_1,z_1)\neq(1,1)$ to ensure convergence [**37**, (0.1)]. (See Table 1 for the first few.) Meanwhile, $\mathbb{Q}$-linear combinations of Hoffman's (alternating) multiple zeta values [**51**] (also known as Euler–Zagier numbers [**79, 90**])

$$\zeta_{A_1,\dots,A_n} := \operatorname{Li}_{|A_1|,\dots,|A_n|}\left(\frac{A_1}{|A_1|},\dots,\frac{A_n}{|A_n|}\right) \tag{1.5}$$

enter the representations of $\pi^2 m_k(1+x_1+x_2+x_3)$ (see Table 2 for examples).

In §2, we verify all the statements in the last paragraph. On the qualitative side, we do so by linking Broadhurst's formula [**22**, (9)] for moments of random walks (§§2.1–2.2) to Brown's homotopy-invariant iterated integrals [**27**, §1.1] on the moduli spaces $\mathfrak{M}_{0,n}$ of genus-zero curves with n marked points (§§2.3–2.4). On the quantitative side, with Panzer's `HyperInt` package [**69**] that implements Brown's algorithm [**25–27**] as well as Au's `MultipleZetaValues` package [**5**] that reduces

[1] While defining $\operatorname{Li}_{a_1,\dots,a_n}(z_1,\dots,z_n)$ and $\zeta_{a_1,\dots,a_n}$, we sort the positive integer parameters $a_1,\dots,a_n$ as in `Maple` 2022's built-in functions `MultiPolylog` and `MultiZeta` (which is Hoffman's [**51**] order—see also [**5, 7, 23, 37, 43–45, 79**] for the same practice). In the `HyperInt` package [**69**] for `Maple`, our $\operatorname{Li}_{a_1,\dots,a_n}(z_1,\dots,z_n)$ is represented by `Mpl`$([a_n,\dots,a_1],[z_n,\dots,z_1])$ (the latter of which follows Zagier's [**90**] order—see also [**25–27, 41, 46, 49, 60**] for the same convention).

TABLE 2. Selected closed forms for the k-Mahler measures $m_k(1+x_1+x_2+x_3)$.

k	$m_k(1+x_1+x_2+x_3)$
1	$\frac{7\zeta_3}{2\pi^2}$
2	$\frac{2^2\cdot 3\zeta_{-3,1}}{\pi^2}+\frac{\pi^2}{2^2\cdot 5}$
3	$-\frac{3\cdot 5\cdot 23\zeta_5}{2^2\pi^2}-\frac{2^2\cdot 3^3\zeta_{-3,1,1}}{\pi^2}+\frac{61\zeta_3}{2^3}$
4	$-\frac{2^4\cdot 3^2\cdot 13\zeta_{-5,1}}{\pi^2}+\frac{2^4\cdot 3^4\zeta_{-3,1,1,1}}{\pi^2}-2\cdot 3\zeta_{-3,1}+\frac{3^3\cdot 29\zeta_3^2}{\pi^2}-\frac{1867\pi^4}{2^4\cdot 3\cdot 5\cdot 7}$
5	$\frac{3\cdot 5\cdot 22973\zeta_7}{2^2\cdot 17\pi^2}+\frac{2^4\cdot 3^2\cdot 5^2\cdot 11\zeta_{-5,1,1}}{17\pi^2}+\frac{2^8\cdot 3\cdot 5\cdot 19\zeta_{-3,3,1}}{17\pi^2}+\frac{2^2\cdot 3\cdot 5\zeta_{-3,1}\zeta_3}{\pi^2}$ $-\frac{2^4\cdot 3^5\cdot 5\zeta_{-3,1,1,1,1}}{\pi^2}-\frac{5^2\cdot 29\cdot 59\zeta_5}{2^3\cdot 17}+2\cdot 3^2\cdot 5\zeta_{-3,1,1}-\frac{26267\pi^2\zeta_3}{2^5\cdot 3\cdot 17}$
6	$-\frac{2^5\cdot 3^3\cdot 5\cdot 317\zeta_{-7,1}}{17\pi^2}+\frac{2\cdot 3^3\cdot 11\cdot 29\cdot 43\zeta_{5,3}}{17\pi^2}+\frac{3^4\cdot 5^3\cdot 11\cdot 113\zeta_5\zeta_3}{2^2\cdot 17\pi^2}$ $-\frac{2^5\cdot 3^4\cdot 5^2\cdot 11\zeta_{-5,1,1,1}}{17\pi^2}-\frac{2^9\cdot 3^3\cdot 5\cdot 19\zeta_{-3,3,1,1}}{17\pi^2}-\frac{2^3\cdot 3^3\cdot 5\zeta_{-3,1,1}\zeta_3}{\pi^2}$ $+\frac{2^5\cdot 3^7\cdot 5\zeta_{-3,1,1,1,1,1}}{\pi^2}+2^2\cdot 3^2\cdot 5\cdot 13\zeta_{-5,1}-2^2\cdot 3^4\cdot 5\zeta_{-3,1,1,1}$ $+\frac{3\cdot 11\pi^2\zeta_{-3,1}}{2^2}-\frac{3^3\cdot 5^2\cdot 1213\zeta_3^2}{2^3\cdot 17}-\frac{26459029\pi^6}{2^6\cdot 3^2\cdot 5^2\cdot 7\cdot 17}$

certain special values of multiple polylogarithms, we generate Tables 1 and 2 as stated.[2]

In §3, we extend our methods to several families of infinite series occurring in recent studies of number theory and high energy physics. While these infinite sums may or may not be directly related to hyper-Mahler measures, they can all be expressed algorithmically in terms of multiple polylogarithms at roots of unity. Our automated evaluations provide either new patterns for infinite series or succinct alternatives to existing treatments of the same objects.

2. Integral representations of k-Mahler measures

2.1. Broadhurst's formula for zeta Mahler measures. Consider the zeta Mahler measure (see [**1**, (1.1)] and [**15**, (2.3)])

(2.1)

$$W_n(s):=\int_0^1\mathrm{d}t_1\cdots\int_0^1\mathrm{d}t_n\left|\sum_{j=1}^n e^{2\pi i t_j}\right|^s=\int_0^1\mathrm{d}t_1\cdots\int_0^1\mathrm{d}t_{n-1}\left|1+\sum_{j=1}^{n-1}e^{2\pi i t_j}\right|^s,$$

[2]Computer codes related to this publication are available as ancillary files for arXiv:2210.17243v4.

the k-th order derivative of which at $s=0$ evaluates the k-Mahler measure:

(2.2)
$$\left.\frac{\mathrm{d}^k W_n(s)}{\mathrm{d}\,s^k}\right|_{s=0}=\int_0^1\mathrm{d}t_1\cdots\int_0^1\mathrm{d}t_{n-1}\,\log^k\left|1+\sum_{j=1}^{n-1}e^{2\pi i t_j}\right|=:m_k\left(1+\sum_{j=1}^{n-1}x_j\right).$$

By definition, we have

$$W_1(s)=1. \tag{2.3}$$

The evaluation

$$W_2(s)=\frac{2^s\Gamma\left(\frac{1+s}{2}\right)}{\sqrt{\pi}\Gamma\left(1+\frac{s}{2}\right)} \tag{2.4}$$

can be found in [**1**, Theorem 1(1)], where $\Gamma(\sigma):=\int_0^\infty t^{\sigma-1}e^{-t}\,\mathrm{d}t$, $\sigma>0$ defines Euler's gamma function. In [**11**, (25)] and [**58**, Theorem 14], the quantity $W_2(s)$ was written as an equivalent binomial coefficient $\binom{s}{s/2}$.

In the next lemma, we recapitulate Broadhurst's integral representation [**22**, (9)] of $W_n(s)$, which will be instrumental in the rest of this section.

LEMMA 2.1 (Broadhurst's formula for $W_n(s)$). *For $n\in\mathbb{Z}_{>0}$, the integral representation*

$$W_n(s)=2^{s+1}\frac{\Gamma\left(1+\frac{s}{2}\right)}{\Gamma\left(-\frac{s}{2}\right)}\int_0^\infty J_0^n(x)\frac{\mathrm{d}\,x}{x^{s+1}} \tag{2.5}$$

is valid for $s\in\left(\max\{-2,-\frac{n}{2}\},0\right)$, where $J_0(x):=\sum_{\ell=0}^\infty\frac{(-1)^\ell}{(\ell!)^2}\left(\frac{x}{2}\right)^{2\ell}$ is the zeroth-order Bessel function of the first kind. □

REMARK 2.2. For $n\in\{1,2\}$, Broadhurst's integral representations are compatible with (2.3) and (2.4), as noted by Borwein–Straub–Wan [**14**, Example 2.2]. For $n=2$, one may also check this compatibility against a critical case of the Weber–Schafheitlin integral [**81**, §13.41(2)].

2.2. Some integral formulae involving Bessel functions. In this subsection, we will reformulate Broadhurst's integral representations of $W_3(s)$ and $W_4(s)$ through the Parseval–Plancherel identity for Hankel transforms [**81**, §14.3(3)]

(2.6)
$$\begin{aligned}&\frac{1}{2}\int_0^\infty\left[\int_0^\infty J_\nu\left(\sqrt{ux}\right)f(x)x\,\mathrm{d}x\right]\left[\int_0^\infty J_\nu\left(\sqrt{ux}\right)g(x)x\,\mathrm{d}x\right]\mathrm{d}u\\&\quad=\int_0^\infty f(x)g(x)x\,\mathrm{d}x,\qquad\nu\geqslant-\frac{1}{2},\end{aligned}$$

where $J_\nu(X):=\sum_{\ell=0}^\infty\frac{(-1)^\ell}{\ell!\Gamma(\nu+\ell+1)}\left(\frac{X}{2}\right)^{2\ell+\nu}$ is the Bessel function of order ν. This motivates the study of the integrals in the next proposition.

PROPOSITION 2.3 (Weber–Schafheitlin integral and its generalizations). *For $s \in (-2, 2)$, we have*

(2.7)

$$\int_0^\infty J_{1+\frac{s}{2}}\left(\sqrt{ux}\right)\frac{J_0(x)}{x^{s/2}}\mathrm{d}x = \begin{cases} 0, & u \in (0,1), \\ \dfrac{1}{u^{\frac{2+s}{4}}\Gamma\left(1+\frac{s}{2}\right)}\left(\dfrac{u-1}{2}\right)^{\frac{s}{2}}, & u \in (1,\infty), \end{cases}$$

(2.8)

$$\int_0^\infty J_{1+\frac{s}{2}}\left(\sqrt{ux}\right)\frac{J_0^2(x)}{x^{s/2}}\mathrm{d}x = \begin{cases} \displaystyle\int_0^{2\arcsin\frac{\sqrt{u}}{2}} \frac{\left(u-4\sin^2\frac{\phi}{2}\right)^{\frac{s}{2}}}{2^{\frac{s}{2}}\pi u^{\frac{2+s}{4}}\Gamma\left(1+\frac{s}{2}\right)}\mathrm{d}\phi, & u \in (0,4), \\ \displaystyle\int_0^{\pi} \frac{\left(u-4\sin^2\frac{\phi}{2}\right)^{\frac{s}{2}}}{2^{\frac{s}{2}}\pi u^{\frac{2+s}{4}}\Gamma\left(1+\frac{s}{2}\right)}\mathrm{d}\phi, & u \in (4,\infty). \end{cases}$$

PROOF. See [**81**, §13.46(1)]. □

Armed with the formulae in the proposition above, as well as the observation that

$$\begin{aligned}
&\int_0^\infty J_{1+\frac{s}{2}}\left(\sqrt{ux}\right)\frac{J_0^2(x)-1}{x^{s/2}}\mathrm{d}x \\
&= \int_0^\infty J_{1+\frac{s}{2}}\left(\sqrt{ux}\right)\frac{J_0^2(x)}{x^{s/2}}\mathrm{d}x + \int_0^\infty \frac{\partial}{\partial x}\frac{J_{\frac{s}{2}}\left(\sqrt{ux}\right)}{\sqrt{u}x^{s/2}}\mathrm{d}x \\
&= \int_0^\infty J_{1+\frac{s}{2}}\left(\sqrt{ux}\right)\frac{J_0^2(x)}{x^{s/2}}\mathrm{d}x - \frac{1}{2^{\frac{s}{2}}u^{\frac{2-s}{4}}\Gamma\left(1+\frac{s}{2}\right)},
\end{aligned} \tag{2.9}$$

we will apply the Parseval–Plancherel identity (2.6) to $W_3(s) - W_1(s)$ and $W_4(s) - W_2(s)$, in the corollary below.

COROLLARY 2.4 (Hankel–Broadhurst representations for $W_3(s)$ and $W_4(s)$). *Set $\varrho = e^{\pi i/3}$ and $\mathsf{Q}(x) := \left(1-\frac{x}{\varrho}\right)(1-x\varrho)$. Define the five rational functions*

(2.10)

$$\mathsf{A}_3(t,\tau) := \frac{\left(1-\frac{t}{\tau}\right)(1-t\tau)}{\mathsf{Q}(t)}, \quad \mathsf{B}_3(\tau) := \frac{\tau^2}{(1+\tau)^2\mathsf{Q}(\tau)}, \quad \mathsf{C}_3(\tau) := \frac{(1+\tau)^2}{3\tau},$$

(2.11)

$$\mathsf{A}_4(t,\tau) := \frac{\left(1-\frac{t}{\tau}\right)(1-t\tau)}{t}, \quad \mathsf{B}_4(\tau) := -\frac{\tau}{(1-\tau)^2}.$$

When $s \in \left(-\frac{1}{2}, 0\right)$, we have

(2.12)

$$\begin{aligned}
&W_3(s) - 1 \\
&= \frac{3^{s+\frac{3}{2}}\sin\frac{\pi s}{2}}{\pi^2}\int_0^1\left[\int_{-1}^{\tau}\mathsf{A}_3^{\frac{s}{2}}(t,\tau)\frac{\mathrm{d}t}{\mathsf{Q}(t)} - \mathsf{C}_3^{\frac{s}{2}}(\tau)\int_{-1}^1\frac{\mathrm{d}t}{\mathsf{Q}(t)}\right]\frac{\tau-1}{\tau+1}\frac{\mathsf{B}_3^{\frac{s}{2}}(\tau)\,\mathrm{d}\tau}{\mathsf{Q}(\tau)} \\
&+ \frac{3^{s+\frac{3}{2}}\sin\frac{\pi s}{2}}{\pi^2}\int_1^{\varrho}\left\{\int_{-1}^1\left[\mathsf{A}_3^{\frac{s}{2}}(t,\tau) - \mathsf{C}_3^{\frac{s}{2}}(\tau)\right]\frac{\mathrm{d}t}{\mathsf{Q}(t)}\right\}\frac{\tau-1}{\tau+1}\frac{\mathsf{B}_3^{\frac{s}{2}}(\tau)\,\mathrm{d}\tau}{\mathsf{Q}(\tau)},
\end{aligned}$$

where all the paths of integration are straight line segments; and when $s \in (-1,0)$, we have

(2.13)

$$\begin{aligned} W_4(s) - \frac{2^s\Gamma\left(\frac{1+s}{2}\right)}{\sqrt{\pi}\Gamma\left(1+\frac{s}{2}\right)} &= \frac{\sin\frac{\pi s}{2}}{\pi^3}\oint_1^{-1}\left[\oint_1^{\tau}\mathsf{A}_4^{\frac{s}{2}}(t,\tau)\frac{\mathrm{d}t}{t}\right]\left[\oint_1^{\tau}\mathsf{A}_4^{\frac{s}{2}}(T,\tau)\frac{\mathrm{d}T}{T} - \pi i\mathsf{B}_4^{-\frac{s}{2}}(\tau)\right]\frac{\tau+1}{\tau-1}\frac{\mathsf{B}_4^{\frac{s}{2}}(\tau)\,\mathrm{d}\tau}{\tau} \\ &+ \frac{\sin\frac{\pi s}{2}}{\pi^3}\int_{-1}^{0}\left[\oint_1^{-1}\mathsf{A}_4^{\frac{s}{2}}(t,\tau)\frac{\mathrm{d}t}{t}\right]\left\{\oint_1^{-1}\left[\mathsf{A}_4^{\frac{s}{2}}(T,\tau) - \mathsf{B}_4^{-\frac{s}{2}}(\tau)\right]\frac{\mathrm{d}T}{T}\right\}\frac{\tau+1}{\tau-1}\frac{\mathsf{B}_4^{\frac{s}{2}}(\tau)\,\mathrm{d}\tau}{\tau}, \end{aligned}$$

where the path of integration for each $\oint$ runs counterclockwise along the unit circle, and the path of integration for τ from -1 to 0 is a straight line segment.

PROOF. For $W_3(s)$, make the variable substitutions $\phi = 2\arctan\frac{1+t}{\sqrt{3}(1-t)}$ [which satisfies $4\sin^2\frac{\phi}{2} = \frac{(1+t)^2}{1-t+t^2}$] and $u = \frac{(1+\tau)^2}{1-\tau+\tau^2}$ after invoking (2.6). For $W_4(s)$, use $\phi = \arg t$ [which satisfies $4\sin^2\frac{\phi}{2} = -\frac{(1-t)^2}{t}$ for $|t| = 1$] and $u = -\frac{(1-\tau)^2}{\tau}$ instead. □

2.3. Period structure of $m_k(1+x_1+x_2)$. As a variation on Goncharov's notation [**47**, §1.2], we define $\mathbb{Q}$-vector spaces[3]

$$\mathfrak{Z}_k(N) := \operatorname{span}_{\mathbb{Q}}\left\{\operatorname{Li}_{a_1,\dots,a_n}(z_1,\dots,z_n)\in\mathbb{C}\,\middle|\, \begin{smallmatrix} a_1,\dots,a_n\in\mathbb{Z}_{>0} \\ z_1^N=\dots=z_n^N=1 \\ \sum_{j=1}^n a_j=k \end{smallmatrix}\right\} \tag{2.14}$$

for positive integers k and N. With the understanding that empty sums are zero (so $\operatorname{span}_{\mathbb{Q}}\varnothing = \{0\}$), we have $\mathfrak{Z}_1(1) = \{0\}$. Retroactively, we set $\mathfrak{Z}_0(N) := \mathbb{Q}$.

A number $\operatorname{Li}_{a_1,\dots,a_n}(z_1,\dots,z_n) \in \mathfrak{Z}_k(N)$ is referred to as a cyclotomic multiple zeta value (CMZV) of weight k and level N [**74**, (1)]. Some authors [**5, 74**] write $\mathsf{CMZV}_{k,N}$ or CMZV_k^N for $\mathfrak{Z}_k(N)$. At level 1, a CMZV $\zeta_{a_1,\dots,a_n} = \operatorname{Li}_{a_1,\dots,a_n}(1,\dots,1) \in \mathfrak{Z}_{a_1+\dots+a_n}(1)$ is simply a multiple zeta value (MZV); at level 2, a CMZV becomes an alternating multiple zeta value (AMZV).

It is clear from the definition (2.14) that we have a set inclusion $\mathfrak{Z}_k(N) \subseteq \mathfrak{Z}_k(M)$ if N divides M. Furthermore, we have an elementary fact that (see [**5**, Lemma 4.1] or [**74**, Remark 1.3])

$$\pi i = \frac{N}{N-2}\left[\operatorname{Li}_1(e^{2\pi i/N}) - \operatorname{Li}_1(e^{-2\pi i/N})\right] \in \mathfrak{Z}_1(N) \tag{2.15}$$

when $N \in \mathbb{Z}_{\geqslant 3}$.

For each fixed N, the $\mathbb{Q}$-algebra $\mathfrak{Z}(N) := \sum_{k\in\mathbb{Z}_{\geqslant 0}}\mathfrak{Z}_k(N)$ is filtered by weight [**47**, §1.2]:

$$\mathfrak{Z}_j(N)\mathfrak{Z}_k(N) \subseteq \mathfrak{Z}_{j+k}(N), \qquad \text{for all } j,k\in\mathbb{Z}_{\geqslant 0}. \tag{2.16}$$

[3] In this article, we write "$\cdots \in \mathbb{C}$" (as opposed to "$\cdots \in \mathbb{C}\cup\{\infty\}$") to indicate that the series or integral in question is convergent. Thus, the condition of convergence $(a_1, z_1) \neq (1,1)$ for (1.4) is imposed tacitly here, despite being absent from the predicate in (2.14).

In what follows, we will refer to members of $\mathfrak{Z}_k(N)$ as Goncharov–Deligne periods,[4] in honor of the pioneers [**37**, **38**, **47**] who worked on them.[5]

THEOREM 2.5 ($\pi i\, m_k(1+x_1+x_2)$ as Goncharov–Deligne periods). *For* $n \in \mathbb{Z}_{\geqslant 0}$, *define*

$$\mathsf{X}_3^{(n)}(\tau) := \int_{-1}^{\tau} \frac{\log^n \mathsf{A}_3(t,\tau)\,\mathrm{d}t}{\mathsf{Q}(t)} - \frac{\pi \log^n \mathsf{C}_3(\tau)}{\sqrt{3}}, \qquad \mathsf{Y}_3^{(n)}(\tau) := \frac{\tau-1}{\tau+1}\frac{\log^n \mathsf{B}_3(\tau)}{\mathsf{Q}(\tau)}, \tag{2.17}$$

$$\Xi_3^{(n)}(\tau) := \int_{-1}^{1} [\log^n \mathsf{A}_3(t,\tau) - \log^n \mathsf{C}_3(\tau)] \frac{\mathrm{d}t}{\mathsf{Q}(t)}, \tag{2.18}$$

using the rational functions in (2.10). *For each positive integer* k, *we have*

$$\begin{aligned}\pi m_k(1+x_1+x_2) &\in \operatorname{span}_{\mathbb{Q}}\left\{\frac{\sqrt{3}\log^{n_1} 3}{\pi^{-2n_2}}\int_0^1 \mathsf{X}_3^{(n_3)}(\tau)\mathsf{Y}_3^{(n_4)}(\tau)\,\mathrm{d}\tau \,\middle|\, \begin{smallmatrix} n_1,n_2,n_3,n_4\in\mathbb{Z}_{\geqslant 0}\\ n_1+(2n_2+1)+n_3+n_4=k\end{smallmatrix}\right\}\\ &\quad + \operatorname{span}_{\mathbb{Q}}\left\{\frac{\sqrt{3}\log^{n_1} 3}{\pi^{-2n_2}}\int_1^{\varrho} \Xi_3^{(n_3)}(\tau)\mathsf{Y}_3^{(n_4)}(\tau)\,\mathrm{d}\tau \,\middle|\, \begin{smallmatrix} n_1,n_2,n_3,n_4\in\mathbb{Z}_{\geqslant 0}\\ n_1+(2n_2+1)+n_3+n_4=k\end{smallmatrix}\right\}\\ &\subseteq i\mathfrak{Z}_{k+1}(6).\end{aligned} \tag{2.19}$$

PROOF. The "$\in$" part is a direct consequence of (2.2) and (2.12).

To prove the "$\subseteq$" part, we follow Brown's algorithmic studies [**25**–**27**] of $L(\mathfrak{M}_{0,n})$, the totality of homotopy-invariant iterated integrals [**27**, §1.1] on the moduli spaces $\mathfrak{M}_{0,n}$ of Riemann spheres with n marked points. Thanks to the filtration property (2.16) and the facts that

$$\log 3 = -\operatorname{Li}_1(-\varrho) - \operatorname{Li}_1\left(-\frac{1}{\varrho}\right) \in \mathfrak{Z}_1(6), \qquad \pi^2 = 6\operatorname{Li}_2(1) \in \mathfrak{Z}_2(6), \tag{2.20}$$

we may reduce our task to the demonstration of

$$\left\{\sqrt{3}\int_0^1 \mathsf{X}_3^{(n_3)}(\tau)\mathsf{Y}_3^{(n_4)}(\tau)\,\mathrm{d}\tau,\quad \sqrt{3}\int_1^{\varrho} \Xi_3^{(n_3)}(\tau)\mathsf{Y}_3^{(n_4)}(\tau)\,\mathrm{d}\tau\right\} \subset i\mathfrak{Z}_{k+1}(6) \tag{2.21}$$

for $n_3, n_4 \in \mathbb{Z}_{\geqslant 0}$ and $n_3 + n_4 = k-1$.

Using the notation of Frellesvig–Tommasini–Wever [**44**, §2], we define the generalized polylogarithm[6] (GPL) by the recursion

$$G(\alpha_1,\dots,\alpha_n;z) := \int_0^z \frac{\mathrm{d}x}{x-\alpha_1} G(\alpha_2,\dots,\alpha_n;x) \tag{2.22}$$

---

[4]Instead of pointing to Kontsevich–Zagier periods [**57**] or Deligne periods [**39**] in a broad sense, our discussions of "period structure" in the rest of this article will pertain exclusively to Goncharov–Deligne periods.

[5]It is worth mentioning that the provable upper bound for $\dim_{\mathbb{Q}} \mathfrak{Z}_k(N)$ in the joint work of Deligne and Goncharov [**38**, Corollaire 5.25] provides guidance to Au's software [**4**, Theorem 5.5] for symbolic reductions of CMZVs.

[6]The GPL $G(\alpha_1,\dots,\alpha_n;z)$ is implemented in `Maple` as `GeneralizedPolylog([`$\alpha_1,\dots,\alpha_n$`], z)` by Frellesvig [**43**]. The same GPL corresponds to `Hlog(z, [`$\alpha_1,\dots,\alpha_n$`])` in Panzer's `HyperInt` package [**69**] for `Maple`, where `Hlog` stands for "hyperlogarithm", another name for GPL. We also note that Panzer used "GPL" as an abbreviation for "Goncharov polylogarithm" [**69**, §1], referring to the same object as our GPL.

for any $\boldsymbol{\alpha} = (\alpha_1, \ldots, \alpha_n) \neq \mathbf{0} \in \mathbb{C}^n$, along with the additional settings that

$$G(\underbrace{0,\ldots,0}_{n};z) := \frac{\log^n z}{n!}, \qquad G(-;z) := 1. \tag{2.23}$$

Here, it is always understood that the integration path in (2.22) is a straight line segment joining 0 to z. For convenience, we will use a short-hand $G(\boldsymbol{\alpha};z) = G(\alpha_1,\ldots,\alpha_{w(\boldsymbol{\alpha})};z)$ and refer to $w(\boldsymbol{\alpha})$ (the number of components in the vector $\boldsymbol{\alpha} \in \mathbb{C}^{w(\boldsymbol{\alpha})}$) as the weight of the GPL $G(\boldsymbol{\alpha};z)$.

Set $\mathfrak{G}_0^{(z)}(N) = \mathbb{Q}$ and $\mathfrak{H}_0^{(z)}(N) = \mathbb{Q}$. For every $k \in \mathbb{Z}_{>0}$, we construct two $\mathbb{Q}$-vector spaces[7]

$$\mathfrak{G}_k^{(z)}(N) := \operatorname{span}_{\mathbb{Q}}\left\{ G(\alpha_1,\ldots,\alpha_k;z) \in \mathbb{C} \left| \begin{smallmatrix} \alpha_1^N,\ldots,\alpha_{k-1}^N \in \{0,1\} \\ \alpha_k^N = 1 \end{smallmatrix} \right.\right\}, \tag{2.24}$$

$$\mathfrak{H}_k^{(z)}(N) := \operatorname{span}_{\mathbb{Q}}\left\{ (\pi i)^\ell G(\alpha_1,\ldots,\alpha_{k-\ell};z) \in \mathbb{C} \left| \begin{smallmatrix} \alpha_1^N,\ldots,\alpha_{k-\ell}^N \in \{0,1\} \\ 0 \leqslant \ell \leqslant k \end{smallmatrix} \right.\right\}. \tag{2.25}$$

The $\mathbb{Q}$-algebras $\mathfrak{G}^{(z)}(N) := \sum_{k\in\mathbb{Z}_{\geqslant 0}} \mathfrak{G}_k^{(z)}(N)$ and $\mathfrak{H}^{(z)}(N) := \sum_{k\in\mathbb{Z}_{\geqslant 0}} \mathfrak{H}_k^{(z)}(N)$ are filtered by weight:

$$\mathfrak{G}_j^{(z)}(N)\mathfrak{G}_k^{(z)}(N) \subseteq \mathfrak{G}_{j+k}^{(z)}(N), \qquad \mathfrak{H}_j^{(z)}(N)\mathfrak{H}_k^{(z)}(N) \subseteq \mathfrak{H}_{j+k}^{(z)}(N), \tag{2.26}$$

thanks to a shuffle identity (see [**10**, §5.4] or [**44**, (2.4)])

$$G(\alpha_1,\ldots,\alpha_{w_1};z)G(\alpha_{w_1+1},\ldots,\alpha_{w_1+w_2};z) = \sum_{\sigma\in Ш_{w_1,w_2}} G(\alpha_{\sigma(1)},\ldots,\alpha_{\sigma(w_1+w_2)};z), \tag{2.27}$$

where $Ш_{w_1,w_2}$ consists of $\frac{(w_1+w_2)!}{w_1!w_2!}$ permutations $\sigma\colon \mathbb{Z}\cap[1,w_1+w_2] \longrightarrow \mathbb{Z}\cap[1,w_1+w_2]$ that satisfy $\sigma^{-1}(m) < \sigma^{-1}(n)$ whenever $1 \leqslant m < n \leqslant w_1$ or $w_1+1 \leqslant m < n \leqslant w_1+w_2$. A multiple polylogarithm (MPL) can be converted to a GPL (see [**44**, (2.5)] or [**69**, (1.3)])

$$\operatorname{Li}_{a_1,\ldots,a_n}(z_1,\ldots,z_n) = (-1)^n G\left(\underbrace{0,\ldots,0}_{a_1-1},\frac{1}{z_1},\underbrace{0,\ldots,0}_{a_2-1},\frac{1}{z_1z_2},\ldots,\underbrace{0,\ldots,0}_{a_n-1},\frac{1}{\prod_{j=1}^n z_j};1\right) \tag{2.28}$$

when $\prod_{j=1}^n z_j \neq 0$, and *vice versa* (see [**44**, (2.6)] or [**69**, (1.3)]), i.e.,

$$G(\underbrace{0,\ldots,0}_{a_1-1},\widetilde{\alpha}_1,\underbrace{0,\ldots,0}_{a_2-1},\widetilde{\alpha}_2,\ldots,\underbrace{0,\ldots,0}_{a_n-1},\widetilde{\alpha}_n;z) = (-1)^n \operatorname{Li}_{a_1,\ldots,a_n}\left(\frac{z}{\widetilde{\alpha}_1},\frac{\widetilde{\alpha}_1}{\widetilde{\alpha}_2},\ldots,\frac{\widetilde{\alpha}_{n-1}}{\widetilde{\alpha}_n}\right) \tag{2.29}$$

when $\prod_{j=1}^n \widetilde{\alpha}_j \neq 0$.[8] As a result, we have $\mathfrak{G}_k^{(1)}(N) = \mathfrak{Z}_k(N)$. Furthermore, the convergence condition for MPLs can be transferred onto GPLs, allowing us to

---

[7] According to Goncharov [**47**, Theorem 2.1], the space $\mathfrak{G}_k^{(z)}(N)$ is spanned by a special class of GPLs that can be converted to CMZVs, when $z^N = 1$. [See also (2.28)–(2.29) below.] In the space $\mathfrak{H}_k^{(z)}(N)$, one allows $\alpha_{k-\ell} = 0$, as in the "harmonic polylogarithms" for Maître's `HPL` [**66,67**] or in the "hyperlogarithms" for Panzer's `HyperInt` [**69**].

[8] The equations (2.28) and (2.29) can also be used as working definitions for an MPL when its series form (1.4) does not converge.

reformulate (2.24) and (2.25) as

$$\mathfrak{G}_k^{(z)}(N) := \operatorname{span}_{\mathbb{Q}}\left\{G(\alpha_1,\dots,\alpha_k;z)\,\middle|\,\begin{smallmatrix}\alpha_1^N,\dots,\alpha_{k-1}^N\in\{0,1\}\\ \alpha_1\neq z;\alpha_k^N=1\end{smallmatrix}\right\}, \tag{2.24$'$}$$

$$\mathfrak{H}_k^{(z)}(N) := \operatorname{span}_{\mathbb{Q}}\left\{(\pi i)^{\ell}G(\alpha_1,\dots,\alpha_{k-\ell};z)\,\middle|\,\begin{smallmatrix}\alpha_1^N,\dots,\alpha_{k-\ell}^N\in\{0,1\}\\ \alpha_1\neq z;0\leqslant\ell\leqslant k\end{smallmatrix}\right\}. \tag{2.25$'$}$$

To facilitate further analysis, we define

(2.30a)

$$\mathfrak{g}_k^{(z)}(N) := \operatorname{span}_{\mathbb{Q}}\left\{G(\boldsymbol{\alpha};1)G(\boldsymbol{\beta};z)(\pi i)^{\ell}(\log z)^{k-\ell-w(\boldsymbol{\alpha})-w(\boldsymbol{\beta})}\,\middle|\,\begin{smallmatrix}G(\boldsymbol{\alpha};1)\in\mathfrak{Z}_{w(\boldsymbol{\alpha})}(N)\\ G(\boldsymbol{\beta};z)\in\mathfrak{G}^{(z)}_{w(\boldsymbol{\beta})}(N)\\ \ell,w(\boldsymbol{\alpha}),w(\boldsymbol{\beta})\in\mathbb{Z}_{\geqslant0}\\ \ell+w(\boldsymbol{\alpha})+w(\boldsymbol{\beta})\leqslant k\end{smallmatrix}\right\}$$

$$\equiv \operatorname{span}_{\mathbb{Q}}\left\{G(\boldsymbol{\alpha};1)G(\boldsymbol{\beta};z)(\pi i)^{\ell}\,\middle|\,\begin{smallmatrix}G(\boldsymbol{\alpha};1)\in\mathfrak{Z}_{w(\boldsymbol{\alpha})}(N)\\ G(\boldsymbol{\beta};z)(\pi i)^{\ell}\in\mathfrak{H}^{(z)}_{\ell+w(\boldsymbol{\beta})}(N)\\ \ell,w(\boldsymbol{\alpha}),w(\boldsymbol{\beta})\in\mathbb{Z}_{\geqslant0}\\ \ell+w(\boldsymbol{\alpha})+w(\boldsymbol{\beta})=k\end{smallmatrix}\right\}, \tag{2.30b}$$

where the equivalence between two definitions stems from the GPL shuffle algebra [**41**, p. 24]. By the shuffle identity (2.27), we have $\mathfrak{g}_j^{(z)}(N)\mathfrak{g}_k^{(z)}(N)\subseteq\mathfrak{g}_{j+k}^{(z)}(N)$ for $j,k\in\mathbb{Z}_{\geqslant0}$. Combining (2.30b) with the recursive definition of GPL in (2.22), we see that if $\alpha^N\in\{0,1\}$, then

$$\int_0^z\frac{g(x)\,\mathrm{d}x}{x-\alpha}\in\mathfrak{g}_{k+1}^{(z)}(N) \tag{2.31}$$

so long as the integral in question converges, and $g(x)\in\mathfrak{g}_k^{(x)}(N)$ on the path of integration.

In view of the facts that

(2.32)

$$\log\mathsf{A}_3(t,\tau) := \log\frac{\left(1-\frac{t}{\tau}\right)(1-t\tau)}{\mathsf{Q}(t)} = G(\tau;t)+G\left(\frac{1}{\tau};t\right)-G(\varrho;t)-G\left(\frac{1}{\varrho};t\right),$$

(2.33)

$$\begin{aligned}\log\mathsf{C}_3(\tau) := \log\frac{(1+\tau)^2}{3\tau} &= 2G(-1;\tau)-G(0;\tau)-G(-\varrho;1)-G\left(-\frac{1}{\varrho};1\right)\\ &\in\mathfrak{g}_1^{(\tau)}(6),\end{aligned}$$

and

$$\frac{1}{\mathsf{Q}(t)}=\frac{1}{i\sqrt{3}}\left(\frac{1}{t-\varrho}-\frac{1}{t-\frac{1}{\varrho}}\right),\quad\frac{\pi i}{3}=G\left(-\frac{1}{\varrho};1\right)-G(-\varrho;1)\in\mathfrak{g}_1^{(\tau)}(6), \tag{2.34}$$

we have the following arguments for $k\in\mathbb{Z}_{>0}$:

$$\begin{aligned}\mathsf{X}_3^{(k-1)}(\tau) &:= \int_{-1}^{\tau}\frac{\mathrm{d}t}{\mathsf{Q}(t)}\log^{k-1}\frac{\left(1-\frac{t}{\tau}\right)(1-t\tau)}{\mathsf{Q}(t)}-\frac{\pi\log^{k-1}\frac{(1+\tau)^2}{3\tau}}{\sqrt{3}}\\ &\in\operatorname{span}_{\mathbb{Q}}\left\{\frac{G(\alpha_1,\dots,\alpha_k;z)}{i\sqrt{3}}\,\middle|\,\begin{smallmatrix}\alpha_1\in\{\varrho,\frac{1}{\varrho}\}\smallsetminus\{\tau\}\\ \alpha_2,\dots,\alpha_k\in\{\varrho,\frac{1}{\varrho},\tau,\frac{1}{\tau}\}\\ z\in\{-1,\tau\}\end{smallmatrix}\right\}+\frac{\mathfrak{g}_k^{(\tau)}(6)}{i\sqrt{3}}\subseteq\frac{\mathfrak{g}_k^{(\tau)}(6)}{i\sqrt{3}}.\end{aligned} \tag{2.35}$$

Here, the "$\in$" step issues from the GPL recursion (2.22) and the shuffle identity (2.27) (along with Panzer's logarithmic regularizations at $t=\tau$ [**69**, §2.3] that preserve the shuffle structure), while the "$\subseteq$" step is a result of the fibration basis in Brown's algorithm [**69**, Lemma 2.14 and Corollary 3.2]. Concretely speaking, through the fibration procedure, one can rewrite a GPL $G(\boldsymbol{\alpha};\tau)$ (in which the components of $\boldsymbol{\alpha}$ belong to $\{\varrho,\frac{1}{\varrho},\tau,\frac{1}{\tau}\}$) as a member of the $\mathbb{Z}$-module[9]

$$\operatorname{span}_{\mathbb{Z}}\left\{G(\beta_1,\dots,\beta_{w(\boldsymbol{\alpha})};\tau)\,\middle|\,\beta_1,\dots,\beta_{w(\boldsymbol{\alpha})}\in\left\{\varrho,\frac{1}{\varrho},-1,0,1\right\}\right\},\tag{2.36}$$

by induction on the weight $w(\boldsymbol{\alpha})$ and applications of the GPL recursion (2.22) (where $z=\tau$) to the GPL differential form[10] [**86**, (8.8)]

$$\begin{aligned}&\mathrm{d}\,G(\alpha_1,\dots,\alpha_n;z)\\&=\sum_{j=1}^n G(\alpha_1,\dots,\widehat{\alpha_j},\dots,\alpha_n;z)[\mathrm{d}\log(\alpha_{j-1}-\alpha_j)-\mathrm{d}\log(\alpha_{j+1}-\alpha_j)],\end{aligned}\tag{2.37}$$

where an element under the caret is removed, and $\alpha_0:=z$, $\alpha_{n+1}:=0$. Thus, it is clear that $G(\boldsymbol{\alpha};\tau)\in\mathfrak{g}^{(\tau)}_{w(\boldsymbol{\alpha})}(6)$. Likewise, one can show inductively that $G(\boldsymbol{\alpha};-1)$ (in which the components of $\boldsymbol{\alpha}$ belong to $\{\varrho,\frac{1}{\varrho},\tau,\frac{1}{\tau}\}$) is a member of

$$\operatorname{span}_{\mathbb{Z}}\left\{G(\beta_1,\dots,\beta_j;\tau)G(\beta_{j+1},\dots,\beta_{w(\boldsymbol{\alpha})};-1)\,\middle|\,\substack{\beta_1,\dots,\beta_{w(\boldsymbol{\alpha})}\in\{\varrho,\frac{1}{\varrho},-1,0,1\}\\0\leqslant j\leqslant w(\boldsymbol{\alpha})}\right\}.\tag{2.38}$$

This justifies the relation $\mathsf{X}_3^{(k-1)}(\tau)\in\frac{\mathfrak{g}^{(\tau)}_k(6)}{i\sqrt{3}}$ for every $k\in\mathbb{Z}_{>0}$.

Reasoning in a similar vein as the last paragraph, one can verify that $\Xi_3^{(k-1)}(\tau)\in\frac{\mathfrak{g}^{(\tau)}_k(6)}{i\sqrt{3}}$ for $k\in\mathbb{Z}_{>0}$.

With the additional observation that

$$\begin{aligned}&\mathsf{Y}_3^{(n)}(\tau)\\&:=\frac13\left(\frac{1}{\tau-\varrho}+\frac{1}{\tau-\frac{1}{\varrho}}-\frac{2}{\tau+1}\right)\log^n\frac{\tau^2}{(1+\tau)^2\mathsf{Q}(\tau)}\\&=\frac13\left(\frac{1}{\tau-\varrho}+\frac{1}{\tau-\frac{1}{\varrho}}-\frac{2}{\tau+1}\right)\left[2G(0;\tau)-2G(-1;\tau)-G(\varrho;\tau)-G\left(\frac{1}{\varrho};\tau\right)\right]^n\end{aligned}\tag{2.39}$$

together with the scaling property [**44**, (2.3)] of GPL $G(\varrho\boldsymbol{\alpha};\varrho z)=G(\boldsymbol{\alpha};z)$ when $\alpha_{w(\boldsymbol{\alpha})}\neq0$, we may complete all the integrations over τ (with Panzer's logarithmic regularizations at $\tau=\varrho$) and arrive at

$$\pi m_k(1+x_1+x_2)\in i\mathfrak{G}^{(1)}_{k+1}(6)+i\mathfrak{g}^{(\varrho)}_{k+1}(6)=i\mathfrak{G}^{(1)}_{k+1}(6)=i\mathfrak{Z}_{k+1}(6)\tag{2.40}$$

for all $k\in\mathbb{Z}_{>0}$. Here, the set inclusion $\mathfrak{g}^{(\varrho)}_{k+1}(6)\subseteq\mathfrak{G}^{(1)}_{k+1}(6)$ descends from the relation $G(\frac{1}{\varrho};1)=\log(1-\varrho)=-\frac{\pi i}{3}\in\mathfrak{G}^{(1)}_1(6)$, and parameter rescaling by a factor of ϱ for members of $\mathfrak{G}^{(\varrho)}(6)$.

---

[9]Such fibration procedures are automated by Panzer's `HyperInt` package [**69**, Lemma 2.14 and Corollary 3.2], where `fibrationBasis(Hlog(`$\tau,[\alpha_1,\dots,\alpha_{w(\boldsymbol{\alpha})}]$`),`$[\tau]$`)` produces a unique representation of $G(\boldsymbol{\alpha};\tau)$ as a member of (2.36).

[10]As in [**86**, §8.1, p. 297], we adopt the convention that the 1-form $\mathrm{d}\log(\alpha_j-\alpha_{j'})$ is zero when $\alpha_j=\alpha_{j'}$. Accordingly, logarithmic regularizations [**69**, §2.3] are implicit when we encounter $G(z,\dots;z)$ and $G(\dots,0;0)$.

While the last paragraph finishes the proof of (2.19), we still have a slightly stronger result:
(2.40′)

$$\pi i\, m_k(1+x_1+x_2)$$
$$\in \operatorname{span}_{\mathbb{Q}}\left\{ (\pi i)^\ell G(\alpha_1,\dots,\alpha_j;\tau)G(\alpha_{j+1},\dots,\alpha_n;z)\log^r 3 \;\middle|\; \begin{smallmatrix} \alpha_1,\dots,\alpha_n,z\in\left\{\varrho,\frac{1}{\varrho},-1,0,1\right\} \\ \tau\in\{1,\varrho\} \\ \ell,j,n,r\in\mathbb{Z}_{\geqslant 0};\ell+n+r=k+1 \end{smallmatrix} \right\}$$
$$\subseteq \mathfrak{g}^{(\varrho)}_{k+1}(2).$$

Here, the "$\in$" step traces back to (2.35)–(2.39), along with all the foregoing fibrations, while the "$\subseteq$" step comes from the observation that

$$\log 3 = G(1;\varrho) + 2G(-1;\varrho) \in \mathfrak{G}^{(\varrho)}_1(2), \tag{2.41}$$

and the fibration of GPLs with respect to ϱ. □

REMARK 2.6. If one defines a multiple zeta value (MZV) $\zeta_{a_1,\dots,a_n} = \mathrm{Li}_{a_1,\dots,a_n}(1,\dots,1)$ via the iterative constructions of GPL on the right-hand side of (2.28), then one gets the Drinfel'd integral representation of MZV (see [**40**, §2] or [**90**, §9]). In general, the representation of an MPL via the iterative integrations for the corresponding GPL is called the Leibniz–Kontsevich formula (see [**48**, Theorem 2.2] or [**49**, §1.2]) by Goncharov, who ascribed it to a private communication with Kontsevich that generalized the Leibniz representation for ζ_n as an n-dimensional integral.

By considering mixed Tate motives [**49**, (2) and (3)], Goncharov–Manin asked [**49**, Conjecture 4.5] if certain integral periods on $\mathfrak{M}_{0,n}$ (generalizations of the Drinfel'd integral representation of MZV) are still representable as $\mathbb{Q}$-linear combinations of MZVs. This deep question has been answered in the positive by Brown [**27**, Theorem 1.1], whose algorithmic approach [**25–27**] to the Goncharov–Manin periods on $\mathfrak{M}_{0,n}$ [along with generalizations to a larger set $L(\mathfrak{M}_{0,n})$] has been implemented in Panzer's `HyperInt` package [**69**] for `Maple`.

In the following, the absolute fibration operation $\mathrm{Fib}(f)$ (where f is an expression involving a formal variable ϱ) consists of four tasks in sequence:

1. Evaluate `fibrationBasis(f)` in `HyperInt`, which (among other things) reduces the alternating multiple zeta values (AMZVs) $\zeta_{A_1,\dots,A_n} \in \mathfrak{Z}_{|A_1|+\dots+|A_n|}(2)$ [see (1.5)] and a special class of multiple polylogarithms [**16**, (2–3)]

$$\mathrm{Li}_{\underbrace{1,\dots,1}_{n}}(z,\underbrace{1,\dots,1}_{n-1}) = \frac{(-1)^n}{n!}\log^n(1-z); \tag{2.42}$$

2. Set $\varrho = \frac{1}{2}+\frac{i\sqrt{3}}{2}$ and $\delta_\varrho := \frac{\operatorname{Im}\varrho}{|\operatorname{Im}\varrho|} = 1$, before evaluating logarithms and simplifying all the complex arguments of (multiple) polylogarithms;
3. Set $\frac{1}{2}+\frac{i\sqrt{3}}{2} = \varrho$, $-\frac{1}{2}+\frac{i\sqrt{3}}{2} = -\frac{1}{\varrho}$, $-\frac{1}{2}-\frac{i\sqrt{3}}{2} = -\varrho$, and $\frac{1}{2}-\frac{i\sqrt{3}}{2} = \frac{1}{\varrho}$, using the `eval` command in `Maple`;
4. Convert to `Hlog` form in `HyperInt`.

Define the k-th Hankel–Broadhurst integral

$$\mathscr{I}^{\mathrm{HB}}_k := \operatorname*{fib}_{\varrho}\mathrm{Fib}\left.\frac{\mathrm{d}^k W_3(s)}{\mathrm{d}s^k}\right|_{s=0} \tag{2.43}$$

by computing the k-th order derivative of (2.12) at $s=0$ with `HyperInt` [where $\mathsf{Q}(x):=\left(1-\frac{x}{\varrho}\right)(1-x\varrho)$ and ϱ is treated as a formal variable], before evaluating its absolute fibration ("Fib") and relative fibration basis ("fib_ϱ") with respect to the formal variable ϱ.

The last relative fibration sends the k-th Hankel–Broadhurst integral to $\mathfrak{g}^{(\varrho)}_{k+1}(2)=\sum_{\ell=0}^{k+1}\mathfrak{H}^{(\varrho)}_{\ell}(2)\mathfrak{Z}_{k+1-\ell}(2)$, where $\mathfrak{H}^{(\varrho)}_{\ell}(2)$ is a $\mathbb{Q}$-vector space generated by the Remiddi–Vermaseren [**71**] harmonic polylogarithms[11] of weight ℓ. Members of the latter $\mathbb{Q}$-vector space can be decomposed into Lyndon words [**70**] by Maître's `HPL` package [**66**] for `Mathematica`. We use "Lyn" to denote this Lyndon word decomposition together with a subsequent reduction by "$\mathrm{fib}_\varrho\,\mathrm{Fib}$".

The processed Lyndon words $\mathrm{Lyn}\,\mathscr{I}^{\mathrm{HB}}_{k}$, $1\leqslant k\leqslant 4$ can now be fed into Au's `Mathematica` package `MultipleZetaValues` (v1.2.0), which performs automated (and provable [**5**, §5]) reductions for members of $\mathfrak{Z}_k(6)$, $1\leqslant k\leqslant 5$.[12] This confirms all the entries in Table 1. It is worth noting that all those tabulated formulae suggest that[13]

$$\pi i\, m_k(1+x_1+x_2)\in\mathfrak{Z}_{k+1}(3),\tag{2.44}$$

which is stronger than the relation (2.19) established in Theorem 2.5.

To get ready for the proof of a sharper statement (2.44) in the next theorem, we recall the multi-dimensional polylogarithm

$$L_{a_1,\dots,a_n}(x)\equiv \mathrm{Li}_{a_1,\dots,a_n}(x,\underbrace{1,\dots,1}_{n-1}):=\sum_{\ell_1>\dots>\ell_n>0}\frac{x^{\ell_1}}{\prod_{j=1}^{n}\ell_j^{a_j}}\tag{2.45}$$

from the work of Borwein–Broadhurst–Kamnitzer [**16**, §2]. For brevity, we write $L_{\boldsymbol A}(x)\equiv L_{a_1,\dots,a_n}(x)$ for a vector $\boldsymbol A=(a_1,\dots,a_{d(\boldsymbol A)})\in\mathbb{Z}^{d(\boldsymbol A)}_{>0}$, while calling $d(\boldsymbol A)$ the depth of $L_{\boldsymbol A}$, and $w(\boldsymbol A):=\sum_{j=1}^{d(\boldsymbol A)}a_j$ the weight of $L_{\boldsymbol A}$. In a trigonometric analysis of the hyper-Mahler measures for $1+x_1+x_2$, Borwein–Borwein–Straub–Wan [**9**, Theorem 26] introduced a $\mathbb{Q}$-linear combination of certain multi-dimensional polylogarithms of weight $n\in\mathbb{Z}_{>1}$ and all possible depths, namely

$$\rho_n(X):=\frac{(-1)^n n!}{4^n}\sum_{\substack{w(\boldsymbol A)=n,a_1=2\\ a_2,\dots,a_{d(\boldsymbol A)}\in\{1,2\}}}4^{d(\boldsymbol A)}L_{\boldsymbol A}\left(X^2\right),\qquad X\in(0,1),\tag{2.46}$$

and defined additionally $\rho_0(X)=1$, $\rho_1(X)=0$.

THEOREM 2.7 ($\pi i\, m_k(1+x_1+x_2)$ as Goncharov–Deligne periods, reprise). *For each positive integer k, we have*

$$\begin{aligned}&\pi m_k(1+x_1+x_2)\\&=2\int_0^{\pi/6}\rho_k(2\sin\vartheta)\,\mathrm{d}\vartheta+2\sum_{j=0}^{k}\binom{k}{j}\int_{\pi/6}^{\pi/2}\rho_j\left(\frac{1}{2\sin\vartheta}\right)\log^{k-j}(2\sin\vartheta)\,\mathrm{d}\vartheta\in i\mathfrak{Z}_{k+1}(3).\end{aligned}\tag{2.47}$$

---

[11] If all the components of $\boldsymbol\alpha=(\alpha_1,\dots,\alpha_k)$ belong to $\{-1,0,1\}$, then `HPL[`$\{\alpha_1,\dots,\alpha_k\}$`, z]` (in Maître's `HPL` package [**66**]) is equal to $(-1)^{n_1(\boldsymbol\alpha)}G(\boldsymbol\alpha;z)$, where $n_1(\boldsymbol\alpha):=\#\{j\in\mathbb{Z}\cap[1,k]|\alpha_j=1\}$ counts the number of 1's in the components of $\boldsymbol\alpha$.

[12] Prior to Au's analytic work on MPLs, the Henn–Smirnov–Smirnov database [**50**] provided empirical reduction formulae for CMZVs in $\mathfrak{Z}_k(6)$, $1\leqslant k\leqslant 6$.

[13] The filtration property of $\mathfrak{Z}(3)$ helps us to identify quickly certain terms as Goncharov–Deligne periods at third roots of unity. For example, the relation $\pi i\zeta_3\in\mathfrak{Z}_4(3)$ builds upon $\pi i=3[\mathrm{Li}_1(\omega)-\mathrm{Li}_1(\omega^2)]\in\mathfrak{Z}_1(3)$ [cf. (2.15)], $\zeta_3=\mathrm{Li}_3(1)\in\mathfrak{Z}_3(3)$, and $\mathfrak{Z}_1(3)\mathfrak{Z}_3(3)\subseteq\mathfrak{Z}_4(3)$.

PROOF. The integral representation in (2.47) is due to Borwein–Borwein–Straub–Wan [**9**, Theorem 17].

We reparametrize the integral over $\vartheta\in\left(0,\frac{\pi}{6}\right)$ with $\vartheta=\arctan\frac{1+t}{\sqrt{3}(1-t)}$, $t\in(-1,0)$, so that

$$\int_0^{\pi/6}\rho_k(2\sin\vartheta)\,\mathrm{d}\vartheta=-\frac{i}{2}\int_{-1}^{0}\left(\frac{1}{t-\varrho}-\frac{1}{t-\frac{1}{\varrho}}\right)\rho_k\left(\frac{1+t}{\sqrt{1-t+t^2}}\right)\mathrm{d}t. \tag{2.48}$$

By induction on the weight $w(\boldsymbol{A})$ and applications of the GPL recursion (2.22) to $z=\frac{(1+t)^2}{1-t+t^2}$, we obtain

$$\begin{aligned}&L_{\boldsymbol{A}}\left(\frac{(1+t)^2}{1-t+t^2}\right)\\&\quad\in\operatorname{span}_{\mathbb{Z}}\left\{(\pi i)^{\ell}G(\alpha_1,\dots,\alpha_j;t)G(\alpha_{j+1},\dots,\alpha_{w(\boldsymbol{A})-\ell};-1)\in\mathbb{C}\;\middle|\;\substack{\alpha_1,\dots,\alpha_{w(\boldsymbol{A})-\ell}\in\left\{0,-1,\varrho,\frac{1}{\varrho}\right\}\\0\leqslant\ell\leqslant w(\boldsymbol{A});0\leqslant j\leqslant w(\boldsymbol{A})-\ell}\right\}\end{aligned} \tag{2.49}$$

for $\varrho=e^{\pi i/3}$ and

$$\begin{aligned}&i\int_0^{\pi/6}\rho_k(2\sin\vartheta)\,\mathrm{d}\vartheta\\&\quad\in\operatorname{span}_{\mathbb{Q}}\left\{(\pi i)^{\ell}G(\alpha_1,\dots,\alpha_{k+1-\ell};-1)\in\mathbb{C}\;\middle|\;\substack{\alpha_1,\dots,\alpha_{k+1-\ell}\in\left\{0,-1,\varrho,\frac{1}{\varrho}\right\}\\0\leqslant\ell\leqslant k+1}\right\}\\&\quad\subseteq\operatorname{span}_{\mathbb{Q}}\left\{(\pi i)^{k+1-n}G(\alpha_1,\dots,\alpha_n;1)\;\middle|\;\substack{\alpha_1,\dots,\alpha_n\in\left\{0,1,-\varrho=\frac{1}{\varrho^2}=\omega^2,-\frac{1}{\varrho}=\varrho^2=\omega\right\}\\0\leqslant n\leqslant k+1;\alpha_1\neq1;\alpha_n\neq0}\right\}\subseteq\mathfrak{Z}_{k+1}(3)\end{aligned} \tag{2.50}$$

for $\omega=e^{2\pi i/3}$.

Arguing as in the last paragraph, we can check that

$$i\int_0^{\pi/6}\log^k(2\sin\vartheta)\,\mathrm{d}\vartheta\in\mathfrak{Z}_{k+1}(3). \tag{2.51}$$

Meanwhile, a constructive proof for the relation [**13**, §2.1]

$$\int_0^{\pi/2}\log^k(2\sin\vartheta)\,\mathrm{d}\vartheta\in\pi\mathfrak{Z}_k(1)\subseteq\pi\mathfrak{Z}_k(3) \tag{2.52}$$

is known. Combining the last two displayed equations with the fact that $\pi i\in\mathfrak{Z}_1(3)$ [see (2.15)], we get

$$i\int_{\pi/6}^{\pi/2}\log^k(2\sin\vartheta)\,\mathrm{d}\vartheta\in\mathfrak{Z}_{k+1}(3). \tag{2.53}$$

Each one of the remaining integrals $\int_{\pi/6}^{\pi/2}\rho_j\left(\frac{1}{2\sin\vartheta}\right)\log^{k-j}(2\sin\vartheta)\,\mathrm{d}\vartheta$ (where $2\leqslant j\leqslant k$) contributes a summand to the right-hand side of a vanishing identity

$$\begin{aligned}0=\lim_{\varepsilon\to0^+}&\left(\int_{i\varepsilon}^{1+i\varepsilon}+\oint_{1-\varepsilon}^{(1-\varepsilon)\varrho}+\oint_{\varrho}^{1/\varrho}\right.\\&\left.+\oint_{(1-\varepsilon)/\varrho}^{1-\varepsilon}+\int_{1-i\varepsilon}^{-i\varepsilon}\right)\rho_j\left(\sqrt{\frac{-z}{(1-z)^2}}\right)\left[\log\frac{(1-z)^2}{-z}\right]^{k-j}\frac{\mathrm{d}z}{iz},\end{aligned} \tag{2.54}$$

where the path of integration for each $\fint$ runs counterclockwise along a circular arc centered at the origin. Concretely speaking, we have

$$\begin{aligned}&\fint_{\varrho}^{1/\varrho}\rho_j\left(\sqrt{\frac{-z}{(1-z)^2}}\right)\left[\log\frac{(1-z)^2}{-z}\right]^{k-j}\frac{\mathrm{d}\,z}{iz}\\&=2^{k-j+2}\int_{\pi/6}^{\pi/2}\rho_j\left(\frac{1}{2\sin\vartheta}\right)\log^{k-j}(2\sin\vartheta)\,\mathrm{d}\vartheta,\end{aligned}\tag{2.55}$$

which is equal to the sum of

$$\begin{aligned}&-\lim_{\varepsilon\to0^+}\left(\int_{i\varepsilon}^{1+i\varepsilon}+\int_{1-i\varepsilon}^{-i\varepsilon}\right)\rho_j\left(\sqrt{\frac{-z}{(1-z)^2}}\right)\left[\log\frac{(1-z)^2}{-z}\right]^{k-j}\frac{\mathrm{d}\,z}{iz}\\&\in\operatorname{span}_{\mathbb{Q}}\left\{\int_0^1\rho_j\left(\sqrt{\frac{-z}{(1-z)^2}}\right)\left[\log\frac{(1-z)^2}{z}\right]^{k-j-n}(\pi i)^n\frac{\mathrm{d}\,z}{iz}\right|\left.0\leqslant n\leqslant k-j\vphantom{\int}\right\}\end{aligned}\tag{2.56}$$

and

$$\begin{aligned}&-\lim_{\varepsilon\to0^+}\left(\fint_{1-\varepsilon}^{(1-\varepsilon)\varrho}+\fint_{(1-\varepsilon)/\varrho}^{1-\varepsilon}\right)\rho_j\left(\sqrt{\frac{-z}{(1-z)^2}}\right)\left[\log\frac{(1-z)^2}{-z}\right]^{k-j}\frac{\mathrm{d}\,z}{iz}\\&\in i\operatorname{span}_{\mathbb{Q}}\left\{\fint_1^{\varrho}(\pi i)^nZ_\ell G\left(\alpha_1,\dots,\alpha_{j-\ell-n};\frac{(1-z)^2}{-z}\right)\right.\\&\qquad\times\left.\left[\log\frac{(1-z)^2}{-z}\right]^{k-j}\frac{\mathrm{d}\,z}{iz}\right|\left.\begin{smallmatrix}Z_\ell\in\mathfrak{Z}_\ell(1)\\ \alpha_1,\dots,\alpha_{j-\ell-n}\in\{0,1\}\\ 0\leqslant n\leqslant j;0\leqslant\ell\leqslant j-n\end{smallmatrix}\right\}\\&+i\operatorname{span}_{\mathbb{Q}}\left\{\fint_{1/\varrho}^{1}(\pi i)^nZ_\ell G\left(\alpha_1,\dots,\alpha_{j-\ell-n};\frac{(1-z)^2}{-z}\right)\right.\\&\qquad\times\left.\left[\log\frac{(1-z)^2}{-z}\right]^{k-j}\frac{\mathrm{d}\,z}{iz}\right|\left.\begin{smallmatrix}Z_\ell\in\mathfrak{Z}_\ell(1)\\ \alpha_1,\dots,\alpha_{j-\ell-n}\in\{0,1\}\\ 0\leqslant n\leqslant j;0\leqslant\ell\leqslant j-n\end{smallmatrix}\right\}.\end{aligned}\tag{2.57}$$

Here, the multi-dimensional polylogarithm behaves like $L_{\boldsymbol{A}}(\varXi\pm i0^+)\in\mathfrak{g}^{(1/\varXi)}_{w(\boldsymbol{A})}(1)+\pi i\mathfrak{g}^{(1/\varXi)}_{w(\boldsymbol{A})-1}(1)$, with a jump discontinuity [**67, 69**] $L_{\boldsymbol{A}}(\varXi+i0^+)-L_{\boldsymbol{A}}(\varXi-i0^+)\in\pi i\mathfrak{g}^{(1/\varXi)}_{w(\boldsymbol{A})-1}(1)$ [see (2.30b) for the definition of $\mathfrak{g}^{(z)}_k(N)$] across its branch cut, when

$\varXi > 1$. To handle the right-hand side of (2.56), we may switch back to the parameter $t = \frac{\varrho z-1}{z-\varrho}$, as follows:

(2.58)

$$\begin{aligned}
&\int_0^1 \rho_j\left(\sqrt{\frac{-z}{(1-z)^2}}\right)\left[\log\frac{(1-z)^2}{z}\right]^{k-j-n}(\pi i)^n\frac{\mathrm{d}\,z}{iz}\\
&= i\int_{-1}^{1/\varrho}\rho_j\left(\frac{\sqrt{1-t+t^2}}{1+t}\right)\left[\log\frac{-(1+t)^2}{1-t+t^2}\right]^{k-j-n}(\pi i)^n\left(\frac{1}{t-\varrho}-\frac{1}{t-\frac{1}{\varrho}}\right)\mathrm{d}t\\
&\in i\,\mathrm{span}_{\mathbb{Q}}\left\{(\pi i)^{k+1-\ell}G(\alpha_1,\dots,\alpha_j;u)G(\alpha_{j+1},\dots,\alpha_\ell;\varrho)\,\middle|\,\begin{smallmatrix}\alpha_1,\dots,\alpha_\ell\in\{0,-1,\varrho,\frac{1}{\varrho}\}\\ u\in\{-1,\frac{1}{\varrho}\}\\ 0\leqslant j\leqslant\ell\leqslant k+1\end{smallmatrix}\right\}\\
&\subseteq i\,\mathrm{span}_{\mathbb{Q}}\left\{(\pi i)^{k+1-\ell}G(\alpha_1,\dots,\alpha_\ell;1)\,\middle|\,\begin{smallmatrix}\alpha_1,\dots,\alpha_\ell\in\{0,1,\omega=-\frac{1}{\varrho}=\varrho^2,\omega^2=-\varrho=\frac{1}{\varrho^2}\}\\ 0\leqslant\ell\leqslant k+1;\alpha_1\neq1;\alpha_\ell\neq0\end{smallmatrix}\right\}\\
&\subseteq i\mathfrak{Z}_{k+1}(3).
\end{aligned}$$

Likewise, by integrating GPLs over $t = \frac{\varrho z-1}{z-\varrho} \in [-1, 0]$, one can identify the right-hand side of (2.57) with a subset of $i\mathfrak{Z}_{k+1}(3)$, as we did in (2.50).

Therefore, we have $\pi m_k(1 + x_1 + x_2) \in i\mathfrak{Z}_{k+1}(3)$ as claimed. □

REMARK 2.8. If we reparametrize all the integrals in (2.47) with $\vartheta = \arctan\frac{1+t}{\sqrt{3}(1-t)}$, $t \in (-1, 1)$, then we have

(2.47′)

$$\begin{aligned}
\mathscr{I}_k^{\mathrm{BBSW}} &:= \underset{\varrho}{\mathrm{fib}}\int_{-1}^{0}\left(\frac{1}{t-\varrho}-\frac{1}{t-\frac{1}{\varrho}}\right)\frac{\rho_k\left(\frac{1+t}{\sqrt{\mathsf{Q}(t)}}\right)}{\pi i}\,dt\\
&\quad+\sum_{j=0}^{k}\binom{k}{j}\underset{\varrho}{\mathrm{fib}}\int_0^1\left(\frac{1}{t-\varrho}-\frac{1}{t-\frac{1}{\varrho}}\right)\frac{\rho_j\left(\frac{\sqrt{\mathsf{Q}(t)}}{1+t}\right)}{2^{k-j}\pi i}\log^{k-j}\frac{(1+t)^2}{\mathsf{Q}(t)}\,dt,
\end{aligned}$$

where $\mathsf{Q}(t) := \left(1-\frac{t}{\varrho}\right)(1-t\varrho)$. Evaluating in `HyperInt` and `HPL`, one can check that $\mathrm{Lyn}\,\mathscr{I}_k^{\mathrm{BBSW}} = \mathrm{Lyn}\,\mathscr{I}_k^{\mathrm{HB}}$, $k \in \{1, 2, 3, 4\}$.

REMARK 2.9. Let

$$\mathrm{Ls}_k(\varphi) := -\int_0^{\varphi}\log^{k-1}\left|2\sin\frac{\vartheta}{2}\right|\mathrm{d}\vartheta \tag{2.59}$$

be Lewin's log-sine integral [**62**, **63**] for $k \in \mathbb{Z}_{>0}$. From Borwein–Straub [**12**, Theorem 3], one sees that $\mathrm{Ls}_k(2\pi/3) \in i\mathfrak{Z}_k(3)$. According to (2.51), we also have $\mathrm{Ls}_k(\pi/3) \in i\mathfrak{Z}_k(3)$, which is not immediate from the Borwein–Straub reduction [**12**, Theorem 3] of log-sine integrals. Borwein–Borwein–Straub–Wan [**9**, (108) and (109)] proposed conjectural expressions for $m_3(1 + x_1 + x_2)$ and $m_4(1 + x_1 + x_2)$, using log-sine integrals at $\pi/3$ and $2\pi/3$. These conjectures are equivalent to the corresponding formulae in Table 1, as one can check with Au's algorithm [**5**, §5].

REMARK 2.10. Let

$$\mathrm{Ls}_k^{(m)}(\varphi) := -\int_0^{\varphi}\vartheta^m\log^{k-1-m}\left|2\sin\frac{\vartheta}{2}\right|\mathrm{d}\vartheta \tag{2.60}$$

be Lewin's generalized log-sine integral [**63**, Appendix A.1(20)] for $k, m \in \mathbb{Z}_{>0}$. By the explicit construction of Borwein–Straub [**12**, Theorem 3], we have $\mathrm{Ls}_k^{(m)}(2\pi/3) \in$

$i^{m+1}\mathfrak{Z}_k(3)$. Meanwhile, we note that the proof of the theorem above can be readily adapted to

$$\begin{aligned}&\mathrm{Ls}_k^{(m)}\left(\frac{\pi}{3}\right)\\ &=-\frac{1}{2^{k-1-m}}\fint_1^{\varrho}\left(\frac{\log z}{i}\right)^m\left[\log\frac{(1-z)^2}{-z}\right]^{k-1-m}\frac{\mathrm{d}z}{iz}\\ &\in i^{m+1}\operatorname{span}_{\mathbb{Q}}\left\{(\pi i)^{k-\ell}G(\alpha_1,\dots,\alpha_\ell;u)\in\mathbb{C}\left|\,{\scriptstyle \alpha_1,\dots,\alpha_\ell\in\{0,1\};u\in\{1,\varrho\} \atop 0\leqslant\ell\leqslant k;\alpha_\ell\neq0}\right.\right\}\\ &\subseteq i^{m+1}\mathfrak{Z}_k(3).\end{aligned}\tag{2.61}$$

Here, the "$\subseteq$" step builds upon a more general result

$$\operatorname{span}_{\mathbb{Q}}\left\{G(\alpha_1,\dots,\alpha_k;z)\left|\,{\scriptstyle \alpha_1,\dots,\alpha_k,z\in\{0,1,\varrho,\frac{1}{\varrho}\} \atop \alpha_1\neq z;\alpha_k\neq0}\right.\right\}\subseteq\mathfrak{Z}_k(3).\tag{2.62}$$

To prove this, use the GPL recursion (2.22) (with an integration path connecting ω^2 and z) to show inductively that

$$\begin{aligned}&\operatorname{span}_{\mathbb{Q}}\left\{G(\alpha_1,\dots,\alpha_k;w)\left|\,{\scriptstyle \alpha_1,\dots,\alpha_k,w\in\left\{0,1,\frac{\omega^2-1}{z-1},\frac{z}{\omega^2}\frac{\omega^2-1}{z-1}\right\} \atop \alpha_1\neq1;\alpha_k\neq0}\right.\right\}\\ &\subseteq\operatorname{span}_{\mathbb{Q}}\left\{(\pi i)^{k-\ell}G(\alpha_1,\dots,\alpha_j;z)G(\alpha_{j+1},\dots,\alpha_\ell;\omega^2)\left|\,{\scriptstyle \alpha_1,\dots,\alpha_\ell\in\{0,1,\omega^2\} \atop 0\leqslant j\leqslant\ell\leqslant k}\right.\right\},\end{aligned}\tag{2.63}$$

before specializing to $z=\omega:=e^{2\pi i/3}$ (with logarithmic regularizations [**69**, §2.3] in case of $\alpha_1=z$ and $\alpha_{j+1}=\omega^2$).

2.4. Period structure of $m_k(1+x_1+x_2+x_3)$. With the $\mathbb{Q}$-vector space $\mathfrak{g}_k^{(z)}(2)$ introduced in (2.30b), we can extend our proof of Theorem 2.5 to the k-Mahler measures of $1+x_1+x_2+x_3$.

THEOREM 2.11 ($\pi^2 m_k(1+x_1+x_2+x_3)$ as Goncharov–Deligne periods). *For $n\in\mathbb{Z}_{\geqslant0}$, define*[14]

$$\mathsf{X}_4^{(n)}(\tau):=\fint_1^{\tau}\frac{[\log\mathsf{A}_4(t,\tau)+\log\mathsf{B}_4(\tau)]^n\,\mathrm{d}t}{t}-\pi i\delta_{n,0},\qquad\widetilde{\mathsf{X}}_4^{(n)}(\tau):=\fint_1^{\tau}\frac{\log^n\mathsf{A}_4(t,\tau)\,\mathrm{d}t}{t},\tag{2.64}$$

$$\Xi_4^{(n)}(\tau):=\fint_1^{-1}\frac{[\log\mathsf{A}_4(t,\tau)+\log\mathsf{B}_4(\tau)]^n-\delta_{n,0}}{t}\,\mathrm{d}t,\qquad\widetilde{\Xi}_4^{(n)}(\tau):=\fint_1^{-1}\frac{\log^n\mathsf{A}_4(t,\tau)}{t}\,\mathrm{d}t,\tag{2.65}$$

using the rational functions in (2.11). *For each positive integer k, we have*

$$\begin{aligned}&\pi^2 m_k(1+x_1+x_2+x_3)-\pi^2 m_k(1+x_1)\\ &\in\operatorname{span}_{\mathbb{Q}}\left\{\frac{1}{\pi^{-2n_1}}\fint_1^{-1}\widetilde{\mathsf{X}}_4^{(n_2)}(\tau)\mathsf{X}_4^{(n_3)}(\tau)\frac{\tau+1}{\tau-1}\frac{\mathrm{d}\tau}{\tau}\left|\,{\scriptstyle n_1,n_2,n_3\in\mathbb{Z}_{\geqslant0} \atop (2n_1+1)+n_2+n_3=k}\right.\right\}\\ &\quad+\operatorname{span}_{\mathbb{Q}}\left\{\frac{1}{\pi^{-2n_1}}\int_{-1}^{0}\widetilde{\Xi}_4^{(n_2)}(\tau)\Xi_4^{(n_3)}(\tau)\frac{\tau+1}{\tau-1}\frac{\mathrm{d}\tau}{\tau}\left|\,{\scriptstyle n_1,n_2,n_3\in\mathbb{Z}_{\geqslant0} \atop (2n_1+1)+n_2+n_3=k}\right.\right\}\\ &\subseteq\mathfrak{Z}_{k+2}(2),\end{aligned}\tag{2.66}$$

which also implies

$$\pi^2 m_k(1+x_1+x_2+x_3)\in\mathfrak{Z}_{k+2}(2).\tag{2.67}$$

[14] We set the Kronecker delta as $\delta_{a,b}=1$ when $a=b$, and $\delta_{a,b}=0$ otherwise.

PROOF. As we may recall, Borwein–Straub [13, (2.4), (2.5), (4.3)] devised a recursion for $m_k(1+x_1), k\in\mathbb{Z}_{>0}$ and showed that

$$\begin{aligned}(2.68)\qquad m_k(1+x_1)&\in\operatorname{span}_{\mathbb{Q}}\left\{\prod_{j=1}^{n}\zeta(a_j)\,\middle|\,a_j\in\mathbb{Z}_{>1},1\leqslant j\leqslant n;\sum_{j=1}^{n}a_j=k\right\}\\&\subseteq\mathfrak{Z}_k(1)\subseteq\mathfrak{Z}_k(2),\end{aligned}$$

which simplified an earlier proof of the same statement by Kurokawa–Lalín–Ochiai [58, Proposition 4]. Thus, it indeed suffices to work out (2.66) for our goal in (2.67).

Again, owing to the filtration property (2.16) and the fact that

$$\pi^2=6\operatorname{Li}_2(1)\in\mathfrak{Z}_2(2),\tag{2.69}$$

we may reduce our task to the demonstration of

(2.70)

$$\left\{⨍_1^{-1}\widetilde{\mathsf{X}}_4^{(n_2)}(\tau)\mathsf{X}_4^{(n_3)}(\tau)\frac{\tau+1}{\tau-1}\frac{d\tau}{\tau},\quad\int_{-1}^{0}\widetilde{\Xi}_4^{(n_2)}(\tau)\Xi_4^{(n_3)}(\tau)\frac{\tau+1}{\tau-1}\frac{d\tau}{\tau}\right\}\subset\mathfrak{Z}_{k+2}(2)$$

for $n_2,n_3\in\mathbb{Z}_{\geqslant0}$ and $n_2+n_3=k-1$.

With

$$\log\mathsf{A}_4(t,\tau):=\log\frac{\left(1-\frac{t}{\tau}\right)(1-t\tau)}{t}=G(\tau;t)+G\left(\frac{1}{\tau};t\right)-G(0;t)\tag{2.71}$$

in hand, we have the following result for $k\in\mathbb{Z}_{>0}$:

$$\begin{aligned}(2.72)\qquad\widetilde{\mathsf{X}}_4^{(k)}(\tau):=&⨍_1^{\tau}\frac{dt}{t}\log^k\frac{\left(1-\frac{t}{\tau}\right)(1-t\tau)}{t}\\&\in\operatorname{span}_{\mathbb{Q}}\left\{G(0,\alpha_2,\dots,\alpha_{k+1};z)\,\middle|\,{}^{\alpha_2,\dots,\alpha_{k+1}\in\{\tau,\frac{1}{\tau},0\}}_{z\in\{1,\tau\}}\right\}\\&\subseteq\mathfrak{g}_{k+1}^{(\tau)}(2).\end{aligned}$$

Here, the rationales behind both the "$\in$" step and the "$\subseteq$" step are identical to what we provided for (2.35). For the "$\subseteq$" step, the fibration output resides in the $\mathbb{Z}$-module

$$\operatorname{span}_{\mathbb{Z}}\left\{(\pi i)^{\ell}G(\beta_1,\dots,\beta_j;\tau)Z_m\in\mathbb{C}\,\middle|\,{}^{\beta_1,\dots,\beta_j\in\{-1,0,1\};Z_m\in\mathfrak{Z}_m(1)}_{j,\ell,m\in\mathbb{Z}_{\geqslant0};j+\ell+m=k+1}\right\},\tag{2.73}$$

as we inductively apply the GPL recursion (2.22) (with $z=\tau$ and $z=1$) to the GPL differential form (2.37) $\left[\text{where } d\log(\alpha_{j-1}-\alpha_j), d\log(\alpha_{j+1}-\alpha_j)\in\left\{d\log\tau,\right.\right.$ $\left.\left.d\log(\tau-1), d\log\left(\frac{1}{\tau}-1\right), d\log\left(\tau-\frac{1}{\tau}\right),0\right\}\right]$. Taking into account

$$\log\mathsf{B}_4(\tau):=\log\frac{-\tau}{(1-\tau)^2}\equiv\pi i+G(0;\tau)-2G(1;\tau)\pmod{2\pi i\mathbb{Z}},\tag{2.74}$$

we may check that $\mathsf{X}_4^{(k)}(\tau)\in\pi i\log^k\mathsf{B}_4(\tau)+\operatorname{span}_{\mathbb{Q}}\{\widetilde{\mathsf{X}}_4^{(j)}(\tau)\log^{k-j}\mathsf{B}_4(\tau)\mid 0\leqslant j\leqslant k\}\subseteq\mathfrak{g}_{k+1}^{(\tau)}(2)$.

Likewise, one can verify that $\{\widetilde{\Xi}_4^{(k)}(\tau),\Xi_4^{(k)}(\tau)\}\subset i\mathfrak{g}_{k+1}^{(\tau)}(2)$ holds for all $k\in\mathbb{Z}_{\geqslant0}$.

Thus far, we see that

$$\widetilde{\mathsf{X}}_4^{(n_2)}(\tau)\mathsf{X}_4^{(n_3)}(\tau)\in\mathfrak{g}_{n_2+n_3+2}^{(\tau)}(2),\tag{2.75}$$

$$\widetilde{\Xi}_4^{(n_2)}(\tau)\Xi_4^{(n_3)}(\tau)\in\mathfrak{g}_{n_2+n_3+2}^{(\tau)}(2).\tag{2.76}$$

As we expand

$$\frac{\tau+1}{\tau-1}\frac{1}{\tau}=\frac{2}{\tau-1}-\frac{1}{\tau} \tag{2.77}$$

and invoke the closure property (2.31) for integrations over τ, we may deduce $\pi^2 m_k(1+x_1+x_2+x_3)\in\mathfrak{g}^{(1)}_{k+2}(2)\cap\mathbb{R}=\mathfrak{Z}_{k+2}(2)$. $\square$

Thanks to the data mine of Blümlein–Broadhurst–Vermaseren [**7**], we can reduce every AMZV (up to weight 8) to a polynomial in $\mathbb{Q}[\log 2,\pi^2,\zeta_3,\zeta_{-3,1},\zeta_5,\zeta_{-3,1,1},\dots]$, using a look-up table that ships with `HyperInt`. This leads us to the evaluations in Table 2.

All the entries in the aforementioned look-up table for $\mathfrak{Z}_k(2), 1\leqslant k\leqslant 8$ are analytically provable by Deligne's standard relations [**93**, Definition 1.1] for $\mathbb{Q}$-linearly dependent multiple polylogarithms.[15]

Before closing this section, we compare our results to some related works on (hyper-)Mahler measures.

In their foundational paper [**58**, Theorem 3], Kurokawa, Lalín, and Ochiai established a remarkable connection between the $\mathbb{Q}$-vector space $\mathfrak{Z}_k(1)$ incorporating MZVs of weight k and the hyper-Mahler measure $m_k(1+x_1)$ for $k\in\mathbb{Z}_{>1}$:

$$m_k(1+x_1)=(-1)^k k!\sum_{\substack{a_1+\dots+a_n=k\\ a_j\geqslant 2,1\leqslant j\leqslant n}}\frac{\zeta_{a_1,\dots,a_n}}{2^{2n}}\in\mathfrak{Z}_k(1). \tag{2.78}$$

The same extends to the $k=1$ case, where $m_1(1+x_1)=0$ and an empty sum over $\mathfrak{Z}_1=\{0\}$ must vanish.

Using Lalín's method [**59**], Kurokawa–Lalín–Ochiai originally demonstrated (2.78) by writing the hyper-Mahler measures $m_k(1+x_1)$ as analogs of the Drinfel'd integral representation of MZVs (see [**40**, §2] or [**90**, §9]), the latter of which are special cases of Goncharov–Manin periods [**49**] on $\mathfrak{M}_{0,n}$.

Our Theorems 2.7 and 2.11 extend the Kurokawa–Lalín–Ochiai sum rule (2.78) to Goncharov–Deligne periods in $\mathfrak{Z}_{k+1}(3)$ and $\mathfrak{Z}_{k+2}(2)$ for $k\in\mathbb{Z}_{>0}$:

$$\pi i\, m_k(1+x_1+x_2)\in\mathfrak{Z}_{k+1}(3), \tag{2.79}$$

$$(\pi i)^2 m_k(1+x_1+x_2+x_3)\in\mathfrak{Z}_{k+2}(2). \tag{2.80}$$

If we set $m_0(1+x_1+x_2)=m_0(1+x_1+x_2+x_3)=1$, then the same is true for $k=0$.

The hyper-Mahler measures for the multivariate family $R_m(x_1,\dots,x_m,z)=z+\prod_{j=1}^m\frac{1-x_j}{1+x_j}$ are expressible as $\mathbb{Q}$-linear combinations of multiple polylogarithms whose arguments are fourth roots of unity, as established by Lalín–Lechasseur [**60**, Theorem 1.1]. The results of Lalín–Lechasseur hinged on a representation of $m_k(a+x_1)$ via multi-dimensional polylogarithms of $|a|^{-2\operatorname{sgn}\log|a|}$ [**60**, Theorem 4.5] (extending a result of Akatsuka [**1**, Theorem 7], in a similar spirit to Borwein–Borwein–Straub–Wan [**9**, Theorem 26]), and subsequent computations of certain motivic periods on $\mathfrak{M}_{0,n}$ [**60**, §§7–8].

---

[15] Some provable non-standard $\mathbb{Q}$-linear relations in $\mathfrak{Z}(N)$ are covered by Au's method [**5**, §5.1], where $N\in\{6,8,10,12\}$.

We are not sure whether there are systematic classifications of all the hyper-Mahler measures $m_k(1+\sum_{j=1}^{\ell}x_j)$, where k and ℓ are positive integers. The conjectural characterizations of $m_1(1+\sum_{j=1}^{4}x_j)$ and $m_1(1+\sum_{j=1}^{5}x_j)$ by Rodríguez-Villegas [**18**, §8] seem to suggest that the moduli spaces $\mathfrak{M}_{g,n}$ for $g>0$ are needed (probably by the routes of elliptic polylogarithms [**6, 72, 73**] and elliptic multiple zeta values [**24, 91, 92**]) to quantify these logarithmic Mahler measures for polynomials in 4 and 5 variables.

At present, we do not have natural generalizations of Proposition 2.3 to

$$\int_0^\infty J_{1+\frac{s}{2}}\left(\sqrt{u}x\right)\frac{J_0^3(x)}{x^{s/2}}\,\mathrm{d}x \quad\text{or}\quad \int_0^\infty J_{1+\frac{s}{2}}\left(\sqrt{u}x\right)\frac{J_0^4(x)}{x^{s/2}}\,\mathrm{d}x \tag{2.81}$$

when s is generic. On a different note, for $s=-2$, Kluyver's random walk integrals [**56**]

$$\int_0^\infty J_0\left(\sqrt{u}x\right)J_0^3(x)x\,\mathrm{d}x \quad\text{and}\quad \int_0^\infty J_0\left(\sqrt{u}x\right)J_0^4(x)x\,\mathrm{d}x \tag{2.82}$$

have been scrutinized by Borwein and coworkers [**11, 14, 15**]. A potential link from these random walk integrals to the two conjectures of Rodríguez-Villegas [**18**, §8] has been explored by Straub–Zudilin [**75**, §§5–7] and Zudilin [**28**, Chapter 6].

3. Some infinite series behind hyper-Mahler measures

3.1. Non-linear Euler sums revisited. We open this section with a non-exhaustive overview of some infinite sums that revolve around the harmonic numbers of order r

$$\mathsf{H}_n^{(r)}:=\sum_{k=1}^\infty\left[\frac{1}{k^r}-\frac{1}{(k+n)^r}\right],\qquad n\in\mathbb{C}\smallsetminus\mathbb{Z}_{<0}, \tag{3.1}$$

which become $\mathsf{H}_n^{(r)}=\sum_{k=1}^n\frac{1}{k^r}$ for $n\in\mathbb{Z}_{>0}$. Aside from their connections to the hyper-Mahler measures (see §§3.2–3.3 for details) investigated in the current work, these infinite series involving harmonic numbers play significant rôles in the quantitative understanding of Feynman diagrams in quantum electrodynamics and quantum chromodynamics [**7, 20, 34, 35, 54**].

Studies of "linear Euler sums" like

$$\sum_{n=1}^\infty\frac{\mathsf{H}_n^{(r)}}{n^s}=\zeta_{s,r}+\zeta_{r+s},\qquad r\in\mathbb{Z}_{>0},\quad s\in\mathbb{Z}_{>1} \tag{3.2}$$

were initiated by Euler [**42**], in response to Goldbach's challenge in 1742 [**17**, §1]. Borwein–Zucker–Boersma [**17**] generalized the Euler–Goldbach double sums to $[p,q](s,t):=\sum_{m>n>0}\frac{\chi_p(m)\chi_q(n)}{m^sn^t}$ for Dirichlet characters p and q. This allows one to convert infinite sums like [**17**, Table 1]

$$\sum_{n=1}^\infty\frac{\mathsf{H}_n^{(r)}}{n^s},\quad\sum_{n=1}^\infty\frac{(-1)^n\mathsf{H}_n^{(r)}}{n^s},\quad\sum_{n=1}^\infty\frac{\mathsf{H}_n^{(r)}}{(2n-1)^s},\quad\sum_{n=1}^\infty\frac{(-1)^n\mathsf{H}_n^{(r)}}{(2n-1)^s} \tag{3.3}$$

into a single integral over polylogarithmic expressions [**17**, Table 2]. Although predating Brown's work [**25–27**], the integral representations of Borwein–Zucker–Boersma [**17**, Table 2] belong to Brown's $L(\mathfrak{M}_{0,n})$ class. A linear Euler sum $\zeta_{-3,1}$ in its polylogarithmic avatar $\zeta_{-3,1}=2\operatorname{Li}_4\left(\frac12\right)-\frac{\pi^4}{48}+\frac{7\zeta_3\log2}{4}-\frac{\pi^2\log^22}{12}+\frac{\log^42}{12}$

[**17**, (5.5)] featured prominently in the Laporta–Remiddi formula [**61**, (5)] for the 3-loop contribution to the magnetic moment of the electron.

The "non-linear Euler sums"

$$\sum_{n=1}^{\infty}\frac{\left[\mathsf{H}_n^{(1)}\right]^2}{n^2},\quad \sum_{n=1}^{\infty}\frac{(-1)^n\left[\mathsf{H}_n^{(1)}\right]^2}{n^2},\quad \sum_{n=1}^{\infty}\frac{\left[\mathsf{H}_n^{(2)}\right]^2}{n^2}\tag{3.4}$$

had been considered by Borwein–Borwein [**8**], De Doelder [**36**], and Mező [**68**] before Panzer's `HyperInt` package [**69**] became available. In retrospect, these authors effectively converted their series of interest into members of $L(\mathfrak{M}_{0,n})$ and referred to the classical treatise on polylogarithms [**63**] for integral evaluations, without regard to the algebraic geometry of the moduli spaces $\mathfrak{M}_{0,n}$. Like their linear brethren, the non-linear Euler sums found their ways into perturbative expansions in high energy physics. While considering on-shell charge renormalization, Broadhurst [**20**, (42)] differentiated a certain hypergeometric expression and encountered non-linear Euler sums in the form of

$$\sum_{n=1}^{\infty}\frac{1}{(2n-1)^a}\prod_{i=1}^{k-1}\sum_{\ell_i=1}^{n-1}\frac{1}{(2\ell_i-1)^{b_i}}.\tag{3.5}$$

More sophisticated analogs of such infinite series showed up in the ε-expansions of Davydychev–Kalmykov [**34**, (1.11)], Kalmykov–Veretin [**54**, (4)], and Kalmykov–Ward–Yost [**55**, Theorems A and B].

For the prototypical non-linear Euler sums like those in (3.4), we have the following characterization of their period structure, which in turn, generalizes the Xu–Wang algorithm [**88**].

THEOREM 3.1 (Non-linear Euler sums, HPLs, and AMZVs).

(a) *For $|z|<1$ and $r_1,\dots,r_M\in\mathbb{Z}_{>0}$, we have*

$$\sum_{n=1}^{\infty}\left[\prod_{j=1}^{M}\mathsf{H}_n^{(r_j)}\right]z^n\in\frac{1}{1-z}\operatorname{span}_{\mathbb{Q}}\left\{L_{\boldsymbol{A}}(z)\,\middle|\,w(\boldsymbol{A})=\sum_{j=1}^{M}r_j\right\}\equiv\frac{\mathfrak{G}^{(z)}_{\sum_{j=1}^{M}r_j}(1)}{1-z},\tag{3.6}$$

where the multi-dimensional polylogarithms $L_{\boldsymbol{A}}(z)$ are defined by (2.45) *and the $\mathbb{Q}$-vector space $\mathfrak{G}_k^{(z)}(1)$ is given in* (2.24)*. By extension, we have* (*alternating*) *MZV representations* (*cf.* [**88**, Theorems 2.2 and 4.2])

$$\sum_{n=1}^{\infty}\left[\prod_{j=1}^{M}\mathsf{H}_n^{(r_j)}\right]\frac{1}{n^{s+1}}\in\mathfrak{Z}_{k+1}(1),\tag{3.7}$$

$$\sum_{n=1}^{\infty}\left[\prod_{j=1}^{M}\mathsf{H}_n^{(r_j)}\right]\frac{(-1)^n}{n^{s}}\in\mathfrak{Z}_{k}(2),\tag{3.8}$$

where $s\in\mathbb{Z}_{>0}$ and $k=s+\sum_{j=1}^{M}r_j$.

(b) *Define*

$$\overline{\mathsf{H}}_n^{(r)}:=\sum_{k=1}^{n}\frac{(-1)^{k-1}}{k^r},\qquad n\in\mathbb{Z}_{>0}.\tag{3.9}$$

For $|z|<1$, $r_1,\dots,r_M\in\mathbb{Z}_{>0}$, *and* $\overline{r}_1,\dots,\overline{r}_{\overline{M}}\in\mathbb{Z}_{>0}$, *we have*

$$\sum_{n=1}^{\infty}\left[\prod_{j=1}^{M}\mathsf{H}_n^{(r_j)}\right]\left[\prod_{\overline{j}=1}^{\overline{M}}\overline{\mathsf{H}}_n^{(\overline{r}_{\overline{j}})}\right]z^n\in\frac{\mathfrak{G}^{(z)}_{\sum_{j=1}^{M}r_j+\sum_{\overline{j}=1}^{\overline{M}}\overline{r}_{\overline{j}}}(2)}{1-z},\tag{3.10}$$

where $\mathfrak{G}_k^{(z)}(2)$ [*defined in* (2.24)] *is a subspace of* $\mathfrak{H}_k^{(z)}(2)$ [*defined in* (2.25)], *the latter being the* $\mathbb{Q}$*-vector space spanned by all the Remiddi–Vermaseren harmonic polylogarithms* (*HPLs*) [**71**] *of weight* k. *By extension, we have AMZV representations* (*cf.* [**88**, Theorem 4.2])

$$\sum_{n=1}^{\infty}\left[\prod_{j=1}^{M}\mathsf{H}_n^{(r_j)}\right]\left[\prod_{\overline{j}=1}^{\overline{M}}\overline{\mathsf{H}}_n^{(\overline{r}_{\overline{j}})}\right]\frac{1}{n^{s+1}}\in\mathfrak{Z}_{k+1}(2),\tag{3.11}$$

$$\sum_{n=1}^{\infty}\left[\prod_{j=1}^{M}\mathsf{H}_n^{(r_j)}\right]\left[\prod_{\overline{j}=1}^{\overline{M}}\overline{\mathsf{H}}_n^{(\overline{r}_{\overline{j}})}\right]\frac{(-1)^n}{n^{s}}\in\mathfrak{Z}_{k}(2),\tag{3.12}$$

where $s\in\mathbb{Z}_{>0}$ *and* $k=s+\sum_{j=1}^{M}r_j+\sum_{\overline{j}=1}^{\overline{M}}\overline{r}_{\overline{j}}$.

PROOF. (a) First, we note that [**68**, §5.1]

$$\sum_{n=1}^{\infty}\mathsf{H}_n^{(r)}z^n=\frac{\operatorname{Li}_r(z)}{1-z}\tag{3.13}$$

fits into (3.6) when $M=1$. Suppose that (3.6) holds for positive integers M up to a certain $M_0\in\mathbb{Z}_{>1}$. In the next paragraph, we show that the same will apply to the M_0+1 case.

According to our induction hypothesis, we have the following relation for $1\leqslant M\leqslant M_0$:

$$\begin{aligned}\sum_{n=1}^{\infty}\left[\prod_{j=1}^{M}\mathsf{H}_n^{(r_j)}\right]\frac{z^n}{n}&\in\operatorname{span}_{\mathbb{Q}}\left\{\int_0^z\frac{L_{\boldsymbol{A}}(x)\,\mathrm{d}x}{x(1-x)}\,\middle|\,w(\boldsymbol{A})=\sum_{j=1}^{M}r_j\right\}\\&\subseteq\operatorname{span}_{\mathbb{Q}}\left\{L_{\boldsymbol{A}}(z)\,\middle|\,w(\boldsymbol{A})=1+\sum_{j=1}^{M}r_j\right\},\end{aligned}\tag{3.14}$$

where the "$\subseteq$" step involves the standard recursion for GPL [see (2.22)] and the conversions between MPL and GPL [see (2.28)–(2.29)]. Since

$$\int_0^z\frac{L_{a_1,a_2,\dots,a_n}(x)}{x}\,\mathrm{d}x=L_{a_1+1,a_2,\dots,a_n}(z),\tag{3.15}$$

it also follows that

$$\sum_{n=1}^{\infty}\left[\prod_{j=1}^{M}\mathsf{H}_n^{(r_j)}\right]\frac{z^n}{n^s}\in\operatorname{span}_{\mathbb{Q}}\left\{L_{\boldsymbol{A}}(z)\,\middle|\,w(\boldsymbol{A})=s+\sum_{j=1}^{M}r_j\right\}\tag{3.16}$$

for $s\in\mathbb{Z}_{>0}$ and $1\leqslant M\leqslant M_0$. As we note that $\mathsf{H}_0^{(r)}=0$ and $\mathsf{H}_n^{(r)}-\mathsf{H}_{n-1}^{(r)}=\frac{1}{n^r}$, we may deduce the following result for $k=\sum_{j=1}^{M_0+1}r_j$:

$$\begin{aligned}(1-z)\sum_{n=1}^\infty\left[\prod_{j=1}^{M_0+1}\mathsf{H}_n^{(r_j)}\right]z^n&=\sum_{n=1}^\infty\left[\prod_{j=1}^{M_0+1}\mathsf{H}_n^{(r_j)}-\prod_{j=1}^{M_0+1}\mathsf{H}_{n-1}^{(r_j)}\right]z^n\\ &\in\mathbb{Q}\operatorname{Li}_k(z)+\operatorname{span}_{\mathbb{Q}}\left\{\sum_{n=1}^\infty\left[\prod_{j=1}^{M'}\mathsf{H}_n^{(r'_j)}\right]\frac{z^n}{n^s}\left|\begin{smallmatrix}k=s+\sum_{j=1}^{M'}r_j\\ s\in\mathbb{Z}_{>0}\\ 1\leqslant M'\leqslant M_0\end{smallmatrix}\right.\right\}\\ &\subseteq\operatorname{span}_{\mathbb{Q}}\left\{L_{\boldsymbol{A}}(z)\,\middle|\,w(\boldsymbol{A})=k\right\},\end{aligned}\tag{3.17}$$

where the "$\subseteq$" step follows from induction.

Therefore, the relation in (3.6) holds for all positive integers M. As a consequence, we may also extend (3.16) to all $M\in\mathbb{Z}_{>0}$.

As we closely examine the derivation of (3.16), we see that on its right-hand side, the first component of the vector $\boldsymbol{A}$ is at least s. This implies that the left-hand side of (3.16) converges at the point $z=1$ (resp. $z=-1$) when $s\in\mathbb{Z}_{>1}$ (resp. $s\in\mathbb{Z}_{>0}$). Specializing (3.16) to the $z=1$ (resp. $z=-1$) scenario, we arrive at (3.7) [resp. (3.8)].

(b) Direct computations reveal that

$$\sum_{n=1}^\infty\overline{\mathsf{H}}_n^{(\overline{r})}z^n=-\frac{\operatorname{Li}_{\overline{r}}(-z)}{1-z}\tag{3.18}$$

holds for $|z|<1$, where

$$-\operatorname{Li}_{\overline{r}}(-z)=G(\underbrace{0,\dots,0}_{\overline{r}-1},1;-z)=G(\underbrace{0,\dots,0}_{\overline{r}-1},-1;z)\in\mathfrak{G}_{\overline{r}}^{(z)}(2).\tag{3.19}$$

Here, in the penultimate step, we have resorted to the scaling property [**44**, (2.3)] for GPLs: $G(a_1,a_2,\dots,a_k;-z)=G(-a_1,-a_2,\dots,-a_k;z)$ if $a_k\neq0$. This scaling property also entails the relation $\mathfrak{G}_k^{(z)}(2)=\mathfrak{G}_k^{(-z)}(2)$ for all $k\in\mathbb{Z}_{>0}$.

After this, repeated invocations of the GPL recursion in (2.22) take us to

$$\sum_{n=1}^\infty\overline{\mathsf{H}}_n^{(\overline{r})}\frac{z^n}{n^s}\in\mathfrak{G}_{\overline{r}+s}^{(z)}(2),\tag{3.20}$$

so it follows that

$$(1-z)\sum_{n=1}^\infty\mathsf{H}_n^{(r)}\overline{\mathsf{H}}_n^{(\overline{r})}z^n=\sum_{n=1}^\infty\overline{\mathsf{H}}_n^{(\overline{r})}\frac{z^n}{n^r}-\sum_{n=1}^\infty\mathsf{H}_n^{(r)}\frac{(-z)^n}{n^{\overline{r}}}\in\mathfrak{G}_{r+\overline{r}}^{(z)}(2),\tag{3.21}$$

which confirms (3.10) for $M=\overline{M}=1$.

Subsequently, we can prove (3.10) by inductions on M and $\overline{M}$, as we did for part (a).

Like what we have encountered before, the relation in (3.10) lifts to

$$\sum_{n=1}^\infty\left[\prod_{j=1}^{M}\mathsf{H}_n^{(r_j)}\right]\left[\prod_{\overline{j}=1}^{\overline{M}}\overline{\mathsf{H}}_n^{(\overline{r}_{\overline{j}})}\right]\frac{z^n}{n^s}\in\mathfrak{G}_w^{(z)}(2)\tag{3.22}$$

for $s\in\mathbb{Z}_{>0}$ and $w=s+\sum_{j=1}^{M}r_j+\sum_{\overline{j}=1}^{\overline{M}}r_{\overline{j}}$. Here, if $s=1$, then the left-hand side of (3.22) is a $\mathbb{Q}$-linear combination of functions of the form

$$\int_0^z\frac{G(\alpha_1,\dots,\alpha_k;x)}{x(1-x)}\,\mathrm{d}x=G(0,\alpha_1,\dots,\alpha_k;z)-G(1,\alpha_1,\dots,\alpha_k;z),\tag{3.23}$$

where $\alpha_1,\dots,\alpha_k\in\{-1,0,1\}$, $k=\sum_{j=1}^{M}r_j+\sum_{\overline{j}=1}^{\overline{M}}r_{\overline{j}}$, so that it converges when $z=-1$; if $s\in\mathbb{Z}_{>1}$, then the left-hand side of (3.22) is a $\mathbb{Q}$-linear combination of functions of the form

$$\int_0^z\frac{G(\alpha_1,\dots,\alpha_k;x)}{x}\,\mathrm{d}x=G(0,\alpha_1,\dots,\alpha_k;z),\tag{3.24}$$

where $\alpha_1,\dots,\alpha_k\in\{-1,0,1\}$, $k=s-1+\sum_{j=1}^{M}r_j+\sum_{\overline{j}=1}^{\overline{M}}r_{\overline{j}}$, so that it converges when $z=\pm1$. Therefore, special cases of (3.22) lead us to (3.11) and (3.12). □

REMARK 3.2. In the statements of the theorem above and its corollaries in two subsections to follow, we only pinpoint the $\mathbb{Q}$-vector spaces in which the infinite sums reside. The accompanying proofs contain effective algorithms to evaluate each individual series explicitly.

For example, in Table 3, we compute a few generating functions by the procedures in the proof of the theorem above, and convert some of them to classical polylogarithms via the Frellesvig–Tommasini–Wever method [**44**]. Here, the generating functions $\mathsf{G}^{(1),2}(x)$ and $\mathsf{G}^{(1),3}(x)$ have been reported by Mező [**68**, Theorem 9] in the form of classical polylogarithms. Our induction step (3.17) is inspired by Mező's derivation of $\mathsf{G}^{(1),2}(x)$ [**68**, (24)].

For $r,\ell\in\mathbb{Z}_{>0}$ and $|z|<1$, the formula $\sum_{n=1}^\infty\mathsf{H}_n^{(r)}\frac{z^n}{n^\ell}=\operatorname{Li}_{\ell+r}(z)+\operatorname{Li}_{\ell,r}(z,1)\in\mathfrak{G}_{\ell+r}^{(z)}(1)$ in Table 3 builds inductively on GPL recursions, as in our proof of (3.16). [As pointed out by an anonymous referee, one can confirm the same by writing $\mathsf{H}_n^{(r)}=\frac{1}{n^r}+\sum_{0<m<n}\frac{1}{m^r}$ and appealing to the definition of MPLs in (1.4).] Likewise, for r, ℓ, and z in the same ranges, one can prove $\sum_{n=1}^\infty\overline{\mathsf{H}}_n^{(r)}\frac{z^n}{n^\ell}=-\operatorname{Li}_{\ell+r}(-z)-\operatorname{Li}_{\ell,r}(z,-1)\in\mathfrak{G}_{\ell+r}^{(z)}(2)$ recursively [or through an application of $\overline{\mathsf{H}}_n^{(r)}=\frac{(-1)^{n-1}}{n^r}+\sum_{0<m<n}\frac{(-1)^{m-1}}{m^r}$ to (1.4)].

REMARK 3.3. From the generating functions given by the theorem above, one can construct moment sequences that express harmonic numbers. For instance, if we pick a generating function

$$\mathsf{G}_s^{(r_1,\dots,r_M)}(z):=\sum_{n=1}^\infty\left[\prod_{j=1}^M\mathsf{H}_n^{(r_j)}\right]\frac{z^n}{n^s}\in\mathfrak{G}^{(z)}_{s+\sum_{j=1}^M r_j}(1)\tag{3.25}$$

for certain positive integers $r_1,\dots,r_M$ and s, then we can extract its Taylor coefficients by residue calculus:

$$\left[\prod_{j=1}^M\mathsf{H}_n^{(r_j)}\right]\frac{1}{n^s}=\oint_{|z|=1}\frac{\mathsf{G}_s^{(r_1,\dots,r_M)}(z)}{z^n}\frac{\mathrm{d}z}{2\pi iz}=\oint_{|w|=1}\frac{\mathsf{G}_s^{(r_1,\dots,r_M)}(1/w)}{(1/w)^n}\frac{\mathrm{d}w}{2\pi iw},\tag{3.26}$$

where $n\in\mathbb{Z}_{>0}$. In view of the jump discontinuities [**67**, **69**] $\mathsf{G}_s^{(r_1,\dots,r_M)}(1/w+i0^+)-\mathsf{G}_s^{(r_1,\dots,r_M)}(1/w-i0^+)\in\pi i\mathfrak{g}^{(w)}_{s-1+\sum_{j=1}^M r_j}(1)$ [cf. (2.30b)] for $0<w<1$, we can

TABLE 3. Some generating functions of Mező type.

r	$\mathsf{G}^{(r),m}(z):=\sum_{n=1}^{\infty}\left[\mathsf{H}_n^{(r)}\right]^m z^n,\quad 0<\lvert z\rvert<1$
1	$\mathsf{G}^{(1),1}(z)=\frac{\mathrm{Li}_1(z)}{1-z}$
	$\mathsf{G}^{(1),2}(z)=\frac{\mathrm{Li}_2(z)+2L_{1,1}(z)}{1-z}=\frac{[\mathrm{Li}_1(z)]^2+\mathrm{Li}_2(z)}{1-z}$
	$\mathsf{G}^{(1),3}(z)=\frac{\mathrm{Li}_3(z)+3L_{2,1}(z)+3L_{1,2}(z)+6L_{1,1,1}(z)}{1-z}$
	$=\frac{\frac{\pi^2}{2}\mathrm{Li}_1(z)+[\mathrm{Li}_1(z)]^3+\frac{3}{2}[\mathrm{Li}_1(z)]^2\log z+3\,\mathrm{Li}_3(1-z)+\mathrm{Li}_3(z)-3\zeta_3}{1-z}$
	$\mathsf{G}^{(1),4}(z)=\frac{\mathrm{Li}_4(z)+4L_{3,1}(z)+6L_{2,2}(z)+4L_{1,3}(z)}{1-z}$
	$+\frac{12L_{2,1,1}(z)+12L_{1,2,1}(z)+12L_{1,1,2}(z)+24L_{1,1,1,1}(z)}{1-z}$
	$=\frac{-4\zeta_3\,\mathrm{Li}_1(z)+\frac{5\pi^2}{3}[\mathrm{Li}_1(z)]^2-\frac{2}{3}[\mathrm{Li}_1(z)]^3[2\log(-z)-3\log z]}{1-z}$
	$+\frac{\frac{1}{3}[\mathrm{Li}_1(z)]^4+\frac{2\pi^4}{9}+[\mathrm{Li}_2(z)]^2-4\,\mathrm{Li}_1(z)\,\mathrm{Li}_3(z)}{1-z}$
	$-\frac{8\,\mathrm{Li}_4\left(\frac{1}{1-z}\right)+8\,\mathrm{Li}_4\left(\frac{z}{z-1}\right)+12\,\mathrm{Li}_4(1-z)+7\,\mathrm{Li}_4(z)}{1-z}$
2	$\mathsf{G}^{(2),1}(z)=\frac{\mathrm{Li}_2(z)}{1-z}$
	$\mathsf{G}^{(2),2}(z)=\frac{\mathrm{Li}_4(z)+2L_{2,2}(z)}{1-z}$
	$=\frac{4\zeta_3\,\mathrm{Li}_1(z)+\frac{\pi^2}{3}[\mathrm{Li}_1(z)]^2-\frac{2}{3}[\mathrm{Li}_1(z)]^3\log(-z)-\frac{1}{3}[\mathrm{Li}_1(z)]^4+\frac{2\pi^4}{45}}{1-z}$
	$+\frac{[\mathrm{Li}_2(z)]^2-4\,\mathrm{Li}_1(z)\,\mathrm{Li}_3(z)-4\,\mathrm{Li}_4\left(\frac{1}{1-z}\right)-4\,\mathrm{Li}_4\left(\frac{z}{z-1}\right)-3\,\mathrm{Li}_4(z)}{1-z}$

(r,m,ℓ)	$\mathsf{G}_\ell^{(r),m}(z):=\sum_{n=1}^{\infty}\left[\mathsf{H}_n^{(r)}\right]^m\frac{z^n}{n^\ell},\quad \lvert z\rvert<1$
$(r,1,\ell)$	$\mathrm{Li}_{\ell+r}(z)+L_{\ell,r}(z),\quad r,\ell\in\mathbb{Z}_{>0}$
$(1,2,1)$	$\mathrm{Li}_3(z)+2L_{2,1}(z)+L_{1,2}(z)+2L_{1,1,1}(z)$
$(1,2,2)$	$\mathrm{Li}_4(z)+2L_{3,1}(z)+L_{2,2}(z)+2L_{2,1,1}(z)$
$(1,3,1)$	$\mathrm{Li}_4(z)+3L_{3,1}(z)+3L_{2,2}(z)+L_{1,3}(z)$ $+6L_{2,1,1}(z)+3L_{1,2,1}(z)+3L_{1,1,2}(z)+6L_{1,1,1,1}(z)$

reformulate the last contour integral in the complex w-plane as the moment relation

$$\left[\prod_{j=1}^{M}\mathsf{H}_n^{(r_j)}\right]\frac{1}{n^s}=\int_0^1 x^{n-1}\mathsf{L}_s^{(r_1,\dots,r_M)}(x)\,\mathrm{d}x,\tag{3.27}$$

where $n \in \mathbb{Z}_{>0}$ and $\mathsf{L}_s^{(r_1,\dots,r_M)}(x) \in \mathfrak{g}^{(x)}_{s-1+\sum_{j=1}^M r_j}(1)$. Since the difference between the two sides of the equation above defines a bounded holomorphic function of n for $\operatorname{Re} n > 0$ that vanishes at all the positive integers, it is in fact identically vanishing for all $\operatorname{Re} n > 0$ [**2**, Theorem 2.8.1]. In particular, for the same $\mathsf{L}_s^{(r_1,\dots,r_M)}(x) \in \mathfrak{g}^{(x)}_{s-1+\sum_{j=1}^M r_j}(1)$ as given above, we have

$$\begin{aligned}(3.28)\qquad \left[\prod_{j=1}^{M}\mathsf{H}^{(r_j)}_{n-\frac{1}{2}}\right]\frac{1}{(2n-1)^s} &= \frac{1}{2^s}\int_0^1 x^{n-\frac{3}{2}}\mathsf{L}_s^{(r_1,\dots,r_M)}(x)\,\mathrm{d}x\\ &= \frac{1}{2^{s-1}}\int_0^1 X^{2n-2}\mathsf{L}_s^{(r_1,\dots,r_M)}(X^2)\,\mathrm{d}X,\end{aligned}$$

a moment relation that will become useful later in Corollary 3.14.

3.2. Infinite series related to $m_k(1+x_1+x_2)$. Following the footsteps of Broadhurst [**21**, §§2 and 6], Davydychev–Kalmykov [**35**, §2], and Borwein–Borwein–Straub–Wan [**9**, §5] at an early stage of the present research, we converted $m_k(1+x_1+x_2)$ into infinite series by differentiating a formula of Borwein–Straub–Wan–Zudilin [**15**, Theorem 7.2] (see also [**13**, Theorem 4.2]):

$$(3.29)\qquad W_3(s) = \frac{3^{s+\frac{3}{2}}}{2\pi}\frac{\left[\Gamma\left(1+\frac{s}{2}\right)\right]^2}{\Gamma(2+s)}\,{}_3F_2\left(\left.\begin{matrix}1+\frac{s}{2},1+\frac{s}{2},1+\frac{s}{2}\\ 1,\frac{3+s}{2}\end{matrix}\right|\frac{1}{4}\right),\quad |s|<2,$$

where ${}_3F_2$ is the generalized hypergeometric series. With $c_n := \frac{(n!)^2}{(2n+1)!} = \frac{1}{(n+1)\binom{2n+2}{n+1}} = \frac{1}{2}\int_0^1 x^n(1-x)^n\,\mathrm{d}x$ and $\mathfrak{h}_n^{(r)} := 3\mathsf{H}_n^{(r)} - \mathsf{H}^{(r)}_{n+\frac{1}{2}} + 2\delta_{1,r}\log\frac{3}{2}$, we found (cf. [**55**, §4])

$$(3.30)\qquad \frac{\pi}{\sqrt{3}}\left.\frac{\mathrm{d}^k W_3(s)}{\mathrm{d}s^k}\right|_{s=0} \in \operatorname{span}_{\mathbb{Q}}\left\{\sum_{n=0}^{\infty}c_n\prod_{j=1}^{M}\mathfrak{h}_n^{(r_j)}\,\middle|\, r_1,\dots,r_M\in\mathbb{Z}_{>0};\ \sum_{j=1}^{M}r_j = k\right\}$$

by slightly modifying the proof of [**9**, Theorem 10].

In the terminology of Kalmykov–Ward–Yost [**55**, §2.1], the infinite series appearing in (3.30) are special cases of "inverse binomial sums". In the next two corollaries, we explore such infinite sums in broader context. As precursors to Goncharov–Deligne periods in $\mathfrak{Z}_k(N)$, we will also need new $\mathbb{Q}$-vector spaces[16]

$$(3.31)\qquad \mathfrak{g}_k[z_1,\dots,z_n;z](N) := \operatorname{span}_{\mathbb{Q}}\left\{(\pi i)^{k-\ell}G(\alpha_1,\dots,\alpha_j;z)Z_{\ell-j}\in\mathbb{C}\,\middle|\,\begin{matrix}\alpha_1,\dots,\alpha_j\in\{0\}\cup\{z_1,\dots,z_n\}\\ Z_{\ell-j}\in\mathfrak{Z}_{\ell-j}(N)\\ 0\leqslant j\leqslant\ell\leqslant k\end{matrix}\right\},$$

$$(3.32)\qquad \mathfrak{g}_{k,N}^{\{z_1,\dots,z_n\}} := \operatorname{span}_{\mathbb{Q}}\left\{\prod_{j=1}^{J}G_j\,\middle|\,\begin{matrix}G_j\in\mathfrak{g}_{k_j}[z_1,\dots,z_n;z](N)\\ z\in\{z_1,\dots,z_n\}\\ \sum_{j=1}^{J}k_j=k\end{matrix}\right\},$$

which extend $\mathfrak{g}_k^{(z)}(N) = \mathfrak{g}_k\left[1,e^{2\pi i/N},\dots,e^{2\pi i(N-1)/N};z\right](N)$ [defined in (2.30b)].

---

[16]For $S\subset\mathbb{C}$, our notation $\mathfrak{g}^S_{k,N}$ is related (but not identical) to Au's MZV_k^S [**5**, §5.3]. The sets $\mathfrak{g}_k[z_1,\dots,z_n;z](N)$ and $\mathfrak{g}_{k,N}^{\{z_1,\dots,z_n\}}$ are filtered by weight just as $\mathfrak{g}_k^{(z)}(N)$, while $\mathfrak{g}_k[z_1,\dots,z_n;z](N)$ also behaves like (2.31) under GPL recursions with respect to the variable z. In the corollaries below, one will encounter the $\mathbb{Q}$-vector spaces in (3.31) and (3.32) as natural outcomes of shuffle, recursion, or fibration of GPLs.

COROLLARY 3.4 (Inverse binomial sums via GPLs). *In the following, we set* $r_1, \dots, r_M \in \mathbb{Z}_{>0}$ *and* $r'_1, \dots, r'_{M'} \in \mathbb{Z}_{>0}$.
(a) *If*

$$\begin{cases} |1-\xi|^2 \leqslant 4|\xi|, \\ s \in \mathbb{Z}_{>0} \end{cases} \quad \text{or} \quad \begin{cases} |1-\xi|^2 < 4|\xi|, \\ s \in \mathbb{Z}_{\geqslant 0}, \end{cases} \tag{3.33}$$

then (*cf.* [**55**, (1.1), (5.1a), and (5.1b)])

$$\sum_{n=1}^{\infty} \left[\prod_{j=1}^{M} \mathsf{H}_{n-1}^{(r_j)}\right] \frac{\left(\frac{1+\xi}{1-\xi}\right)^{\delta_{0,s}}}{n^{s+1}\binom{2n}{n}} \left[-\frac{(1-\xi)^2}{\xi}\right]^n \in \mathfrak{g}_{k+1}^{(\xi)}(2) \cap \mathfrak{g}_{k+1,1}^{\left\{1,\frac{1}{1-\xi},\frac{\xi}{\xi-1}\right\}}, \tag{3.34}$$

for $k = s + \sum_{j=1}^{M} r_j$. *More generally, for any moment sequence*[17] $a_{n,M'} = \int_0^1 x^{n-1}(1-x)^{n-1} f_{M'}(x)\,\mathrm{d}x$, $n \in \mathbb{Z}_{>0}$, *satisfying* $f_{M'}(x) \in \mathfrak{H}_{M'}^{(x)}(1)$, *we have*

$$\sum_{n=1}^{\infty} \left[\prod_{j=1}^{M} \mathsf{H}_{n-1}^{(r_j)}\right] \frac{\left(\frac{1+\xi}{1-\xi}\right)^{\delta_{0,s}}}{n^{s}} \left[-\frac{(1-\xi)^2}{\xi}\right]^n a_{n,M'} \in \mathfrak{g}_{k+1}^{(\xi)}(2) \cap \mathfrak{g}_{k+1,1}^{\left\{1,\frac{1}{1-\xi},\frac{\xi}{\xi-1}\right\}} \tag{3.34*}$$

for $k = s + M' + \sum_{j=1}^{M} r_j$, *so long as the infinite series are convergent.*
(b) *If*

$$\begin{cases} |1+\tau^2| \leqslant 2|\tau|, \\ s \in \mathbb{Z}_{>0}, \end{cases} \quad \text{or} \quad \begin{cases} |1+\tau^2| < 2|\tau|, \\ s \in \mathbb{Z}_{\geqslant 0}, \end{cases} \tag{3.35}$$

then we have

$$\begin{aligned} &\sum_{n=1}^{\infty} \left[\prod_{j=1}^{M} \mathsf{H}_{n-1}^{(r_j)}\right] \left\{\prod_{j'=1}^{M'} \left[\mathsf{H}_{2n-1}^{(r'_{j'})} + \overline{\mathsf{H}}_{2n-1}^{(r'_{j'})}\right]\right\} \frac{\left(\frac{1-\tau^2}{1+\tau^2}\right)^{\delta_{0,s}}}{n^{s+1}\binom{2n}{n}} \left(\frac{1+\tau^2}{\tau}\right)^{2n} \\ &\qquad \in \mathfrak{g}_{k+1}^{(\tau)}(4) \cap \mathfrak{g}_{k+1,2}^{\left\{i,-i,\tau,-\frac{1}{\tau},-\tau,\frac{1}{\tau}\right\}} \end{aligned} \tag{3.36}$$

for $k = s + \sum_{j=1}^{M} r_j + \sum_{j'=1}^{M'} r'_{j'}$. *More generally, for any moment sequence* $b_{n,M''} = \int_0^1 x^{n-1}(1-x)^{n-1} g_{M''}(x)\,\mathrm{d}x$, $n \in \mathbb{Z}_{>0}$, *satisfying* $g_{M''}(x) \in \mathfrak{G}_{M''}^{(x)}(1)$, *we have*

$$\begin{aligned} &\sum_{n=1}^{\infty} \left[\prod_{j=1}^{M} \mathsf{H}_{n-1}^{(r_j)}\right] \left\{\prod_{j'=1}^{M'} \left[\mathsf{H}_{2n-1}^{(r'_{j'})} + \overline{\mathsf{H}}_{2n-1}^{(r'_{j'})}\right]\right\} \frac{\left(\frac{1-\tau^2}{1+\tau^2}\right)^{\delta_{0,s}}}{n^{s}} \left(\frac{1+\tau^2}{\tau}\right)^{2n} b_{n,M''} \\ &\qquad \in \mathfrak{g}_{k+1}^{(\tau)}(4) \cap \mathfrak{g}_{k+1,2}^{\left\{i,-i,\tau,-\frac{1}{\tau},-\tau,\frac{1}{\tau}\right\}} \end{aligned} \tag{3.36*}$$

for $k = s + M'' + \sum_{j=1}^{M} r_j + \sum_{j'=1}^{M'} r'_{j'}$, *so long as the infinite series are convergent.*

PROOF. From the asymptotic behavior $\frac{1}{n\binom{2n}{n}} = \frac{1}{4^n}\sqrt{\frac{\pi}{n}}\left[1 + O\left(\frac{1}{n}\right)\right]$ as $n \to \infty$, we see that all the infinite sums of interest indeed converge for the prescribed ranges of s, ξ, and τ.

[17] For example, the function $f_1(x) = \log x \in \mathfrak{H}_1^{(x)}(1)$ produces a sequence $a_{n,1} = \frac{2}{n\binom{2n}{n}}[\mathsf{H}_{n-1}^{(1)} - \mathsf{H}_{2n-1}^{(1)}]$, while $f_2(x) = \log^2 x \in \mathfrak{H}_2^{(x)}(1)$ brings us another sequence $a_{n,2} = \frac{2}{n\binom{2n}{n}}\left\{[\mathsf{H}_{n-1}^{(1)} - \mathsf{H}_{2n-1}^{(1)}]^2 + \mathsf{H}_{n-1}^{(2)} - \mathsf{H}_{2n-1}^{(2)}\right\}$.

(a) First, we consider the $s=0$ case in (3.34). By (3.6), we have

(3.37)

$$\begin{aligned}&\sum_{n=1}^{\infty}\left[\prod_{j=1}^{M}\mathsf{H}_{n-1}^{(r_j)}\right]\frac{1}{n\binom{2n}{n}}\left[-\frac{(1-\xi)^2}{\xi}\right]^n\\&=-\frac{(1-\xi)^2}{2\xi}\sum_{n=1}^{\infty}\left[\prod_{j=1}^{M}\mathsf{H}_{n-1}^{(r_j)}\right]\left[-\frac{(1-\xi)^2}{\xi}\right]^{n-1}\int_0^1 x^{n-1}(1-x)^{n-1}\,\mathrm{d}x\\&\in-\frac{(1-\xi)^2}{2\xi}\operatorname{span}_{\mathbb{Q}}\left\{\int_0^1 L_{\boldsymbol{A}}\left(-\frac{(1-\xi)^2}{\xi}x(1-x)\right)\frac{\mathrm{d}x}{1+\frac{(1-\xi)^2}{\xi}x(1-x)}\,\middle|\, w(\boldsymbol{A})=\sum_{j=1}^{M}r_j\right\}.\end{aligned}$$

Like the proof of Theorem 2.5, we will compute the integrals over multi-dimensional polylogarithms using Brown's method. By direct applications of the GPL recursion (2.22) $\left[\text{with } z=-\frac{(1-\xi)^2}{\xi}x(1-x)\right]$, we can show inductively that

$$\begin{aligned}&L_{\boldsymbol{A}}\left(-\frac{(1-\xi)^2}{\xi}x(1-x)\right)\\&\in\operatorname{span}_{\mathbb{Z}}\left\{G(\beta_1,\dots,\beta_{w(\boldsymbol{A})};x)\,\middle|\,\begin{smallmatrix}\beta_1,\dots,\beta_{w(\boldsymbol{A})}\in\left\{0,1,\frac{1}{1-\xi},\frac{\xi}{\xi-1}\right\}\\ \beta_{w(\boldsymbol{A})}\neq0\end{smallmatrix}\right\}.\end{aligned}\tag{3.38}$$

Furthermore, the partial fraction expansion

$$\frac{-\frac{(1-\xi)^2}{2\xi}}{1+\frac{(1-\xi)^2}{\xi}x(1-x)}=\frac{1-\xi}{1+\xi}\left(\frac{1}{x-\frac{1}{1-\xi}}-\frac{1}{x-\frac{\xi}{\xi-1}}\right)\tag{3.39}$$

tells us that

$$\begin{aligned}&-\frac{1+\xi}{1-\xi}\frac{(1-\xi)^2}{2\xi}\int_0^1 L_{\boldsymbol{A}}\left(-\frac{(1-\xi)^2}{\xi}x(1-x)\right)\frac{\mathrm{d}x}{1+\frac{(1-\xi)^2}{\xi}x(1-x)}\\&\in\operatorname{span}_{\mathbb{Q}}\left\{G(\beta_1,\dots,\beta_{w(\boldsymbol{A})+1};1)\,\middle|\,\begin{smallmatrix}\beta_1,\dots,\beta_{w(\boldsymbol{A})+1}\in\left\{0,1,\frac{1}{1-\xi},\frac{\xi}{\xi-1}\right\}\\ \beta_1\in\left\{\frac{1}{1-\xi},\frac{\xi}{\xi-1}\right\},\beta_{w(\boldsymbol{A})+1}\neq0\end{smallmatrix}\right\}\\&\subseteq\mathfrak{g}_{w(\boldsymbol{A})+1}^{(\xi)}(2)\cap\mathfrak{g}_{w(\boldsymbol{A})+1,1}^{\left\{1,\frac{1}{1-\xi},\frac{\xi}{\xi-1}\right\}},\end{aligned}\tag{3.40}$$

where the "$\in$" step comes from the GPL recursion in (2.22), and the "$\subseteq$" step draws on another round of fibration with respect to ξ [which is deducible from repeated applications of the $z=1$ and $z=\xi$ cases of the GPL recursion (2.22) to the GPL differential form (2.37) for $\mathrm{d}\log(\alpha_{j-1}-\alpha_j),\mathrm{d}\log(\alpha_{j+1}-\alpha_j)\in\left\{\mathrm{d}\log\frac{1}{1-\xi},\right.$ $\left.\mathrm{d}\log\frac{\xi}{\xi-1},\mathrm{d}\log\frac{1+\xi}{1-\xi},0\right\}$].

Then, we work with positive integers s. By the GPL recursion in (2.22), we have [cf. (3.14)]

$$\begin{aligned}&\sum_{n=1}^{\infty}\left[\prod_{j=1}^{M}\mathsf{H}_{n-1}^{(r_j)}\right]\frac{z^n}{n}=\sum_{n=1}^{\infty}\left[\prod_{j=1}^{M}\mathsf{H}_{n-1}^{(r_j)}\right]\int_0^z x^{n-1}\,\mathrm{d}x\\&\in\operatorname{span}_{\mathbb{Q}}\left\{\int_0^z\frac{L_{\boldsymbol{A}}(x)}{1-x}\,\mathrm{d}x\,\middle|\,w(\boldsymbol{A})=\sum_{j=1}^{M}r_j\right\}\\&\subseteq\operatorname{span}_{\mathbb{Q}}\left\{L_{\boldsymbol{A}}(z)\,\middle|\,w(\boldsymbol{A})=1+\sum_{j=1}^{M}r_j\right\}.\end{aligned}\tag{3.41}$$

Subsequent inductions in the spirit of (3.15) bring us

$$\sum_{n=1}^{\infty}\left[\prod_{j=1}^{M}\mathsf{H}_{n-1}^{(r_j)}\right]\frac{z^n}{n^s}\in\operatorname{span}_{\mathbb{Q}}\left\{L_{\boldsymbol{A}}(z)\,\middle|\,w(\boldsymbol{A})=s+\sum_{j=1}^{M}r_j\right\}\tag{3.42}$$

for all $s\in\mathbb{Z}_{>0}$. Accordingly, we arrive at

$$\begin{aligned}&\sum_{n=1}^{\infty}\left[\prod_{j=1}^{M}\mathsf{H}_{n-1}^{(r_j)}\right]\frac{1}{n^{s+1}\binom{2n}{n}}\left[-\frac{(1-\xi)^2}{\xi}\right]^n\\&=\sum_{n=1}^{\infty}\left[\prod_{j=1}^{M}\mathsf{H}_{n-1}^{(r_j)}\right]\frac{1}{n^{s}}\left[-\frac{(1-\xi)^2}{\xi}\right]^n\int_0^1x^n(1-x)^n\frac{\mathrm{d}x}{x(1-x)}\\&\in\operatorname{span}_{\mathbb{Q}}\left\{\int_0^1L_{\boldsymbol{A}}\left(-\frac{(1-\xi)^2}{\xi}x(1-x)\right)\frac{\mathrm{d}x}{x(1-x)}\,\middle|\,w(\boldsymbol{A})=s+\sum_{j=1}^{M}r_j\right\},\end{aligned}\tag{3.43}$$

where

$$\begin{aligned}&\int_0^1L_{\boldsymbol{A}}\left(-\frac{(1-\xi)^2}{\xi}x(1-x)\right)\frac{\mathrm{d}x}{x(1-x)}\\&\in\operatorname{span}_{\mathbb{Q}}\left\{G(\beta_1,\dots,\beta_{w(\boldsymbol{A})+1};1)\,\middle|\,\substack{\beta_1,\dots,\beta_{w(\boldsymbol{A})+1}\in\left\{0,1,\frac{1}{1-\xi},\frac{\xi}{\xi-1}\right\}\\ \beta_1\in\{0,1\},\beta_{w(\boldsymbol{A})+1}\neq0}\right\}\\&\subseteq\mathfrak{g}_{w(\boldsymbol{A})+1}^{(\xi)}(2)\cap\mathfrak{g}_{w(\boldsymbol{A})+1,1}^{\left\{1,\frac{1}{1-\xi},\frac{\xi}{\xi-1}\right\}}\end{aligned}\tag{3.44}$$

follows from recursions and fibrations of GPLs (along with logarithmic regularizations [**69**, §2.3] when $\beta_1=1$).

This completes the proof of (3.34). To generalize it to (3.34*), simply note that

$$\begin{aligned}&L_{\boldsymbol{A}}\left(-\frac{(1-\xi)^2}{\xi}x(1-x)\right)f_{M'}(x)\\&\in\operatorname{span}_{\mathbb{Q}}\left\{G(\beta_1,\dots,\beta_{w(\boldsymbol{A})};x)\,\middle|\,\beta_1,\dots,\beta_{w(\boldsymbol{A})}\in\left\{0,1,\frac{1}{1-\xi},\frac{\xi}{\xi-1}\right\}\right\}\end{aligned}\tag{3.45}$$

follows from the condition $f_{M'}(x)\in\mathfrak{H}_{M'}^{(x)}(1)$ and the GPL shuffle algebra.

(b) Combining (3.10) with the observation that

$$\mathsf{H}_{n-1}^{(r)}=2^{r-1}\left[\mathsf{H}_{2n-1}^{(r)}-\overline{\mathsf{H}}_{2n-1}^{(r)}\right],\quad r\in\mathbb{Z}_{>0},\tag{3.46}$$

we may put down

$$\begin{aligned}&\sum_{n=1}^{\infty}\left[\prod_{j=1}^{M}\mathsf{H}_{n-1}^{(r_j)}\right]\left\{\prod_{j'=1}^{M'}\left[\mathsf{H}_{2n-1}^{(r'_{j'})}+\overline{\mathsf{H}}_{2n-1}^{(r'_{j'})}\right]\right\}z^{2n-1}\\&\in\operatorname{span}_{\mathbb{Q}}\left\{\frac{g(z)}{1-z}-\frac{g(-z)}{1+z}\,\middle|\,g(z)\in\mathfrak{G}_{\sum_{j=1}^{M}r_j+\sum_{j'=1}^{M'}r'_{j'}}^{(z)}(2)\right\}.\end{aligned}\tag{3.47}$$

When $s = 0$, to evaluate the integral representation for the left-hand side of (3.36), we need

$$\begin{aligned}
&\frac{1-\tau^2}{1+\tau^2}\frac{1+\tau^2}{\tau}\int_0^1 \frac{G\left(\alpha_1,\dots,\alpha_w;\pm\frac{1+\tau^2}{\tau}\sqrt{X(1-X)}\right)\mathrm{d}X}{\sqrt{X(1-X)}\left[1\mp\frac{1+\tau^2}{\tau}\sqrt{X(1-X)}\right]}\\
&=\frac{2(1-\tau^2)}{\tau}\int_{1/2}^1 \frac{G\left(\alpha_1,\dots,\alpha_w;\pm\frac{1+\tau^2}{\tau}\sqrt{x(1-x)}\right)\mathrm{d}x}{\sqrt{x(1-x)}\left[1\mp\frac{1+\tau^2}{\tau}\sqrt{x(1-x)}\right]}\\
&=4\int_0^1 G\left(\alpha_1,\dots,\alpha_w;\pm\frac{(1+\tau^2)t}{(1+t^2)\tau}\right)\left(\frac{1}{\tau\mp t}-\frac{1}{\frac{1}{\tau}\mp t}\right)\mathrm{d}t,
\end{aligned}\tag{3.48}$$

where $x=\frac{1}{1+t^2}$, $\alpha_1,\dots,\alpha_w\in\{-1,0,1\},\alpha_w\neq 0$ and $w=\sum_{j=1}^M r_j+\sum_{j'=1}^{M'}r'_{j'}$. Again, by recursive applications of (2.22) for $z=\pm\frac{(1+\tau^2)t}{(1+t^2)\tau}$, we can show that

$$\begin{aligned}
&G\left(\alpha_1,\dots,\alpha_w;\pm\frac{(1+\tau^2)t}{(1+t^2)\tau}\right)\\
&\in\operatorname{span}_{\mathbb{Z}}\left\{G(\beta_1,\dots,\beta_w;t)\;\middle|\;{\scriptstyle\beta_1,\dots,\beta_w\in\left\{0,\tau,i,-\frac{1}{\tau},-\tau,-i,\frac{1}{\tau}\right\}\atop\scriptstyle\beta_w\neq 0}\right\}.
\end{aligned}\tag{3.49}$$

Therefore, the left-hand side of (3.36) belongs to the $\mathbb{Q}$-vector space

$$\operatorname{span}_{\mathbb{Q}}\left\{G(\beta_1,\dots,\beta_{k+1};1)\;\middle|\;\beta_1,\dots,\beta_{k+1}\in\left\{0,\tau,i,-\frac{1}{\tau},-\tau,-i,\frac{1}{\tau}\right\}\right\},\tag{3.50}$$

when $s=0$. Fibrating the GPLs in (3.50) with respect to τ, we realize that (3.50) is a subspace of $\mathfrak{g}_{k+1}^{(\tau)}(4)$.

Then we handle (3.36) for $s\in\mathbb{Z}_{>0}$. By the GPL recursion in (2.22), we may promote (3.47) to

$$\begin{aligned}
&\sum_{n=1}^{\infty}\left[\prod_{j=1}^{M}\mathsf{H}_{n-1}^{(r_j)}\right]\left\{\prod_{j'=1}^{M'}\left[\mathsf{H}_{2n-1}^{(r'_{j'})}+\overline{\mathsf{H}}_{2n-1}^{(r'_{j'})}\right]\right\}\frac{z^{2n}}{n^s}\\
&\in\operatorname{span}_{\mathbb{Q}}\left\{g(z)+g(-z)\;\middle|\;g(z)\in\mathfrak{G}^{(z)}_{s+\sum_{j=1}^{M}r_j+\sum_{j'=1}^{M'}r'_{j'}}(2)\right\}
\end{aligned}\tag{3.51}$$

for all $s\in\mathbb{Z}_{>0}$. Accordingly, the integral representation of

$$\sum_{n=1}^{\infty}\left[\prod_{j=1}^{M}\mathsf{H}_{n-1}^{(r_j)}\right]\left\{\prod_{j'=1}^{M'}\left[\mathsf{H}_{2n-1}^{(r'_{j'})}+\overline{\mathsf{H}}_{2n-1}^{(r'_{j'})}\right]\right\}\frac{1}{n^{s+1}\binom{2n}{n}}\left(\frac{1+\tau^2}{\tau}\right)^{2n}\tag{3.52}$$

requires us to evaluate

$$\begin{aligned}
&\int_0^1\frac{G\left(\widetilde{\alpha}_1,\dots,\widetilde{\alpha}_{w+s};\pm\frac{1+\tau^2}{\tau}\sqrt{X(1-X)}\right)}{X(1-X)}\mathrm{d}X\\
&=4\int_0^1 G\left(\widetilde{\alpha}_1,\dots,\widetilde{\alpha}_{w+s};\pm\frac{(1+\tau^2)t}{(1+t^2)\tau}\right)\frac{\mathrm{d}t}{t},
\end{aligned}\tag{3.53}$$

where $\widetilde{\alpha}_1,\dots,\widetilde{\alpha}_{w+s}\in\{-1,0,1\},\alpha_w\neq 0$ and $w=\sum_{j=1}^M r_j+\sum_{j'=1}^{M'}r'_{j'}$. Therefore, the left-hand side of (3.36) is still in the $\mathbb{Q}$-vector space $\mathfrak{g}_{k+1}^{(\tau)}(4)$.

This completes the proof for a weaker version of (3.36). The same arguments extend to the $\mathfrak{g}^{(\tau)}_{k+1}(4)$ part of (3.36*), since $g_{M''}(x) \in \mathfrak{G}^{(x)}_{M''}(1)$ entails $g_{M''}(\frac{1}{1+t^2}) + g_{M''}(1-\frac{1}{1+t^2}) \in \mathfrak{g}_{M''}[i,-i;t](2)$.

To verify (3.36) and (3.36*) in their entirety, we need contour deformations. Concretely speaking, if we have $f_k(z) \in \mathfrak{G}^{(z)}_k(2)$ and $g_{M''}(z) \in \mathfrak{G}^{(z)}_{M''}(1)$ for $|z|<1$, then we can deform the contour of

(3.54)
$$\begin{aligned}\int_0^\infty & f_k\left(\pm\frac{(1+\tau^2)t}{(1+t^2)\tau}\right) g_{M''}\left(\frac{1}{1+t^2}\right) h(t)\,dt\\ &= \left(\int_{-i0^+}^{\infty-i0^+} - \int_{i0^+}^{\infty+i0^+}\right) f_k\left(\pm\frac{(1+\tau^2)t}{(1+t^2)\tau}\right) g_{M''}\left(\frac{1}{1+t^2}\right) h(t)\frac{\log(-t)\,dt}{2\pi i},\\ &\text{where } h(t) \in \left\{\frac{1}{\tau\mp t} - \frac{1}{\frac{1}{\tau}\mp t}, \frac{1}{t}\right\}\end{aligned}$$

to tight loops enveloping the branch cuts, so that the net contribution to the integral results from the jump discontinuities [**67**, **69**]

$$\text{(3.55)}\qquad \begin{cases} f_k(\pm(\varXi+i0^+)) - f_k(\pm(\varXi-i0^+)) \in \pi i \mathfrak{g}^{(\pm 1/\varXi)}_{k-1}(2),\\ g_{M''}(\varXi+i0^+) - g_{M''}(\varXi-i0^+) \in \pi i \mathfrak{g}^{(1/\varXi)}_{M''-1}(1)\end{cases}$$

for $\varXi > 1$ (with the understanding that $\mathfrak{g}^{(z)}_{-1}(1) = \{0\}$). Here, in the complex t-plane, the end points of these branch cuts belong to the set $\{i,-i,\tau,-\frac{1}{\tau},-\tau,\frac{1}{\tau}\}$. Recursive applications of (2.22) with $z=i$ and $z=t$ lead us to

$$\text{(3.56)}\qquad \mathfrak{g}^{(\pm 1/\varXi)}_{k-1}(2) \subseteq \mathfrak{g}^{(t)}_{k-1}(4) \quad\text{and}\quad \mathfrak{g}^{(1+t^2)}_{M''-1}(1) \subseteq \mathfrak{g}^{(t)}_{M''-1}(4)$$

for $\varXi = \frac{(1+\tau^2)t}{(1+t^2)\tau}$, hence the full characterization stated in (3.36) and (3.36*). □

COROLLARY 3.5 (Inverse binomial sums via Goncharov–Deligne periods). *Let $\boldsymbol{r} = (r_1,\dots,r_M)$ and $\boldsymbol{r}' = (r'_1,\dots,r'_{M'})$ be vectors whose components are positive integers. Define their 1-norms by $\|\boldsymbol{r}\|_1 := \sum_{j=1}^M r_j$ and $\|\boldsymbol{r}'\|_1 := \sum_{j=1}^{M'} r'_j$. We will examine the inverse binomial sums*

$$\text{(3.57)}\qquad \mathsf{s}_{\boldsymbol{r}}(u,s) := \sum_{n=1}^\infty \left[\prod_{j=1}^M \mathsf{H}^{(r_j)}_{n-1}\right] \frac{u^n}{n^{s+1}\binom{2n}{n}},$$

$$\text{(3.58)}\qquad \mathsf{S}_{\boldsymbol{r};\boldsymbol{r}'}(U,s) := \sum_{n=1}^\infty \left[\prod_{j=1}^M \mathsf{H}^{(r_j)}_{n-1}\right] \left\{\prod_{j'=1}^{M'} \left[\mathsf{H}^{(r'_{j'})}_{2n-1} + \overline{\mathsf{H}}^{(r'_{j'})}_{2n-1}\right]\right\} \frac{U^n}{n^{s+1}\binom{2n}{n}}$$

for certain special values of $u = -\frac{(1-\xi)^2}{\xi}$ and $U = \left(\frac{1+\tau^2}{\tau}\right)^2$.

(a) *For* $s\in\mathbb{Z}_{\geqslant0}$, $k=s+\|\boldsymbol{r}\|_1$, *we have the following particular cases of* (3.34), *where* $\xi\in\{\varrho:=e^{\pi i/3},\,1-\rho:=\frac{3-\sqrt5}{2}=\rho^2,\,i,\,\omega:=e^{2\pi i/3}\}$:[18]

$$(i\sqrt3)^{\delta_{0,s}}\mathsf{s}_{\boldsymbol{r}}(1,s)\in\mathfrak{g}^{(\varrho)}_{k+1}(2)\cap\mathfrak{Z}_{k+1}(3),\tag{3.59}$$

$$(\sqrt5)^{\delta_{0,s}}\mathsf{s}_{\boldsymbol{r}}(-1,s)\in\mathfrak{g}^{(\rho^2)}_{k+1}(2)\cap\mathfrak{Z}_{k+1}(10),\tag{3.60}$$

$$i^{\delta_{0,s}}\mathsf{s}_{\boldsymbol{r}}(2,s)\in\mathfrak{g}^{(i)}_{k+1}(2)\cap\mathfrak{Z}_{k+1}(4),\tag{3.61}$$

$$(i\sqrt3)^{\delta_{0,s}}\mathsf{s}_{\boldsymbol{r}}(3,s)\in\mathfrak{g}^{(\omega)}_{k+1}(2)\cap\mathfrak{Z}_{k+1}(6).\tag{3.62}$$

For $s\in\mathbb{Z}_{>0}$, $k=s+\|\boldsymbol{r}\|_1$, *we have*

$$\mathsf{s}_{\boldsymbol{r}}(4,s)=\sum_{n=1}^{\infty}\left[\prod_{j=1}^{M}\mathsf{H}^{(r_j)}_{n-1}\right]\frac{2^{2n}}{n^{s+1}\binom{2n}{n}}\in\mathfrak{Z}_{k+1}(2),\tag{3.63}$$

along with the quadratic analog of inverse binomial sums:

$$\sum_{n=1}^{\infty}\left[\prod_{j=1}^{M}\mathsf{H}^{(r_j)}_{n-1}\right]\frac{2^{4n}}{n^{s+2}\binom{2n}{n}^2}\in\mathfrak{Z}_{k+2}(4).\tag{3.64}$$

(b) *For* $s\in\mathbb{Z}_{\geqslant0}$, $k=s+\|\boldsymbol{r}\|_1+\|\boldsymbol{r}'\|_1$, *we have the following particular cases of* (3.36), *where* $\tau\in\{\varrho:=e^{\pi i/3},\,i\rho:=i\frac{\sqrt5-1}{2},\,\theta:=e^{\pi i/4},\,i/\varrho=e^{\pi i/6}\}$:

$$(i\sqrt3)^{\delta_{0,s}}\mathsf{S}_{\boldsymbol{r};\boldsymbol{r}'}(1,s)\in\mathfrak{g}^{(\varrho)}_{k+1}(4)\cap\mathfrak{Z}_{k+1}(12),\tag{3.65}$$

$$(\sqrt5)^{\delta_{0,s}}\mathsf{S}_{\boldsymbol{r};\boldsymbol{r}'}(-1,s)\in\mathfrak{g}^{(i\rho)}_{k+1}(4)\cap\mathfrak{Z}_{k+1}(60),\tag{3.66}$$

$$i^{\delta_{0,s}}\mathsf{S}_{\boldsymbol{r};\boldsymbol{r}'}(2,s)\in\mathfrak{g}^{(\theta)}_{k+1}(4)\cap\mathfrak{Z}_{k+1}(8),\tag{3.67}$$

$$(i\sqrt3)^{\delta_{0,s}}\mathsf{S}_{\boldsymbol{r};\boldsymbol{r}'}(3,s)\in\mathfrak{g}^{(i/\varrho)}_{k+1}(4)\cap\mathfrak{Z}_{k+1}(6).\tag{3.68}$$

For $s\in\mathbb{Z}_{>0}$, $k=s+\|\boldsymbol{r}\|_1+\|\boldsymbol{r}'\|_1$, *we have*

(3.69)

$$\mathsf{S}_{\boldsymbol{r};\boldsymbol{r}'}(4,s)=\sum_{n=1}^{\infty}\left[\prod_{j=1}^{M}\mathsf{H}^{(r_j)}_{n-1}\right]\left\{\prod_{j'=1}^{M'}\left[\mathsf{H}^{(r'_{j'})}_{2n-1}+\overline{\mathsf{H}}^{(r'_{j'})}_{2n-1}\right]\right\}\frac{2^{2n}}{n^{s+1}\binom{2n}{n}}\in\mathfrak{Z}_{k+1}(4),$$

as well as the quadratic analog of inverse binomial sums:

$$\sum_{n=1}^{\infty}\left[\prod_{j=1}^{M}\mathsf{H}^{(r_j)}_{n-1}\right]\left\{\prod_{j'=1}^{M'}\left[\mathsf{H}^{(r'_{j'})}_{2n-1}+\overline{\mathsf{H}}^{(r'_{j'})}_{2n-1}\right]\right\}\frac{2^{4n}}{n^{s+2}\binom{2n}{n}^2}\in\mathfrak{Z}_{k+2}(4).\tag{3.70}$$

PROOF. (a) With $\xi=\varrho$ in (3.34), we obtain $\{1,\frac{1}{1-\xi},\frac{\xi}{\xi-1}\}=\{1,\varrho,\frac{1}{\varrho}\}$. This hearkens back to the scenario in (2.62), so the left-hand side of (3.59) is in $\mathfrak{Z}_{k+1}(3)$.

Setting $\xi=1-\rho=\frac{3-\sqrt5}{2}$ in (3.34), we get $\{1,\frac{1}{1-\xi},\frac{\xi}{\xi-1}\}=\{1,\frac{1+\sqrt5}{2},\frac{1-\sqrt5}{2}\}$. By a variation on the arguments for (2.62), we may use GPL fibration in the

[18]In this work, the Greek letters in the notations for roots of unity $\omega:=e^{2\pi i/3}$, $\varsigma:=e^{2\pi i/5}$, $\varrho:=e^{2\pi i/6}$, and $\theta:=e^{2\pi i/8}$ are chosen for their visual resemblance to the corresponding Arabic numerals. The notation $\rho:=\frac{\sqrt5-1}{2}$ is standard in modern literature on polylogarithms [**16**, **64**].

TABLE 4. Selected closed forms for inverse binomial sums, where $\omega := e^{2\pi i/3}$, $\varrho := e^{\pi i/3}$, $\rho := \frac{\sqrt{5}-1}{2}$, $\varsigma := e^{2\pi i/5}$, $G := \operatorname{Im}\operatorname{Li}_2(i)$, $\lambda := \log 2$, and $\varLambda := \log 3$.

(r,m,ξ)	$\sum_{n=1}^{\infty}\left[\mathsf{H}_{n-1}^{(r)}\right]^m\frac{1}{n\binom{2n}{n}}\left[-\frac{(1-\xi)^2}{\xi}\right]^{n-1}=\frac{1}{2}\int_0^1\mathsf{G}^{(r),m}\left(-\frac{(1-\xi)^2}{\xi}x(1-x)\right)\mathrm{d}x$
$(1,1,\varrho)$	$\frac{2\operatorname{Im}\operatorname{Li}_2(\omega)}{\sqrt{3}}-\frac{\pi\varLambda}{3\sqrt{3}}$
$(1,2,\varrho)$	$-\frac{2^3\operatorname{Im}L_{2,1}(\omega)}{\sqrt{3}}-\frac{2^2\varLambda\operatorname{Im}\operatorname{Li}_2(\omega)}{\sqrt{3}}+\frac{\pi\varLambda^2}{3\sqrt{3}}+\frac{\pi^3}{2\cdot 3^3\sqrt{3}}$
$(1,3,\varrho)$	$-\frac{3^2\cdot 13\operatorname{Im}\operatorname{Li}_4(\omega)}{2^2\sqrt{3}}+\frac{2^4\cdot 3\operatorname{Im}L_{2,1,1}(\omega)}{\sqrt{3}}+\frac{7^2\pi\zeta_3}{3^2\sqrt{3}}+\frac{2^3\cdot 3\varLambda\operatorname{Im}L_{2,1}(\omega)}{\sqrt{3}}$ $+\left(\frac{5\pi^2}{2\cdot 3}+2\cdot 3\varLambda^2\right)\frac{\operatorname{Im}\operatorname{Li}_2(\omega)}{\sqrt{3}}-\left(\frac{\pi^2}{2\cdot 3}+\varLambda^2\right)\frac{\pi\varLambda}{3\sqrt{3}}$
$(1,4,\varrho)$	$-\frac{2^2\cdot 3^2\cdot 79\operatorname{Im}L_{4,1}(\omega)}{13\sqrt{3}}-\frac{2\cdot 3^2\cdot 109\operatorname{Im}L_{3,2}(\omega)}{13\sqrt{3}}-\frac{2^7\cdot 3\operatorname{Im}L_{2,1,1,1}(\omega)}{\sqrt{3}}+\frac{3^2\cdot 13\varLambda\operatorname{Im}\operatorname{Li}_4(\omega)}{\sqrt{3}}$ $+\frac{2^5\pi\operatorname{Re}L_{3,1}(\omega)}{\sqrt{3}}-\frac{2^6\cdot 3\varLambda\operatorname{Im}L_{2,1,1}(\omega)}{\sqrt{3}}+\frac{2^3\zeta_3\operatorname{Im}\operatorname{Li}_2(\omega)}{\sqrt{3}}-\frac{2^2\cdot 7^2\pi\zeta_3\varLambda}{3^2\sqrt{3}}$ $-\frac{2^2(5\pi^2+2^2\cdot 3^2\varLambda^2)\operatorname{Im}L_{2,1}(\omega)}{3\sqrt{3}}-\frac{3\pi[\operatorname{Im}\operatorname{Li}_2(\omega)]^2}{2\sqrt{3}}-\left(\frac{5\pi^2}{3}+2^2\varLambda^2\right)\frac{2\varLambda\operatorname{Im}\operatorname{Li}_2(\omega)}{\sqrt{3}}$ $+\left(\frac{\pi^2}{3}+\varLambda^2\right)\frac{\pi\varLambda^2}{3\sqrt{3}}-\frac{2267\pi^5}{2^2\cdot 3^4\cdot 5\cdot 13\sqrt{3}}$
$(2,1,\varrho)$	$\frac{\pi^3}{2\cdot 3^4\sqrt{3}}$
$(2,2,\varrho)$	$\frac{2^2\cdot 3^3\operatorname{Im}L_{4,1}(\omega)}{13\sqrt{3}}+\frac{2^2\cdot 3^2\operatorname{Im}L_{3,2}(\omega)}{13\sqrt{3}}+\frac{2^2\zeta_3\operatorname{Im}\operatorname{Li}_2(\omega)}{\sqrt{3}}-\frac{3\pi[\operatorname{Im}\operatorname{Li}_2(\omega)]^2}{2\sqrt{3}}-\frac{5^2\pi^5}{2^2\cdot 3^6\cdot 13\sqrt{3}}$
$(1,1,\rho^2)$	$\frac{2^2}{\sqrt{5}}\operatorname{Re}\operatorname{Li}_{1,1}\left(-\frac{1}{\varsigma^2}=e^{\pi i/5},\varsigma\right)+\frac{11\pi^2}{2\cdot 5^2\sqrt{5}}$
$(1,2,\rho^2)$	$-\frac{2^3}{\sqrt{5}}\left[2^2\operatorname{Re}\operatorname{Li}_{1,1,1}\left(\varsigma,1,-\frac{1}{\varsigma^2}\right)+\operatorname{Re}\operatorname{Li}_{1,1,1}\left(-\varsigma,1,-\frac{1}{\varsigma^2}\right)\right]-\frac{5\cdot 11}{3\sqrt{5}}\operatorname{Re}\operatorname{Li}_3(\varsigma)-\frac{2^6\zeta_3}{5\sqrt{5}}$ $+\frac{2^2\pi[\operatorname{Im}\operatorname{Li}_2(\varsigma)+2\operatorname{Im}\operatorname{Li}_2(\varsigma^2)]}{\sqrt{5}}+\frac{2^4}{\sqrt{5}}[\operatorname{Re}\operatorname{Li}_1(\varsigma)]^2\operatorname{Re}[\operatorname{Li}_1(\varsigma)-\operatorname{Li}_1(\varsigma^2)]$ $-\frac{2\pi^2}{3\cdot 5^2\sqrt{5}}\operatorname{Re}[37\operatorname{Li}_1(\varsigma)-11\operatorname{Li}_1(\varsigma^2)]$
$(2,1,\rho^2)$	$\frac{2^2}{3\sqrt{5}}\left\{\operatorname{Re}[\operatorname{Li}_1(\varsigma^2)-\operatorname{Li}_1(\varsigma)]\right\}^3$
$(1,1,i)$	$G-\frac{\pi\lambda}{2^2}$
$(1,2,i)$	$-2^2\operatorname{Im}L_{2,1}(i)-(2^3G-\pi\lambda)\frac{\lambda}{2^2}+\frac{\pi^3}{2^5}$
$(1,3,i)$	$-3\cdot 5\operatorname{Im}\operatorname{Li}_4(i)+2^3\cdot 3\operatorname{Im}L_{2,1,1}(i)+\frac{7\cdot 11\pi\zeta_3}{2^5}+2^2\cdot 3\lambda\operatorname{Im}L_{2,1}(i)+G(\pi^2+3\lambda^2)$ $-\left(\frac{3\pi^2}{2^3}+\lambda^2\right)\frac{\pi\lambda}{2^2}$
$(1,4,i)$	$-\frac{2\cdot 3^3\cdot 17\operatorname{Im}L_{4,1}(i)}{7}-\frac{2\cdot 293\operatorname{Im}L_{3,2}(i)}{7}-2^6\cdot 3\operatorname{Im}L_{2,1,1,1}(i)+2^2\cdot 3\cdot 5\lambda\operatorname{Im}\operatorname{Li}_4(i)$ $+\frac{3\cdot 5\pi\zeta_{-3,1}}{2}-2^5\cdot 3\lambda\operatorname{Im}L_{2,1,1}(i)-\left(\frac{5\cdot 7\pi^2}{2^2}+2^3\cdot 3\lambda^2\right)\operatorname{Im}L_{2,1}(i)$ $+\frac{7\cdot 11\zeta_3}{2^4}(3G-2\pi\lambda)-\frac{5\pi G^2}{2}-2^2G\lambda(\pi^2+\lambda^2)+\frac{\pi\lambda^2}{2^2}\left(\frac{3\pi^2}{2^2}+\lambda^2\right)-\frac{13^2\pi^5}{2^7\cdot 3\cdot 7}$
$(2,1,i)$	$\frac{\pi^3}{2^5\cdot 3}$
$(2,2,i)$	$2\cdot 3\operatorname{Im}L_{4,1}(i)+2\operatorname{Im}L_{3,2}(i)+\frac{5\cdot 7\zeta_3G}{2^4}+\frac{\pi^2\operatorname{Im}L_{2,1}(i)}{2^2}-\frac{\pi G^2}{2}-\frac{\pi^5}{2^7\cdot 3^2}$

variable z to prove[19]

$$\mathfrak{g}_{k+1,1}^{\left\{1,-\frac{(1+\varsigma)\varsigma(\varsigma+z)}{\varsigma^4+z},-\frac{(1+\varsigma)\varsigma(\varsigma^3+z)}{\varsigma^6+z}\right\}}\subseteq\mathfrak{g}_{k+1}\left[1,-\varsigma,-\varsigma^2,-\varsigma^3,-\varsigma^4;z\right](10),\tag{3.71}$$

[19]The rational functions in (3.71) paraphrase [**5**, Example 4.7]. We also note that the $M=1$ case of (3.60) has already been covered by [**5**, Theorem 7.2]. For the fibration here, the GPL recursion integrals run from $-\varsigma$ (or $-\varsigma^4$) to z, and the GPL differential forms involve terms like $\mathrm{d}\log\left(1+\frac{(1+\varsigma)\varsigma(\varsigma+z)}{\varsigma^4+z}\right)=\mathrm{d}\log\frac{\varsigma^2+z}{\varsigma^4+z}$ and $\mathrm{d}\log\left(1+\frac{(1+\varsigma)\varsigma(\varsigma^3+z)}{\varsigma^6+z}\right)=\mathrm{d}\log\frac{\varsigma^4+z}{\varsigma^6+z}$.

for $\varsigma = e^{2\pi i/5}$, while checking that

$$(3.72)\qquad -\frac{(1+\varsigma)\varsigma(\varsigma+z)}{\varsigma^4+z}=\frac{1+\sqrt{5}}{2},\qquad -\frac{(1+\varsigma)\varsigma(\varsigma^3+z)}{\varsigma^6+z}=\frac{1-\sqrt{5}}{2}$$

for $z=1$. This identifies the left-hand side of (3.60) as a CMZV of level 10 (cf. [**16**, §6]).

Specializing (3.34) to $\xi\in\{i,\omega,-1\}$, while noting that $\pi i\in\mathfrak{Z}_1(N), N\in\mathbb{Z}_{\geqslant 3}$ [cf. (2.15)] and $\mathfrak{g}^{(-1)}_{k+1}(2)\cap\mathbb{R}=\mathfrak{Z}_{k+1}(2)$, one can verify (3.61)–(3.63). Here, the relation (3.61) [resp. (3.62)] is consistent with [**4**, Theorem 4.1] (resp. [**5**, Proposition 7.11] involving $a_{n,1}=\frac{2}{n\binom{2n}{n}}[\mathsf{H}^{(1)}_{n-1}-\mathsf{H}^{(1)}_{2n-1}]$) for the $M=1$ case. The Euler–Apéry-type series in (3.63) (cf. [**80**, (1.1)]) were previously known to be AMZVs (namely CMZVs of level 2).

Combining the following integral along a straight line contour (on which $|1-z^2|\leqslant 2|z|$ holds)

$$(3.73)\qquad \int_1^i\left(\frac{1-z^2}{iz}\right)^{2n-1}\frac{\mathrm{d}\,z}{iz}=\int_1^i\left[-\frac{(1-z^2)^2}{z^2}\right]^n\frac{\mathrm{d}z}{1-z^2}=-\frac{2^{4n-2}}{n\binom{2n}{n}}$$

with the relation $\mathfrak{g}^{(z^2)}_{k+1}(2)\subseteq\mathfrak{g}^{(z)}_{k+1}(4)$ (provable by variable substitutions in the integrals defining GPLs), we can confirm (3.64) by integrating (3.34).

It is clear from the analysis so far that (3.34) and (3.59)–(3.64) extend retroactively to the cases where $M=0$ (thus covering Zucker's sums [**95**]

$$(3.74)\qquad \sum_{n=1}^{\infty}\frac{1}{n^{s+1}\binom{2n}{n}}$$

and the like), and that the designated $\mathbb{Q}$-vector spaces are not affected if we replace any single factor of $\mathsf{H}^{(r)}_{n-1}=\mathsf{H}^{(r)}_{n}-\frac{1}{n^r}$ by $\mathsf{H}^{(r)}_{n}$.

(b) With $\pi i\in\mathfrak{Z}_1(N), N\in\mathbb{Z}_{\geqslant 3}$ [cf. (2.15)] in mind, we can deduce (3.65)–(3.69) from (3.36), except that we need the next two paragraphs to account for the CMZV portrayal in (3.66) and (3.68).

Setting $R_n(z):=i\big(ze^{n\pi i/5}+\frac{1}{ze^{n\pi i/5}}\big)$, we consider

$$(3.75)\qquad \mathfrak{g}^{\{i,-i,\tau,-\frac{1}{\tau},-\tau,\frac{1}{\tau}\}}_{k+1,2}\bigg|_{\tau=i\rho}=\mathfrak{g}^{\{i,-i,R_0(z),R_1(z),R_2(z),R_3(z)\}}_{k+1,2}\bigg|_{z=e^{\pi i/5}}.$$

The right-hand side of the equation above is a $\mathbb{Q}$-linear subspace of $\mathfrak{Z}_{k+1}(60)$, as one can check by the fibration basis with respect to z.[20]

A naïve set inclusion $\mathfrak{g}^{(i/\varrho)}_{k+1}(4)\subseteq\mathfrak{Z}_{k+1}(12)$ [based on a weaker form of (3.36)] is sharpened by the following invocation of the scaling property [**44**, (2.3)] for GPLs:

$$(3.76)\qquad \mathfrak{g}^{\{i,-i,\tau,-\frac{1}{\tau},-\tau,\frac{1}{\tau}\}}_{k+1,2}\bigg|_{\tau=i/\varrho}\subseteq\mathfrak{g}^{\{1,\varrho,\varrho^2,-1,-\varrho,-\varrho^2\}}_{k+1,2}\subseteq\mathfrak{Z}_{k+1}(6),$$

so the statement in (3.68) is true.

[20] To show that $\mathfrak{g}^{\{iz,-iz,zR_0(z),zR_1(z),zR_2(z),zR_3(z)\}}_{k+1,2}\Big|_{z=e^{\pi i/5}}\subseteq\mathfrak{Z}_{k+1}(60)$, one may apply the GPL recursion (2.22) to the GPL differential form (2.37), while noting that $\mathrm{d}\log(iz+zR_1(z))=\mathrm{d}\log(z-e^{7\pi i/15})+\mathrm{d}\log(z+e^{2\pi i/15})$ and so forth.

TABLE 5. Selected closed forms for quadratic analogs of inverse binomial sums, where $G := \operatorname{Im}\operatorname{Li}_2(i)$ and $\lambda := \log 2$.

(r,m,ℓ)	$\sum_{n=1}^{\infty}\left[\mathsf{H}_n^{(r)}\right]^m \frac{2^{4n}}{n^{\ell+2}\binom{2n}{n}^2} = -2\int_1^i\left[\int_0^1 \mathsf{G}_\ell^{(r),m}\left(-\frac{(1-z^2)^2}{z^2}x(1-x)\right)\frac{\mathrm{d}x}{x(1-x)}\right]\frac{\mathrm{d}z}{1-z^2}$
$(1,1,1)$	$2^4\cdot 7\boldsymbol{\zeta}_{-3,1} - 2^9\operatorname{Re}L_{2,1,1}(i) + 2^5G(G+\pi\lambda) - 2^3\cdot 7\zeta_3\lambda - \frac{13\pi^4}{3^2\cdot 5}$
$(1,1,2)$	$\frac{5\cdot 211\zeta_5}{2^2} + 2^5\cdot 3^2\boldsymbol{\zeta}_{-3,1,1} + 2^9\operatorname{Re}L_{2,2,1}(i) - 2^5\cdot 3^2\cdot 5\pi\operatorname{Im}\operatorname{Li}_4(i) + 2^7\boldsymbol{\zeta}_{-3,1}\lambda + 2^9\cdot 3\pi\operatorname{Im}L_{2,1,1}(i) + \zeta_3(11\cdot 23\pi^2 - 2^4\cdot 7\lambda^2) + 2^9(G+\pi\lambda)\operatorname{Im}L_{2,1}(i) + 2^7G^2\lambda + 2^4\pi G(3\pi^2+2^2\lambda^2) + \frac{2\cdot 7\pi^4\lambda}{3^2\cdot 5}$
$(1,1,3)$	$2^4\cdot 3^3\cdot 17\boldsymbol{\zeta}_{-5,1} + 2^{10}\cdot 3\operatorname{Re}L_{4,2}(i) + 2^7\cdot 3^2\boldsymbol{\zeta}_{-3,1,1,1} + 2^{12}\operatorname{Re}L_{2,2,1,1}(i) + 5\cdot 211\zeta_5\lambda + \frac{2^9\cdot 3\cdot 83\pi\operatorname{Im}L_{4,1}(i)}{7} + \frac{2^8\cdot 13\cdot 23\pi\operatorname{Im}L_{3,2}(i)}{7} + 2^7\cdot 3^2\boldsymbol{\zeta}_{-3,1,1}\lambda + 2^{11}\lambda\operatorname{Re}L_{2,2,1}(i) + 2^{14}\pi\operatorname{Im}L_{2,1,1,1}(i) - 2^7\cdot 3^2\cdot 5\pi\lambda\operatorname{Im}\operatorname{Li}_4(i) - 2^4\boldsymbol{\zeta}_{-3,1}(3^2\cdot 5\pi^2 - 2^4\lambda^2) + 2^{11}(2G+3\pi\lambda)\operatorname{Im}L_{2,1,1}(i) - 4271\zeta_3^2 + 2^2\zeta_3\lambda\left(11\cdot 23\,\pi^2 - \frac{2^4\cdot 7\lambda^2}{3}\right) + 2^9\pi\left(\frac{2^2G\lambda}{\pi} + \pi^2 + 2\lambda^2\right)\operatorname{Im}L_{2,1}(i) + 2^6G^2(\pi^2+2^2\lambda^2) + \frac{2^6\pi G\lambda}{3}(3^2\pi^2+2^2\lambda^2) + \frac{2^2\cdot 7\pi^4\lambda^2}{3^2\cdot 5} + \frac{118171\pi^6}{2^2\cdot 3^4\cdot 5\cdot 7}$
$(1,2,1)$	$\frac{7\cdot 313\zeta_5}{2} + 2^6\cdot 17\boldsymbol{\zeta}_{-3,1,1} + 2^{10}\cdot 3\operatorname{Re}L_{2,2,1}(i) - 2^5\cdot 3^3\cdot 5\pi\operatorname{Im}\operatorname{Li}_4(i) + 2^9\cdot 7\pi\operatorname{Im}L_{2,1,1}(i) + 2^7\boldsymbol{\zeta}_{-3,1}\lambda + 2^2\zeta_3\left(\frac{601\pi^2}{3} - 2^2\cdot 7\lambda^2\right) + 2^8(2^3G+3\pi\lambda)\operatorname{Im}L_{2,1}(i) + 2^8G^2\lambda + 2^2\pi G\left(\frac{5\cdot 17\pi^2}{3} + 2^4\lambda^2\right) + \frac{2\cdot 7\pi^4\lambda}{3^2\cdot 5}$
$(1,2,2)$	$2^7\cdot 13\cdot 29\boldsymbol{\zeta}_{-5,1} + 2^9\cdot 37\operatorname{Re}L_{4,2}(i) + 2^8\cdot 17\boldsymbol{\zeta}_{-3,1,1,1} + 2^{13}\cdot 3\operatorname{Re}L_{2,2,1,1}(i) + 2\cdot 7\cdot 313\zeta_5\lambda + \frac{2^8\cdot 5^3\cdot 11\pi\operatorname{Im}L_{4,1}(i)}{7} + \frac{2^9\cdot 11\cdot 43\pi\operatorname{Im}L_{3,2}(i)}{7} + 2^8\cdot 17\boldsymbol{\zeta}_{-3,1,1}\lambda + 2^{12}\cdot 3\lambda\operatorname{Re}L_{2,2,1}(i) + 2^{12}\cdot 13\pi\operatorname{Im}L_{2,1,1,1}(i) - 2^7\cdot 3^3\cdot 5\pi\lambda\operatorname{Im}\operatorname{Li}_4(i) - 2^8\boldsymbol{\zeta}_{-3,1}(2\cdot 5\pi^2 - \lambda^2) + 2^{11}(2^2\cdot 3G+7\pi\lambda)\operatorname{Im}L_{2,1,1}(i) - 2\cdot 3^2\cdot 1511\zeta_3^2 + \frac{2^4\zeta_3\lambda}{3}(601\pi^2 - 2^2\cdot 7\lambda^2) - 2^{12}[\operatorname{Im}L_{2,1}(i)]^2 + 2^5\left(\frac{157\pi^3}{3} + 2^8G\lambda + 2^4\cdot 3\pi\lambda^2\right)\operatorname{Im}L_{2,1}(i) + 2^7G^2(3\pi^2+2^2\lambda^2) + \frac{2^4\pi G\lambda}{3}(5\cdot 17\pi^2+2^4\lambda^2) + \frac{2^2\cdot 7\pi^4\lambda^2}{3^2\cdot 5} + \frac{11\cdot 263\cdot 409\pi^6}{2^3\cdot 3^4\cdot 5\cdot 7}$
$(1,3,1)$	$2^4\cdot 29\cdot 263\boldsymbol{\zeta}_{-5,1} + 2^9\cdot 3\cdot 31\operatorname{Re}L_{4,2}(i) + 2^7\cdot 79\boldsymbol{\zeta}_{-3,1,1,1} + 2^{12}\cdot 3\cdot 5\operatorname{Re}L_{2,2,1,1}(i) + 10321\zeta_5\lambda + \frac{2^{10}\cdot 3^2\cdot 103\pi\operatorname{Im}L_{4,1}(i)}{7} + \frac{2^8\cdot 2503\pi\operatorname{Im}L_{3,2}(i)}{7} + 2^7\cdot 79\boldsymbol{\zeta}_{-3,1,1}\lambda + 2^{11}\cdot 3\cdot 5\lambda\operatorname{Re}L_{2,2,1}(i) + 2^{13}\cdot 17\pi\operatorname{Im}L_{2,1,1,1}(i) - 2^8\cdot 3(5^2G+47\pi\lambda)\operatorname{Im}\operatorname{Li}_4(i) - 2^4\boldsymbol{\zeta}_{-3,1}(5\cdot 83\pi^2 - 2^4\lambda^2) + 2^{11}(2\cdot 3\cdot 7G+13\pi\lambda)\operatorname{Im}L_{2,1,1}(i) - 3\cdot 5\cdot 4561\zeta_3^2 + 2^5\cdot 109\pi\zeta_3G + \frac{2^2\zeta_3\lambda}{3}(5113\pi^2 - 2^4\cdot 7\lambda^2) - 2^{13}\cdot 3[\operatorname{Im}L_{2,1}(i)]^2 + 2^6\pi\left(\frac{2^5\cdot 3^2G\lambda}{\pi} + 3\cdot 23\pi^2 + 2^5\lambda^2\right)\operatorname{Im}L_{2,1}(i) + 2^6\cdot 3G^2(3^2\pi^2+2^2\lambda^2) + \frac{2^4\pi G\lambda}{3}(3\cdot 53\pi^2+2^4\lambda^2) + \frac{2^2\cdot 7\pi^4\lambda^2}{3^2\cdot 5} + \frac{13\cdot 77213\pi^6}{2^3\cdot 3^3\cdot 5\cdot 7}$
$(2,1,1)$	$3\cdot 7^2\zeta_5 + 2^7\boldsymbol{\zeta}_{-3,1,1} - 2^5\cdot 13\pi\operatorname{Im}\operatorname{Li}_4(i) + 2^7\boldsymbol{\zeta}_{-3,1}\lambda + 2^9\pi\operatorname{Im}L_{2,1,1}(i) + \frac{2\zeta_3}{3}(89\pi^2 - 2^3\cdot 3\cdot 7\lambda^2) + 2^8\pi\lambda\operatorname{Im}L_{2,1}(i) + 2^2\pi G\left(\frac{13\pi^2}{3} + 2^4\lambda^2\right) + \frac{2\cdot 7\pi^4\lambda}{3^2\cdot 5}$
$(2,1,2)$	$-2^6\cdot 13\boldsymbol{\zeta}_{-5,1} + 2^9\boldsymbol{\zeta}_{-3,1,1,1} + 2^2\cdot 3\cdot 7^2\zeta_5\lambda + \frac{2^8\cdot 5\cdot 31\pi\operatorname{Im}L_{4,1}(i)}{7} + \frac{2^9\cdot 41\pi\operatorname{Im}L_{3,2}(i)}{7} + 2^9\boldsymbol{\zeta}_{-3,1,1}\lambda + 2^{12}\pi\operatorname{Im}L_{2,1,1,1}(i) - 2^7\cdot 13\pi\lambda\operatorname{Im}\operatorname{Li}_4(i) - 2^5\boldsymbol{\zeta}_{-3,1}(5\pi^2 - 2^3\lambda^2) + 2^{11}\pi\lambda\operatorname{Im}L_{2,1,1}(i) + 2\cdot 151\zeta_3^2 + \frac{2^3\zeta_3\lambda}{3}(89\pi^2 - 2^3\cdot 7\lambda^2) + \frac{2^5\pi}{3}(13\pi^2 + 2^4\cdot 3\lambda^2)\operatorname{Im}L_{2,1}(i) + \frac{2^4\pi G\lambda}{3}(13\pi^2+2^4\lambda^2) + \frac{2^2\cdot 7\pi^4\lambda^2}{3^2\cdot 5} + \frac{197\pi^6}{2^2\cdot 3^2\cdot 7}$

With

$$\int_1^i\left(\frac{1+\tau^2}{\tau}\right)^{2n}\frac{i\,\mathrm{d}\tau}{1+\tau^2} = -\frac{2^{4n-2}}{n\binom{2n}{n}} \tag{3.77}$$

TABLE 6. Continuation of Table 4, with $\omega := e^{2\pi i/3}$, $\varrho := e^{\pi i/3}$, $\theta = e^{\pi i/4}$, $G := \operatorname{Im}\operatorname{Li}_2(i)$, $\lambda := \log 2$, and $\varLambda := \log 3$.

(r,m,τ)	$\sum_{n=1}^{\infty}\left[\mathsf{H}_{2n-1}^{(r)}\right]^m\frac{1}{n\binom{2n}{n}}\left(\frac{1+\tau^2}{\tau}\right)^{2n-1} = \int_0^1\left[\mathsf{G}^{(r),m}\left(\frac{(1+\tau^2)t}{(1+t^2)\tau}\right)-\mathsf{G}^{(r),m}\left(-\frac{(1+\tau^2)t}{(1+t^2)\tau}\right)\right]\frac{\mathrm{d}t}{1+t^2}$
$(1,1,\varrho)$	$\frac{7}{2\sqrt3}\operatorname{Im}\operatorname{Li}_2(\omega)-\frac{\pi\varLambda}{3\sqrt3}$
$(1,2,\varrho)$	$-\frac{2\cdot7}{\sqrt3}\operatorname{Im}L_{2,1}(\omega)-\frac{2^5}{3\sqrt3}\operatorname{Im}L_{2,1}(i)+\frac{2^3}{\sqrt3}\operatorname{Im}\operatorname{Li}_{1,1,1}\left(\frac{i}{\varrho},1,-1\right)-\frac{7\varLambda}{\sqrt3}\operatorname{Im}\operatorname{Li}_2(\omega)$ $-\frac{2^4G\lambda}{3\sqrt3}+\frac{2\pi\lambda^2}{\sqrt3}+\frac{\pi\varLambda^2}{3\sqrt3}+\frac{2^2\pi\lambda}{\sqrt3}\operatorname{Re}\operatorname{Li}_1\left(\frac{i}{\varrho}\right)-\frac{\pi}{3\sqrt3}\left[\operatorname{Re}\operatorname{Li}_1\left(\frac{i}{\varrho}\right)\right]^2-\frac{97\pi^3}{2^4\cdot3^4\sqrt3}$
$(2,1,\varrho)$	$\frac{2^5}{3\sqrt3}\operatorname{Im}L_{2,1}(i)-\frac{2^3}{\sqrt3}\operatorname{Im}\operatorname{Li}_{1,1,1}\left(\frac{i}{\varrho},1,-1\right)+\frac{2^4G\lambda}{3\sqrt3}-\frac{2\pi\lambda^2}{\sqrt3}-\frac{2^2\pi\lambda}{\sqrt3}\operatorname{Re}\operatorname{Li}_1\left(\frac{i}{\varrho}\right)$ $+\frac{\pi}{3\sqrt3}\left[\operatorname{Re}\operatorname{Li}_1\left(\frac{i}{\varrho}\right)\right]^2+\frac{7\cdot13\pi^3}{2^4\cdot3^3\sqrt3}$
$(1,1,\theta)$	$\frac{3G}{\sqrt2}-\frac{\pi\lambda}{2^2\sqrt2}$
$(1,2,\theta)$	$-\frac{7}{\sqrt2}\operatorname{Im}L_{2,1}(i)+\frac{2}{\sqrt2}\operatorname{Im}\operatorname{Li}_{2,1}(\theta,\theta)-\frac{2\cdot3}{\sqrt2}\operatorname{Im}L_{2,1}(\theta)+\frac{2}{\sqrt2}[\lambda+3\operatorname{Re}\operatorname{Li}_1(\theta)]\operatorname{Im}\operatorname{Li}_2(\theta)$ $-\frac{7G\lambda}{2\sqrt2}+\frac{\pi\lambda^2}{2^3\sqrt2}-\frac{\pi}{2^2\sqrt2}[\operatorname{Re}\operatorname{Li}_1(\theta)-\operatorname{Re}\operatorname{Li}_1(i\theta)]^2+\frac{3\cdot11\pi^3}{2^8\sqrt2}$
$(2,1,\theta)$	$\frac{1}{\sqrt2}\operatorname{Im}L_{2,1}(i)-\frac{2}{\sqrt2}\operatorname{Im}\operatorname{Li}_{2,1}(\theta,\theta)+\frac{2\cdot3}{\sqrt2}\operatorname{Im}L_{2,1}(\theta)-\frac{2}{\sqrt2}[\lambda+3\operatorname{Re}\operatorname{Li}_1(\theta)]\operatorname{Im}\operatorname{Li}_2(\theta)$ $+\frac{G\lambda}{2\sqrt2}+\frac{\pi}{2^2\sqrt2}[\operatorname{Re}\operatorname{Li}_1(\theta)-\operatorname{Re}\operatorname{Li}_1(i\theta)]^2+\frac{61\pi^3}{2^8\cdot3\sqrt2}$
$(1,1,\frac{i}{\varrho})$	$2^2\operatorname{Im}\operatorname{Li}_2(\omega)+\frac{\pi\varLambda}{3}$
$(1,2,\frac{i}{\varrho})$	$13\operatorname{Im}L_{2,1}(\omega)+\frac{13\varLambda}{2}\operatorname{Im}\operatorname{Li}_2(\omega)+\frac{\pi\varLambda^2}{2^2\cdot3}+\frac{59\pi^3}{2^2\cdot3^4}$
$(2,1,\frac{i}{\varrho})$	$-5\operatorname{Im}L_{2,1}(\omega)-\frac{5\varLambda}{2}\operatorname{Im}\operatorname{Li}_2(\omega)+\frac{\pi\varLambda^2}{2^2\cdot3}+\frac{5\cdot7\pi^3}{2^2\cdot3^4}$

in hand, we can prove (3.70) by integrating (3.36) along a straight line segment (on which $|1+\tau^2|\leqslant 2|\tau|$) and applying Panzer's logarithmic regularization [**69**, §2.3] at $\tau = i$.

It is clear that the $\mathbb{Q}$-vector spaces appearing on the right-hand sides of (3.65)–(3.70) incorporate their counterparts in (3.59)–(3.64) as subsets. Therefore, in view of (3.46), we can replace any occurrence of $\mathsf{H}_{2n-1}^{(r)}+\overline{\mathsf{H}}_{2n-1}^{(r)}$ in (3.65)–(3.70) by $\mathsf{H}_{2n-1}^{(r)}$, without affecting the corresponding CMZV characterizations. □

REMARK 3.6. With Panzer's `HyperInt` [**69**] and Au's `MultipleZetaValues` [**5**], one can verify all the entries of Tables 4–6.

REMARK 3.7. In [**55**, (2.1)], Kalmykov–Ward–Yost considered

$$\sum_{n=1}^{\infty}\left[\prod_{j=1}^{M}\mathsf{H}_{n-1}^{(r_j)}\right]\left[\prod_{j'=1}^{M'}\mathsf{H}_{2n-1}^{(r'_{j'})}\right]\frac{\left(\frac{1-\tau^2}{1+\tau^2}\right)^{\delta_{0,s}}}{n^{s+1}\binom{2n}{n}}\left(\frac{1+\tau^2}{\tau}\right)^{2n} \tag{3.78}$$

among other things. In their capacities as $\mathbb{Q}$-linear combinations of the left-hand side in (3.65), these Kalmykov–Ward–Yost series are also explicitly representable as members of $\mathfrak{g}_{k+1}^{(\tau)}(4)$ for $s\in\mathbb{Z}_{\geqslant0}$ and $k = s+\sum_{j=1}^{M}r_j+\sum_{j'=1}^{M'}r'_{j'}$. For $\tau=\varrho$ and small values of k, Kalmykov–Veretin [**54**, Appendix B] proposed closed-form formulae for some special cases of (3.78), based on numerical evidence. Diligent readers may check these experimental findings analytically, using Au's package [**5**]. For generic τ and small values of k, Davydychev–Kalmykov [**35**] evaluated (3.78) in terms of classical polylogarithms and their associates.

REMARK 3.8. We note that

$$\begin{aligned}&\mathsf{H}^{(r)}_{n-\frac12}+2\delta_{1,r}\log 2-2^{r-1}\left[\mathsf{H}^{(r)}_{2n-1}+\overline{\mathsf{H}}^{(r)}_{2n-1}\right]\\ &=\begin{cases}0, & r=1,\\ 2(1-2^{r-1})\zeta_r, & r\in\mathbb{Z}_{>1}\end{cases}\end{aligned}\tag{3.79}$$

is always a member of $\mathfrak{g}^{(\varrho)}_r(2)$. Therefore, the relation in (3.65) can be rewritten as

(3.65′)

$$\sum_{n=1}^\infty\left[\prod_{j=1}^M\mathsf{H}^{(r_j)}_{n-1}\right]\left\{\prod_{j'=1}^{M'}\left[\mathsf{H}^{(r'_{j'})}_{n-\frac12}+2\delta_{1,r'_{j'}}\log 2\right]\right\}\frac{(i\sqrt3)^{\delta_{0,s}}}{n^{s+1}\binom{2n}{n}}\in\mathfrak{g}^{(\varrho)}_{k+1}(4)\cap\mathfrak{Z}_{k+1}(12),$$

the $s=0$ case of which combines with (2.41) into

$$i\sqrt3\sum_{n=0}^\infty c_n\prod_{j=1}^M\mathfrak{h}^{(r_j)}_n\in\mathfrak{g}^{(\varrho)}_{k+1}(4)\cap\mathfrak{Z}_{k+1}(12).\tag{3.80}$$

Unfortunately, this result is not sharp enough [cf. (3.120) in §3.4] to recover the MPLs at sixth roots exhibited in Theorem 2.5, let alone the third roots sharpened by Theorem 2.7.

We initially confirmed the sufficiency of third roots in the MPL characterizations of $m_k(1+x_1+x_2)$ for small positive integers k by checking manually that these series representations [cf. (3.30)] of hyper-Mahler measures fit into the framework of (3.34*) and (3.59). Later on, we performed variable substitutions and contour deformations on the Borwein–Borwein–Straub–Wan integrals (Theorem 2.7) that led us to $\pi i\, m_k(1+x_1+x_2)\in\mathfrak{Z}_{k+1}(3)$ as a whole, for all positive integers k.

In parallel to the inverse binomial sums treated in Corollary 3.4, we will soon present period characterizations for the binomial sums, the latter of which played their part in some hyper-Mahler measures studied by Kurokawa–Lalín–Ochiai [**58**, Theorem 19].

COROLLARY 3.9 (Binomial sums via GPLs). *As before, we set* $r_1,\dots,r_M\in\mathbb{Z}_{>0}$ *and* $r'_1,\dots,r'_{M'}\in\mathbb{Z}_{>0}$.
(a) *If* $4|\chi|\leqslant|1+\chi|^2$ *and* $s\in\mathbb{Z}_{>0}$, *then we have* (*cf.* [**55**, (1.1) and (5.2b)])

$$\sum_{n=1}^\infty\left[\prod_{j=1}^M\mathsf{H}^{(r_j)}_{n-1}\right]\frac{\binom{2n}{n}}{n^s}\left[\frac{\chi}{(1+\chi)^2}\right]^n\in\mathfrak{g}^{(\chi)}_k(2)\cap\mathfrak{g}^{\{-1,\chi,\frac1\chi\}}_{k,1},\tag{3.81}$$

where $k=s+\sum_{j=1}^M r_j$.
(b) *If* $2|v|\leqslant|1+v^2|$ *and* $s\in\mathbb{Z}_{>0}$, *then we have*

$$\begin{aligned}&\sum_{n=1}^\infty\left[\prod_{j=1}^M\mathsf{H}^{(r_j)}_{n-1}\right]\left\{\prod_{j'=1}^{M'}\left[\mathsf{H}^{(r'_{j'})}_{2n-1}+\overline{\mathsf{H}}^{(r'_{j'})}_{2n-1}\right]\right\}\frac{\binom{2n}{n}}{n^s}\left(\frac{v}{1+v^2}\right)^{2n}\\ &\in\mathfrak{g}^{(v)}_k(4)\cap\mathfrak{g}^{\{i,-i,v,-\frac1v,-v,\frac1v\}}_{k,2},\end{aligned}\tag{3.82}$$

where $k=s+\sum_{j=1}^{M}r_j+\sum_{j'=1}^{M'}r'_{j'}$.

PROOF. All the infinite series of our concern are convergent, thanks to the asymptotic behavior $\binom{2n}{n}=\frac{4^n}{\sqrt{n\pi}}\left[1+O\left(\frac{1}{n}\right)\right]$ as $n\to\infty$.

(a) When $s\in\mathbb{Z}_{>0}$, the left-hand side of (3.81) is

$$\begin{aligned}\sum_{n=1}^{\infty}\left[\prod_{j=1}^{M}\mathsf{H}_{n-1}^{(r_j)}\right]\left[\frac{16\chi}{(1+\chi)^2}\right]^n\int_0^1\frac{x^n(1-x)^n}{\pi n^s}\frac{\mathrm{d}x}{\sqrt{x(1-x)}}\\ \in\operatorname{span}_{\mathbb{Q}}\left\{\int_0^1 L_{\boldsymbol{A}}\left(\frac{16\chi x(1-x)}{(1+\chi)^2}\right)\frac{\mathrm{d}x}{\pi\sqrt{x(1-x)}}\,\middle|\, w(\boldsymbol{A})=s+\sum_{j=1}^{M}r_j\right\}.\end{aligned}\tag{3.83}$$

Setting $x=\frac{1}{2}-\frac{1}{4i}\left(\sqrt{t}-\frac{1}{\sqrt{t}}\right)$ for t on the unit circle, we have

$$\int_0^1 L_{\boldsymbol{A}}\left(\frac{16\chi x(1-x)}{(1+\chi)^2}\right)\frac{\mathrm{d}x}{\pi\sqrt{x(1-x)}}=\oint_{|t|=1}L_{\boldsymbol{A}}\left(\Xi(\chi,t)\right)\frac{\mathrm{d}t}{2\pi it},\tag{3.84}$$

where $\Xi(\chi,t):=\frac{(1+t)^2\chi}{(1+\chi)^2t}$.

The following fibration of the multi-dimensional polylogarithm

$$\begin{aligned}&L_{\boldsymbol{A}}\left(\Xi(\chi,t)\right)\\ &\in\operatorname{span}_{\mathbb{Q}}\left\{\left(\pi i\frac{\operatorname{Im}\Xi(\chi,t)}{|\operatorname{Im}\Xi(\chi,t)|}\right)^{\ell}Z_j\log^r\frac{1}{\Xi(\chi,t)}\,\middle|\,\begin{smallmatrix}\ell,j,r\in\mathbb{Z}_{\geqslant0}\\ Z_j\in\mathfrak{Z}_j(1)\\ \ell+j+r=w(\boldsymbol{A})\end{smallmatrix}\right\}\\ &\quad+\operatorname{span}_{\mathbb{Q}}\left\{\left(\pi i\frac{\operatorname{Im}\Xi(\chi,t)}{|\operatorname{Im}\Xi(\chi,t)|}\right)^{\ell}Z_jL_{\boldsymbol{B}}\left(\frac{1}{\Xi(\chi,t)}\right)\log^r\frac{1}{\Xi(\chi,t)}\,\middle|\,\begin{smallmatrix}\ell,j,r\in\mathbb{Z}_{\geqslant0};w(\boldsymbol{B})\in\mathbb{Z}_{>0}\\ Z_j\in\mathfrak{Z}_j(1)\\ \ell+j+w(\boldsymbol{B})+r=w(\boldsymbol{A})\end{smallmatrix}\right\}\end{aligned}\tag{3.85}$$

highlights its branch cuts [**67**, **69**], on which $\Xi(\chi,t)>1$. If $f(\Xi(\chi,t))$ belongs to the first summand on the right-hand side of (3.85), then

$$\oint_{|t|=1}f\left(\Xi(\chi,t)\right)\frac{\mathrm{d}t}{2\pi it}\in\mathfrak{g}_{w(\boldsymbol{A})}^{(\chi)}(2)\cap\mathfrak{g}_{w(\boldsymbol{A}),1}^{\{-1,\chi,\frac{1}{\chi}\}}\tag{3.86}$$

follows from the Borwein–Straub relation [the "$\in$" part of (2.52)] and the fact that $\log\frac{\chi}{(1+\chi)^2}=G(0;\chi)-2G(-1;\chi)+2\pi i\mathbb{Z}\in\mathfrak{g}_1^{(\chi)}(2)\cap\mathfrak{g}_{1,1}^{\{-1,\chi,\frac{1}{\chi}\}}$. If $f(\Xi(\chi,t))$ belongs to the second summand on the right-hand side of (3.85), then (3.86) remains valid, for two reasons: (1) One can shrink the contour to tight loops wrapping around the branch cuts inside the unit circle, whose boundary points belong to $\{0,-1,\chi,\frac{1}{\chi}\}$; (2) Across the branch cuts, the integrand $f(\Xi(\chi,t))$ has jumps in the $\mathbb{Q}$-vector space $\pi i\mathfrak{g}_{w(\boldsymbol{A})-1}^{(1/\Xi(\chi,t))}(1)$, so (3.86) results from fibrations of GPLs with respect to t and integrations of GPLs along the branch cuts.

The arguments in the last paragraph take us to the the right-hand side of (3.81).

(b) When $s\in\mathbb{Z}_{>0}$, the left-hand side of (3.82) is

(3.87)
$$\sum_{n=1}^\infty\left[\prod_{j=1}^M\mathsf{H}_{n-1}^{(r_j)}\right]\left\{\prod_{j'=1}^{M'}\left[\mathsf{H}_{2n-1}^{(r'_{j'})}+\overline{\mathsf{H}}_{2n-1}^{(r'_{j'})}\right]\right\}\frac{1}{n^s}\oint_{|\tau|=1}\left[\frac{(1+\tau^2)\upsilon}{(1+\upsilon^2)\tau}\right]^{2n}\frac{\mathrm{d}\,\tau}{2\pi i\tau}$$
$$\in\operatorname{span}_{\mathbb{Q}}\left\{\oint_{|\tau|=1}\frac{G\left(\widetilde\alpha_1,\dots,\widetilde\alpha_w;\frac{(1+\tau^2)\upsilon}{(1+\upsilon^2)\tau}\right)\mathrm{d}\,\tau}{2\pi i\tau}\middle|\begin{smallmatrix}\widetilde\alpha_1,\dots,\widetilde\alpha_{w-1}\in\{-1,0,1\};\widetilde\alpha_w\in\{-1,1\}\\ w=s+\sum_{j=1}^M r_j+\sum_{j'=1}^{M'}r'_{j'}\end{smallmatrix}\right\},$$
in light of (3.51).

To verify the claim in (3.82), use GPL fibrations and contour deformations as in our proof of (3.81). □

REMARK 3.10. For the treatment of the $s=0$ case in binomial sums, see [**94**, §2]. The contour deformation techniques in the proof above can also be generalized to infinite sums involving $\binom{3n}{n}$ and $\binom{4n}{2n}$ [**94**, §3]. For further developments in series related to $\binom{3n}{n}$ and $\binom{4n}{2n}$, see [**30, 78**].

3.3. Infinite series related to $m_k(1+x_1+x_2+x_3)$. Akin to the descriptions at the beginning of the last subsection, we once evaluated $m_k(1+x_1+x_2+x_3)$, $1\leqslant k\leqslant 6$, by taking k-th order derivatives of Crandall's hypergeometric representation [**15**, (6.8)] $W_4(s)=U_4(s)+V_4(s)$, where

$$U_4(s):=\frac{2^s}{\cot\frac{\pi s}{2}}\left[\frac{\Gamma\left(1+\frac{s}{2}\right)}{\sqrt{\pi}\Gamma\left(\frac{3+s}{2}\right)}\right]^3{}_4F_3\left(\left.\begin{matrix}\frac12,\frac12,\frac12,1+\frac{s}{2}\\ \frac{3+s}{2},\frac{3+s}{2},\frac{3+s}{2}\end{matrix}\right|1\right),\tag{3.88}$$

$$V_4(s):=\frac{2^s\Gamma\left(\frac{1+s}{2}\right)}{\sqrt{\pi}\Gamma\left(1+\frac{s}{2}\right)}{}_4F_3\left(\left.\begin{matrix}\frac12,-\frac{s}{2},-\frac{s}{2},-\frac{s}{2}\\ 1,1,\frac{1-s}{2}\end{matrix}\right|1\right).\tag{3.89}$$

During these evaluations, the derivative

$$\left.\frac{\mathrm{d}^k\,U_4(s)}{\mathrm{d}\,s^k}\right|_{s=0},\quad\text{resp.}\quad\left.\frac{\mathrm{d}^k\,V_4(s)}{\mathrm{d}\,s^k}\right|_{s=0}\tag{3.90}$$

produced a $\mathbb{Q}$-linear combination of

$$\pi^{2(r_0-1)}\sum_{n=1}^\infty\frac{\prod_{j=1}^\ell\left[\mathsf{H}_{n-1}^{(r_j)}-3\mathsf{H}_{n-\frac12}^{(r_j)}+2\delta_{1,r_j}\log2\right]}{(2n-1)^3},\tag{3.91}$$

$$\text{resp.}\quad\pi^{2\widetilde r_0}\sum_{n=1}^\infty\frac{\prod_{j=1}^{\widetilde\ell}\left[3\mathsf{H}_{n-1}^{(\widetilde r_j)}-\mathsf{H}_{n-\frac12}^{(\widetilde r_j)}-2\delta_{1,\widetilde r_j}\log2\right]}{n^3},\tag{3.92}$$

where $2r_0+\sum_{j=1}^\ell r_j=k-1$, $2\widetilde r_0+\sum_{j=1}^{\widetilde\ell}\widetilde r_j=k+1$, and $r_0,\widetilde r_0\in\mathbb{Z}_{\geqslant0}$. (Here, both ℓ and $\widetilde\ell$ may be zero, whereupon the empty products become equal to 1.)

The next corollary will allow us to strengthen Theorem 2.11 as

$$\left.\frac{\mathrm{d}^k\,U_4(s)}{\mathrm{d}\,s^k}\right|_{s=0}\in\frac{1}{\pi^2}\mathfrak{Z}_{k+2}(2),\qquad\left.\frac{\mathrm{d}^k\,V_4(s)}{\mathrm{d}\,s^k}\right|_{s=0}\in\mathfrak{Z}_k(2).\tag{3.93}$$

Such AMZV characterizations will also apply to Broadhurst's series in (3.5).

COROLLARY 3.11 (Euler sums, HPLs, and Goncharov–Deligne periods).

(a) *For* $s \in \mathbb{Z}_{\geqslant 0}$, $r_1, \dots, r_M \in \mathbb{Z}_{>0}$, *and* $r'_1, \dots, r'_{M'} \in \mathbb{Z}_{>0}$, *we have*

$$\sum_{n=1}^{\infty}\left[\prod_{j=1}^{M}\mathsf{H}_{n-1}^{(r_j)}\right]\left\{\prod_{j'=1}^{M'}\left[\mathsf{H}_{2n-1}^{(r'_{j'})}+\overline{\mathsf{H}}_{2n-1}^{(r'_{j'})}\right]\right\}\frac{z^{2n-1}}{(2n-1)^{s+1}}\in\mathfrak{G}_{k+1}^{(z)}(2),\tag{3.94}$$

where $|z|<1$, $k=s+\sum_{j=1}^{M}r_j+\sum_{j'=1}^{M'}r'_{j'}$. *For* $s \in \mathbb{Z}_{\geqslant 0}$, $r_1, \dots, r_M \in \mathbb{Z}_{>0}$, *and* $r'_1, \dots, r'_{M'} \in \mathbb{Z}_{>0}$, *the result above extends to*

$$\sum_{n=1}^{\infty}\left[\prod_{j=1}^{M}\mathsf{H}_{n-1}^{(r_j)}\right]\left\{\prod_{j'=1}^{M'}\left[\mathsf{H}_{2n-1}^{(r'_{j'})}+\overline{\mathsf{H}}_{2n-1}^{(r'_{j'})}\right]\right\}\frac{1}{(2n-1)^{s+2}}\in\mathfrak{Z}_{k+2}(2),\tag{3.95}$$

$$\sum_{n=1}^{\infty}\left[\prod_{j=1}^{M}\mathsf{H}_{n-1}^{(r_j)}\right]\left\{\prod_{j'=1}^{M'}\left[\mathsf{H}_{2n-1}^{(r'_{j'})}+\overline{\mathsf{H}}_{2n-1}^{(r'_{j'})}\right]\right\}\frac{(-1)^n}{(2n-1)^{s+1}}\in i\mathfrak{Z}_{k+1}(4),\tag{3.96}$$

where $k=s+\sum_{j=1}^{M}r_j+\sum_{j'=1}^{M'}r'_{j'}$.

(b) *For* $s \in \mathbb{Z}_{\geqslant 0}$, $r_1, \dots, r_M \in \mathbb{Z}_{>0}$, *and* $r'_1, \dots, r'_{M'} \in \mathbb{Z}_{>0}$, *we have*

$$\sum_{n=1}^{\infty}\left[\prod_{j=1}^{M}\mathsf{H}_{n-1}^{(r_j)}\right]\left\{\prod_{j'=1}^{M'}\left[\mathsf{H}_{2n-1}^{(r'_{j'})}+\overline{\mathsf{H}}_{2n-1}^{(r'_{j'})}\right]\right\}\frac{z^{2n}}{n^{s+1}}\in\mathfrak{G}_{k+1}^{(z)}(2),\tag{3.97}$$

where $|z|<1$, $k=s+\sum_{j=1}^{M}r_j+\sum_{j'=1}^{M'}r'_{j'}$. *For* $s \in \mathbb{Z}_{\geqslant 0}$, $r_1, \dots, r_M \in \mathbb{Z}_{>0}$, *and* $r'_1, \dots, r'_{M'} \in \mathbb{Z}_{>0}$, *the result above extends to*

$$\sum_{n=1}^{\infty}\left[\prod_{j=1}^{M}\mathsf{H}_{n-1}^{(r_j)}\right]\left\{\prod_{j'=1}^{M'}\left[\mathsf{H}_{2n-1}^{(r'_{j'})}+\overline{\mathsf{H}}_{2n-1}^{(r'_{j'})}\right]\right\}\frac{1}{n^{s+2}}\in\mathfrak{Z}_{k+2}(2),\tag{3.98}$$

$$\sum_{n=1}^{\infty}\left[\prod_{j=1}^{M}\mathsf{H}_{n-1}^{(r_j)}\right]\left\{\prod_{j'=1}^{M'}\left[\mathsf{H}_{2n-1}^{(r'_{j'})}+\overline{\mathsf{H}}_{2n-1}^{(r'_{j'})}\right]\right\}\frac{(-1)^n}{n^{s+1}}\in\mathfrak{Z}_{k+1}(4),\tag{3.99}$$

where $k=s+\sum_{j=1}^{M}r_j+\sum_{j'=1}^{M'}r'_{j'}$.

PROOF. (a) To prove (3.94) for $s=0$, divide both sides of (3.47) by z and integrate. For progress onto all $s \in \mathbb{Z}_{>0}$, apply the operations $f(z)\mapsto\int_0^z f(x)\frac{\mathrm{d}x}{x}$ iteratively. In fact, this method allows us to deduce a slightly stronger statement:

$$\begin{aligned}&\sum_{n=1}^{\infty}\left[\prod_{j=1}^{M}\mathsf{H}_{n-1}^{(r_j)}\right]\left\{\prod_{j'=1}^{M'}\left[\mathsf{H}_{2n-1}^{(r'_{j'})}+\overline{\mathsf{H}}_{2n-1}^{(r'_{j'})}\right]\right\}\frac{z^{2n-1}}{(2n-1)^{s+1}}\\&\quad\in\operatorname{span}_{\mathbb{Q}}\left\{g(z)-g(-z)\,\middle|\,g(z)\in\mathfrak{G}^{(z)}_{s+1+\sum_{j=1}^{M}r_j+\sum_{j'=1}^{M'}r'_{j'}}(2)\right\}.\end{aligned}\tag{3.100}$$

When $s\in\mathbb{Z}_{\geqslant 0}$ (resp. $s\in\mathbb{Z}_{\geqslant 1}$), the integral representation for the left-hand side of (3.94) converges at $z=i$ (resp. $z=1$), according to variations on the proof of Theorem 3.1. Thus, the period characterizations in (3.95) and (3.96) immediately follow.

(b) These are collateral results of (3.51). □

REMARK 3.12. We note that Xu [**87**] (resp. Xu–Wang [**89**]) studied the $M'=0$ (resp. $M=0$) cases of the series (3.95) and (3.96), using different methods based on Hoffman's multiple t-values [**52**]. A recent result of Au [**3**, Theorem 1] extends (3.95) and (3.96) to $\mathfrak{Z}(N)$ with $N\in\mathbb{Z}_{\geqslant2}$.

REMARK 3.13. Since $\log 2\in\mathfrak{Z}_1(2)$, one can confirm (3.93) by combining (3.95) and (3.98) with (3.79). Unlike the discrepancy between Theorem 2.7 and (3.80), the AMZV characterizations in Theorem 2.11 remain effective down to the individual non-linear Euler sums [see (3.91) and (3.92)] that constitute the k-th order derivatives in (3.93).

The techniques in the last four corollaries can be combined to take some recently proven series evaluations [**31, 33**] a fair distance.

COROLLARY 3.14 (GPL and CMZV representations of generalized Sun series). *If* $r_1,\dots,r_M\in\mathbb{Z}_{>0}$, $r_1',\dots,r_{M'}'\in\mathbb{Z}_{>0}$, $|1+\tau^2|\leqslant 2|\tau|$, *and* $s\in\mathbb{Z}_{\geqslant0}$, *then we have*

$$\begin{aligned}\sum_{n=1}^\infty\left[\prod_{j=1}^M\mathsf{H}_{n-1}^{(r_j)}\right]\left\{\prod_{j'=1}^{M'}\left[\mathsf{H}_{2n-1}^{(r_{j'}')}+\overline{\mathsf{H}}_{2n-1}^{(r_{j'}')}\right]\right\}\frac{\binom{2n-2}{n-1}\left(\frac{1+\tau^2}{\tau}\right)^{2n-1}}{2^{4(n-1)}(2n-1)^{s+1}}\\ \in i\left(\mathfrak{g}_{k+1}^{(\tau)}(4)\cap\mathfrak{g}_{k+1,2}^{\{i,-i,\tau,-\frac1\tau,-\tau,\frac1\tau\}}\right),\end{aligned}\tag{3.101}$$

where $k=s+\sum_{j=1}^M r_j+\sum_{j'=1}^{M'}r_{j'}'$.

For $\tau=\varrho:=e^{\pi i/3}$, *we have*

$$\sum_{n=1}^\infty\left[\prod_{j=1}^M\mathsf{H}_{n-1}^{(r_j)}\right]\left\{\prod_{j'=1}^{M'}\left[\mathsf{H}_{2n-1}^{(r_{j'}')}+\overline{\mathsf{H}}_{2n-1}^{(r_{j'}')}\right]\right\}\frac{\binom{2n-2}{n-1}}{2^{4(n-1)}(2n-1)^{s+1}}\in i\left[\mathfrak{g}_{k+1}^{(\varrho)}(4)\cap\mathfrak{Z}_{k+1}(12)\right],\tag{3.102}$$

where $k=s+\sum_{j=1}^M r_j+\sum_{j'=1}^{M'}r_{j'}'$. *It is possible to use CMZVs at a lower level to represent a special subclass of* (3.102), *namely*

$$\sum_{n=1}^\infty\left[\prod_{j=1}^M\mathsf{H}_{n-\frac12}^{(r_j)}\right]\frac{\binom{2n-2}{n-1}}{2^{4(n-1)}(2n-1)^{s+1}}\in i\mathfrak{Z}_{k+1}(3),\tag{3.102$'$}$$

where $k=s+\sum_{j=1}^M r_j$.

For $\tau=i\rho:=i\frac{\sqrt5-1}{2}$, *we have*

$$\sum_{n=1}^\infty\left[\prod_{j=1}^M\mathsf{H}_{n-1}^{(r_j)}\right]\left\{\prod_{j'=1}^{M'}\left[\mathsf{H}_{2n-1}^{(r_{j'}')}+\overline{\mathsf{H}}_{2n-1}^{(r_{j'}')}\right]\right\}\frac{\binom{2n-2}{n-1}(-1)^{n-1}}{2^{4(n-1)}(2n-1)^{s+1}}\in\mathfrak{g}_{k+1}^{(i\rho)}(4)\cap\mathfrak{Z}_{k+1}(60),\tag{3.103}$$

where $k=s+\sum_{j=1}^M r_j+\sum_{j'=1}^{M'}r_{j'}'$. *If* $k+1\leqslant3$, *then the left-hand side of* (3.103) *belongs to* $\mathfrak{Z}_{k+1}(10)$.

PROOF. Noting that

$$\frac{\binom{2n-2}{n-1}}{2^{4(n-1)}}=\int_0^1\left[\sqrt{X(1-X)}\right]^{2n-1}\frac{\mathrm{d}X}{\pi X(1-X)},\tag{3.104}$$

we may identify the left-hand side of (3.101) with a member of

$$\operatorname{span}_{\mathbb{Q}}\left\{\int_0^{\infty}\left[g\left(\frac{(1+\tau^2)t}{(1+t^2)\tau}\right)-g\left(-\frac{(1+\tau^2)t}{(1+t^2)\tau}\right)\right]\frac{\mathrm{d}\,t}{\pi t}\,\middle|\, g(z)\in\mathfrak{G}^{(z)}_{k+1}(2)\right\},\tag{3.105}$$

by virtue of the symmetry displayed in (3.100) and a variable substitution $X=\frac{1}{1+t^2}$. This effectively reduces our target to an integral

$$\int_{-\infty}^{\infty} G\left(\alpha_1,\dots,\alpha_{k+1};\frac{(1+\tau^2)t}{(1+t^2)\tau}\right)\frac{\mathrm{d}\,t}{\pi t},\qquad \alpha_1,\dots,\alpha_{k+1}\in\{-1,0,1\},\tag{3.106}$$

the contour of which can be deformed to loops wrapping around the branch cuts in the upper half-plane, whereon $\frac{(1+\tau^2)t}{(1+t^2)\tau}\in(-\infty,-1)\cup(1,\infty)\subset\mathbb{R}$. The end points of these branch cuts are in $\{i,\tau,\frac{1}{\tau},-\tau,-\frac{1}{\tau}\}\cap\{w\in\mathbb{C}\mid\operatorname{Im}w>0\}$. As we fibrate with respect to t (resp. τ) before (resp. after) integration over branch cuts, we arrive at the claim in the right-hand side of (3.101).

The rationales behind the right-hand sides of (3.102) and (3.103) are similar to those for (3.65) and (3.66).

Enlisting help from (3.28), we may equate the left-hand side of (3.102′) with[21]

$$\begin{aligned}&\frac{1}{2^s}\sum_{n=1}^{\infty}\frac{\binom{2n-2}{n-1}}{2^{4(n-1)}}\int_0^1\frac{\mathsf{L}^{(r_1,\dots,r_M)}_{s+1}(x^2)}{2n-1}\,\mathrm{d}x^{2n-1}\\&\quad=\frac{1}{2^{s-1}}\int_{\tau=i}^{\tau=\varrho}\mathsf{L}^{(r_1,\dots,r_M)}_{s+1}\left(\frac{(1+\tau^2)^2}{\tau^2}\right)\mathrm{d}\arcsin\frac{1+\tau^2}{2\tau},\end{aligned}\tag{3.107}$$

where $\mathsf{L}^{(r_1,\dots,r_M)}_{s+1}(X)\in\mathfrak{g}^{(X)}_k(1)$ and $k=s+\sum_{j=1}^M r_j$. Introducing a new variable $u=-\tau^2$, one can reveal the expression above as member of

(3.108)

$$\begin{aligned}&\operatorname{span}_{\mathbb{Q}}\left\{(\pi i)^{k-\ell}Z_{\ell-j}\int_1^{1/\varrho}G\left(\alpha_1,\dots,\alpha_j;-\frac{(1-u)^2}{u}\right)\frac{\mathrm{d}\,u}{iu}\,\middle|\,\begin{smallmatrix}\alpha_1,\dots,\alpha_j\in\{0,1\}\\ Z_{\ell-j}\in\mathfrak{Z}_{\ell-j}(1)\\ 0\leqslant j\leqslant\ell\leqslant k\end{smallmatrix}\right\}\\&\subseteq\operatorname{span}_{\mathbb{Q}}\left\{(\pi i)^{k-\ell}Z_{\ell-j}\int_1^{1/\varrho}G(\alpha_1,\dots,\alpha_j;u)\frac{\mathrm{d}\,u}{iu}\,\middle|\,\begin{smallmatrix}\alpha_1,\dots,\alpha_j\in\left\{0,1,\varrho=\frac{\omega^2-1}{z-1},\frac{1}{\varrho}=\frac{z}{\omega^2}\frac{\omega^2-1}{z-1}\right\}\\ Z_{\ell-j}\in\mathfrak{Z}_{\ell-j}(1)\\ z=\omega=e^{2\pi i/3};0\leqslant j\leqslant\ell\leqslant k\end{smallmatrix}\right\}\\&\subseteq\left.\frac{\mathfrak{g}_{k+1}[1,\omega,\omega^2;z](3)}{i}\right|_{z=\omega}\subseteq i\mathfrak{Z}_{k+1}(3).\end{aligned}$$

If the weight $k+1$ does not exceed 3 in the case of (3.103), then we are effectively working with $\mathfrak{g}^S_{k+1,2}$, where the subset S of $\{i,-i,\tau,-\frac{1}{\tau},-\tau,\frac{1}{\tau}\}=\{i,-i,i\rho,-\frac{1}{i\rho},-i\rho,\frac{1}{i\rho}\}$ contains no more than 3 members. By virtue of GPL rescaling and the fact that $\log\rho=\operatorname{Re}[\operatorname{Li}_1(e^{4\pi i/5})-\operatorname{Li}_1(e^{2\pi i/5})]\in\mathfrak{Z}_1(5)$, we only need to check that $G(\boldsymbol{\alpha};1)\in\mathfrak{Z}_{w(\boldsymbol{\alpha})}(10)$ when the components of $\boldsymbol{\alpha}$ involve no more than 2 different numbers from one of the following sets:

(3.109)

$$\{1,-1,\rho,\tfrac{1}{\rho},-\rho,-\tfrac{1}{\rho}\},\quad\{\tfrac{1}{\rho},-\tfrac{1}{\rho},1,\tfrac{1}{\rho^2},-1,-\tfrac{1}{\rho^2}\},\quad\{\rho,-\rho,\rho^2,1,-\rho^2,-1\}.$$

This can be done automatically by Au's `IterIntDoableQ` function. □

[21]To construct Tables 7 and 8, we need the following special cases of $\mathsf{L}^{(1)}_s(x)=\frac{1}{2\pi i}\left[\mathsf{G}^{(1),1}_s\left(\frac{1}{x-i0^+}\right)-\mathsf{G}^{(1),1}_s\left(\frac{1}{x+i0^+}\right)\right]$ for $0<x<1$: $\mathsf{L}^{(1)}_1(x)=\operatorname{Li}_1(x)$, $\mathsf{L}^{(1)}_2(x)=-\operatorname{Li}_2(x)+\frac{\pi^2}{6}$, $\mathsf{L}^{(1)}_3(x)=\operatorname{Li}_3(x)-\frac{\pi^2\log x}{6}-\zeta_3$, and $\mathsf{L}^{(1)}_4(x)=-\operatorname{Li}_4(x)+\zeta_3\log x+\frac{\pi^2\log^2 x}{12}+\frac{\pi^4}{90}$.

TABLE 7. Selected closed forms for binomial sums of Sun's type, where $\omega := e^{2\pi i/3}$, $\varrho := e^{\pi i/3}$, $\rho := \frac{\sqrt{5}-1}{2}$, $\varsigma := e^{2\pi i/5}$, $G := \operatorname{Im}\operatorname{Li}_2(i)$, and $\lambda := \log 2$.

s	$\sum_{n=1}^{\infty}\frac{\binom{2n-2}{n-1}}{2^{4(n-1)}(2n-1)^{s+1}} = \frac{1}{2^s s!}\int_1^{\varrho}\left[\log\left(-\frac{u}{(1-u)^2}\right)\right]^s\frac{\mathrm{d}u}{iu}$	$\sum_{n=1}^{\infty}\mathsf{H}^{(1)}_{n-\frac{1}{2}}\frac{\binom{2n-2}{n-1}}{2^{4(n-1)}(2n-1)^{s}} = \frac{1}{2^{s-1}}\int_1^{\varrho}\mathsf{L}^{(1)}_s\left(-\frac{(1-u)^2}{u}\right)\frac{\mathrm{d}u}{iu}$
1	$\frac{3\operatorname{Im}\operatorname{Li}_2(\omega)}{2}$	$\operatorname{Im}\operatorname{Li}_2(\omega)$
2	$\frac{7\pi^3}{2^3\cdot 3^3}$	$3\operatorname{Im}L_{2,1}(\omega)+\frac{\pi^3}{3^3}$
3	$\frac{3^3\operatorname{Im}\operatorname{Li}_4(\omega)}{2^5}+\frac{\pi\zeta_3}{2^2\cdot 3}$	$\frac{3\operatorname{Im}\operatorname{Li}_4(\omega)}{2^3}+\frac{\pi\zeta_3}{2^2\cdot 3}$
4	$-\frac{3^4\operatorname{Im}L_{4,1}(\omega)}{2^3\cdot 13}-\frac{3^3\operatorname{Im}L_{3,2}(\omega)}{2^3\cdot 13}+\frac{229\pi^5}{2^7\cdot 3^2\cdot 5\cdot 13}$	$-\frac{3^4\operatorname{Im}L_{4,1}(\omega)}{2^2\cdot 13}-\frac{3^2\cdot 5^2\operatorname{Im}L_{3,2}(\omega)}{2^4\cdot 13}+\frac{5\cdot 19\pi^5}{2^6\cdot 3^4\cdot 13}$

s	$\sum_{n=1}^{\infty}\mathsf{H}^{(1)}_{2n-1}\frac{\binom{2n-2}{n-1}}{2^{4(n-1)}(2n-1)^{s}} = 2\int_{\varrho}^{i}\mathsf{L}^{(1)}_s\left(\frac{1+t^2}{t}\right)\frac{\mathrm{d}t}{it}$
1	$\frac{2^3G}{3}-2\operatorname{Im}\operatorname{Li}_2(\omega)$
2	$3\operatorname{Im}L_{2,1}(\omega)-\frac{2^5\operatorname{Im}L_{2,1}(i)}{3}+2^3\operatorname{Im}L_{2,1}\left(\frac{i}{\varrho}\right)+2^3\operatorname{Im}\operatorname{Li}_{1,1,1}\left(\frac{i}{\varrho},1,-1\right)$ $-\frac{2^4G}{3}\left[\lambda+\operatorname{Re}\operatorname{Li}_1\left(\frac{i}{\varrho}\right)\right]+\pi\operatorname{Re}\operatorname{Li}_1\left(\frac{i}{\varrho}\right)\left[\frac{1}{3}\operatorname{Re}\operatorname{Li}_1\left(\frac{i}{\varrho}\right)+2^2\lambda\right]+2\pi\lambda^2-\frac{11\cdot 19\pi^3}{2^4\cdot 3^4}$

s	$\sum_{n=1}^{\infty}\frac{\binom{2n-2}{n-1}(-1)^{n-1}}{2^{4(n-1)}(2n-1)^{s+1}} = \frac{2}{s!}\int_{i\rho}^{i}\log^s\left(\frac{t}{i(1+t^2)}\right)\frac{\mathrm{d}t}{t}$	$\sum_{n=1}^{\infty}\mathsf{H}^{(1)}_{2n-1}\frac{\binom{2n-2}{n-1}(-1)^{n-1}}{2^{4(n-1)}(2n-1)^{s}} = 2\int_{i\rho}^{i}\mathsf{L}^{(1)}_s\left(\frac{i(1+t^2)}{t}\right)\frac{\mathrm{d}t}{t}$
1	$\frac{\pi^2}{2\cdot 5}$	$-2^2\operatorname{Re}\operatorname{Li}_{1,1}\left(-\frac{1}{\varsigma^2},\varsigma\right)+3\left\{\operatorname{Re}[\operatorname{Li}_1(\varsigma^2)-\operatorname{Li}_1(\varsigma)]\right\}^2-2^2\left\{[\operatorname{Re}\operatorname{Li}_1(\varsigma^2)]^2-[\operatorname{Re}\operatorname{Li}_1(\varsigma)]^2\right\}-\frac{2^2\pi^2}{3\cdot 5^2}$
2	$\frac{5^2\operatorname{Re}\operatorname{Li}_3(\varsigma)}{2^2\cdot 3}+\frac{\zeta_3}{2}$	$-5\operatorname{Re}\operatorname{Li}_3(\varsigma)+\frac{2^3\zeta_3}{5}$

REMARK 3.15. As revealed by Table 7, two $\mathfrak{Z}_{k+1}(10)$ cases of (3.103) cover an infinite series evaluated in [**33**, §4]:

$$\sum_{n=0}^{\infty}\frac{\binom{2n}{n}\left[5\mathsf{H}^{(1)}_{2n+1}+\frac{12}{2n+1}\right]}{(2n+1)^2(-16)^n}=14\zeta_3. \tag{3.110}$$

Charlton–Gangl–Lai–Xu–Zhao [**33**, §§2–3] verified another infinite series due to Zhi-Wei Sun, namely

$$\sum_{n=0}^{\infty}\frac{\binom{2n}{n}\left[9\mathsf{H}^{(1)}_{2n+1}+\frac{32}{2n+1}\right]}{(2n+1)^3 16^n}=40\operatorname{Im}\operatorname{Li}_4(i)+\frac{5\pi\zeta_3}{12} \tag{3.111}$$

by maneuvering identities of polylogarithms. Au [**3**, Corollary 12] achieved the same by an application of the Wilf–Zeilberger method. At the time of this writing, we do not have an automated proof of (3.111), since Au's `MultipleZetaValues` package currently does not simplify CMZVs in $\mathfrak{Z}_4(12)$.

COROLLARY 3.16 (Generalized Wang–Chu series as Goncharov–Deligne periods). *For $s\in\mathbb{Z}_{\geqslant0}$, $r_1,\dots,r_M\in\mathbb{Z}_{>0}$, and $r'_1,\dots,r'_{M'}\in\mathbb{Z}_{>0}$, we have* (*cf.* [**31**, (17), (18), (27)–(29)])

$$\sum_{n=1}^{\infty}\left[\prod_{j=1}^{M}\mathsf{H}_{n-1}^{(r_j)}\right]\left\{\prod_{j'=1}^{M'}\left[\mathsf{H}_{2n-1}^{(r'_{j'})}+\overline{\mathsf{H}}_{2n-1}^{(r'_{j'})}\right]\right\}\frac{\binom{2n-2}{n-1}^2}{2^{4n}(2n-1)^{s+1}}\in\frac{\mathfrak{Z}_{k+2}(4)}{\pi i},\tag{3.112}$$

$$\sum_{n=1}^{\infty}\left[\prod_{j=1}^{M}\mathsf{H}_{n-1}^{(r_j)}\right]\left\{\prod_{j'=1}^{M'}\left[\mathsf{H}_{2n-1}^{(r'_{j'})}+\overline{\mathsf{H}}_{2n-1}^{(r'_{j'})}\right]\right\}\frac{\binom{2n}{n}^2}{2^{4n}n^{s+1}}\in\frac{\mathfrak{Z}_{k+2}(4)}{\pi i},\tag{3.113}$$

where $k=s+\sum_{j=1}^{M}r_j+\sum_{j'=1}^{M'}r'_{j'}$.

PROOF. Observe that

$$\frac{\pi}{2}\binom{2n-2}{n-1}=\int_0^{\pi/2}(2\cos\phi)^{2n-2}\,\mathrm{d}\phi=\int_1^i\left(\frac{1+\tau^2}{\tau}\right)^{2n-1}\frac{\mathrm{d}\tau}{i(1+\tau^2)},\tag{3.114}$$

where the integral contour in the complex τ-plane is a straight line segment, on which $|1+\tau^2|\leqslant2|\tau|$. Hence, the last corollary allows us to equate the left-hand side of (3.112) with a member of the $\mathbb{Q}$-vector space

$$\frac{1}{\pi}\operatorname{span}_{\mathbb{Q}}\left\{\int_1^i\frac{(\pi i)^{\ell}G(\alpha_1,\dots,\alpha_{k+1-\ell};\tau)\,\mathrm{d}\tau}{1+\tau^2}\;\middle|\;\substack{\alpha_1,\dots,\alpha_{k+1-\ell}\in\{0,1,i,-1,-i\}\\0\leqslant\ell\leqslant k+1}\right\}.\tag{3.115}$$

Appealing to the GPL recursion in (2.22) and applying Panzer's logarithmic regularization [**69**, §2.3] at $\tau=i$, we may recognize the expression above as a subspace of $\frac{1}{\pi i}\mathfrak{Z}_{k+2}(4)$.

To prove (3.113), first verify

$$\int_0^1\left(\frac{v}{1+v^2}\right)^{2n}\frac{\mathrm{d}v}{1+v^2}=\frac12\int_0^{1/2}[x(1-x)]^{n-\frac12}\,\mathrm{d}x=\frac{\pi\binom{2n}{n}}{2^{2(2n+1)}}\tag{3.116}$$

by a substitution $v=\sqrt{x}/\sqrt{1-x}$, then integrate the relation (3.82) in Corollary 3.9(b). □

REMARK 3.17. Some special cases of (3.112) and (3.113) have been investigated by Cantarini–D'Aurizio [**32**, §§2–3] in the Fourier–Legendre framework. See Table 8 for more examples.

3.4. An outlook on ε-expansions of hypergeometric expressions. As pointed out in [**55**, §3], a common source for non-linear Euler sums and related series in high energy physics is an ε-expansion in the form

$$\begin{aligned}&\left.\frac{\partial^{\ell}}{\partial\varepsilon^{\ell}}\right|_{\varepsilon=0}{}_pF_{p-1}\left(\substack{a_1+r_1\varepsilon,\dots,a_p+r_p\varepsilon\\b_1+s_1\varepsilon,\dots,b_{p-1}+s_{p-1}\varepsilon}\;\middle|\;z\right)\\&:=\left.\frac{\partial^{\ell}}{\partial\varepsilon^{\ell}}\right|_{\varepsilon=0}\left[1+\sum_{n=1}^{\infty}\frac{\prod_{j=1}^{p}(a_j+r_j\varepsilon)_n}{\prod_{k=1}^{p-1}(b_k+s_k\varepsilon)_n}\frac{z^n}{n!}\right]\in\mathbb{C},\qquad\ell\in\mathbb{Z}_{\geqslant0},\end{aligned}\tag{3.117}$$

TABLE 8. Selected closed forms for quadratic analogs of binomial sums, where $G := \operatorname{Im}\operatorname{Li}_2(i)$ and $\lambda := \log 2$.

s	$\sum_{n=1}^{\infty}\frac{\binom{2n-2}{n-1}^2}{2^{4n}(2n-1)^{s+1}} = \frac{1}{2^2\pi s!}\int_1^i\left[\int_i^\tau \log^s\frac{(1+\tau^2)t}{(1+t^2)\tau}\frac{\mathrm{d}t}{t}\right]\frac{\mathrm{d}\tau}{1+\tau^2}$	$\sum_{n=1}^{\infty}\frac{\binom{2n}{n}^2}{2^{4n}n^{s+1}} = \frac{2^{s+4}}{\pi s!}\int_0^1\left[\int_0^v \log^s\frac{(1+t^2)v}{(1+v^2)t}\frac{t\,\mathrm{d}t}{1+t^2}\right]\frac{\mathrm{d}v}{1+v^2}$
0	$\frac{G}{2^2\pi}$	$-\frac{2^3G}{\pi}+2^2\lambda$
1	$\frac{\operatorname{Im}L_{2,1}(i)}{\pi}+\frac{G\lambda}{2\pi}$	$\frac{2^6\operatorname{Im}L_{2,1}(i)}{\pi}+\frac{2^5G\lambda}{\pi}-2^3\lambda^2$
2	$-\frac{3\operatorname{Im}\operatorname{Li}_4(i)}{\pi}+\frac{2^2\operatorname{Im}L_{2,1,1}(i)}{\pi}+\frac{5\cdot7\zeta_3}{2^6}$ $+\frac{2\lambda\operatorname{Im}L_{2,1}(i)}{\pi}+\frac{G}{2\pi}\left(\frac{\pi^2}{2^2}+\lambda^2\right)$	$\frac{2^7\cdot3\operatorname{Im}\operatorname{Li}_4(i)}{\pi}-\frac{2^9\operatorname{Im}L_{2,1,1}(i)}{\pi}-2\cdot3\cdot11\zeta_3$ $-\frac{2^8\lambda\operatorname{Im}L_{2,1}(i)}{\pi}-\frac{2^6G}{\pi}\left(\frac{\pi^2}{2^2}+\lambda^2\right)+\frac{2^5\lambda^3}{3}$
3	$\frac{2^4\cdot3^2\operatorname{Im}L_{4,1}(i)}{7\pi}+\frac{2^2\cdot19\operatorname{Im}L_{3,2}(i)}{7\pi}$ $+\frac{2^4\operatorname{Im}L_{2,1,1,1}(i)}{\pi}-\frac{2\cdot3\lambda\operatorname{Im}\operatorname{Li}_4(i)}{\pi}-\frac{5\zeta_{-3,1}}{2^3}$ $+\frac{2^3\lambda\operatorname{Im}L_{2,1,1}(i)}{\pi}+\frac{5\cdot7\zeta_3\lambda}{2^5}$ $+\left(\frac{\pi^2}{2^2}+\lambda^2\right)\frac{2\operatorname{Im}L_{2,1}(i)}{\pi}+\frac{G\lambda}{\pi}\left(\frac{\pi^2}{2^2}+\frac{\lambda^2}{3}\right)$ $+\frac{73\pi^4}{2^8\cdot3^2\cdot7}$	$\frac{2^{12}\cdot3^2\operatorname{Im}L_{4,1}(i)}{7\pi}+\frac{2^{10}\cdot19\operatorname{Im}L_{3,2}(i)}{7\pi}$ $+\frac{2^{12}\operatorname{Im}L_{2,1,1,1}(i)}{\pi}-\frac{2^9\cdot3\lambda\operatorname{Im}\operatorname{Li}_4(i)}{\pi}$ $-2^5\cdot5\zeta_{-3,1}+\frac{2^{11}\lambda\operatorname{Im}L_{2,1,1}(i)}{\pi}$ $+2^3\cdot3\cdot11\zeta_3\lambda+\left(\frac{\pi^2}{2^2}+\lambda^2\right)\frac{2^9\operatorname{Im}L_{2,1}(i)}{\pi}$ $+\frac{2^8G\lambda}{\pi}\left(\frac{\pi^2}{2^2}+\frac{\lambda^2}{3}\right)-\frac{2^5\lambda^4}{3}+\frac{73\pi^4}{3^2\cdot7}$

s	$\sum_{n=1}^{\infty}\frac{\binom{2n-2}{n-1}^2\mathsf{H}^{(1)}_{n-\frac12}}{2^{4n}(2n-1)^{s}} = \frac{1}{2^{s+1}\pi}\int_1^i\left[\int_i^\tau \mathsf{L}^{(1)}_s\left(\frac{(1+t^2)^2\tau^2}{(1+\tau^2)^2t^2}\right)\frac{\mathrm{d}t}{t}\right]\frac{\mathrm{d}\tau}{1+\tau^2}$	$\sum_{n=1}^{\infty}\frac{\binom{2n}{n}^2\mathsf{H}^{(1)}_{n}}{2^{4n}n^{s}} = \frac{2^{4}}{\pi}\int_0^1\left[\int_0^v \mathsf{L}^{(1)}_s\left(\frac{(1+v^2)^2t^2}{(1+t^2)^2v^2}\right)\frac{t\,\mathrm{d}t}{1+t^2}\right]\frac{\mathrm{d}v}{1+v^2}$
1	$\frac{2^2\operatorname{Im}L_{2,1}(i)}{\pi}+\frac{G\lambda}{\pi}$	$-\frac{2^6\operatorname{Im}L_{2,1}(i)}{\pi}-\frac{\pi^2}{2\cdot3}$
2	$-\frac{37\operatorname{Im}\operatorname{Li}_4(i)}{2\pi}+\frac{2^3\cdot3\operatorname{Im}L_{2,1,1}(i)}{\pi}+\frac{3\cdot5\cdot7\zeta_3}{2^5}$ $+\frac{2^3\lambda\operatorname{Im}L_{2,1}(i)}{\pi}+\frac{3G}{\pi}\left(\frac{\pi^2}{2^2}+\frac{\lambda^2}{3}\right)$	$-\frac{2^5\cdot5^2\operatorname{Im}\operatorname{Li}_4(i)}{\pi}+\frac{2^{10}\operatorname{Im}L_{2,1,1}(i)}{\pi}+3\cdot47\zeta_3$ $+\frac{2^8\lambda\operatorname{Im}L_{2,1}(i)}{\pi}+2^5\pi G+\frac{2\pi^2\lambda}{3}$
3	$\frac{2^3\cdot149\operatorname{Im}L_{4,1}(i)}{7\pi}+\frac{2\cdot313\operatorname{Im}L_{3,2}(i)}{7\pi}$ $+\frac{2^7\operatorname{Im}L_{2,1,1,1}(i)}{\pi}-\frac{37\lambda\operatorname{Im}\operatorname{Li}_4(i)}{\pi}-5\zeta_{-3,1}$ $+\frac{2^4\cdot3\lambda\operatorname{Im}L_{2,1,1}(i)}{\pi}+\frac{3\cdot5\cdot7\zeta_3\lambda}{2^4}$ $+(\pi^2+2\lambda^2)\frac{2^2\operatorname{Im}L_{2,1}(i)}{\pi}+\frac{G\lambda}{\pi}\left(\frac{3\pi^2}{2}+\frac{2\lambda^2}{3}\right)$ $+\frac{5\cdot233\pi^4}{2^9\cdot3^2\cdot7}$	$-\frac{2^{10}\cdot113\operatorname{Im}L_{4,1}(i)}{7\pi}-\frac{2^8\cdot3\cdot79\operatorname{Im}L_{3,2}(i)}{7\pi}$ $-\frac{2^{12}\cdot3\operatorname{Im}L_{2,1,1,1}(i)}{\pi}+\frac{2^7\cdot5^2\lambda\operatorname{Im}\operatorname{Li}_4(i)}{\pi}$ $+2^5\cdot3\cdot5\zeta_{-3,1}-\frac{2^{12}\lambda\operatorname{Im}L_{2,1,1}(i)}{\pi}$ $-2^2\cdot3\cdot47\zeta_3\lambda-(3\pi^2+2^2\lambda^2)\frac{2^7\operatorname{Im}L_{2,1}(i)}{\pi}$ $-2^7\pi G\lambda-\frac{2^2\pi^2\lambda^2}{3}-\frac{29\cdot151\pi^4}{2^2\cdot3^2\cdot5\cdot7}$

s	$\sum_{n=1}^{\infty}\frac{\binom{2n-2}{n-1}^2\mathsf{H}^{(1)}_{2n-1}}{2^{4n}(2n-1)^{s}} = \frac{1}{2^{2}\pi}\int_1^i\left[\int_i^\tau \mathsf{L}^{(1)}_s\left(\frac{(1+t^2)\tau}{(1+\tau^2)t}\right)\frac{\mathrm{d}t}{t}\right]\frac{\mathrm{d}\tau}{1+\tau^2}$	$\sum_{n=1}^{\infty}\frac{\binom{2n}{n}^2\mathsf{H}^{(1)}_{2n}}{2^{4n}n^{s}} = \frac{2^{s+3}}{\pi}\int_0^1\left[\int_0^v \mathsf{L}^{(1)}_s\left(\frac{(1+v^2)t}{(1+t^2)v}\right)\frac{t\,\mathrm{d}t}{1+t^2}\right]\frac{\mathrm{d}v}{1+v^2}$
1	$\frac{5\operatorname{Im}L_{2,1}(i)}{2\pi}+\frac{G\lambda}{2\pi}+\frac{\pi^2}{2^7}$	$-\frac{2^4\cdot3\operatorname{Im}L_{2,1}(i)}{\pi}-\frac{\pi^2}{2^2\cdot3}$
2	$-\frac{5^2\operatorname{Im}\operatorname{Li}_4(i)}{2\pi}+\frac{2^4\operatorname{Im}L_{2,1,1}(i)}{\pi}+\frac{5\cdot7\zeta_3}{2^4}$ $+\frac{5\lambda\operatorname{Im}L_{2,1}(i)}{\pi}+\frac{G}{2\pi}(\pi^2+\lambda^2)+\frac{\pi^2\lambda}{2^6}$	$-\frac{2^5\cdot19\operatorname{Im}\operatorname{Li}_4(i)}{\pi}+\frac{2^8\cdot3\operatorname{Im}L_{2,1,1}(i)}{\pi}+109\zeta_3$ $+\frac{2^6\cdot3\lambda\operatorname{Im}L_{2,1}(i)}{\pi}+2^3\cdot3\pi G+\frac{\pi^2\lambda}{3}$
3	$\frac{2^6\cdot13\operatorname{Im}L_{4,1}(i)}{7\pi}+\frac{2^2\cdot109\operatorname{Im}L_{3,2}(i)}{7\pi}$ $+\frac{2^3\cdot11\operatorname{Im}L_{2,1,1,1}(i)}{\pi}-\frac{5^2\lambda\operatorname{Im}\operatorname{Li}_4(i)}{\pi}$ $-\frac{5\cdot11\zeta_{-3,1}}{2^4}+\frac{2^5\lambda}{\pi}\operatorname{Im}L_{2,1,1}(i)+\frac{5\cdot7\zeta_3\lambda}{2^3}$ $+\left(\frac{11\pi^2}{2^2}+5\lambda^2\right)\frac{\operatorname{Im}L_{2,1}(i)}{\pi}+\frac{G\lambda}{\pi}\left(\pi^2+\frac{\lambda^2}{3}\right)$ $+\frac{\pi^2\lambda^2}{2^6}+\frac{821\pi^4}{2^9\cdot3^2\cdot7}$	$-\frac{2^{11}\cdot43\operatorname{Im}L_{4,1}(i)}{7\pi}-\frac{2^{10}\cdot3^2\cdot5\operatorname{Im}L_{3,2}(i)}{7\pi}$ $-\frac{2^{10}\cdot3^2\operatorname{Im}L_{2,1,1,1}(i)}{\pi}+\frac{2^7\cdot19\lambda}{\pi}\operatorname{Im}\operatorname{Li}_4(i)$ $+2^3\cdot3^2\cdot5\zeta_{-3,1}-\frac{2^{10}\cdot3\lambda}{\pi}\operatorname{Im}L_{2,1,1}(i)$ $-2^2\cdot109\zeta_3\lambda-2^5\cdot3\left(3\pi^2+2^2\lambda^2\right)\frac{\operatorname{Im}L_{2,1}(i)}{\pi}$ $-2^5\cdot3\pi G\lambda-\frac{2\pi^2\lambda^2}{3}-\frac{29\cdot113\pi^4}{2^2\cdot3^2\cdot5\cdot7}$

where the (generalized) hypergeometric series ${}_pF_{p-1}$ is defined with

$$\begin{cases}2a_j\in\mathbb{Z},\ r_j\in\mathbb{Q}, & 1\leqslant j\leqslant p,\\ 2b_k\in\mathbb{Z},\ s_k\in\mathbb{Q}, & 1\leqslant k\leqslant p-1,\end{cases}\tag{3.118}$$

and $(a)_n:=\Gamma(a+n)/\Gamma(a)$.

Up to this point, the studies of non-linear Euler sums [together with their (inverse) binomial variations] in this section allow us to evaluate some (but not all) ε-expansions via multiple polylogarithms. We emphasize "some (but not all)", because we have only handled certain series whose summands involve powers of binomial coefficients $\binom{2n}{n}^{\mathscr{N}}$ for $\mathscr{N}\in\{-2,-1,0,1,2\}$. These closed-form series are further subdivided into two classes:

- If $\mathscr{N}\in\{-1,0,1\}$ and $|z|\leqslant 1$, then we have polylogarithmic evaluations of (3.117), provided that the choice of parameters in (3.118) turns the right-hand side of (3.117) into a $\mathbb{Q}$-linear combination of (3.6), (3.10), (3.34), (3.36), (3.81), (3.82), (3.94), (3.97), and (3.101), while ensuring its convergence at $z=1$;
- If $\mathscr{N}\in\{-2,2\}$ and $z=1$, then we have polylogarithmic reductions of (3.117) when the parameters in (3.118) are orchestrated in such a way that the right-hand side of (3.117) becomes a $\mathbb{Q}$-linear combination of (3.64), (3.70), (3.112), and (3.113).

Instead of a microscopic dissection of (3.117) into a $\mathbb{Q}$-linear combination of [(inverse) binomial variations on] non-linear Euler sums, one can compute ε-expansions while treating the (generalized) hypergeometric series ${}_pF_{p-1}$ as a whole.

To illustrate the holistic perspective of ${}_pF_{p-1}$, we consider a double integral representation[22] of a zeta Mahler measure for $s\in(-2,0)$,

$$W_3(s)=-\frac{3^{s+\frac32}\sin\frac{\pi s}{2}}{\pi^2}\int_0^1\mathrm{d}t\int_0^1\mathrm{d}u\,\frac{2\left[\frac{1-u^2}{4-(1-u^2)t}\frac{t}{1-t}\right]^{s/2}}{(1-t)\left[4-(1-u^2)t\right]},\tag{3.119}$$

which descends from a formula of Borwein–Borwein–Straub–Wan [**9**, (24), with $z=\sqrt{1-u^2}$] and Euler's integral representation for ${}_2F_1$ [**2**, Theorem 2.2.1]. Exploiting a beta integral $\int_0^1\left(\frac{t}{1-t}\right)^{s/2}\frac{\mathrm{d}t}{1-t}=-\frac{\pi}{\sin\frac{\pi s}{2}}$, we can extend (3.119) to

(3.119′)

$$W_3(s)=\frac{3^{s+\frac32}}{\pi}\left(\int_0^1\frac{2\left(\frac{1-u^2}{3+u^2}\right)^{s/2}}{3+u^2}\mathrm{d}u-\frac{\sin\frac{\pi s}{2}}{\pi}\int_0^1\mathrm{d}t\int_0^1\mathrm{d}u\,\frac{2\left[\frac{1-u^2}{4-(1-u^2)t}\frac{t}{1-t}\right]^{s/2}}{\left(t-\frac{4}{1-u^2}\right)(3+u^2)}\right.$$
$$\left.-\frac{\sin\frac{\pi s}{2}}{\pi}\int_0^1\mathrm{d}t\int_0^1\mathrm{d}u\,\frac{2\left(\frac{t}{1-t}\right)^{s/2}\left\{\left[\frac{1-u^2}{4-(1-u^2)t}\right]^{s/2}-\left(\frac{1-u^2}{3+u^2}\right)^{s/2}\right\}}{(1-t)(3+u^2)}\right)$$

[22] The author thanks an anonymous referee for suggesting this approach.

for $s \in (-2,2)$. Differentiating (3.119′) under the integral sign, and fibrating with respect to u after integrating over t, we find

(3.120)

$$\left.\frac{\mathrm{d}^k W_3(s)}{\mathrm{d}\, s^k}\right|_{s=0} \in \frac{\sqrt{3}}{\pi}\operatorname{span}_{\mathbb{Q}}\left\{\pi^{2\ell}(\log 3)^m\int_0^1\frac{G(\boldsymbol{\alpha};u)G(\boldsymbol{\beta};1)\,\mathrm{d}u}{3+u^2}\left|\begin{smallmatrix}\alpha_1,\dots,\alpha_{w(\boldsymbol{\alpha})}\in\{0,1,-1,i\sqrt{3},-i\sqrt{3}\}\\ \beta_1,\dots,\beta_{w(\boldsymbol{\beta})}\in\{0,1,4\}\\ \ell,m,w(\boldsymbol{\alpha}),w(\boldsymbol{\beta})\in\mathbb{Z}_{\geqslant 0}\\ 2\ell+m+w(\boldsymbol{\alpha})+w(\boldsymbol{\beta})=k\end{smallmatrix}\right.\right\} \subseteq \frac{\mathfrak{Z}_{k+1}(6)}{\pi i}.$$

Here, to confirm the "$\in$" step, build

$$\begin{aligned}&\operatorname{span}_{\mathbb{Q}}\left\{G(\widetilde{\alpha}_1,\dots,\widetilde{\alpha}_n;1)\left|\,\widetilde{\alpha}_1,\dots,\widetilde{\alpha}_n\in\left\{0,1,\frac{4}{1-u^2}\right\}\right.\right\}\\ &\subseteq\operatorname{span}_{\mathbb{Q}}\left\{G(\boldsymbol{\alpha};u)G(\boldsymbol{\beta};1)\left|\begin{smallmatrix}\alpha_1,\dots,\alpha_{w(\boldsymbol{\alpha})}\in\{0,1,-1,i\sqrt{3},-i\sqrt{3}\}\\ \beta_1,\dots,\beta_{w(\boldsymbol{\beta})}\in\{0,1,4\}\\ w(\boldsymbol{\alpha}),w(\boldsymbol{\beta})\in\mathbb{Z}_{\geqslant 0};w(\boldsymbol{\alpha})+w(\boldsymbol{\beta})=n\end{smallmatrix}\right.\right\}\end{aligned}\tag{3.121}$$

inductively on (2.37), with logarithmic regularization [**69**, §2.3] when $\widetilde{\alpha}_1 = 1$; to confirm the "$\subseteq$" step, simply note that $\pi^2 \in \mathfrak{Z}_2(1)$, $\log 2 \in \mathfrak{Z}_1(2)$, and $\log 3 \in \mathfrak{Z}_1(3)$, before evaluating `IterIntDoableQ[{0, 1, -1, I Sqrt[3], -I Sqrt[3]}]` and `IterIntDoableQ[{0, 1, 4}]` in Au's `MultipleZetaValues` package [**5**]. This recovers Theorem 2.5.

For $|z| \leqslant 1$, an ε-expansion in (3.117) may also give rise to summands involving $\binom{2n}{n}^{\mathscr{N}}$ for $\mathscr{N} \in \mathbb{Z} \smallsetminus \{-1,0,1\}$. Such binomial variations on non-linear Euler sums may or may not be representable by CMZVs, according to Zhi-Wei Sun's recent experimental discoveries (see [**76**, §§2–4] and [**77**, §6]). Some of Sun's new identities have been verified analytically by Campbell [**29**], Hou–Sun [**53**], Wei [**82, 83**], Li–Chu [**65**], and Wei–Xu [**84, 85**]. For future research, it is possibly worth studying (multiple) integral representations for these newly discovered ε-expansions and their generalizations, in search of algorithmic reductions to CMZVs and non-CMZVs.

Acknowledgments

The current research grew out of my unpublished notes on polylogarithms and multiple zeta values prepared in 2012 for Prof. Weinan E's working seminar on mathematical problems in quantum field theory at Princeton University. Some parts of this paper were written during my visits to Prof. E at Peking University in summers of 2017 and 2022. I dedicate this work to his 60th birthday and his vision for automating mathematical research.

I am deeply indebted to three anonymous referees, whose perceptive comments helped me improve the presentation of this paper in many ways.

References

[1] H. Akatsuka, *Zeta Mahler measures*, J. Number Theory **129** (2009), no. 11, 2713–2734, DOI 10.1016/j.jnt.2009.05.007. MR2549527

[2] G. E. Andrews, R. Askey, and R. Roy, *Special functions*, Encyclopedia of Mathematics and its Applications, vol. 71, Cambridge University Press, Cambridge, 1999, DOI 10.1017/CBO9781107325937. MR1688958

[3] K. C. Au, *Multiple zeta values, WZ-pairs and infinite sums computations*, Ramanujan J. **66** (2025), no. 1, Paper No. 3, 35 pp., DOI 10.1007/s11139-024-00987-3. MR4843119

[4] K. C. Au. Evaluation of one-dimensional polylogarithmic integral, with applications to infinite series. `arXiv:2007.03957v3` [math.NT], 2022.

[5] K. C. Au. Iterated integrals and multiple polylogarithm at algebraic arguments. `arXiv:2201.01676v2` [math.NT] (software available at `https://www.researchgate.net/publication/357601353_Mathematica_package_MultipleZetaValues` and `https://www.wolframcloud.com/obj/pisco125/MultipleZetaValues-1.2.0.paclet`), 2022.

[6] S. Bloch, M. Kerr, and P. Vanhove, *A Feynman integral via higher normal functions*, Compos. Math. **151** (2015), no. 12, 2329–2375, DOI 10.1112/S0010437X15007472. MR3433889

[7] J. Blümlein, D. J. Broadhurst, and J. A. M. Vermaseren, *The multiple zeta value data mine*, Comput. Phys. Comm. **181** (2010), no. 3, 582–625, DOI 10.1016/j.cpc.2009.11.007. MR2578167

[8] D. Borwein and J. M. Borwein, *On an intriguing integral and some series related to* $\zeta(4)$, Proc. Amer. Math. Soc. **123** (1995), no. 4, 1191–1198, DOI 10.2307/2160718. MR1231029

[9] D. Borwein, J. M. Borwein, A. Straub, and J. Wan, *Log-sine evaluations of Mahler measures, II*, Integers **12** (2012), no. 6, 1179–1212, DOI 10.1515/integers-2012-0035. MR3011556

[10] J. M. Borwein, D. M. Bradley, D. J. Broadhurst, and P. Lisoněk, *Special values of multiple polylogarithms*, Trans. Amer. Math. Soc. **353** (2001), no. 3, 907–941, DOI 10.1090/S0002-9947-00-02616-7. MR1709772

[11] J. M. Borwein, D. Nuyens, A. Straub, and J. Wan, *Some arithmetic properties of short random walk integrals*, Ramanujan J. **26** (2011), no. 1, 109–132, DOI 10.1007/s11139-011-9325-y. MR2837721

[12] J. M. Borwein and A. Straub, *Special values of generalized log-sine integrals*, ISSAC 2011—Proceedings of the 36th International Symposium on Symbolic and Algebraic Computation, ACM, New York, 2011, pp. 43–50, DOI 10.1145/1993886.1993899. MR2895193

[13] J. M. Borwein and A. Straub, *Log-sine evaluations of Mahler measures*, J. Aust. Math. Soc. **92** (2012), no. 1, 15–36, DOI 10.1017/S1446788712000067. MR2945674

[14] J. M. Borwein, A. Straub, and J. Wan, *Three-step and four-step random walk integrals*, Exp. Math. **22** (2013), no. 1, 1–14, DOI 10.1080/10586458.2013.748379. MR3038778

[15] J. M. Borwein, A. Straub, J. Wan, and W. Zudilin, *Densities of short uniform random walks*, Canad. J. Math. **64** (2012), no. 5, 961–990, DOI 10.4153/CJM-2011-079-2. With an appendix by Don Zagier. MR2979573

[16] J. M. Borwein, D. J. Broadhurst, and J. Kamnitzer, *Central binomial sums, multiple Clausen values, and zeta values*, Experiment. Math. **10** (2001), no. 1, 25–34. MR1821569

[17] J. M. Borwein, I. J. Zucker, and J. Boersma, *The evaluation of character Euler double sums*, Ramanujan J. **15** (2008), no. 3, 377–405, DOI 10.1007/s11139-007-9083-z. MR2390277

[18] D. Boyd, D. Lind, F. Rodríguez-Villegas, and C. Deninger. The many aspects of Mahler's measure. Final report of the Banff workshop 03w5035 (April 26–May 1, 2003), 2003.

[19] D. W. Boyd, *Speculations concerning the range of Mahler's measure*, Canad. Math. Bull. **24** (1981), no. 4, 453–469, DOI 10.4153/CMB-1981-069-5. MR644535

[20] D. J. Broadhurst. On the enumeration of irreducible k-fold Euler sums and their roles in knot theory and field theory. `arXiv:hep-th/9604128`, 1996.

[21] D. J. Broadhurst, *Massive 3-loop Feynman diagrams reducible to* SC* *primitives of algebras of the sixth root of unity*, Eur. Phys. J. C Part. Fields **8** (1999), no. 2, 313–333, DOI 10.1007/s100529900935. MR1739510

[22] D. Broadhurst. Bessel moments, random walks and Calabi–Yau equations. preprint, `https://carma.newcastle.edu.au/jon/Preprints/Papers/Submitted%20Papers/%4step-walks/walk-broadhurst.pdf`, 2009.

[23] D. Broadhurst, *Multiple zeta values and modular forms in quantum field theory*, Computer algebra in quantum field theory, Texts Monogr. Symbol. Comput., Springer, Vienna, 2013, pp. 33–73. MR3616746

[24] J. Broedel, O. Schlotterer, and F. Zerbini, *From elliptic multiple zeta values to modular graph functions: open and closed strings at one loop*, J. High Energy Phys. **1** (2019), 155, front matter+67, DOI 10.1007/jhep01(2019)155. MR3919340

[25] F. Brown, *The massless higher-loop two-point function*, Comm. Math. Phys. **287** (2009), no. 3, 925–958, DOI 10.1007/s00220-009-0740-5. MR2486668

[26] F. Brown. On the periods of some Feynman integrals. `arXiv:0910.0114v2` [math.AG], 2010.

[27] F. C. S. Brown, *Multiple zeta values and periods of moduli spaces* $\overline{\mathfrak{M}}_{0,n}$ (English, with English and French summaries), Ann. Sci. Éc. Norm. Supér. (4) **42** (2009), no. 3, 371–489, DOI 10.24033/asens.2099. MR2543329

[28] F. Brunault and W. Zudilin, *Many variations of Mahler measures—a lasting symphony*, Australian Mathematical Society Lecture Series, vol. 28, Cambridge University Press, Cambridge, 2020. MR4382435

[29] J. M. Campbell. On a conjecture on a series of convergence rate $\frac{1}{2}$. arXiv:2304.00360 [math.NT], 2023.

[30] J. M. Campbell, M. L. Glasser, and Y. Zhou, *New evaluations of inverse binomial series via cyclotomic multiple zeta values*, SIGMA Symmetry Integrability Geom. Methods Appl. **20** (2024), Paper No. 079, 14 pp., DOI 10.3842/SIGMA.2024.079. MR4843368

[31] J. M. Campbell, P. Levrie, C. Xu, and J. Zhao, *On a problem involving the squares of odd harmonic numbers*, Ramanujan J. **63** (2024), no. 2, 387–408, DOI 10.1007/s11139-023-00765-7. MR4694513

[32] M. Cantarini and J. D'Aurizio, *On the interplay between hypergeometric series, Fourier-Legendre expansions and Euler sums*, Boll. Unione Mat. Ital. **12** (2019), no. 4, 623–656, DOI 10.1007/s40574-019-00198-5. MR4022298

[33] S. Charlton, H. Gangl, L. Lai, C. Xu, and J. Zhao, *On two conjectures of Sun concerning Apéry-like series*, Forum Math. **35** (2023), no. 6, 1533–1547, DOI 10.1515/forum-2022-0325. MR4661542

[34] A. I. Davydychev and M. Yu. Kalmykov, *New results for the ϵ-expansion of certain one-, two- and three-loop Feynman diagrams*, Nuclear Phys. B **605** (2001), no. 1-3, 266–318, DOI 10.1016/S0550-3213(01)00095-5. MR1845851

[35] A. I. Davydychev and M. Yu. Kalmykov, *Massive Feynman diagrams and inverse binomial sums*, Nuclear Phys. B **699** (2004), no. 1-2, 3–64, DOI 10.1016/j.nuclphysb.2004.08.020. MR2095786

[36] P. J. de Doelder, *On some series containing $\psi(x) - \psi(y)$ and $(\psi(x) - \psi(y))^2$ for certain values of x and y*, J. Comput. Appl. Math. **37** (1991), no. 1-3, 125–141, DOI 10.1016/0377-0427(91)90112-W. MR1136919

[37] P. Deligne, *Le groupe fondamental unipotent motivique de $\mathbf{G}_m - \mu_N$, pour $N = 2, 3, 4, 6$ ou 8* (French), Publ. Math. Inst. Hautes Études Sci. **112** (2010), 101–141, DOI 10.1007/s10240-010-0027-6. MR2737978

[38] P. Deligne and A. B. Goncharov, *Groupes fondamentaux motiviques de Tate mixte* (French, with English and French summaries), Ann. Sci. École Norm. Sup. (4) **38** (2005), no. 1, 1–56, DOI 10.1016/j.ansens.2004.11.001. MR2136480

[39] C. Deninger, *Deligne periods of mixed motives, K-theory and the entropy of certain $\mathbf{Z}^n$-actions*, J. Amer. Math. Soc. **10** (1997), no. 2, 259–281, DOI 10.1090/S0894-0347-97-00228-2. MR1415320

[40] V. G. Drinfel′d, *On quasitriangular quasi-Hopf algebras and on a group that is closely connected with* $\mathrm{Gal}(\overline{\mathbf{Q}}/\mathbf{Q})$ (Russian), Algebra i Analiz **2** (1990), no. 4, 149–181; English transl., Leningrad Math. J. **2** (1991), no. 4, 829–860. MR1080203

[41] C. Duhr and F. Dulat, *PolyLogTools—polylogs for the masses*, J. High Energy Phys. **8** (2019), 135, 56, DOI 10.1007/jhep08(2019)135. MR4013407

[42] L. Euler. Meditationes circa singulare serierum genus. *Novi Commun. Acad. Sci. Petropol.*, 20:140–186, 1775. (see also *Opera Omnia*, Series 1, Vol. 15, pp. 217–267).

[43] H. Frellesvig. Generalized polylogarithms in Maple. arXiv:1806.02883 [hep-th], 2018.

[44] H. Frellesvig, D. Tommasini, and C. Wever. On the reduction of generalized polylogarithms to Li_n and $\mathrm{Li}_{2,2}$ and on the evaluation thereof. *J. High Energy Phys.*, 2016(3):189, 2016. arXiv:1601.02649v3 [hep-ph].

[45] J. Fresán. Une introduction aux périodes. arXiv:2210.03407v1 [math.AG], 2022.

[46] A. B. Goncharov, *The double logarithm and Manin's complex for modular curves*, Math. Res. Lett. **4** (1997), no. 5, 617–636, DOI 10.4310/MRL.1997.v4.n5.a1. MR1484694

[47] A. B. Goncharov, *Multiple polylogarithms, cyclotomy and modular complexes*, Math. Res. Lett. **5** (1998), no. 4, 497–516, DOI 10.4310/MRL.1998.v5.n4.a7. MR1653320

[48] A. B. Goncharov. Multiple polylogarithms and mixed Tate motives. arXiv:math/0103059 [math.AG], 2001.

[49] A. B. Goncharov and Yu. I. Manin, *Multiple ζ-motives and moduli spaces* $\overline{\mathcal{M}}_{0,n}$, Compos. Math. **140** (2004), no. 1, 1–14, DOI 10.1112/S0010437X03000125. MR2004120

[50] J. M. Henn, A. V. Smirnov, and V. A. Smirnov, *Evaluating multiple polylogarithm values at sixth roots of unity up to weight six*, Nuclear Phys. B **919** (2017), 315–324, DOI 10.1016/j.nuclphysb.2017.03.026. MR3648336

[51] M. E. Hoffman, *Multiple harmonic series*, Pacific J. Math. **152** (1992), no. 2, 275–290. MR1141796

[52] M. E. Hoffman, *An odd variant of multiple zeta values*, Commun. Number Theory Phys. **13** (2019), no. 3, 529–567. MR4013729

[53] Q.-H. Hou and Z.-W. Sun. Taylor coefficients and series involving harmonic numbers. arXiv:2310.03699v2 [math.NT], 2023.

[54] M. Yu. Kalmykov and O. Veretin, *Single-scale diagrams and multiple binomial sums*, Phys. Lett. B **483** (2000), no. 1-3, 315–323, DOI 10.1016/S0370-2693(00)00574-8. MR1768598

[55] M. Y. Kalmykov, B. F. L. Ward, and S. A. Yost, *On the all-order ϵ-expansion of generalized hypergeometric functions with integer values of parameters*, J. High Energy Phys. **11** (2007), 009, 13 pp., DOI 10.1088/1126-6708/2007/11/009. MR2362140

[56] J. C. Kluyver. A local probability problem. *Nederl. Acad. Wetensch. Proc.*, 8:341–350, 1905. http://www.dwc.knaw.nl/DL/publications/PU00013859.pdf.

[57] M. Kontsevich and D. Zagier, *Periods*, Mathematics unlimited—2001 and beyond, Springer, Berlin, 2001, pp. 771–808. MR1852188

[58] N. Kurokawa, M. Lalín, and H. Ochiai, *Higher Mahler measures and zeta functions*, Acta Arith. **135** (2008), no. 3, 269–297, DOI 10.4064/aa135-3-5. MR2457199

[59] M. N. Lalín, *Some examples of Mahler measures as multiple polylogarithms*, J. Number Theory **103** (2003), no. 1, 85–108, DOI 10.1016/S0022-314X(03)00101-X. MR2008068

[60] M. N. Lalín and J.-S. Lechasseur, *Higher Mahler measure of an n-variable family*, Acta Arith. **174** (2016), no. 1, 1–30, DOI 10.4064/aa8111-3-2016. MR3517530

[61] S. Laporta and E. Remiddi. The analytical value of the electron $(g-2)$ at order α^3 in QED. *Physics Letters B*, 379(1):283–291, 1996. arXiv:hep-ph/9602417.

[62] L. Lewin. On the evaluation of log-sine integrals. *Math. Gaz.*, 42(340):125–128, 1958.

[63] L. Lewin, *Polylogarithms and associated functions*, North-Holland Publishing Co., New York-Amsterdam, 1981. MR618278

[64] L. Lewin (ed.), *Structural properties of polylogarithms*, Mathematical Surveys and Monographs, vol. 37, American Mathematical Society, Providence, RI, 1991, DOI 10.1090/surv/037. MR1148371

[65] C. Li and W. Chu, *Infinite series about harmonic numbers inspired by Ramanujan-like formulae*, Electron. Res. Arch. **31** (2023), no. 8, 4611–4636, DOI 10.3934/era.2023236. MR4611610

[66] D. Maître. HPL, a Mathematica implementation of the harmonic polylogarithms. *Comput. Phys. Comm.*, 174(3):222–240, 2006. arXiv:hep-ph/0507152v2 (software available at https://www.physik.uzh.ch/data/HPL/).

[67] D. Maître. Extension of HPL to complex arguments. *Comput. Phys. Comm.*, 183(3):846, 2012. arXiv:hep-ph/0703052v2.

[68] I. Mező, *Nonlinear Euler sums*, Pacific J. Math. **272** (2014), no. 1, 201–226, DOI 10.2140/pjm.2014.272.201. MR3270178

[69] E. Panzer. Algorithms for the symbolic integration of hyperlogarithms with applications to Feynman integrals. *Comput. Phys. Commun.*, 188:148–166, 2015. (software available at https://bitbucket.org/PanzerErik/hyperint/wiki/Home).

[70] D. E. Radford, *A natural ring basis for the shuffle algebra and an application to group schemes*, J. Algebra **58** (1979), no. 2, 432–454, DOI 10.1016/0021-8693(79)90171-6. MR540649

[71] E. Remiddi and J. A. M. Vermaseren, *Harmonic polylogarithms*, Internat. J. Modern Phys. A **15** (2000), no. 5, 725–754, DOI 10.1142/S0217751X00000367. MR1746652

[72] M. Rogers, *Hypergeometric formulas for lattice sums and Mahler measures*, Int. Math. Res. Not. IMRN **17** (2011), 4027–4058, DOI 10.1093/imrn/rnq240. MR2836402

[73] D. Samart, *On a noncritical symmetric square L-value of the congruent number elliptic curves*, Bull. Aust. Math. Soc. **101** (2020), no. 1, 13–22, DOI 10.1017/s000497271900056x. MR4052904

[74] J. Singer and J. Zhao, *Finite and symmetrized colored multiple zeta values*, Finite Fields Appl. **65** (2020), 101676, 20 pp., DOI 10.1016/j.ffa.2020.101676. MR4082312

[75] A. Straub and W. Zudilin, *Short walk adventures*, From analysis to visualization, Springer Proc. Math. Stat., vol. 313, Springer, Cham, 2020, pp. 423–439, DOI 10.1007/978-3-030-36568-4_27. MR4100648

[76] Z.-W. Sun. Series with summands involving harmonic numbers. In Melvyn B. Nathanson, editor, *Combinatorial and Additive Number Theory VI*. Springer, 2024. (to appear) arXiv:2210.07238v8 [math.NT].

[77] Z.-W. Sun. New series involving binomial coefficients. arXiv:2307.03086v5 [math.NT], 2023.

[78] Z.-W. Sun and Y. Zhou. Evaluations of $\sum_{k=1}^{\infty}\frac{x^k}{k^2\binom{3k}{k}}$ and related series. arXiv:2401.12083v1 [math.CO], 2024.

[79] M. Waldschmidt, *Multiple polylogarithms: an introduction*, Number theory and discrete mathematics (Chandigarh, 2000), Trends Math., Birkhäuser, Basel, 2002, pp. 1–12. MR1952273

[80] W. Wang and C. Xu, *Alternating multiple zeta values, and explicit formulas of some Euler-Apéry-type series*, European J. Combin. **93** (2021), Paper No. 103283, 25 pp., DOI 10.1016/j.ejc.2020.103283. MR4186623

[81] G. N. Watson, *A Treatise on the Theory of Bessel Functions*, Cambridge University Press, Cambridge; The Macmillan Company, New York, 1944. MR10746

[82] C. Wei. On some conjectural series containing binomial coefficients and harmonic numbers. arXiv:2306.02641v1 [math.CO], 2023.

[83] C. Wei. On some conjectures of Z.-W. Sun involving harmonic numbers. arXiv:2304.09753v1 [math.CO], 2023.

[84] C. Wei and C. Xu. On some conjectural series containing harmonic numbers of order 3. arXiv:2308.06440v1 [math.CO], 2023.

[85] C. Wei and C. Xu. On two conjectural series involving Riemann zeta function. arXiv:2310.04642v1 [math.CO], 2023.

[86] S. Weinzierl, *Feynman integrals—a comprehensive treatment for students and researchers*, Unitext for Physics, Springer, Cham, 2022, DOI 10.1007/978-3-030-99558-4. MR4461090

[87] C. Xu, *Explicit evaluations for several variants of Euler sums*, Rocky Mountain J. Math. **51** (2021), no. 3, 1089–1106, DOI 10.1216/rmj.2021.51.1089. MR4298827

[88] C. Xu and W. Wang, *Explicit formulas of Euler sums via multiple zeta values*, J. Symbolic Comput. **101** (2020), 109–127, DOI 10.1016/j.jsc.2019.06.009. MR4109711

[89] C. Xu and W. Wang, *Two variants of Euler sums*, Monatsh. Math. **199** (2022), no. 2, 431–454, DOI 10.1007/s00605-022-01683-4. MR4480817

[90] D. Zagier, *Values of zeta functions and their applications*, First European Congress of Mathematics, Vol. II (Paris, 1992), Progr. Math., vol. 120, Birkhäuser, Basel, 1994, pp. 497–512. MR1341859

[91] D. Zagier and F. Zerbini, *Genus-zero and genus-one string amplitudes and special multiple zeta values*, Commun. Number Theory Phys. **14** (2020), no. 2, 413–452, DOI 10.4310/CNTP.2020.v14.n2.a4. MR4084140

[92] F. Zerbini. *Elliptic multiple zeta values, modular graph functions and genus 1 superstring scattering amplitudes*. PhD thesis, Rheinischen Friedrich-Wilhelms-Universität Bonn, 2017. https://arxiv.org/abs/1804.07989v1 [math-ph].

[93] J. Zhao, *Standard relations of multiple polylogarithm values at roots of unity*, Doc. Math. **15** (2010), 1–34. MR2628849

[94] Y. Zhou, *Sun's series via cyclotomic multiple zeta values*, SIGMA Symmetry Integrability Geom. Methods Appl. **19** (2023), Paper No. 074, 20 pp., DOI 10.3842/SIGMA.2023.074. MR4653197

[95] I. J. Zucker, *On the series* $\sum_{k=1}^{\infty}\binom{2k}{k}^{-1}k^{-n}$ *and related sums*, J. Number Theory **20** (1985), no. 1, 92–102, DOI 10.1016/0022-314X(85)90019-8. MR777973

Program in Applied and Computational Mathematics (PACM), Princeton University, Princeton, NJ 08544

Current address: Academy of Advanced Interdisciplinary Studies (AAIS), Peking University, Beijing 100871, People's Republic of China

Email address: yajunz@math.princeton.edu

Email address: yajun.zhou.1982@pku.edu.cn

Selected Published Titles in This Series

818 **Howard S. Cohl, Roberto S. Costas-Santos, and Robert S. Maier, Editors,** Classical Hypergeometric Functions and Generalizations, 2025

814 **Susan Montgomery and Siu-Hung Ng, Editors,** Quantum Groups, Hopf Algebras, and Applications, 2025

812 **Ross Geoghegan, Craig R. Guilbault, and Kim Ruane, Editors,** Topology at Infinity of Discrete Groups, 2025

811 **Krishnendu Gongopadhyay and Shashikant A. Katre, Editors,** Geometry, Groups and Mathematical Philosophy, 2025

810 **Robert J. Berman and Yanir A. Rubinstein, Editors,** Convex and Complex: Perspectives on Positivity in Geometry, 2025

809 **Fabrice Baudoin and Luca Rizzi, Editors,** New Trends in Sub-Riemannian Geometry, 2025

808 **Loring W. Tu, Editor,** Group Actions and Equivariant Cohomology, 2024

807 **Galina Filipuk, Editor,** Recent Progress in Special Functions, 2024

806 **Maximilian Engel, Hildeberto Jardón-Kojakhmetov, and Cinzia Soresina, Editors,** Topics in Multiple Time Scale Dynamics, 2024

805 **Anthony Iarrobino, Pedro Macias Marques, Maria Evelina Rossi, and Jean Vallès, Editors,** Deformation of Artinian Algebras and Jordan Type, 2024

804 **Mahir Bilen Can and Jörg Feldvoss, Editors,** A Glimpse into Geometric Representation Theory, 2024

803 **Peter Gothen, Margarida Melo, and Montserrat Teixidor i Bigas, Editors,** Moduli Spaces and Vector Bundles—New Trends, 2024

802 **Ralph M. Kaufmann, Martin Markl, and Alexander A. Voronov, Editors,** Higher Structures in Topology, Geometry, and Physics, 2024

801 **K. A. Brown, T. J. Hodges, M. Vancliff, and J. J. Zhang, Editors,** Recent Advances in Noncommutative Algebra and Geometry, 2024

800 **Avinoam Mann, Louis H. Rowen, David J. Saltman, Aner Shalev, Lance W. Small, and Uzi Vishne, Editors,** Amitsur Centennial Symposium, 2024

799 **Alberto A. Condori, Elodie Pozzi, William T. Ross, and Alan A. Sola, Editors,** Recent Progress in Function Theory and Operator Theory, 2024

798 **Marat V. Markin, Igor V. Nikolaev, and Carsten Trunk, Editors,** Advances in Functional Analysis and Operator Theory, 2024

797 **Sangita Jha, Mrinal Kanti Roychowdhury, and Saurabh Verma, Editors,** Recent Developments in Fractal Geometry and Dynamical Systems, 2024

796 **John Cremona, John Jones, Jennifer Paulhus, Andrew V. Sutherland, and John Voight, Editors,** LuCaNT: LMFDB, Computation, and Number Theory, 2024

795 **Michael A. Jones, David McCune, and Jennifer M. Wilson, Editors,** Mathematical Analyses of Decisions, Voting and Games, 2024

794 **Jean-Pierre Magnot, Editor,** Recent Advances in Diffeologies and Their Applications, 2024

793 **Abba Gumel, Editor,** Mathematical and Computational Modeling of Phenomena Arising in Population Biology and Nonlinear Oscillations, 2024

792 **Shaoming Guo, Zane Kun Li, and Brian Street, Editors,** Recent Developments in Harmonic Analysis and its Applications, 2024

791 **Mee Seong Im, Bach Nguyen, and Arik Wilbert, Editors,** Algebraic and Topological Aspects of Representation Theory, 2024

790 **Peter Crooks and Alexandru I. Suciu, Editors,** Compactifications, Configurations, and Cohomology, 2023

For a complete list of titles in this series, visit the
AMS Bookstore at **www.ams.org/bookstore/conmseries/**.

Printed and bound by CPI Group (UK) Ltd, Croydon, CR0 4YY
15/09/2026
14962376-0001